Proceedings of the Steklov Institute of Mathematics

edited by
S. M. Nikol′skiĭ

Number 101 (1972)

The theory of extremal problems for univalent functions of class S

by

K. I. Babenko

AMERICAN MATHEMATICAL SOCIETY
Providence, Rhode Island
1975

Академия Наук
Союза Советских Социалистических Республик

ТРУДЫ
ордена Ленина
МАТЕМАТИЧЕСКОГО ИНСТИТУТА
имени В. А. СТЕКЛОВА
CI

К ТЕОРИИ ЭКСТРЕМАЛЬНЫХ ЗАДАЧ ДЛЯ ОДНОЛИСТНЫХ ФУНКЦИЙ КЛАССА S

К. И. БАБЕНКО

Ответственный редактор
академик С. М. НИКОЛЬСКИЙ

Издательство "Наука"
Москва 1972

Translated by J. W. Noonan

AMS (*MOS*) *subject classifications* (1970). Primary 30A38; Secondary 30A34.

Abstract. This monograph is devoted to the theory of extremal problems involving univalent functions of class S. The class S is an example of a nonlinear closed family in which, at the present time, the construction of variational techniques is still an active area of interest. In the present monograph we discuss some local aspects of the theory of extremal problems; we present some introductory material on variations of functions in S and on functionals defined on this class. The second variation of a functional is introduced, and the structure of the boundary of the coefficient region is examined in terms of this variation. Considerable attention is given to the question of the smoothness of the boundary of the coefficient region. The singularities of this boundary are investigated and, in particular, the dimension of the set of singularities is determined. We follow the well-developed recent theories for a solution to the question of sufficient conditions for the local extremum of a functional. In addition, some questions on the global structure of the boundary of the coefficient region are examined.

Library of Congress Cataloging in Publication Data CIP

Babenko, Konstantin Ivanovich.
The theory of extremal problems for univalent functions of class S.

(Proceedings of the Steklov Institute of Mathematics ; no. 101 (1972))
Translation of K teorii ėkstremal′nykh zadach dli͡a odnolistnykh funkt͡siĭ klassa S.
Bibliography: p.
1. Univalent functions. 2. Maxima and minima. I. Title. II. Series: Akademii͡a nauk SSSR. Matematicheskiĭ institut. Proceedings ; no. 101.
QA1.A413 no. 101 [QA331] 515'.64 74-23425
ISBN 0-8218-3001-5

PROCEEDINGS OF THE STEKLOV INSTITUTE OF MATHEMATICS IN THE ACADEMY OF SCIENCES OF THE USSR

(Труды математического института им. В. А. Стеклова, т. CI, 1972)

TABLE OF CONTENTS

Introduction

Variational methods have been widely used in the theory of extremal problems for univalent functions. Many striking and profound results have been obtained by these methods, although with regard to the application of these methods to specific classical problems we can do little more than hope for good luck. In our opinion, this situation is due to the following two reasons.

In the first place, the development itself of the theory of variational methods is not nearly as advanced as the development of methods for the classical problems of the variational calculus. For example, until recently we had been limited to only one necessary condition for a local extremum, the fact that the extremal function must satisfy a certain differential equation. Thus, if $f(z)$ is a function in S, $f(z) = \sum_0^\infty a_k z^{k+1}$, $a_0 = 1$, then with regard to the problem of determining

$$\sup_{f \in S} \operatorname{Re} a_n$$

we know only that the extremal function must satisfy the equation

$$Q\,[f_0(z)] \left(\frac{z f_0'(z)}{f_0(z)}\right)^2 + P(z) = 0, \tag{1}$$

a fact originally proved by M. Schiffer [1]. In equation (1), $Q(w)$ and $P(z)$ are rational functions whose coefficients are easily defined in terms of the coefficients $a_1^0, \cdots, a_n^0$ of the power series for $f_0(z)$. If $n > 1$, it is impossible to determine from equation (1) the least upper bound for the above functional. At the same time, it is clear that equation (1) gives only a single necessary condition for a local extremum for the problem under consideration. Since equation (1) reflects only the fact that the functional has a critical point at $f_0(z)$, it is necessary to have at least a second condition reflecting the fact that the second variation of the functional must be nonpositive.

In the Proceedings of the Fourth All-Union Mathematical Congress, held in Leningrad in the summer of 1961, the author introduced the second variation of the functional considered above and showed how this variation could be used to obtain more significant information concerning the extremal function than could be obtained from equation (1) alone.

Immediately the vital question arises as to whether or not these necessary conditions are also sufficient for a local extreme of the functional. While for the traditional problems of the variational calculus it is relatively simple to answer this question, it is far from simple with regard to the extremal problems encountered in the theory of univalent functions. The fundamental difficulty is the nonlinearity of the class S.

In the second place, even if the variational methods were to be sufficiently developed, they cannot, by their very nature, give the value of the maximum; these methods determine only local extrema, and it is then necessary to select the absolute maximum from among the local extrema. This is clearly a problem for a global variational calculus, and for the class S we do not really have even the beginnings of such a theory.

From the above remarks it is clear that the development of a variational calculus for the nonlinear class S is an important undertaking. S is metrizable, and is a complete metric space. This can be proved in various ways, and in particular we can define a metric by the formula

$$d_r(f, g) = \max_{|z|=r} |f(z) - g(z)|, \quad r < 1.$$

With this metric, any convergent sequence $\{f_n(z)\} \subset S$ converges uniformly on compact subsets of $\{z : |z| < 1\}$ to a function $f(z) \in S$. This follows from the inequality

$$\log d_{r_2}(f, g) \leqslant \frac{\log \frac{1+r_2}{2r_2}}{\log \frac{1+r_1}{2r_1}} \log d_{r_1}(f, g) + \frac{\log \frac{r_2}{r_1}}{\log \frac{1+r_2}{2r_1}} \log \frac{4(1+r_2)}{(1-r_2)^2}, \qquad r_1 < r_2 < 1,$$

which follows trivially from Bieberbach's inequality and Hadamard's three circles theorem. From Montel's theorem it follows that S is compact. In view of the complexity of the structure of the compact family S, we limit ourselves to a relatively narrow class of functionals, namely those connected with the coefficient problem for the class S. These functionals will all have the form

$$F(a_1, \cdots, a_n), \tag{2}$$

where F is a sufficiently smooth function of all its arguments, and where F also satisfies the condition $\operatorname{grad} F \neq 0$ inside a sphere $|a_1|^2 + \cdots + |a_n|^2 \leq R^2$ of sufficiently large radius. It is clear that the investigation of the extrema of functionals of the form (2) reduces to an elementary problem of analysis—the investigation of the extrema of a function of several variables on some set in $2n$-dimensional space. This compact set arises via a transformation

$$\Phi : S \to R^{2n}$$

defined so that $\Phi(f) = (a_1, \cdots, a_n)$, where $(a_1, \cdots, a_n)$ is a complex n-tuple corresponding to a vector in the real space R^{2n}. Clearly Φ is continuous, and so the image of S, henceforth denoted by $\mathscr{V}_n$, is compact. This compact set $\mathscr{V}_n$ is investigated in the monograph [2] of Schaeffer and Spencer. A series of important results is obtained in [2], and we shall later refer to many of them. It is clear that investigation of the extrema of the function (2) on the compact set $\mathscr{V}_n$ can be carried out only in those cases when we have detailed information concerning the structure of $\mathscr{V}_n$ and, in particular, the differentiability properties of the boundary of $\mathscr{V}_n$. As we shall see later on, any investigation of the metric properties of $\partial\mathscr{V}_n$ is connected with the theory of the second variation of a functional of the form (2), and, in particular, with sufficient conditions for a local extremum. Therefore the well-known problem concerning the structure of $\mathscr{V}_n$ is in some sense a reformulation of the problem of the construction of a variational calculus for functionals of the form (2), and both problems should be considered at the same time.

The nonlinearity of the space S causes difficulty when one attempts to describe constructively a neighborhood of a function $f_0(z) \in S$ which yields a local extremum for a particular functional. This situation does not pose any great obstacle when one is concerned with necessary conditions for an extremum, since in any given neighborhood U of an element f_0, we can construct a sufficiently large set U_0, $U_0 \subset U$, so that every element $g \in U_0$ is fairly simply related to $f_0(z)$. By then comparing f_0 to functions $g \in U_0$ we obtain necessary conditions for an extremum. For functionals of the form (2) we find that the extremal function satisfies an equation of the form (1), a result which is well known. In the course of this work we shall also introduce and calculate the second variation of the functional (2).

If we seek sufficient conditions for an extremum, it does not suffice to limit ourselves to elements of U_0. Rather we must consider the entire neighborhood U. It is at this point that the nonlinearity of S comes to the fore, and to proceed further with regard to a functional of the form (2) we must overcome many individual difficulties during our investigation of the metric properties of the boundary of $\mathscr{V}_n$.

Equation (1), which is satisfied by any function $f_0(z)$ yielding a local maximum for the functional (2), has a special structure, since

$$Q(w) = \sum_{1}^{n} A_k w^{-k}, \quad P(z) = \sum_{-n}^{n} B_k z^{-k}, B_k = \overline{B}_{-k}, \quad 0 \leqslant k \leqslant n,$$

and in addition $P(z)$ possesses the very important property that

$$P(z) \geqq 0 \quad \text{if} \quad |z| = 1. \tag{3}$$

It will prove useful to study that subset of S consisting of those univalent functions which satisfy (1) with $Q(w)$ and $P(z)$ satisfying the above requirements (this point of view is systematically undertaken in [**2**]). Functions in this subset will be called extremal, and the subset itself will be denoted by S_n. The justification for this terminology is as follows. Let S_x be the preimage under the mapping Φ of the point $x \in \mathscr{V}_n$. Then, as is shown in [**2**], for every point $x \in \partial\mathscr{V}_n$,

$$S_x \cap S_n \neq \emptyset. \tag{4}$$

On the other hand, we have the very striking and deep result of Teichmüller [**3**], which in combination with (4) gives that S_x contains exactly one function, and, as is clear, that this function will be extremal. In this way S_n is homeomorphic to $\partial \mathscr{V}_n$. It will be shown that if we neglect a set of small dimension, then for $x \in \partial \mathscr{V}_n$ it is possible to define a functional (2) which has an absolute maximum at the function $f(z) \in S_x$. Thus, we can say roughly that S_n consists of those, and only those, functions which give an extremum for some functional of the form (2).

Equation (1) clearly reveals the fundamental role of the quadratic differential $Q(w)(dw/w)^2$ in the theory we shall be discussing. The previously mentioned result of Teichmüller was proved by a method which is fundamental to the theory of the quadratic differential and which is also connected with differential geometry; Teichmüller's original proof remains the most clear-cut exposition available. Detailed discussions of these points can be found in the interesting monograph [**4**] of Jenkins.

We shall not go into a detailed analysis of the contents of the individual chapters of this book, but we shall quickly indicate some of the basic results; before each chapter we shall give a short introduction, in which we shall also review the relevant points made above.

One of the basic questions on which we must concentrate is the question of the structure and construction of the set U_0. There exist many different methods for the construction of univalent functions (interior variation, boundary variation, etc.); although these methods are similar, they are not equivalent. We shall follow the method of Goluzin [**5**], though not introducing a small parameter explicitly. Our Theorem 1 is a refinement and, in some sense, a generalization of a theorem of Goluzin.

On the basis of Theorem 1 it is possible to establish various formulas for the variation of a given univalent function $f(z)$ and to derive formulas giving the first and second variations of functionals; these formulas depend on the construction of the set U_0. A widely used construction of U_0 is obtained by the application of Theorem 1 to the function

$$f(z) + \varepsilon \int_{\mathscr{E}} \frac{f^2(z)}{f(z) - f(t)} \left(\frac{tf'(t)}{f(t)} \right)^2 \nu(dt), \tag{5}$$

where $\nu(dt)$ is an arbitrary complex measure with support contained in the disc $|z| \leq r < 1$, and where ϵ is an arbitrary small real constant. The resulting formula for the second variation is a bit cumbersome. Clearly the second variation is a quadratic functional relative to the measure $\nu(dz)$. If $f(z) \in S_n$, then it is possible to derive several equivalent formulations of the second variation, and, in particular, it is possible to represent the second variation as a quadratic functional of the first variation of $f(z)$. Such a representation is completely independent of the method of construction of U_0, and the formula for the second variation continues to hold even for the most general types of variation of $f(z)$. From this follows the remarkable fact that the second variation naturally induces a scalar product on the set of first variations of $f(z)$. The completion of the space of first variations with respect to the induced metric is the real Hilbert space of first variations. This Hilbert space, H_f, in our opinion, most completely characterizes the extremal function $f(z)$. If x is the image in H_f of $\partial\mathscr{V}_n$, then the properties of H_f in a neighborhood of x are closely related to the smoothness of $\partial\mathscr{V}_n$.

The second variation was introduced independently of the present work by Duren and Schiffer [**6**]. The basic result obtained in [**6**] is a formula for the second variation of a linear functional derived under some relatively strong assumptions on the boundary structure of the image of $|z| < 1$ under the extremal function. The class of variations considered by Duren and Schiffer consists essentially of those obtained by applying Theorem 1 to (5). The formula of Duren and Schiffer is formally different from ours, since these two authors do not represent the second variation as a quadratic functional of the first variation of the extremal function. In essence, however, the formula of Duren and Schiffer is a special case of one of our formulas. At the same time, we reemphasize that the fact that the second variation is a quadratic functional of the first variation is very important to our whole theory.

The classical treatment by the variational calculus of the second variation can be applied to our situation; this theory assumes in this case a distinctive form. The problem under consideration takes the following form: to find a least upper bound for the second variation, under the condition that the first variation of the first n coefficients in the power series for $f(z)$ remains fixed. It turns out that the upper bound is always attained at some element of H_f which, as expected, satisfies a differential equation corresponding to (1). The upper bound itself has in this case a simple geometric interpretation, and it is expressed in terms of various smoothness properties of $\mathscr{V}_n$. By this means the theory of the second variation is connected with the theory of the metric properties of the boundary of $\mathscr{V}_n$. With regard to the condition which guarantees the nonpositivity of the second variation, we use here purely algebraic conditions in place of the well-known Jacobi criteria. In practice it is extremely difficult to verify these conditions, and even in a simple case, when the extremal function is $f_0(z) = z(1-z)^{-2}$, we must proceed in a somewhat roundabout manner.

Up to now we have considered only functionals of the form (2). For the coefficient problem, the most interesting functional is a linear functional of the form

$$L(f) = \operatorname{Re}\sum_1^n \overline{\lambda}_k a_k, \tag{6}$$

where $\lambda = (\lambda_1, \cdots, \lambda_k)$ is a fixed vector. Immediately there arises the question of which properties of the extremal functions are "singled out" by functionals of the form (6). Any extremal function maps $|z| < 1$ onto a domain whose boundary consists of finitely many analytic arcs. The difference between the case of a linear functional and a nonlinear functional arises first and foremost in the properties of these arcs. For a function giving a local extremum for a linear functional, the arc is quite easily described. If the function $f(z)$ gives rise to the absolute extremum of a linear functional, then its boundary consists of a single arc, and the presence of the quadratic differential $Q(w)\,(dw/w)^2$ implies that $w=\infty$ is a first order pole. This follows from the well-known result of Schiffer [7] and from results to be presented here.

The boundary of the coefficient region $\mathscr{V}_n$ has quite a complex structure; with the exception of a set of small dimension, at each point of $\partial\mathscr{V}_n$ there exists a normal vector satisfying a Lipschitz condition. The separate parts of the boundary are quite smooth, and, in particular, on the boundary there exists a connected open set such that at each point of this set the boundary is an analytic hypersurface. It is an established fact that the image of any function which gives the absolute extremum of a linear functional either lies on one of the above-mentioned analytic hypersurfaces or belongs to the set of limit points of these hypersurfaces. This fact will be very important to us; as we shall see, it will play an essential role in our treatment of global extremes of linear functionals.

One can easily understand the interest displayed in the Koebe function $f_0(z)=z(1-z)^{-2}$ with regard to the theory of extremal problems. It is easily seen that the functional (6) will have a critical point at $f_0(z)$ if $\operatorname{Im}\lambda_k=0$, $1\leqq k\leqq n$. The set of all functionals with nonpositive second variation at $f_0(z)$ forms a convex cone. The interior of this cone consists of all those functionals for which a sufficient condition for a local extreme is satisfied. We have shown that the functional $L_k(f)=\operatorname{Re}a_k$, $1\leqq k\leqq n$, will be an interior point of this cone if k is odd, and will belong to the boundary of the cone if k is even. The question of sufficient conditions for a local extremum for the Koebe function was also considered in [8]. A paper[1)] of E. Bombieri, in which the second variation of $L_k(f)$ at $f_0(z)$ is discussed, should also be mentioned.

This Introduction is not intended to be a complete survey, and therefore we have mentioned only those papers which have an immediate bearing on our own work or which overlap and complement our results.

For the convenience of the reader we have occasionally included some known results such as the formula for the first variation, the differential equation for the extremal function, etc. We apologize for this, but these results are absolutely necessary for the derivation of the formulas for the second variation. We have also included complete statements of all theorems referred to but not proved in the text.

[1)] Written in 1967, the text of this paper has not yet been revised.

CHAPTER 1

VARIATION OF FUNCTIONALS AND OF UNIVALENT FUNCTIONS OF CLASS S

The class S can be made into a metric space by the following definition of the distance between two functions $f(z), g(z) \in S$:

$$\rho(f, g) = \max_{|z| \leqslant 1/2} |f(z) - g(z)|.$$

With this metric S is a compact complete metric space. The complexity of variational problems within S is due to the fact that S is not a linear space. In order to investigate the theory of variational problems, we must first of all decide how we can construct a sufficiently small neighborhood of a given extremal function. Taking as a neighborhood of an element $f \in S$ the closed sphere $U = \{\varphi : \varphi \in S, \rho(\varphi, f) \leqq \epsilon\}$, we have for any element $\varphi \in U$ the relation $\varphi = f + \epsilon v$, where all that we know about $v(z)$ is that v is regular in $|z| < 1$,

$$\max_{|z| \leqslant 1/2} |v(z)| \leqslant 1,$$

and $f + \epsilon v \in S$. This type of characterization is not very useful to us with regard to the formulation of a specific variational problem.

If the compact set S were an infinite-dimensional smooth manifold, we could represent $v(z, \epsilon)$ in the form $v(z, \epsilon) = v_0(z) + o(1)$, where $v_0(z)$ belongs to the tangent space, and hence we would have

$$\varphi(z) = f(z) + \epsilon v_0(z) + o(\epsilon).$$

But, as it happens, the case of the compact set S is somewhat different from this. In order to overcome the difficulties which arise in the theory of variational problems in the class S, we must use a very precise approach, consisting of the following. We shall not consider the entire neighborhood U, but only a sufficiently large subset U_0 of U. U_0 will be constructed so that for any function $\varphi \in U_0$ the above relation holds. We must also choose a sufficiently general range of functions $v_0(z)$ so that it is always possible to obtain necessary conditions for the extremum of the functional.

Eventually, since we shall be interested in the question of sufficient conditions for a local extremum, we shall be forced to consider the entire neighborhood U. For some classes of functionals we shall succeed in overcoming this difficulty by using a method which seems to us to be both quite instructive and quite elegant; we shall have more to say about this later.

The theory of variational problems on the compact set S is in a rather peculiar state. Although at the present time there are relatively simple theorems giving criteria for local extrema in the case of a linear metric space, the corresponding theorems for the class S would be both very difficult to prove and very powerful. The set U_0 is defined differently by different authors; also, in the literature on extremal problems in S, univalent functions are varied by several different methods, and the use of these different methods is really equivalent to different constructions of the set U_0.

Below we give one of the several available methods for the construction of the set U_0.

§1. DEFINITION 1. The function $g(z)$ will be called *admissible* with respect to a given $f(z) \in S$ if there exists an annulus $r_0 \leqq |z| < 1$ in which the function

$$F(z) = f(z) + g(z) \tag{1.1}$$

is regular and univalent.

The construction of a wide class of admissible functions is really the same as the construction of a variational formula for a given $f(z)$, since it is not difficult to obtain from $F(z)$ a function univalent in $|z| < 1$, provided $g(z)$ is "small". The precise meaning of "small" will be given below. With the aid of a transformation $z = \psi(\zeta)$ which is univalent in some annulus $r_1 \leqq |\zeta| < 1$ and which maps the unit circle $|\zeta| = 1$ onto itself, it is possible to eliminate the singularities of $F(z)$ and to obtain a function $F_*(\zeta) = F[\psi(\zeta)]$ univalent in $|\zeta| < 1$ and satisfying

$$F_*(0) = 0. \tag{1.2}$$

Let $K_0 = K(r_0, \Delta) = \{z : r_0 \leqq |z| \leqq r_0 + \Delta < 1\}$, and set

$$\max_{z \in K_0} |g(z)| = \varepsilon. \tag{1.3}$$

The following theorem is a refinement of a result of G. M. Goluzin.

THEOREM 1. *Let $f(z) \in S$, and let $g(z)$ be an admissible function such that* (1.3) *holds. There exists a constant ϵ_0 depending only on Δ and r_0 such that if $\epsilon \leqq \epsilon_0$, there exists a function $\varphi(z)$ with the following properties*:

1. *$\varphi(z)$ is regular in the annulus $r_0 + \Delta/4 \leqq |z| \leqq (r_0 + \Delta/4)^{-1}$, and* $\operatorname{Re} \varphi(z) = 0$ *for $|z| = 1$.*

2. *The function $z \exp\{\varphi(z)\}$ is univalent in the annulus $r_1 \leqq |z| < 1$, and its image domain lies in the annulus $r_0 \leqq |z| < 1$.*

3. *The function $F_*(z) = F[z \exp(\varphi(z))]$ is univalent in $|z| < 1$ and satisfies condition* (1.2).

The proof of the theorem requires three lemmas.

LEMMA 1.1. *The function $\xi_l(x) = \sum_1^\infty k^l x^k$, $0 < x < 1$, where l is a natural number, satisfies the inequality*

$$\xi_l(x) < l!\,(1 - x)^{-l-1}.$$

PROOF. Applying the Abel transform to $\sum_1^\infty k^l x^k$, we find that

$$\xi_l(x) = \sum_1^\infty [k^l - (k-1)^l]\, S_k(x),$$

where $S_k(x) = \sum_k^\infty x^\nu$. But $k^l - (k-1)^l < lk^{l-1}$ and $S_k(x) = x^k(1-x)^{-1}$. Therefore

$$\xi_l(x) < \frac{l}{(1-x)} \sum_1^\infty k^{l-1} x^k = \frac{l}{1-x}\, \xi_{l-1}(x).$$

From this it follows that $\xi_l(x) < l!\,(1-x)^{-l-1}$.

LEMMA 1.2. *If* $0 < x < 1$, *then*

$$\frac{1}{2\pi} \int_0^{2\pi} \Big| \sum_1^\infty k^l x^k e^{ik\theta} \Big| \, d\theta < l!\,(1+x)^{-1}(1-x)^{-l}.$$

PROOF. Set $z = xe^{i\theta}$, $S_k(z) = z^k(1-z)^{-1}$, and $\sigma_k(z) = z^k(1-z)^{-2}$. Applying Abel's transform twice to $\sum_1^\infty k^l z^k$, we find that

$$\sum_1^\infty k^l z^k = \sigma_1(z) + \sum_2^\infty [k^l - 2(k-1)^l + (k-2)^l]\, \sigma_k(z).$$

From this we get

$$\frac{1}{2\pi} \int_0^{2\pi} \Big| \sum_1^\infty k^l x^k e^{ik\theta} \Big| \, d\theta < \frac{1}{2\pi} \int_0^{2\pi} \Big\{ |\sigma_1(z)| + \sum_2^\infty [k^l - 2(k-1)^l + (k-2)^l]$$

$$\times |\sigma_k(z)| \Big\} \, d\theta < (1-x^2)^{-1} \Big\{ x + \sum_2^\infty [k^l - 2(k-1)^l + (k-2)^l]\, x^k \Big\}$$

$$= \frac{1-x}{1+x}\, \xi_l(x) < \frac{l!}{(1+x)(1-x)^l},$$

where we have used Lemma 1.1. Q.E.D.

LEMMA 1.3. *If* $\eta(x, \vartheta) = \eta(x) = [1 - \vartheta \sum_1^\infty x^k/k^2]^{-1}$ *for* $|x| < 1$ *and* $0 < \vartheta < 6/\pi^2$, *then*

$$\frac{1}{n!}\, \eta^{(n)}(0) < \vartheta \left(1 + \frac{2}{3}\pi^2 \vartheta\right) \eta^3(1)\, n^{-2}, \qquad n = 1, 2, \ldots.$$

PROOF. Applying Cauchy's formula and integrating twice by parts, we see that

$$\frac{1}{n!}\, \eta^{(n)}(0) = \frac{1}{2\pi i n^2} \int_{|z|=\rho} D^2 \eta(z) \frac{dz}{z^{n+1}}, \qquad \rho < 1,$$

where D signifies the operator $z\,d/dz$. We shall use this notation from now on. Computation gives that

$$D^2 \eta(z) = 2\vartheta^2 \eta^3(z) \Big(\sum_1^\infty \frac{z^k}{k} \Big)^2 + \vartheta \eta^2(z)(1-z)^{-1}.$$

and we also note that

$$\eta^2(z) = \eta^2(1) + \vartheta\,[\eta(z) + \eta(1)]\,\eta(z)\,\eta(1)\sum_1^\infty \frac{z^k - 1}{k^2}.$$

Substituting these formulas into the above and changing the path of integration, we find that

$$\frac{\eta^{(n)}(0)}{n!} = n^{-2}\left[\vartheta\eta^2(1) + \frac{\vartheta^2}{\pi i}\int_{|z|=1} \eta^3(z)\left(\sum_1^\infty \frac{z^k}{k}\right)^2 \frac{dz}{z^{k+1}}\right.$$

$$\left. + \vartheta^2\eta(1)\frac{1}{2\pi i}\int_{|z|=1} [\eta(z) + \eta(1)]\,\eta(z)\sum_1^\infty \frac{z^k - 1}{(1-z)\,k^2}\,\frac{dz}{z^{n+1}}\right].$$

Therefore

$$\frac{\eta^{(n)}(0)}{n!} < n^{-2}\left[\vartheta\eta^2(1) + \frac{\pi^2}{3}\vartheta^2\eta^3(1) + 2\vartheta^2\eta^3(1)\left(\frac{1}{2\pi}\int_{|z|=1}\left|\sum_1^\infty \frac{z^k-1}{(z-1)\,k^2}\right|^2 |dz|\right)^{1/2}\right]$$

$$< n^{-2}\left[\vartheta\eta^2(1) + \frac{5}{6}\pi^2\vartheta^2\eta^3(1)\right] < \vartheta\left(1 + \frac{2}{3}\pi^2\vartheta\right)\eta^3(1)\,n^{-2}.$$

Proof of the Theorem. Set $h(z) = \epsilon^{-1}g(z)$ and

$$F(z,\xi) = f(z) + \xi h(z), \quad F(z,\epsilon) = F(z),$$

where ξ is a parameter. Define the function $\varphi(z,\xi) = \sum_1^\infty \xi^k\varphi_k(z)$ so that the function $F\{z\exp[\varphi(z,\xi)],\xi\}$ is regular in $|z| < 1$. We note that we have the formal expansion

$$F\{z\exp[\varphi(z,\xi)],\ \xi\} = \sum_{k=0}^\infty \xi^k F_k(z).$$

Clearly

$$F_n(z) = \frac{1}{n!}\left\{\frac{d^n}{d\xi^n} F\,[z\exp(\varphi(z,\xi)),\ \xi]\right\}_{\xi=0}$$

$$= \frac{1}{n!}\left\{\frac{d^n}{d\xi^n} f\,[z\exp(\varphi(z,\xi))]\right\}_{\xi=0} + \frac{1}{(n-1)!}\left\{\frac{d^{n-1}}{d\xi^{n-1}} h\,[z\exp(\varphi(z,\xi))]\right\}_{\xi=0}.$$

We note that

$$\frac{d}{d\xi} f\,[z\exp(\varphi(z,\xi))] = Df\,[z\exp(\varphi(z,\xi))]\,\varphi'(z,\xi),$$

where $'$ denotes differentiation with respect to ξ. Repeating the above process, we arrive at the following formula, which is analogous to the formula of Faà di Bruno:

$$\frac{d^n}{d\xi^n} f\,[z\exp(\varphi(z,\xi))]$$

$$= \sum_{\nu_1+2\nu_2+\cdots=n} k(\nu_1,\nu_2,\ldots)\,D^{\nu_1+\nu_2+\cdots} f\,[z\exp(\varphi(z,\xi))]\,[\varphi']^{\nu_1}\,[\varphi'']^{\nu_2}\cdots,$$

where $k(\nu_1,\nu_2,\cdots)$ are nonnegative coefficients as in di Bruno's formula. From this it follows that

$$(1.4)\quad \begin{aligned} F_n(z) &= \frac{1}{n!} \sum_{\nu_1+2\nu_2+\ldots=n} k(\nu_1, \nu_2, \ldots)(1!)^{\nu_1}(2!)^{\nu_2}\ldots D^{\nu_1+\nu_2+\ldots} f(z)\, \varphi_1^{\nu_1}\varphi_2^{\nu_2} \\ &+ \frac{1}{(n-1)!} \sum_{\nu_1+2\nu_2+\ldots=n-1} k(\nu_1, \nu_2, \ldots)(1!)^{\nu_1}(2!)^{\nu_2}\ldots D^{\nu_1+\nu_2+\ldots} h(z)\, \varphi_1^{\nu_1}\varphi_2^{\nu_2}\ldots \end{aligned}$$

In the case $n = 1$ we have

$$(1.5)\quad F_1(z) = Df(z)\,\varphi_1(z) + h(z) = Df(z)\left[\varphi_1(z) + \frac{h(z)}{Df(z)}\right].$$

If

$$(1.6)\quad \psi_1(z) = \frac{h(z)}{Df(z)} = \sum_{k=-\infty}^{\infty} b_{1k} z^k,$$

then we set

$$(1.7)\quad \varphi_1(z) = -\sum_{k=-\infty}^{-1} (b_{1k}z^k - \bar{b}_{1k}z^{-k}) + ic_1,$$

where c_1 is an arbitrary real constant.

Assuming that $\varphi_1, \cdots, \varphi_{n-1}$ have been defined, we now represent (1.4) in the form

$$(1.8)\quad F_n(z) = Df(z)\,[\varphi_n(z) + \psi_n(z)],$$

where

$$\begin{aligned} \psi_n(z) &= \frac{1}{n!} \sum_{\substack{\nu_1+2\nu_2+\ldots=n \\ \nu_1+\nu_2+\ldots>1}} k(\nu_1, \nu_2, \ldots)(1!)^{\nu_1}(2!)^{\nu_2}\ldots \frac{D^{\nu_1+\nu_2+\ldots}}{Df(z)}\, \varphi_1^{\nu_1}\varphi_2^{\nu_2}\ldots\varphi_{n-1}^{\nu_{n-1}} \\ &+ \frac{1}{(n-1)!} \sum_{\nu_1+2\nu_2+\ldots=n-1} k(\nu_1, \nu_2, \ldots)(1!)^{\nu_1}(2!)^{\nu_2}\ldots \frac{D^{\nu_1+\nu_2+\ldots}\, h(z)}{Df(z)}\, \varphi_1^{\nu_1}\varphi_2^{\nu_2}\ldots\varphi_{n-1}^{\nu_{n-1}}. \end{aligned}$$

Set

$$\psi_n(z) = \sum_{k=-\infty}^{\infty} b_{nk} z^k;$$

then

$$\varphi_n(z) = -\sum_{-\infty}^{-1} (b_{nk}z^k - \bar{b}_{nk}z^{-k}) + ic_n,$$

where c_n is an arbitrary real constant. We have thus defined $\varphi_n(z)$ for all $n = 1, 2, \cdots$, and it is clear that $\operatorname{Re}\varphi_n(z) = 0$ for $|z| = 1$.

We now consider the question of the convergence of the series. Set

$$K_1 = \left\{z\colon\ r_0 + \tfrac{1}{4}\Delta \leqslant |z| \leqslant r_0 + \tfrac{3}{4}\Delta\right\}.$$

Define a norm for a function $\chi(z) = \sum_{-\infty}^{\infty} \alpha_k z^k$ regular in K_1 by

$$\|\chi\| = \max_{z\in K_1} \sum_{-\infty}^{\infty} |\alpha_k z^k|.$$

Clearly

$$\|\chi_1+\chi_2\|\leqslant\|\chi_1\|+\|\chi_2\|,\quad \|\chi_1\chi_2\|\leqslant\|\chi_1\|\cdot\|\chi_2\|.$$

and so $\|\varphi_n\| \leqq 2\|\psi_n\| + |c_n|$. We now estimate the norm of $\psi_1(z)$. We note that $\|\chi\| = \sum_{-\infty}^{\infty} \alpha_k e^{i\theta_k} z^k$ for some $z \in K_1$, and so, if $\chi(z)$ is regular in the annulus K_0, we have

$$\|\chi\| = \frac{1}{2\pi i}\int_{|t|=r_2} \chi(t) \sum_{-\infty}^{-1} \left(\frac{z}{t}\right)^k e^{i\theta_k}\frac{dt}{t} + \frac{1}{2\pi i}\int_{|t|=r_3} \chi(t)\sum_0^{\infty}\left(\frac{z}{t}\right)^k e^{i\theta_k}\frac{dt}{t}$$
$$\leqslant \frac{1}{2\pi}\int_{|t|=r_2} |\chi(t)| \frac{|dt|}{|z|-|t|} + \frac{1}{2\pi}\int_{|t|=r_3} |\chi(t)|\frac{|dt|}{|t|-|z|},$$

where $r_2 = r_0 + \Delta/8$ and $r_3 = r_0 + 7\Delta/8$. Hence

$$\|\chi\| < \frac{4}{\pi\Delta}\left(\int_{|t|=r_2} |\chi(t)||dt| + \int_{|t|=r_3} |\chi(t)||dt|\right). \tag{1.9}$$

Applying this inequality to the function $\psi_1(z)$ and recalling that $|h(z)| \leqq 1$ if $z \in K_0$, we find that $\|\psi_1\| < 4J/\pi\Delta$, where

$$J = \int_{|t|=r_2} \frac{|dt|}{|Df(t)|} + \int_{|t|=r_3} \frac{|dt|}{|Df(t)|}.$$

From Bieberbach's inequality

$$\frac{1}{|Df(z)|} \leqslant \frac{(1+|z|)^3}{|z|(1-|z|)} < \frac{8}{|z|(1-|z|)},$$

it follows that $J < 4\pi 8^2 \Delta^{-1}$ and

$$\|\psi_1\| < 2\cdot 8^3 \Delta^{-2}. \tag{1.10}$$

Hence $\|\varphi_1\| < 4\cdot 8^3\Delta^{-2} + |c_1|$. From this point on we shall choose the arbitrary real constant so that

$$|c_n| \leqq \|\psi_n\|. \tag{1.11}$$

Then $\|\varphi_1\| < 48(8/\Delta)^2$. Set

$$A = 48(8/\Delta)^2. \tag{1.12}$$

We now assume that the estimate $\|\varphi_j\| \leqq AB^{j-1}j^{-2}$ holds for $j = 1, \cdots, n-1$, with some constant B. We shall show that this estimate also holds for $\|\varphi_n\|$, and we shall give an estimate for the constant B.

Toward this end we must first estimate $\|D^l h/Df\|$ and $\|D^l f/Df\|$. In the annulus $r_2 \leqq |z| \leqq r_3$ we have

$$D^l h(z) = -\frac{1}{2\pi i}\int_{|t|=r_0} h(t)\, D^l\left(\frac{1}{t-z}\right)dt + \frac{1}{2\pi i}\int_{|t|=r_0+\Delta} h(t)\, D^l\left(\frac{1}{t-z}\right)dt.$$

From this and Lemma 1.2 we find that

$$|D^l h(z)| < \frac{1}{2\pi}\int_0^{2\pi}\left|\sum_0^\infty k^l\left(\frac{r_0}{|z|}\right)^k e^{ik\theta}\right| d\theta + \frac{1}{2\pi}\int_0^{2\pi}\left|\sum_1^\infty k^l\left(\frac{|z|}{r_0+\Delta}\right)^k e^{ik\theta}\right| d\theta$$
$$< l!\left[\left(\frac{|z|}{|z|-r_0}\right)^l + \left(\frac{r_0+\Delta}{r_0+\Delta-|z|}\right)^l\right] < 2\left(\frac{8}{\Delta}\right)^l l!$$

Applying inequality (1.9) to the function $\chi = D^l h/Df$ and using the last inequality, we see that $\|D^l h/Df\| < (8/\Delta)^{l+1} Jl!/\pi$, and so

$$\left\|\frac{D^l h(z)}{Df(z)}\right\| < 32l!\,(8\Delta^{-1})^{l+2}. \tag{1.13}$$

We next estimate $\|D^l f/Df\|$. The well-known result of Bieberbach gives that

$$\left|\frac{D^2 f(z)}{Df(z)}\right| < \frac{1+4r+r^2}{1-r^2} < \frac{3}{1-r}, \qquad r = |z|.$$

If the function $\chi(z)$ is regular in $|z| < r_0 + \Delta$, then in place of (1.9) we clearly have

$$\|\chi\| < \frac{4}{\pi\Delta}\int_{|t|=r_3} |\chi(t)||dt|. \tag{1.14}$$

Thus

$$\left\|\frac{D^2 f}{Df}\right\| < 3\left(\frac{8}{\Delta}\right)^2.$$

Since $D^3 f/Df = D(D^2 f/Df) + (D^2 f/Df)^2$, it follows that

$$\left|\frac{D^3 f}{Df}\right| < \left|D\left(\frac{D^2 f}{Df}\right)\right| + \left|\frac{D^2 f}{Df}\right|^2.$$

If $|z| \leqq r_3$, then, setting $\rho = r_0 + 15\Delta/16$, we deduce that

$$\left|D\left(\frac{D^2 f(z)}{f(z)}\right)\right| = \left|\frac{1}{2\pi i}\int_{|t|=\rho} \frac{D^2 f(t)}{f(t)}\,\frac{z}{(t-z)^2}\,dt\right| < 6\left(\frac{8}{\Delta}\right)^2.$$

Hence for $|z| \leqq r_3$

$$\left|\frac{D^3 f}{Df}\right| < 15\left(\frac{8}{\Delta}\right)^2,$$

and from (1.14) we get

$$\left\|\frac{D^3 f}{Df}\right\| < 15\left(\frac{8}{\Delta}\right)^3 < 3\cdot 3!\left(\frac{8}{\Delta}\right)^3.$$

We shall not repeat the calculations for the derivation of estimates for $\|D^l f/Df\|$ with $l = 4, 5$, but we merely note that methods similar to the above yield that

$$\left\|\frac{D^l f}{Df}\right\| < \alpha_l l!\left(\frac{8}{\Delta}\right)^l, \quad l = 4, 5,$$

with $\alpha_4 = 9$ and $\alpha_5 = 81$. For $l \geqq 6$ we shall derive a rougher estimate. For $f(z) = \sum_1^\infty a_k z^k$ and $\rho = r_0 + 3\Delta/4$ we have

$$\| D^l f \| < \sum_1^\infty k^l |a_k| \rho^k,$$

Littlewood's inequality states that $|a_k| < ke$, and so it follows from Lemma 1.1 that

$$\| D^l f(z) \| < e(l+1)! \left(\frac{4}{\Delta}\right)^{l+2}.$$

From (1.9) we have

$$\left\| \frac{1}{Df(z)} \right\| < \frac{4}{\pi\Delta} J < 16\left(\frac{8}{\Delta}\right)^2,$$

and so

$$\left\| \frac{D^l f(z)}{Df(z)} \right\| < 64e(l+1)! \left(\frac{4}{\Delta}\right)^{l+4}.$$

Using the inductive hypothesis for the norms of the functions φ_j, $j = 1, \cdots, n-1$, we find from the formula for $\psi_n(z)$ that

$$\| \psi_n \| < \frac{B^n}{n!} \sum_{\substack{\nu_1 + 2\nu_2 + \ldots = n \\ \nu_1 + \nu_2 + \ldots > 1}} k(\nu_1, \nu_2, \ldots) \frac{(1!)^{\nu_1}(2!)^{\nu_2}\ldots}{1^{2\nu_1}2^{2\nu_2}} \left\| \frac{D^{\nu_1+\nu_2+\ldots} f}{Df} \right\| \left(\frac{A}{B}\right)^{\nu_1+\nu_2+\ldots}$$
$$+ \frac{B^{n-1}}{(n-1)!} \sum_{\nu_1+2\nu_2+\ldots=n-1} k(\nu_1, \nu_2, \ldots) \frac{(1!)^{\nu_1}(2!)^{\nu_2}\ldots}{1^{2\nu_1}2^{2\nu_2}} \left\| \frac{D^{\nu_1+\nu_2+\ldots} h}{Df} \right\| \left(\frac{A}{B}\right)^{\nu_1+\nu_2+\ldots}.$$

We choose the constant B so that the inequality

$$\left\| \frac{D^l f}{Df} \right\| \left(\frac{A}{B}\right)^{l-1} \leqslant \vartheta^l l!, \qquad l = 2, 3, \ldots, \tag{1.16}$$

is satisfied, where ϑ is an absolute constant to be defined below.

Upon using the estimates for $D^l f / Df$ for $l = 2, \cdots, 5$, one easily sees that inequality (1.16) will hold with

$$B = \frac{3}{2}\left(\frac{8}{\Delta}\right)^2 \vartheta^{-2} A. \tag{1.17}$$

We shall show that this choice of B also suffices for $l \geqq 6$ in (1.16). We have

$$\left\| \frac{D^l f}{Df} \right\| \left(\frac{A}{B}\right)^{l-1} < 64e\left(\frac{4}{\Delta}\right)^{l+4} (l+1)! \left(\frac{3}{2}\vartheta^2\right)^{l-1} \left(\frac{\Delta}{8}\right)^{2l-2}$$
$$= \vartheta^l l!\, 64e(l+1)\vartheta^{l-2} \left(\frac{\Delta}{4}\right)^{l-6} 6^{-l+1} < \vartheta^l l!.$$

The above choice of B leads to the inequality

$$\left\| \frac{D^l h}{Df} \right\| \left(\frac{A}{B}\right)^l < 32\left(\frac{8}{\Delta}\right)^{l+2} \left(\frac{2}{3}\vartheta^2\right)^l \left(\frac{\Delta}{8}\right)^{2l} l! = A\vartheta^{l+1} l! \left(\frac{\Delta}{12}\right)^l \frac{2}{3}\vartheta^{l-1}$$
$$< A\vartheta^{l+1} l!, \qquad l = 1, 2, \ldots.$$

Thus

$$\|\psi_n\| < \frac{AB^{n-1}}{n!} \sum_{\substack{\nu_1+2\nu_2+\dots=n\\ \nu_1+\nu_2+\dots>1}} k(\nu_1, \nu_2, \dots) \frac{(1!)^{\nu_1}(2!)^{\nu_2}\dots}{1^{2\nu_1}2^{2\nu_2}\dots}(\nu_1+\nu_2+\dots)!\,\vartheta^{\nu_1+\nu_2+\dots}$$

$$+\frac{AB^{n-1}}{(n-1)!}\vartheta \sum_{\nu_1+2\nu_2+\dots=n-1} k(\nu_1, \nu_2, \dots) \frac{(1!)^{\nu_1}(2!)^{\nu_2}\dots}{1^{2\nu_1}2^{2\nu_2}\dots}(\nu_1+\nu_2+\dots)!\,\vartheta^{\nu_1+\nu_2+\dots}.$$

It is clear that the second sum in the first part equals $\eta^{(n-1)}(0)$, where $\eta(x)$ is the function defined in Lemma 1.3. Thus if we strengthen the inequality by summing in the first summation over all possible values of the indices, we have from Lemma 1.3 that

$$\|\psi_n\| < AB^{n-1}\left[\frac{\eta^{(n)}(0)}{n!} + \vartheta\frac{\eta^{(n-1)}(0)}{(n-1)!}\right] < \vartheta\left(1+\frac{2}{3}\pi^2\vartheta\right)\eta^3(1)\,AB^{n-1} \times\left(\frac{1}{n^2}+\frac{\vartheta}{(n-1)^2}\right).$$

As ϑ we take the smallest positive root of the equation

$$6\vartheta\left(1+\frac{2}{3}\pi^2\vartheta\right)\left(1-\frac{\pi^2}{6}\vartheta\right)^{-3} = 1.$$

Then

$$\|\psi_n\| < \frac{1}{6}AB^{n-1}n^{-2}\left[1+\vartheta\frac{n^2}{(n-1)^2}\right] < \frac{1}{6}AB^{n-1}n^{-2}(1+4\vartheta) < \frac{1}{3n^2}AB^{n-1},$$

since $\vartheta < 1/6$. Thus, recalling (1.11), we have

$$\|\varphi_n\| < AB^{n-1}n^{-2},$$

and this inequality holds for $n = 2, 3, \cdots$. Noting that $|x| \leqq \|x\|$, we see from (1.8) that

$$|F_n(z)| < \frac{4}{3}|Df(z)|\,AB^{n-1}n^{-2}, \tag{1.18}$$

if $z \in K_1$. Thus for $|z| \leqq r_0 + 3\Delta/4$ the series

$$\sum_0^\infty \xi^k F_k(z)$$

converges for $|\xi| \leqq B^{-1}$ and is a regular function of z in this disc. From the construction of the functions $F_k(z)$ it is clear that at $z = 0$ the sum of this series is zero. If $r_0 + \Delta/4 \leqq |z| < (r_0 + \Delta/4)^{-1}$, then

$$|\varphi(z, \xi)| < \sum_1^\infty |\xi|^k |\varphi_k(z)| < \frac{A}{B}\sum_1^\infty (|\xi|B)^k k^{-2} \leqslant \frac{\pi^2}{6}\frac{A}{B} \tag{1.19}$$

for $|\xi B| \leqq 1$. We shall show that under this condition the function $\psi(z, \xi) = z\exp(\varphi(z, \xi))$ is univalent in the annulus $r_0 + \Delta/2 \leqq |z| \leqq 1$. First of all note that from (1.19) and Cauchy's formula it follows trivially that

$$|\varphi_z(z, \xi)| < \frac{2\pi^2}{3}\frac{A}{B\Delta}, \quad r_0 + \frac{\Delta}{2} \leqslant |z| \leqslant 1.$$

Therefore, if $r_0+\Delta/2 \leqq |z_j| \leqq 1$, $j=1,2$, then

$$|\varphi(z_1,\xi)-\varphi(z_2,\xi)|<\frac{2\pi^3}{3}\cdot\frac{A}{B\Delta}|z_1-z_2|,$$

and hence from (1.17) and the fact that $\vartheta<6/\pi^2$ we have that

$$|\varphi(z_1,\xi)-\varphi(z_2,\xi)|<\frac{\Delta}{4\pi}|z_1-z_2|.$$

From this estimate it follows that

$$|\psi(z_1,\xi)-\psi(z_2,\xi)|>|z_1-z_2|\exp\{\operatorname{Re}\varphi(z_1)\}\left[1-\left|\frac{\exp\{\varphi(z_2)-\varphi(z_1)\}-1}{z_2-z_1}\right|\right]$$

$$>|z_1-z_2|\exp\{\operatorname{Re}\varphi(z_1)\}\left[1-\left|\frac{\varphi(z_1)-\varphi(z_2)}{z_1-z_2}\right|\exp\{|\varphi(z_1)|+|\varphi(z_2)|\}\right]$$

$$>|z_1-z_2|\exp\{\operatorname{Re}\varphi(z_1)\}\left[1-\frac{\Delta}{4\pi}\exp\left(\frac{\pi^2A}{3B}\right)\right].$$

But

$$\frac{\Delta}{4\pi}\exp\left\{\frac{\pi^2A}{3B}\right\}=\frac{\Delta}{4\pi}\exp\left\{2\left(\frac{\pi\Delta\vartheta}{24}\right)^2\right\}<\frac{\Delta e}{4\pi}<\frac{1}{4},$$

and so

$$|\psi(z_1,\xi)-\psi(z_2,\xi)|>\frac{3}{4}|z_1-z_2|\exp\{\operatorname{Re}\varphi(z_1)\}.$$

If ξ is real and $|z|=1$, we have $|\psi(z,\xi)|=1$ by construction. For $|z|=r_0+\Delta/2$ we have

$$|\psi(z,\xi)|<\left(r_0+\frac{1}{2}\Delta\right)\exp\left(\frac{\pi^2A}{6B}\right)<\left(r_0+\frac{1}{2}\Delta\right)\exp\left\{\left(\frac{\pi\Delta\vartheta}{24}\right)^2\right\}$$

$$<r_0+\frac{1}{2}\Delta+2\left(\frac{\pi\Delta\vartheta}{24}\right)^2<r_0+\frac{1}{2}\Delta+\frac{1}{8}\Delta<1.$$

Similarly we obtain the inequality $|\psi(z,\xi)|>r_0+\Delta/2-\Delta/8>r_0$. Hence for real ξ with $|\xi B|\leqq 1$ the image of the annulus $r_0+\Delta/2\leqq|z|\leqq 1$ under the mapping $t=\psi(z,\xi)$ lies in the annulus $r_0\leqq|z|\leqq 1$. Thus $F[\psi(z,\xi),\xi]$ is a regular function in the annulus $r_0+\Delta/2\leqq|z|\leqq 1$. On the other hand, we have shown that this composite function can be expanded in a convergent series if $|\xi B|\leqq 1$, and the sum of this series is a regular function for $|z|<r_0+3\Delta/4$. Thus for $r_0+\Delta/2\leqq|z|<r_0+3\Delta/4$ we have

$$F[\psi(z,\xi),\xi]=\sum_0^\infty \xi^k F_k(z),\qquad F_0(z)=f(z),$$

and hence $F[\psi(z,\xi),\xi]$ is regular for $|z|<1$. Taking

$$\varepsilon_0=B^{-1}=\frac{\vartheta^2}{72}\left(\frac{\Delta}{8}\right)^2=\alpha\Delta^4 \tag{1.20}$$

and $\xi=\epsilon$, we find that $F[\psi(z,\epsilon),\epsilon]=F_*(z)$ is regular and univalent in $|z|<1$, if $\epsilon\leqq\epsilon_0$. In addition, $F_*(0)=0$, and the proof of the theorem is complete.

In general, $F_*(z)$ will not be normalized, so we define

$$f_*(z) = \frac{F_*(z)}{F'_*(0)}.$$

It is clear that

$$F'_*(0) = 1 + \sum_1^\infty \varepsilon^k F'_k(0).$$

From (1.8) we have $F'_k(0) = b_{k0} + ic_k$, where b_{k0} is the constant term in the Laurent expansion of $\psi_k(z)$. Since $|b_{k0}| \leqq \|\psi_k\|$, from (1.11) we have

$$|F'_k(0)| < 2\|\psi_k\| < \frac{2}{3} AB^{k-1} k^{-2}. \tag{1.21}$$

The role of the arbitrary constants c_k will be very important, as will be seen below.

LEMMA 1.4. *The constants c_k can be defined so that $F'_*(0) > 0$.*

PROOF. Setting $c_k = -\operatorname{Im} b_{k0}$, we have $F'_k(0) = \operatorname{Re} b_{k0}$. Since $|b_{k0}| \leqq \|\psi_k\|$, we have that $|c_k| \leqq \|\psi_k\|$, and thus the above choice of c_k is consistent with (1.11). Thus

$$F'_*(0) = 1 + \sum_1^\infty \varepsilon^k \operatorname{Re} b_{k0} \geqslant 1 - \sum_1^\infty \varepsilon^k |\operatorname{Re} b_{k0}| \geqslant 1 - \sum_1^\infty \varepsilon^k \|\psi_k\|.$$

It follows that

$$F'_*(0) \geqslant 1 - \frac{1}{3}\sum_1^\infty \varepsilon^k AB^{k-1}k^{-2} \geqslant 1 - \frac{\pi^2}{18}\frac{A}{B} > 0.$$

Set $G = \{w : w = F(z), r_0 \leqq |z| < 1\}$, and note that one of the components of the complement (with respect to the w-sphere) of the domain G will contain $w = 0$. If we denote this domain by G_0, then G_0 is bounded by the Jordan curve $\Gamma_0 = \{w : w = F(z), |z| = r_0\}$; we also put $G_* = G \cup \overline{G_0}$, so that $F_*(z)$ maps $|z| < 1$ onto G_*.

LEMMA 1.5. *Suppose that the conformal radius of G_* at $w = 0$ is 1. Then, with an appropriate choice of the arbitrary constants c_k, we have that $F_*(z) \in S$.*

PROOF. From Lemma 1.4 it follows that we can choose the arbitrary constants so that $F'_*(0) > 0$. But by hypothesis there exists $g(z) \in S$ mapping $|z| < 1$ onto G_*, and the uniqueness theorem yields that $F_*(z) \equiv g(z)$. Q.E.D.

Let the normalized function f_* be given by

$$f_*(z) = \sum_0^\infty \varepsilon^k f_k(z), \qquad f_0(z) = f(z).$$

Set

$$[F'_*(0)]^{-1} = 1 + \sum_1^\infty \varepsilon^k d_k.$$

We shall now estimate $|d_k|$. Applying the standard techniques of majorizing series, we find from (1.21) that

$$\left[1+\sum_1^\infty \varepsilon^k F_k'(0)\right]^{-1} \ll \left(1-\frac{2A}{3B}\sum_1^\infty \frac{(\varepsilon B)^k}{k^2}\right)^{-1} = \eta(B\varepsilon, \vartheta_0),$$

where $\vartheta_0 = 2A/3B < 6/\pi^2$. But according to Lemma 1.3

$$\eta(B\varepsilon, \vartheta_0) \ll 1+\vartheta_0\left(1+\frac{2\pi^2}{3}\vartheta_0\right)\eta^3(1, \vartheta_0)\sum_1^\infty \frac{(\varepsilon B)^k}{k^2},$$

and so

$$|d_k| < \vartheta_0\left(1+\frac{2\pi^2}{3}\vartheta_0\right)\eta^3(1, \vartheta_0)B^k k^{-2} < AB^{k-1}k^{-2},$$

where we have used (1.17).

Since

$$f_k(z) = F_k(z) + d_k f(z) + \sum_{l=1}^{k-1} d_{k-l}F_l(z),$$

it follows from (1.18) that for $|z| \leqq r_0 + 3\Delta/4$,

$$|f_k(z)| < AB^{k-1}k^{-2}\left[|f(z)| + \frac{4}{3}|Df(z)|\right] + \frac{4}{3}A^2B^{k-2}|Df(z)|\sum_1^{k-1} l^{-2}(k-l)^{-2}.$$

Therefore

$$|f_k(z)| < [|f(z)| + 2|Df(z)|]\, AB^{k-1}k^{-2}. \tag{1.22}$$

We shall require below a somewhat stronger form of Theorem 1. At first glance the new form appears very much stronger, but in reality it is a trivial extension of Theorem 1. We suppose that $g(z)$ has the form

$$g(z) = g(z, \varepsilon) = \sum_1^m \varepsilon^k g_k(z) + \varepsilon^{m+1}g_{m+1}(z, \varepsilon),$$

where $g_k(z)$, $k = 1, \cdots, m$, and $g_{m+1}(z, \epsilon)$ are regular in the annulus $r_0 \leqq |z| < 1$. We also assume that for $0 < \epsilon \leqq \epsilon_0$ the function $g(z, \epsilon)$ is admissible, and that in the annulus K_0

$$|g_k(z)| \leqq AB^{k-1}k^{-\beta}, \qquad k = 1, \cdots, m, \tag{1.23}$$

$$|g_{m+1}(z, \epsilon)| \leqq AB_1^m(m+1)^{-\beta}, \tag{1.24}$$

where A, B and B_1 are constants (with $B_1 \geqq B$), and where $\beta \geqq 2$ is a natural number. We set

$$F(z, \epsilon) = f(z) + g(z, \epsilon),$$

and we assume that the constant BA^{-1} satisfies

$$BA^{-1} \geqq \alpha_0\Delta^{-4}\vartheta^{-2}, \tag{1.25}$$

where α_0 is an absolute constant and ϑ is a constant to be specified below.

Lemma 1.6. *If* $0 < \epsilon \leqq \epsilon_0(\Delta, B_1)$ *and if condition* (1.25) *holds, then there exists a function* $\varphi(z, \epsilon)$ *satisfying conditions* 1) *and* 2) *of Theorem* 1 *such that*

$$F[z\exp(\varphi(z,\epsilon)),\epsilon] = F_*(z,\epsilon)$$

is regular and univalent for $|z| < 1$, *and such that* $F_*(0,\epsilon) = 0$ *and* $F_*'(0,\epsilon) > 0$. *For* $|z| \leqq r_0 + 3\Delta/4$,

$$F_*(z, \varepsilon) = \sum_{k=0}^{m} \varepsilon^k F_k(z) + \varepsilon^{m+1} F_{m+1}(z, \varepsilon), \qquad F_0(z) = f(z),$$

and

$$(1.26) \qquad |F_k(z)| \leqslant \frac{4}{3} |Df(z)| A_0 B^{k-1} k^{-\beta}, \qquad k = 1, \ldots, m,$$

$$(1.27) \qquad |F_{m+1}(z, \varepsilon)| \leqslant \frac{8}{3} |Df(z)| A_0 B_1^m m^{-\beta},$$

where $A_0 = 24 \cdot 8^3 \Delta^{-2} A$.

For the proof of this lemma we shall require a slight modification of Lemma 1.3.

Lemma 1.7. *The derivatives of* $\eta_\beta(x, \vartheta) = [1 - \vartheta \sum_1^\infty k^{-\beta} x^k]^{-1}$, *where* $|x| < 1$ *and* $0 < \vartheta \leqq 1/2\zeta(\beta)$, *satisfy*

$$\frac{1}{n!} \eta_\beta^{(n)}(0, \vartheta) < \gamma \vartheta n^{-\beta}, \qquad n = 1, 2, \ldots,$$

where the constant γ *depends only on* β.

Since the proof of this lemma is essentially the same as the proof of Lemma 1.3, we shall omit the details. In addition, the lemma will be applied only in the case of $\beta = 3$, and in this case the changes necessary in the proof of Lemma 1.3 are trivial.

Proof of Lemma 1.6. We consider the function

$$F(z, \varepsilon; \xi) = f(z) + \sum_1^m \xi^k g_k(z) + \xi^{m+1} g_{m+1}(z, \varepsilon).$$

It is clear that $F(z, \epsilon; \xi) = F(z, \epsilon)$. The construction of the function $\varphi(z, \xi) = \sum_1^\infty \xi^k \varphi_k(z)$ is exactly the same as the construction given in the course of the proof of Theorem 1. We choose the constants $c_1, c_2, \cdots$ as in Lemma 1.4. It follows that

$$F[z\exp(\varphi(z, \xi)), \varepsilon; \xi] = \sum_{k=0}^\infty \xi^k F_k(z).$$

For $n \leqq m$ we have

$$F_n(z) = \frac{1}{n!} \sum_{\nu_1 + 2\nu_2 + \cdots = n} k(\nu_1, \nu_2, \ldots)(1!)^{\nu_1}(2!)^{\nu_2} \ldots D^{\nu_1+\nu_2+\cdots} f(z) \varphi_1^{\nu_1} \ldots \varphi_n^{\nu_n}$$

$$+ \sum_{r=1}^{n-1} \frac{1}{(n-r)!} \sum_{\nu_1+2\nu_2+\cdots = n-r} k(\nu_1, \nu_2, \ldots)(1!)^{\nu_1}(2!)^{\nu_2} \ldots D^{\nu_1+\nu_2+\cdots} g_r(z) \varphi_1^{\nu_1}\varphi_2^{\nu_2} \cdots$$

$$\cdots + g_n(z).$$

For $n > m$ trivial changes must be made in this formula. We set

$$\psi_n(z) = \frac{1}{n!} \sum_{\substack{\nu_1+2\nu_2+\cdots=n \\ \nu_1+\nu_2+\cdots>1}} k(\nu_1, \nu_2, \ldots)(1!)^{\nu_1}(2!)^{\nu_2}$$

$$\ldots \frac{D^{\nu_1+\nu_2+\cdots} f(z)}{Df(z)} \varphi_1^{\nu_1}\varphi_2^{\nu_2} \cdots \varphi_{n-1}^{\nu_{n-1}} + \sum_{r=1}^{n-1} \frac{1}{(n-r)!} \sum_{\nu_1+2\nu_2+\cdots=n-r} k(\nu_1, \nu_2, \ldots)$$

$$\times (1!)^{\nu_1}(2!)^{\nu_2} \ldots \frac{D^{\nu_1+\nu_2+\cdots} g_r(z)}{Df(z)} \varphi_1^{\nu_1}\varphi_2^{\nu_2} \cdots + \frac{g_n(z)}{Df(z)}. \tag{1.28}$$

Then

$$F_n(z) = Df(z)\left[\varphi_n(z) + \psi_n(z)\right].$$

Since $\psi_1(z) = g_1(z)/Df(z)$, it follows from (1.10) and (1.23) that $\|\psi_1(z)\| < 2 \cdot 8^3\Delta^{-2}A$. In view of the choice of the arbitrary constants, we have that $\|\varphi_n\| < 3\|\psi_n\|$, and so

$$\|\varphi_1\| < A_0.$$

We suppose that the inequality

$$\|\varphi_j\| < A_0 B^{j-1} j^{-\beta}$$

holds for $j \leq n-1$, $n \leq m$, and we shall show that this inequality also holds for $j = n$. We write the formula for $\psi_n(z)$ as follows:

$$\psi_n(z) = \widetilde{\psi}_n(z) + \frac{g_n(z)}{Df(z)}. \tag{1.29}$$

As in the proof of Theorem 1, the previous estimates for $\|D^k f/Df\|$ and (1.25) imply that

$$\left\|\frac{D^k f}{Df}\right\| \left(\frac{A_0}{B}\right)^{k-1} \leqslant k!\, \vartheta^k, \qquad k = 2, 3, \ldots.$$

Similarly, using (1.23) and (1.13) we find that

$$\left\|\frac{D^k g_r}{Df}\right\| \left(\frac{A_0}{B}\right)^k < k!\, \vartheta^{k+1} A_0 B^{r-1} r^{-\beta}, \qquad k = 1, 2, \ldots.$$

Thus, recalling the definition of $\eta_\beta(x, \vartheta)$, we see from (1.28) and the induction hypothesis that

$$\|\widetilde{\psi}_n\| < A_0 B^{n-1} \left[\frac{\eta_\beta^{(n)}(0, \vartheta)}{n!} + \vartheta \sum_{r=1}^{n-1} \frac{\eta_\beta^{(n-r)}(0, \vartheta)}{(n-r)!} r^{-\beta}\right].$$

Upon combining Lemma 1.7 with this latter inequality, we see that

$$\|\widetilde{\psi}_n\| < A_0 B^{n-1} \gamma\vartheta \left[n^{-\beta} + \vartheta \sum_{r=1}^{n-1} (n-r)^{-\beta} r^{-\beta}\right].$$

But

$$\sum_{r=1}^{n-1} (n-r)^{-\beta} r^{-\beta} < 2\left(\frac{2}{n}\right)^\beta \zeta(\beta).$$

and so

$$\|\tilde{\psi}_n\| < \gamma(2^\beta + 1)\vartheta A_0 B^{n-1} n^{-\beta}. \tag{1.30}$$

We set $\vartheta = 1/4\gamma(2^\beta + 1)$. Then

$$\|\tilde{\psi}_n\| < \frac{1}{4} A_0 B^{n-1} n^{-\beta}.$$

Above we stated inequality (1.10) for the function $h(z)/Df(z)$. If this inequality is applied to the function $g_n(z)/Df(z)$, then in view of (1.23) we find that

$$\left\| \frac{g_n(z)}{Df(z)} \right\| < \frac{1}{12} A_0 B^{n-1} n^{-\beta},$$

and so $\|\psi_n\| < (1/3) A_0 B^{n-1} n^{-\beta}$, which in turn gives

$$\|\varphi_n\| < A_0 B^{n-1} n^{-\beta}. \tag{1.31}$$

Hence inequality (1.31) holds for $n \leqq m$. If however $n > m$, then it is clear that the same inequality holds, provided the constant B is replaced by B_1. This also follows as in the proof of Theorem 1. Therefore, for $0 < \epsilon \leqq \epsilon_0(\Delta, B_1)$, the function $F[z\exp(\varphi(z,\epsilon)),\epsilon]$ is regular and univalent in $|z| < 1$. For $|z| \leqq r_0 + 3\Delta/4$

$$F_*(z, \varepsilon) = \sum_{k=0}^{\infty} \varepsilon^k F_k(z),$$

where for $k \geqq m+1$ we have suppressed (for notational ease) the dependence of the function $F_k(z)$ on the parameter ϵ. From previously stated estimates it follows that for $k \leqq m$

$$|F_k(z)| \leqslant \frac{4}{3} |Df(z)| A_0 B^{k-1} k^{-\beta}.$$

If however $k > m$, then the constant B must be changed to B_1. We put

$$F_{m+1}(z, \varepsilon) = \sum_{k=m+1}^{\infty} \varepsilon^{k-m-1} F_k(z).$$

Therefore

$$|F_{m+1}(z, \varepsilon)| < \frac{4}{3} |Df(z)| A_0 B^m \sum_{k=m+1}^{\infty} (\varepsilon B_1)^{k-m-1} k^{-\beta},$$

and since we may assume that $\epsilon \leqq 1/2B_1$, (1.27) follows from this latter inequality. Also, as in the proof of Lemma 1.4, it is easily checked that $F'_*(0,\epsilon) > 0$. This completes the proof of the lemma.

§2. Usually one limits oneself at first to that rather narrow subclass of the class of admissible functions consisting of functions of the form

$$g(z, \varepsilon) = \varepsilon g(z) = \frac{\varepsilon}{2\pi i} \int_{\mathscr{C}} \frac{f^2(z)}{f(z) - f(t)} \left(\frac{Df(t)}{f(t)} \right)^2 h(t) \frac{dt}{t}, \tag{2.1}$$

where $\mathscr{C}$ is a smooth Jordan curve or arc in $|z| < 1$, $h(t)$ is an arbitrary integrable function, and ϵ is quite small. For technical reasons it is more convenient to consider admissible functions of the form

$$\varepsilon g(z) = \varepsilon \int_{\mathscr{C}} \frac{f^2(z)}{f(z) - f(t)} \left(\frac{Df(t)}{f(t)} \right)^2 \nu(dt), \tag{2.2}$$

where $\nu(dt)$ is a complex measure with support $\mathscr{C}$ lying in $|z| \leqq \rho < 1$. We shall now define the manner in which ϵ can be varied subject to the requirement that the function in (2.2) remains admissible. Set $C_r = \{w : w = f(z), |z| = r\}$. We first estimate the distance between C_ρ and $C_{(1+\rho)/2}$. According to a theorem of G. M. Goluzin [5],

$$\left| \frac{f(z_1) - f(z_2)}{z_1 - z_2} \right| \geqslant (1 - r_1^2)^{1/2} (1 - r_2^2)^{1/2} \left| \frac{f(z_1) f(z_2)}{z_1 z_2} \right|, \qquad r_j = |z_j|, \qquad j = 1, 2.$$

It now follows from a result of Bieberbach that

$$\left| \frac{f(z_1) - f(z_2)}{z_1 - z_2} \right| \geqslant \frac{(1 - r_1)^{1/2} (1 - r_2)^{1/2}}{(1 + r_1)^{3/2} (1 + r_2)^{3/2}},$$

so with $r_1 = \rho$ and $r_2 = \frac{1}{2}(1 + \rho)$ we see that

$$|f(z_1) - f(z_2)| \geqslant \frac{1}{\sqrt{2}} \left(\frac{1 - \rho}{4} \right)^2. \tag{2.3}$$

Set $M = \max\{\max_{z \in \mathscr{C}} |f(z)|, 1\}$. We shall use the identity

$$\frac{w_1^2}{w_1 - u} - \frac{w_2^2}{w_2 - u} = (w_1 - w_2) \frac{w_1 w_2 - u(w_1 + w_2)}{(w_1 - u)(w_2 - u)}.$$

Set $w_j = f(z_j)$, $j = 1, 2$, $u = f(t)$, and suppose $|z_j| \geqq \frac{1}{2}(1 + \rho)$. Then, using (2.3) and the inequality $M < (1 - \rho)^{-2}$, one can easily show that

$$\left| \frac{w_1 w_2 - u(w_1 + w_2)}{(w_1 - u)(w_2 - u)} \right| < 12 \left(\frac{2}{1 - \rho} \right)^8.$$

Recalling the inequality

$$\left| \frac{Df(z)}{f(z)} \right| < 2(1 - |z|)^{-1}, \tag{2.4}$$

we see that

$$|g(z_1) - g(z_2)| < 12 \left(\frac{2}{1 - \rho} \right)^{10} v(\nu, \mathscr{C}) |f(z_1) - f(z_2)|,$$

where $v(\nu, \mathscr{C})$ is the total variation of the measure $\nu(dz)$. Thus the function $\epsilon g(z)$ will be admissible in the annulus $\frac{1}{2}(1 + \rho) \leqq |z| < 1$ if

$$12 \left(\frac{2}{1 - \rho} \right)^{10} v(\nu, \mathscr{C}) |\varepsilon| < 1.$$

With the notation of Theorem 1, take $r_0 \geqq \frac{1}{2}(1 + \rho)$ and let Δ satisfy the inequality $r_0 + \Delta < 1$. We shall derive an upper estimate for $|\epsilon g(z)|$, $z \in K_0$. If $|f(z)| \geqq 2M$, then

$$|\varepsilon g(z)| < 2|\varepsilon| \left(\frac{2}{1 - \rho} \right)^2 v(\nu, \mathscr{C}) |f(z)| < 8(1 - r_0 - \Delta)^{-2} (1 - \rho)^{-2} v(\nu, \mathscr{C}) |\varepsilon|.$$

If $|f(z)| \leqq 2M$, then

$$|\epsilon g(z)| < 2^{8.5}(1-\rho)^{-8}v(\nu,\mathscr{C})|\epsilon|.$$

Set

$$\Omega = 8(1-\rho)^{-2}v(\nu,\mathscr{C})\max\{(1-r_0-\Delta)^{-2},\ 2^{5.5}(1-\rho)^{-6}\}. \tag{2.5}$$

From Theorem 1 we have that if

$$\Omega|\epsilon| \leqq \epsilon_0, \tag{2.6}$$

where ϵ_0 is defined by (1.20), then there exists a change of variable $z = \zeta\exp[\varphi(\zeta,\epsilon)]$ leading as before to the function

$$F(z,\epsilon) = f(z) + \epsilon g(z).$$

The following lemma is an immediate consequence of Theorem 1 and the estimate (1.22).

Lemma 2.1. *Suppose $f(z) \in S$ and that we have an admissible function of the form* (2.2). *Then, if the real number ϵ satisfies condition* (2.6), *the function $F(z\exp(\varphi(z,\epsilon)),\epsilon) = F_*(z,\epsilon)$ will be regular and univalent in $|z| < 1$. For $|z| \leqq r_0 + 3\Delta/4$ the normalized function can be expanded in the convergent series*

$$f_*(z,\varepsilon) = \frac{F_*(z,\varepsilon)}{F_*'(0,\varepsilon)} = \sum_0^\infty \varepsilon^k f_k(z), \qquad f_0(z) = f(z),$$

and in this disc

$$|f_k(z)| < AB^{k-1}\Omega^k k^{-2}[|f(z)| + 2|Df(z)|], \qquad k \geqslant 1, \tag{2.7}$$

where A and B are defined by (1.12) *and* (1.17).

With regard to our class of admissible functions, it is relatively easy to introduce the concepts of the variation of a univalent function, of the variation of the coefficients, and so forth. Naturally, if we change the class of admissible functions, the corresponding variational formulas will be changed, and so each time we give a variational formula we shall also indicate the corresponding class of admissible functions. Below we give a definition concerned with the class of admissible functions of the form (2.2).

Definition 2.1. The function $k!f_k(z)$ will be called the *kth variation* of the univalent function $f(z)$, and will be denoted by $\delta^k f(z) = \delta^k f(z;\nu)$.

For the first variation we can immediately give a specific formula, first stated by M. Schiffer for a particular choice of the measure $\nu(dz)$. From (1.5) we have

$$F_1(z) = Df(z)\varphi_1(z) + g(z).$$

But from (2.2) it follows immediately that

$$\varphi_1(z) = \int_{\mathscr{C}} \frac{t}{t-z}\,\nu(dt) + \int_{\mathscr{C}} \frac{\bar{t}z}{1-\bar{t}z}\,\overline{\nu(dt)} + ic_1. \tag{2.8}$$

Thus

$$F_1(z) = \int_{\mathscr{C}} \left[\frac{f^2(z)}{f(z)-f(t)}\left(\frac{Df(t)}{f(t)}\right)^2 + \frac{Df(z)\,t}{t-z}\right] \nu(dt) \tag{2.9}$$
$$+ Df(z)\int_{\mathscr{C}} \frac{\bar{t}z}{1-\bar{t}z}\overline{\nu(dt)} + c_1 Df(z).$$

From this it is easily seen that

$$F_1'(0) = \int_{\mathscr{C}} \nu(dt) + ic_1, \tag{2.10}$$

and so

$$\delta f(z) = \int_{\mathscr{C}} \left[\frac{f^2(z)}{f(z)-f(t)}\left(\frac{Df(t)}{f(t)}\right)^2 + \frac{Df(z)\,t}{t-z} - f(z)\right] \nu(dt) \tag{2.11}$$
$$+ Df(z)\int_{\mathscr{C}} \frac{\bar{t}z}{1-\bar{t}z}\overline{\nu(dt)} + ic_1(Df(z) - f(z)).$$

Without loss of generality we assume that no open neighborhood of the origin is contained in the support of the measure $\nu(dz)$. Since the functions $g(z)$ and $\varphi_1(z)$ are regular on that set in $|z|<1$ which lies outside $\mathscr{C}$, we may consider each term separately in (2.11), provided $|z|$ is small.

DEFINITION 2.2. The coefficients of the power series for the function $\delta^k f(z)$ are called the *kth variations* of the coefficients of $f(z)$.

Using the above remarks, it is easy to compute the first coefficient variation. Set

$$f(z) = \sum_0^\infty a_n z^{n+1}, \qquad a_0 = 1.$$

The coefficient variations will be denoted by $\delta^k a_n = \delta^k a_n(f;\nu)$. If $t \neq 0$, then the function $f^2(z)[f(z)-f(t)]^{-1}$ is regular at $z=0$. Set $w=f(t)$. For small $|z|$ we have

$$\frac{f^2(z)}{f(z)-w} = \sum_1^\infty q_n(w)\, z^{n+1}.$$

It is easy to see that $q_n(w)$ is a polynomial of degree n in $1/w$. We also note that

$$q_n(w) = \frac{1}{2\pi i}\int_{|z|=\rho} \frac{f^2(z)}{f(z)-w}\frac{dz}{z^{n+2}} = -\frac{f^2(t)}{Df(t)}t^{-n-1} + \frac{1}{2\pi i}\int_K \frac{f^2(z)}{f(z)-w}\frac{dz}{z^{n+2}},$$

where $\rho < |t|$, and K is a smooth Jordan curve containing t in its interior. If we choose K so that $|w/f(z)|<1$ for $z \in K$, then

$$\frac{f^2(z)}{f(z)-w} = f(z) + w + w\sum_{l=1}^\infty \left[\frac{w}{f(z)}\right]^l.$$

Therefore

$$q_n[f(t)] = -\frac{f^2(t)}{Df(t)}t^{-n-1} + a_n + \sum_{l=1}^\infty B_l[f(t)]^{l+1}. \tag{2.12}$$

The series is obviously convergent for small $|t|$. From (2.11) it follows immediately that

$$\delta a_n = \int_{\mathscr{C}} \mathfrak{A}_n(z)\, \nu(dz) + \int_{\mathscr{C}} \mathfrak{B}_n(\bar{z})\, \overline{\nu(dz)} + ic_1 na_n, \tag{2.13}$$

where

$$\mathfrak{A}_n(z) = \left(\frac{Df(z)}{f(z)}\right)^2 q_n[f(z)] + \sum_{l=1}^{n} (n-l+1)\, a_{n-l} z^{-l} + na_n \tag{2.14}$$

and

$$\mathfrak{B}_n(z) = \sum_{l=1}^{n} (n-l+1)\, a_{n-l} z^l. \tag{2.15}$$

We now require the following very simple lemma.

Lemma 2.2. *The functions $\mathfrak{A}_n(z)$ and $\mathfrak{B}_n(z)$ are regular in $|z|<1$, and $\mathfrak{A}_n(0) = \mathfrak{B}_n(0) = 0$.*

Proof. From (2.12)

$$\left(\frac{Df(z)}{f(z)}\right)^2 q_n[f(z)] = -Df(z)\, z^{-n-1} + a_n \left(\frac{Df(z)}{f(z)}\right)^2 + \mathfrak{P}(z),$$

where $\mathfrak{P}(z)$ is a series convergent in a neighborhood of $z=0$. Therefore

$$\left(\frac{Df(z)}{f(z)}\right)^2 q_n[f(z)] = -\sum_{l=1}^{n} (n-l+1)\, a_{n-l} z^{-l} - na_n + O(z).$$

The regularity of $\mathfrak{A}_n(z)$ at $z=0$ follows from this, and clearly $\mathfrak{A}_n(0)=0$. It is obvious that $\mathfrak{B}_n(z)$ possesses the required properties.

We next determine a formula for the second variation of the function $f(z)$. This formula will be a bit more cumbersome than the formula for $\delta f(z)$, and we shall not be able to present it in so complete a form as the formula for $\delta f(z)$. From (1.4) we have

$$F_2(z) = \frac{1}{2} D^2 f \varphi_1^2 + Df\varphi_2 + Dg\varphi_1.$$

In place of the function φ_2 we introduce the function $\psi(z) = \varphi_2(z) - \varphi_1(z) D\varphi_1(z)$. The condition $\operatorname{Re}\varphi_2(z) = 0$ for $|z|=1$ implies that $\operatorname{Re}\psi(z)=0$ for $|z|=1$. Thus

$$F_2(z) = -\frac{1}{2} D^2 f \varphi_1^2 + DF_1\varphi_1 + Df\psi. \tag{2.16}$$

We represent the function $\psi(z)$ in the form $\psi(z) = \psi_1(z) + \psi_2(z) + ic_2$, where $\psi_1(z)$ basically represents the singularities of the function $-\frac{1}{2}(D^2f/D)\varphi_1^2$, and where $\psi_2(z)$ similarly represents the singularities of $(DF_1/Df)\varphi_1$. It is clear that

$$\psi_2(z) = -\int_{\mathscr{C}} \frac{DF_1(t)}{Df(t)} \frac{t}{t-z}\, \nu(dt) - \int_{\mathscr{C}} \overline{\frac{DF_1(t)}{Df(t)}} \frac{\bar{t}z}{1-\bar{t}z}\, \overline{\nu(dt)}.$$

The formula for $\psi_1(z)$ is quite involved; we have that

$$\psi_1(z) = \frac{1}{2}\int_{\mathscr{C}}\int_{\mathscr{C}} \frac{\zeta t}{\zeta - t}\left[\frac{\dfrac{D^2 f(\zeta)}{Df(\zeta)}}{\zeta - z} - \frac{\dfrac{D^2 f(t)}{Df(t)}}{t - z}\right] \nu(d\zeta)\,\nu(dt)$$

$$+ \frac{1}{2}\int_{\mathscr{C}}\int_{\mathscr{C}} \frac{\bar\zeta \bar t}{\bar\zeta - \bar t}\left[\frac{\overline{\dfrac{D^2 f(\zeta)}{Df(\zeta)}}\, z}{1 - \bar\zeta z} - \frac{\overline{\dfrac{D^2 f(t)}{Df(t)}}\, z}{1 - \bar t z}\right] \overline{\nu(d\zeta)\nu(dt)}$$

$$+ \int_{\mathscr{C}} \frac{\zeta}{\zeta - z}\,\frac{D^2 f(\zeta)}{Df(\zeta)}\left[\int_{\mathscr{C}} \frac{\bar t \zeta}{1 - \bar t \zeta}\,\overline{\nu(dt)} + ic_1\right] \nu(d\zeta)$$

$$+ \int_{\mathscr{C}} \frac{\bar\zeta z}{1 - \bar\zeta z}\,\overline{\frac{D^2 f(\zeta)}{f(\zeta)}}\left[\int_{\mathscr{C}} \frac{t\bar\zeta}{1 - t\bar\zeta}\,\nu(dt) - i\bar c_1\right] \overline{\nu(d\zeta)}.$$

Since we are assuming that the point $z = 0$ is not on $\mathscr{C}$, it follows that all the terms which go to make up $F_2(z)$ are regular in a neighborhood of $z = 0$. Thus

$$F_2'(0) = -\frac{1}{2}\varphi_1^2(0) + F_1'(0)\,\varphi_2(0) + \psi(0).$$

It is easily seen that

$$\delta^2 f(z) = 2\,[F_2(z) - F_2'(0) f(z) - F_1'(0) F_1(z) + (F_1'(0))^2 f(z)].$$

If we are given a function $\chi(z)$, regular in a neighborhood of $z = 0$, and with $\chi(z) = \sum_0^\infty \alpha_n z^{n+1}$ there, we define the following functions:

$$\mathfrak{B}_n^*(z;\, \chi) = \sum_{l=1}^{n} (n - l + 1)\,\alpha_{n-l} z^{-l} + n\alpha_n$$

and

$$\mathfrak{B}_n(z;\, \chi) = \sum_{l=1}^{n} (n - l + 1)\,\alpha_{n-l} z^{l}.$$

If

$$\eta(z) = \int_{\mathscr{C}} \frac{\zeta}{\zeta - z}\,\mu(d\zeta) + \int_{\mathscr{C}} \frac{\bar\zeta z}{1 - \bar\zeta z}\,\overline{\mu(d\zeta)} + ic,$$

then the function $D\chi(z)\eta(z) - \chi(z)\eta(0)$ will have as the nth coefficient of its power series

$$\int_{\mathscr{C}} \mathfrak{B}_n^*(z;\, \chi)\,\mu(dz) + \int_{\mathscr{C}} \mathfrak{B}_n(\bar z;\, \chi)\,\overline{\mu(dz)} + icn\alpha_n.$$

It is now easy to compute the coefficients of the power series for $Df(z)\psi(z) - f(z)\psi(0)$. The nth coefficient for the series for $Df(z)\psi_1(z) - f(z)\psi_1(0)$ will be

$$\frac{1}{2}\int_{\mathscr{C}}\int_{\mathscr{C}} \frac{1}{t - \zeta}\left[t\,\frac{D^2 f(\zeta)}{Df(\zeta)}\,\mathfrak{B}_n^*(\zeta;\, f) - \zeta\,\frac{D^2 f(t)}{Df(t)}\,\mathfrak{B}_n^*(t;\, f)\right] \nu(d\zeta)\,\nu(dt)$$

(2.17)

$$+ \frac{1}{2}\int_{\mathscr{C}}\int_{\mathscr{C}} \frac{1}{\bar t - \bar\zeta}\left[\bar t\,\overline{\frac{D^2 f(\zeta)}{Df(\zeta)}}\,\mathfrak{B}_n^*(\bar\zeta;\, f) - \bar\zeta\,\overline{\frac{D^2 f(t)}{Df(t)}}\,\mathfrak{B}_n^*(\bar t;\, f)\right] \overline{\nu(d\zeta)\,\nu(dt)}$$

$$+\int_{\mathscr{C}}\int_{\mathscr{C}}\frac{D^2f(\zeta)}{Df(\zeta)}\mathfrak{B}_n^*(\zeta;f)\left[\int_{\mathscr{C}}\frac{\bar{t}\zeta}{1-\bar{t}\zeta}\overline{\nu(dt)}+ic_1\right]\nu(d\zeta)$$
$$+\int_{\mathscr{C}}\int_{\mathscr{C}}\overline{\frac{D^2f(\zeta)}{Df(\zeta)}}\mathfrak{B}_n(\bar{\zeta};f)\left[\int_{\mathscr{C}}\frac{t\bar{\zeta}}{1-t\bar{\zeta}}\nu(dt)-ic_1\right]\overline{\nu(d\zeta)}.$$

The coefficient of $Df(z)\psi_2(z)-f(z)\psi_2(0)$ is clearly equal to

$$-\int_{\mathscr{C}}\frac{DF_1(t)}{Df(t)}\mathfrak{B}_n^*(t;f)\nu(dt)-\int_{\mathscr{C}}\overline{\frac{DF_1(t)}{Df(t)}}\mathfrak{B}_n(\bar{t};f)\overline{\nu(dt)}. \tag{2.18}$$

The coefficient of $DF_1(z)\varphi_1(z)-F_1(z)\varphi_1(0)$ is

$$\int_{\mathscr{C}}\mathfrak{B}_n^*(z;F_1)\nu(dz)+\int_{\mathscr{C}}\mathfrak{B}_n(\bar{z};F_1)\overline{\nu(dz)}+ic_1na_n(F_1), \tag{2.19}$$

where $a_n(F_1)$ is the nth coefficient of the power series of $F_1(z)$. Thus to determine the coefficients of the function $F_2(z)-F_2'(0)f(z)$, we must find the coefficients of the function

$$-\frac{1}{2}D^2f(z)\varphi_1^2(z)+\frac{1}{2}f(z)\varphi_1^2(0)+F_1(z)\varphi_1(0)-F_1'(0)\varphi_1(0)f(z)+ic_2(Df(z)-f(z)).$$

Set

$$-\frac{1}{2}D^2f(z)\varphi_1^2(z)+\frac{1}{2}f(z)\varphi_1^2(0)=\sum_0^\infty\beta_nz^{n+1}.$$

Then the function $F_2(z)-F_2'(0)f(z)-F_1(z)\varphi_1(0)+F_1'(0)\varphi_1(0)f(z)$ has as its coefficient the sum of the following: β_n of the formulas (2.17)–(2.19), and the nth coefficient of the function $ic_2(Df(z)-f(z))$. Instead of computing δ^2a_n, we shall compute the sum $\operatorname{Re}\sum_1^n\bar{\lambda}_k\delta^2a_k$, since this will be more convenient for our purposes. We set

$$\begin{aligned}P(z;\chi)&=\sum_{l=1}^n[\bar{\lambda}_l\mathfrak{B}_l^*(z;\chi)+\lambda_l\overline{\mathfrak{B}_l(\bar{z};\chi)}]\\&=\sum_{l=1}^n\left[\bar{\lambda}_l\mathfrak{B}_l\left(\frac{1}{z};\chi\right)+\lambda_l\overline{\mathfrak{B}_l(\bar{z};\chi)}+\bar{\lambda}_l la_l\right].\end{aligned} \tag{2.20}$$

Then it is easily seen that

$$\begin{aligned}\operatorname{Re}\sum_{l=1}^n\bar{\lambda}_l\delta^2a_l=2\operatorname{Re}\Big\{&\frac{1}{2}\int_{\mathscr{C}}\int_{\mathscr{C}}\frac{1}{z-\zeta}\left[\frac{D^2f(\zeta)}{Df(\zeta)}zP_n(\zeta;f)-\frac{D^2f(z)}{Df(z)}\zeta P_n(z;f)\right]\\&\times\nu(dz)\nu(d\zeta)+\int_{\mathscr{C}}\frac{D^2f(z)}{Df(z)}P_n(z;f)\left[\int_{\mathscr{C}}\frac{\bar{\zeta}z}{1-\bar{\zeta}z}\overline{\nu(d\zeta)}+ic_1\right]\nu(dz)\\&+\int_{\mathscr{C}}\left[P_n(z;F_1)-\frac{DF_1(z)}{Df(z)}P_n(z;f)\right]\nu(dz)+ic_1\sum_{l=1}^n\bar{\lambda}_lla_l(F_1)\\&+\sum_{l=1}^n\bar{\lambda}_l\beta_l+ic_2\sum_{l=1}^n\bar{\lambda}_lla_l\Big\}.\end{aligned} \tag{2.21}$$

We shall alter this formula somewhat a bit later on.

§3. In the theory of extremal problems in the class S a very important role is played by the two systems of functions $\{\mathfrak{A}_k(z)\}$ and $\{\mathfrak{B}_k(z)\}$, and by various linear combinations of these functions. We now set $\overline{\mathfrak{B}}_k(z) = \overline{\mathfrak{C}_k(\bar{z})}$, and introduce the following functions:

$$\varphi_{2k-1}(z) = \mathfrak{A}_k(z) + \overline{\mathfrak{B}}_k(z), \qquad \varphi_{2k}(z) = \frac{1}{i}[\mathfrak{A}_k(z) - \overline{\mathfrak{B}}_k(z)], \qquad k = 1, 2, \cdots.$$

The functions $\varphi_k(z)$, $k = 1, 2, \cdots$, will be called the functions *associated* with the univalent function $f(z)$, and $\{\varphi_k(z)\}$ will be called the *system of associated functions*. From Lemma 2.2 it follows that the functions $\varphi_k(z)$ are regular in $|z| < 1$. The properties of $f(z)$ are closely related to properties of the system of associated functions, and among the most important properties of that system is the property of being either linearly independent or dependent over the reals. Let $\mathscr{C} = \{z : |z| = r^* < 1\}$. As a measure of the linear independence of the functions $\varphi_1, \cdots, \varphi_{2n}$ we can take

$$\inf_{|y|=1} \frac{1}{2\pi} \int_{\mathscr{C}} \left| \sum_{k=1}^{2n} y_k \varphi_k(z) \right|^2 \left| \frac{dz}{z} \right|, \tag{3.1}$$

where $y = (y_1, \cdots, y_{2n}) \in R^{2n}$. We denote (3.1) by $d_n^2(f, r^*) = d_n^2(f) = d_n^2$. It is clear that if $d_n(f, r^*) = 0$ for some r^*, then for $|z| < 1$

$$\sum_1^{2n} y_k \varphi_k(z) \equiv 0$$

and $d_n(f, r) = 0$ for $0 < r < 1$. Thus, as a measure of linear independence, $d_n(f, r^*)$ is independent of the choice of r^*, and so in the following we shall not indicate the dependence of $d_n(f)$ on r^*. In addition to the quantity d_n, another characteristic of the system $\varphi_1, \cdots, \varphi_{2n}$ will be very useful to us. From these functions we select one function, for example φ_j, and with the remaining functions we define a lower bound analogous to (3.1). The maximum value of these lower bounds for $1 \leqq j \leqq 2n$ will be denoted by $d_{n-1/2}^2(f) = d_{n-1/2}^2$.

The following problem is important in the theory of variational techniques. Suppose we are given $f(z) \in S$ and a function $g(z) \in S$ which lies in a small neighborhood of $f(z)$. When is it possible to find a varied function $f_*(z, \epsilon; \nu)$ whose first n coefficients coincide with the first n coefficients, respectively, of $g(z)$? It is easy to state a sufficient condition, and indeed the answer to the question is intuitively clear: if it is possible to choose a measure $\nu(dz)$ so that $\delta a_k(f; \nu) = x_k$, $1 \leqq k \leqq n$, where $x_1, \cdots, x_n$ are arbitrary complex numbers, then it is possible to find a varied function with the desired property. If we insist that this should be possible for any function $g(z)$ in a sufficiently small neighborhood, then the condition is also necessary. In the case of a fixed function $g(z)$ the question is not so simple, and only in some particular cases can we answer it completely.

We shall now discuss the question of when it is possible to select a measure $\nu(dz)$ so that the equations

$$\delta a_k(f;\nu) = x_k, \qquad 1 \leqq k \leqq n, \tag{3.2}$$

are satisfied, where $x_1, \cdots, x_n$ are arbitrary complex numbers. It will prove convenient to consider $(x_1, \cdots, x_n)$ as an element of R^{2n} written in a complex form. We set $x_k = \xi_{2k-1} + i\xi_{2k}$, $1 \leqq k \leqq n$, and so we have $x = (\xi_1, \cdots, \xi_{2n}) = (\xi_1 + i\xi_2, \cdots, \xi_{2n-1} + i\xi_{2n})$. We shall use the usual notation for the norm of the vector. We shall also use formula (2.13) in the special case $c_1 = 0$.

Lemma 3.1. *If $d_n > 0$, then equations* (3.2) *have a solution for any $x_1, \cdots, x_n$, and the measure $\nu(dz)$ can be taken to be of the form $\nu(dz) = \zeta(z)\,dz/2\pi iz$, $z \in \mathscr{C}$, where*

$$\left(\frac{1}{2\pi}\int_{\mathscr{C}} |\zeta(z)|^2 \left|\frac{dz}{z}\right|\right)^{1/2} \leqslant \frac{|x|}{d_n}.$$

Proof. It is clear that the infimum is attained in (3.1) and that the minimizing vector can be determined by elementary analysis. We define a scalar product for functions defined on $\mathscr{C}$ as follows:

$$(\varphi, \psi) = \operatorname{Re}\int_{\mathscr{C}} \varphi(z)\,\overline{\psi(z)}\,\frac{|dz|}{2\pi|z|}. \tag{3.3}$$

Then the equations for the minimizing vector are

$$\sum_{l=1}^{2n} (\varphi_k, \varphi_l)\,y_l - d_n^2 y_k = 0, \qquad k = 1, \ldots, 2n.$$

The matrix $A = ((\varphi_k, \varphi_l))_{k,l=1}^{2n}$ is symmetric, and its smallest eigenvalue is d_n^2. Denote by $\omega_1, \cdots, \omega_{2n}$ the eigenvalues of A, and denote the corresponding eigenvectors by $\eta^1, \cdots, \eta^{2n}$. These eigenvectors form an orthonormal basis. Consider the system of functions

$$\psi_l(z) = \frac{1}{\sqrt{\omega_l}} \sum_{k=1}^{n} \eta_{lk}\varphi_k(z), \qquad l = 1, \ldots, 2n, \tag{3.4}$$

where $\eta_{l1}, \cdots, \eta_{l2n}$ are the components of η^l. We shall show that the system (3.4) is orthonormal with respect to the scalar product (3.3). We have

$$\begin{aligned}(\psi_l, \psi_m) &= (\omega_l\omega_m)^{-\frac{1}{2}} \sum_{p,\,q=1}^{2n} \eta_{lp}\eta_{mq}(\varphi_p, \varphi_q) \\ &= (\omega_l\omega_m)^{-\frac{1}{2}} \sum_{p=1}^{2n} \eta_{lp} \sum_{q=1}^{2n} (\varphi_p, \varphi_q)\,\eta_{mq} = \left(\frac{\omega_m}{\omega_l}\right)^{\frac{1}{2}} \sum_{p=1}^{2n} \eta_{lp}\eta_{mp} = \delta_{lm}.\end{aligned}$$

Trusting that no confusion will arise, we denote the scalar product of x, $y \in R^{2n}$ by (x, y). Let the vector x in the basis $\eta^1, \cdots, \eta^{2n}$ have components

$u_1, \cdots, u_{2n}$, $u_j = (x, \eta^j)$, $1 \leqq j \leqq 2n$. It is easily seen that the desired function $\zeta(z)$ can be written as

$$\zeta(z) = \sum_1^{2n} \omega_l^{-1/2} u_l \overline{\psi_l(z)}.$$

To see this, set $v_j = (\varphi_j, \overline{\zeta})$, $1 \leqq j \leqq 2n$. Computing the scalar product of $v = (v_1, \cdots, v_{2n})$ and η^k, we find that

$$(v, \eta^k) = \operatorname{Re} \frac{1}{2\pi} \int_{\mathscr{C}} \sum_{j=1}^{2n} \eta_{kj} \varphi_j(z) \sum_{l=1}^{2n} \omega_l^{-1/2} u_l \overline{\psi_l(z)} \left| \frac{dz}{z} \right|$$
$$= \sqrt{\omega_k} \left(\psi_k \sum_{l=1}^{2n} \omega_l^{-1/2} u_l \psi_l \right) = u_k, \qquad 1 \leqslant k \leqslant 2n.$$

Thus $v = x$. It follows that $\xi_{2k-1} = (\varphi_{2k-1}, \overline{\zeta})$, $\xi_{2k} = (\varphi_{2k}, \overline{\zeta})$ and

$$x_k = \xi_{2k-1} + i\xi_{2k} = (\varphi_{2k-1}, \overline{\zeta}) + i(\varphi_{2k}, \overline{\zeta}) = \operatorname{Re} \frac{1}{2\pi} \int_{\mathscr{C}} [\mathfrak{A}_k(z) + \overline{\mathfrak{B}}_k(z)] \zeta(z) \left| \frac{dz}{z} \right|$$
$$+ i \operatorname{Re} \frac{1}{2\pi i} \int_{\mathscr{C}} [\mathfrak{A}_k(z) - \overline{\mathfrak{B}}_k(z)] \zeta(z) \left| \frac{dz}{z} \right|$$
$$= \frac{1}{2\pi} \int_{\mathscr{C}} [\mathfrak{A}_k(z) \zeta(z) + \mathfrak{B}_k(\bar{z}) \overline{\zeta(z)}] \left| \frac{dz}{z} \right| = \delta a_k(f; \nu), \qquad 1 \leqslant k \leqslant n.$$

Finally

$$\frac{1}{2\pi} \int_{\mathscr{C}} |\zeta(z)|^2 \left| \frac{dz}{z} \right| = \sum_{k,l=1}^{2n} [\omega_k \omega_l]^{-1/2} u_k u_l (\varphi_k, \varphi_l) = \sum_{k=1}^{2n} \frac{u_k^2}{\omega_k} \leqslant \frac{|x|^2}{d_n^2}.$$

Remark 3.1. From the proof it is clear that $\zeta(z; x)$ is a linear function of the components of x.

Theorem 2. *Let $f(z) \in S$, $f(z) = \sum_0^\infty a_k z^{k+1}$ and $a_0 = 1$. Let*

$$h(z) \in S, \ d_n(h) > 0, \ h(z) = \sum_{k=0}^{\infty} b_k z^{k+1}, \ b_0 = 1.$$

Then there exist absolute constants α, β and β_0 such that if

$$\left(\sum_1^n |a_k - b_k|^2 \right)^{1/2} < \alpha \min [n^{-\beta} d_n^2, \ n^{-\beta_0} d_n],$$

it is possible to find a variation $h_*(z, \epsilon)$ of $h(z)$ such that

$$h_*(z, \varepsilon) = \sum_{k=0}^{n} a_k z^{k+1} + \sum_{k=n+1}^{\infty} b_k^*(\varepsilon) z^{k+1}.$$

Proof. Throughout the proof we shall use the letters α and β to denote absolute constants. Let $x = (\xi_1, \cdots, \xi_{2n})$ be any vector such that $|\xi_j| \leqq 1$, $j = 1, \cdots, 2n$. Corresponding to x, we construct, as described in Lemma 3.1, a measure $\nu(dz) = \zeta(z) dz / 2\pi i z$ with support $\mathscr{C}$. From this lemma we have

$$v(\nu;\mathscr{C})<\frac{|x|}{d_n(h)}\leqslant\frac{\sqrt{2n}}{d_n(h)}.$$

We apply Lemma 2.1 to that admissible function of the form (2.2) defined by the function $h(z)$ and the measure $\nu(dz)$. In the notation of §2 we have $\rho=r^*$, and we take $r^*\leqq 3/8$, $r_0=1-5/2(n+2)$ and $\Delta=2/(n+2)$. In this case we have the estimate

$$\Omega<\alpha n^\beta d_n^{-1}(h)=\Omega_0 \tag{3.5}$$

for the quantity Ω defined by (2.5), where α and β are absolute constants. If $\epsilon>0$ satisfies

$$\alpha\epsilon n^\beta d_n^{-1}(h)\leqq\tfrac{1}{2}\epsilon_0, \tag{3.6}$$

then we may apply Lemma 2.1 to obtain the function

$$h_*(z,\varepsilon)=\sum_{k=0}^{\infty}\varepsilon^k h_k(z),\qquad h_0(z)=h(z), \tag{3.7}$$

where the series converges in $|z|\leqq r_0+3\Delta/4=1-1/(n+2)$. According to (2.7), for z in this disc we have that

$$|h_k(z)|\leqslant AB^{k-1}\Omega^k k^{-2}[|h(z)|+2|Dh(z)|]. \tag{3.8}$$

From (1.12) $A=48(4n+8)^2$, and from (1.17) $B=72(4n+8)^4\vartheta^{-2}$. We take all the arbitrary constants $c_1,c_2,\cdots$ to be zero. Then the function $h_k(z)$ will be a homogeneous polynomial of the coordinate vector x of degree k. For $k=1$ this is clear, and in the general case it follows by induction on k upon using (1.4) and the formulas for $\psi_n(z)$ established in §1. We set

$$h_*(z,\varepsilon)=\sum_0^{\infty}b_k^*(\varepsilon)z^{k+1},\quad b_0=1.$$

From (3.7) we have

$$b_l^*=\sum_{k=0}^{\infty}\varepsilon^k\frac{1}{2\pi i}\int_{|z|=1-\frac{1}{n+2}}h_k(z)z^{-l-2}dz=\sum_{k=0}^{\infty}\varepsilon^k b_{lk}(x).$$

It follows from the above remarks that $b_{lk}(x)$ is a homogeneous polynomial of degree k, and it is also clear that

$$b_{l0}=b_l,\quad b_{l1}=\delta b_l=x_l,\qquad 1\leqq l\leqq n.$$

From the inequality (2.4) and the well-known result

$$\frac{1}{2\pi}\int_{|z|=r}|h(z)|\left|\frac{dz}{z}\right|<r(1-r)^{-1}$$

we find that

$$\frac{1}{2\pi}\int_{|z|=1-\frac{1}{n+2}}[|h(z)|+2|Dh(z)|]\frac{|dz|}{|z|^{l+2}}<\left(1-\frac{1}{n+2}\right)^{-l}[4(n+2)^2+n+2].$$

Thus it follows from (3.8) that for $l \leqq n$

(3.9) $$|b_{lk}(x)| \leqq \alpha_0 AB^{k-1}\Omega^k k^{-2} n^2.$$

Set

$$b_l^* = b_l + \varepsilon x_l + \sum_{k=2}^{\infty} \varepsilon^k b_{lk}(x) = b_l + \varepsilon x_l + \varepsilon \mathcal{P}_l(x, \varepsilon), \qquad 1 \leqslant l \leqslant n.$$

Since $x_k = \xi_{2k-1} + i\xi_{2k}$, $b_{lk}(x)$ is homogeneous with respect to ξ_j of degree at most k, and for $|\xi_j| \leqq 1$ the inequality (3.9) holds. Thus

$$\left|\frac{\partial b_{lk}}{\partial \xi_j}\right| \leqslant \alpha_0 n^2 AB^{k-1}\Omega^k, \qquad |\xi_j| \leqslant 1, \qquad 1 \leqslant j \leqslant 2n,$$

by a theorem of V. A. Markov. Hence

$$\left|\frac{\partial \mathcal{P}_l(x)}{\partial \xi_j}\right| < \alpha_0 n^2 A\Omega \sum_{k=1}^{\infty} (\varepsilon B\Omega)^k < 2\alpha_0 n^2 AB\Omega^2 \varepsilon,$$

since $\epsilon B\Omega < \frac{1}{2}$ (which follows in turn from the definition of the constants α and β in (3.5)). From this latter result it follows that

$$|\operatorname{grad} \mathcal{P}_l(x)| < 2\sqrt{2}\alpha_0 n^{5/2} AB\Omega^2 \varepsilon,$$

and so

(3.10) $$\left[\sum_{l=1}^{n} |\operatorname{grad} \mathcal{P}_l(x)|^2\right]^{1/2} < 4\alpha_0 n^3 AB\Omega^2 \varepsilon.$$

Set $b_l^* = a_l$, $l = 1, \cdots, n$, and put

$$\varepsilon = 2\left(\sum_1^n |a_l - b_l|^2\right)^{1/2}.$$

Then, having put $a_l - b_l = y_l$, $|y_l| \leqq \frac{1}{2}$, $l = 1, \cdots, n$, we arrive at the system of equations

$$y_l = x_l + \mathcal{P}_l(x, \varepsilon), \qquad l = 1, \ldots, n.$$

Since we require that

(3.11) $$4\alpha_0 n^3 AB\Omega^2 \varepsilon < \frac{1}{2},$$

we see from (3.10) that

(3.12) $$\left[\sum_{l=1}^{n} |\operatorname{grad} \mathcal{P}_l(x, \varepsilon)|^2\right]^{1/2} < \frac{1}{2}.$$

If in the left-hand side of (3.11) we have in place of Ω the quantity Ω_0 defined in (3.5), and if we require that the final inequality still hold, then we see that $\epsilon < 2\alpha n^{-\beta} d_n^2$ must hold. In this case, when $\Omega_0 < 1$, (3.11) is obviously true, since $4\alpha_0 n^3 AB\Omega^2 \epsilon < 4\alpha_0 n^3 AB\Omega\epsilon < \frac{1}{2}$ (from (3.6) and the corresponding choices of α and β). From (3.6) we have $\epsilon < 2\alpha n^{-\beta_0} d_n$, and so

(3.13) $$\epsilon < 2\alpha \min(n^{-\beta} d_n^2, n^{-\beta_0} d_n).$$

We denote by $\mathscr{P}(x)$ the vector function $\mathscr{P}(x) = \{\mathscr{P}_1(x), \cdots, \mathscr{P}_n(x)\}$. Then the system of equations for the vector x can be written in the form

(3.14) $$y = x + \mathscr{P}(x).$$

From (3.12) it follows that the transformation (3.14) is a homeomorphism in the cube $Q = \{x : |\xi_j| \leq 1,\ 1 \leq j \leq 2n\}$. Furthermore, from (3.9) with $l = 1, \cdots, n$ we obtain

$$|\mathscr{P}_l(x)| < \alpha_0 n^2 A \Omega \sum_{k=1}^{\infty} \frac{(\varepsilon \Omega B)^k}{(k+1)^2} < \frac{1}{2} \alpha_0 n^2 A B \Omega^2 \varepsilon < \frac{1}{16n},$$

where we have used (3.11). Thus $|y - x| < 1/16\sqrt{n}$ for $x \in Q$, and so the image of ∂Q lies outside the cube $\{x : |\xi_j| \leq \frac{1}{2},\ 1 \leq j \leq 2n\}$. It follows from this that the equations (3.14) have a solution, and for this special value of x we have $|y_j| \leq \frac{1}{2}$, $1 \leq j \leq n$, as required. Q.E.D.

Remark 3.2. One might feel that the choice of the variation of $h(z)$ based on the class of admissible functions of the form (3.2) strongly affects the result of Theorem 2. If $d_n(h) > 0$, then an investigation of a variation based on a wider class cannot give any more information than is given by Theorem 2. If however $d_n(h) = 0$, then with the aid of the class of admissible functions (2.2) it is in general impossible to construct the desired variation, even in those cases when the solution to the problem in question might be a positive answer. In these cases one must consider a much wider class of admissible functions.

The remainder of this section will be devoted to a discussion of some questions related to the quantities d_n and $d_{n-1/2}$ introduced above in connection with the associated system of functions. Suppose that $d_n(f) = 0$. Let $(\xi_1, \cdots, \xi_{2n})$ be one of the vectors for which

$$\frac{1}{2\pi} \int_{\mathscr{C}} \left| \sum_1^{2n} \xi_k \varphi_k(z) \right|^2 \left| \frac{dz}{z} \right| = 0.$$

Then for $|z| < 1$ we have $\sum_1^{2n} \xi_k \varphi_k(z) \equiv 0$, or

$$\sum_{k=1}^{n} [(\xi_{2k-1} - i\xi_{2k}) \mathfrak{A}_k(z) + (\xi_{2k-1} + i\xi_{2k}) \overline{\mathfrak{B}}_k(z)] = 0.$$

Put $\lambda_k = \xi_{2k-1} + i\xi_{2k}$ and $\lambda = (\lambda_1, \cdots, \lambda_n)$, where λ is a vector in R^{2n} written in complex form. Then the latter equation can be written as

(3.15) $$\sum_{k=1}^{n} [\overline{\lambda}_k \mathfrak{A}_k(z) + \lambda_k \overline{\mathfrak{B}}_k(z)] = 0.$$

Upon using relations (2.14) and (2.15), we find that

$$Q_n(w;\, f) = Q_n(w) = Q(w) = \sum_{k=1}^{n} \bar{\lambda}_k q_k(w). \tag{3.16}$$

Thus (3.15) can be written in the form

$$\left(\frac{Df(z)}{f(z)}\right)^2 Q[f(z)] + P(z) = 0, \tag{3.17}$$

where $P(z) = P(z; f)$ is as in (2.20). It is clear that

$$Q(w) = \sum_{k=1}^{n} A_k w^{-k}.$$

It is easy to see that $q_k(w) = -w^{-k} +$ lower powers of w^{-1}. Therefore

$$A_n = -\bar{\lambda}_n \tag{3.18}$$

With regard to $P(z)$, it follows from (2.20) and the definitions of the functions $\mathfrak{B}_k(z; f)$ and $\mathfrak{B}_k^*(z; f)$ that

$$P(z) = \sum_{k=-n}^{n} B_k z^{-k},$$

where

$$B_0 = \sum_{k=1}^{n} \bar{\lambda}_k k a_k, \qquad B_j = \bar{B}_{-j}, \qquad 1 \leqslant j \leqslant n, \tag{3.19}$$

and

$$B_n = \bar{\lambda}_n. \tag{3.20}$$

The vector λ will be called an *associated vector*. Clearly for any real constant α the vector $\alpha\lambda$ will also be an associated vector, and the sum of two associated vectors will also be an associated vector. Thus the set of all associated vectors is a linear manifold, which we denote by Λ_n. Of course, for a given $f(z)$, Λ_n might be empty. If however Λ_n is not empty, then Λ_n can give us much information about $f(z)$, and we shall be especially interested in the dimension of Λ_n. Clearly if $d_n(f) = 0$ and $d_{n-1/2}(f) > 0$, then Λ_n is one-dimensional.

We shall usually assume that the associated vector λ has been normalized so that $|\lambda| = 1$; whenever this normalization is not used, we shall mention this fact explicitly. We now consider the question of the dependence of λ on the given function $f(z)$. Toward this end the following lemma is useful.

LEMMA 3.2. *If $d_n = 0$ and $d_{n-1} > 0$, then any two associated vectors λ and λ^1 satisfy an equation $\lambda^1 = \omega\lambda$ with $\omega^2 = 1$.*

PROOF. Suppose that the lemma is false, and choose two vectors λ and λ^1 such that $\lambda^1 \neq \omega\lambda$. Since $f(z)$ satisfies

$$\left(\frac{Df(z)}{f(z)}\right)^2 Q[f(z);\lambda] + P(z;\lambda) = 0, \tag{3.17}$$

and

$$\left(\frac{Df(z)}{f(z)}\right)^2 Q[f(z);\lambda^1] + P(z;\lambda^1) = 0, \tag{3.17'}$$

it is clear that

$$\frac{Q(f;\lambda)}{Q(f;\lambda^1)} = \frac{P(z;\lambda)}{P(z;\lambda^1)}\,.$$

and moreover the assumption $\lambda^1 \neq \omega\lambda$ implies that $P(z;\lambda)/P(z;\lambda^1) \neq$ const. It therefore follows that the function $f(z)$ is algebraic. From (3.17) it follows that any branch of $f(z)$ tends to zero as $z \to 0$. Indeed, $P(z;\lambda) = \overline{\lambda}_n z^{-n} + O(z^{-n+1})$ for small $|z|$. Thus if in a neighborhood of the origin we have $f(z) = w_0 + b_1 z^\alpha + \cdots, \alpha > 0$, then

$$\left(\frac{Df(z)}{f(z)}\right)^2 Q[f(z);\lambda] = O(z^{2\alpha}),$$

which is a contradiction. In the same way, it is clear that $f(z)$ does not tend to infinity. Then from (3.17) and (3.18) it follows that $f(z) = \eta z + \cdots$, where $\eta^n = 1$. If $z \to \infty$, then $P(z;\lambda) = \lambda_n z^n + O(z^{n-1})$, and from (3.17) it follows again that $f(z) \to 0$ as $z \to \infty$ and $f(z) = \theta/z + \cdots$, where θ is defined by the equation $\theta^n = \overline{\lambda}_n \lambda_n^{-1}$. We take that linear combination of equations (3.17) and (3.17′) which eliminates the term containing w^{-n} from the sum $\gamma Q(w;\lambda) + W(w;\lambda^1)$, where γ is defined by the equation $\gamma\overline{\lambda}_n + \overline{\lambda}_n^1 = 0$. We are thus led to the equation

$$\left(\frac{Df}{f}\right)^2 \sum_{k=1}^{m} A_k^* f^{-k} + \sum_{k=-r}^{s} B_k^* z^{-k} = 0, \tag{3.17''}$$

where $m \leqq n-1$ and $A_m^* \neq 0$ for $r, s \leqq n$. Taking into account the behavior of $f(z)$ in a neighborhood of $z = 0$ and of $z = \infty$, we conclude that $r = s = m$. Thus $\gamma B_{-n} + B_{-n}^1 = 0$, so $\gamma\lambda_n + \lambda_n^1 = 0$, and therefore γ is real. It follows easily from (3.19) that $B_k^* = \overline{B}_{-k}^*$ for $k \neq 0$. At least one branch of $f(z)$ has at $z = 0$ the form $f(z) = z + \cdots$, and so $A_m^* = -B_m^*$; subtracting from the left-hand side of (3.17) the appropriate combination $\overline{\mu}_m \mathfrak{A}_m(z) + \mu_m \overline{\mathfrak{B}}_m(z)$, we arrive at

$$\left(\frac{Df}{f}\right)^2 \sum_{k=1}^{m-1} A_k' f^{-k} + \sum_{k=-m+1}^{m-1} B_k' z^{-k} = -(\overline{\mu}_m \mathfrak{A}_m(z) + \mu_m \overline{\mathfrak{B}}_m(z)),$$

in which $B_k' = \overline{B}_{-k}'$, $k \neq 0$. But the right-hand side is regular at $z = 0$, so $A_{m-1}' = -B_{m-1}'$, and we can now repeat the above process. Equation (3.17″) thus takes the form

$$\sum_{k=1}^{m} (\overline{\mu}_k \mathfrak{A}_k(z) + \mu_k \overline{\mathfrak{B}}_k(z)) = C,$$

and consequently $C=0$ and $d_{n-1}(f)=0$, which is a contradiction. The lemma must therefore be true.

Remark 3.3. From the lemma it follows that if $d_n=0$ and $d_{n-1}>0$, then $d_{n-1/2}>0$. If $d_n=0$ and $d_{n-1/2}>0$, then the normalized associated vector is defined uniquely up to ± 1.

We shall now investigate the dependence of the vector $\lambda(f)$ on the function f. It is clear that if the univalent function $g(z)$ lies in a sufficiently small neighborhood of $f(z)$ and if $d_n(g)=0$, then, with the appropriate choice of the multiplier ω, the vector $\lambda(g)$ must be close to the vector $\lambda(f)$. For the proof of this fact, we require the following technical proposition.

Proposition 3.1. *Let* $f,g\in S$ *and* $r_1<r_2<1$. *Then*

$$\max_{|z|=r_1}|\mathfrak{A}_n(z;f)-\mathfrak{A}_n(z;g)|\leqslant\alpha^n\Delta,\qquad \max_{|z|=r_1}|\mathfrak{B}_n(z;f)-\mathfrak{B}_n(z;g)|\leqq\alpha^n\Delta,$$

where $\Delta=\max_{|z|=r_2}|f(z)-g(z)|$. *The constant* α *depends on* r_1 *and* r_2.

Proof. From (2.14) and (2.15) it is easily seen that the most difficult part of the proof will be to estimate

$$j=\max_{|z|=r_1}\left|\left(\frac{Df(z)}{f(z)}\right)^2 q_n[f(z);f]-\left(\frac{Dg(z)}{g(z)}\right)^2 q_n[g(z);g]\right|.$$

By the definition of $q_n(w;f)$ we have

$$q_n[f(z);f]-q_n[g(z);g]=\frac{1}{2\pi i}\int_{|t|=r_2}\left[\frac{f^2(t)}{f(t)-f(z)}-\frac{g^2(t)}{g(t)-g(z)}\right]\frac{dt}{t^{n+2}}.$$

If $|z|=r_1$ and $|t|=r_2$, then the distortion theorem and the theorem of Goluzin give that

$$|f(t)-f(z)|\geqslant|t-z|(1-r_1^2)^{1/2}(1-r_2^2)^{1/2}\left|\frac{f(z)f(t)}{tz}\right|\geqslant\alpha_0,$$

where α_0 is a constant depending only on r_1 and r_2. A similar inequality holds for $g(z)$. From the identity

$$\frac{f^2(t)}{f(t)-f(z)}-\frac{g^2(t)}{g(t)-g(z)}=[f(t)-f(z)]^{-1}[g(t)-g(z)]^{-1}\{[f(t)-g(t)][f(t)g(t)$$
$$-f(z)(f(t)+g(t))]+f^2(t)[f(z)-g(z)]\},$$

we conclude that

$$|q_n[f(z);f]-q_n[g(z);g]|<\alpha_1^n\Delta.$$

It is clear that $|q_n[f(z);f]|<\alpha_2^n$ and $|q_n[g(z);g]|<\alpha_2^n$ if $|z|=r_1$. From Cauchy's integral formula and the distortion theorems it follows easily that for $|z|=r_1$

$$\left|\left(\frac{Df(z)}{f(z)}\right)^2-\left(\frac{Dg(z)}{g(z)}\right)^2\right|<\alpha_3\Delta.$$

Thus $j<\alpha_4^n\Delta$. The estimates for the other terms which go to make up $\mathfrak{A}_n(z)$

are elementary, and thus we have proved the first inequality. The second inequality, involving $\mathfrak{B}_n(z)$, is trivial, and so the proposition is proved.

Without loss of generality we shall assume that

$$d_{n-1/2}(f)=\inf_{|y|=1}\frac{1}{2\pi}\int_{\mathscr{C}}\left|\sum_{1}^{2n-1}y_k\varphi_k(z)\right|^2\left|\frac{dz}{z}\right|.$$

We now consider the Gram matrix:

$$A_f=((\varphi_k,\varphi_l))_{k,l=1}^{2n-1}.$$

Since the system $\varphi_1,\cdots,\varphi_{2n}$ is linearly dependent, $\varphi_{2n}=\sum_1^{2n-1}u_k\varphi_k$, and so we can derive a system of linear equations

$$A_fu=h_f,$$

for the quantities $u_1,\cdots,u_{2n-1}$, where u and h_f are vectors with the elements u_k and (φ_{2n},φ_k), respectively. Then $u=A_f^{-1}h_f$. The vector λ which satisfies (3.15) will have the form

$$\lambda(f)=(u,-1)\,[|u|^2+1]^{-1/2},$$

where for convenience we have written $(u_1,\cdots,u_{2n-1},-1)=(u,-1)$.

Lemma 3.3. *Let $f,g\in S$, $d_n(f)=d_n(g)=0$, $d_{n-1/2}(f)>0$ and $d_{n-1/2}(g)>0$, and suppose $d_{n-1/2}(f)\geqq d_{n-1/2}(g)$. Then*

$$|\lambda(f)-\lambda(g)|<\alpha_0^n d_{n-1/2}^{-2}(f)\max_{|z|\leqslant 2r^*}|f(z)-g(z)|,$$

where α_0 is an absolute constant.

Proof. Without loss of generality we assume that $d_{n-1/2}(f)$ is defined by the system of functions $\varphi_1(z;f),\cdots,\varphi_{2n-1}(z;f)$. If $\max_{|z|=r^*}|f(z)-g(z)|$ is sufficiently small, then it is clear that the system $\varphi_1(z;g),\cdots,\varphi_{2n-1}(z;g)$ is linearly independent. Let A_g be the Gram matrix for this system. Then the system $A_gu_g=h_g$ defines a corresponding vector $\lambda(g)$, where h_g is a vector with coordinates $(\varphi_{2n}(z;g),\varphi_k(z;g))$, $1\leqq k\leqq 2n-1$. For notational ease we shall write $u_f=u$, $u_g=v$, $\lambda(f)=\lambda$ and $\lambda(g)=\mu$. With this notation we see from the above that $\mu=(v,-1)\,[|v|^2+1]^{-1/2}$. We set $a^2=|u|^2+1$ and $b^2=|v|^2+1$, and we note that

$$|a-b|<|u-v|.$$

Furthermore $\lambda-\mu=(1/ab)(ub-va,a-b)$, and so

$$|\lambda-\mu|<\frac{1}{ab}[(ub-va)^2+(a-b)^2]^{1/2}.$$

But

$$|ub-va|<a|u-v|+|u|\,|a-b|<(a+|u|)|u-v|.$$

Therefore

$$|\lambda-\mu|<\frac{|u-v|}{ab}[2a^2+2|u|^2+1]^{1/2}<2b^{-1}|u-v|.$$

Hence we must estimate $|u-v|$. Clearly we have

$$h_f-h_g=A_fu-A_gv=A_f(u-v)+(A_f-A_g)v$$

and so

$$u-v=A_f^{-1}(h_f-h_g)-A_f^{-1}(A_f-A_g)v,$$

and

$$|u-v|<|A_f^{-1}|\,|h_f-h_g|+|A_f^{-1}|\,|A_f-A_g|\,|v|.$$

But $|A_f^{-1}|=d_{n-1/2}^{-2}(f)$, and $|A_f-A_g|\leqq[\mathrm{tr}(A_f-A_g)^2]^{1/2}$. Therefore

$$|\lambda-\mu|<2d_{n-1/2}^{-2}(f)\{b^{-1}|h_f-h_g|+[\mathrm{tr}\,(A_f-A_g)^2]^{1/2}\}.$$

Now we estimate $|h_f-h_g|$ and $\mathrm{tr}(A_f-A_g)^2$.

Applying Proposition 3.1 in the case when $r_1=r^*$, $r_2=2r^*$ and $r^*<1/2$, where r^* will be an absolute constant, we see that for $|z|=r^*$ we have

$$\varphi_k(z;f)-\varphi_k(z;g)|\leqslant\alpha^{k/2}\Delta,\qquad|\varphi_k(z;f)|<\alpha^{k/2},$$
$$|\varphi_k(z;g)|<\alpha^{k/2},\qquad 1\leqslant k\leqslant 2n,$$

where α is an absolute constant and $\Delta=\max_{|z|=2r^*}|f(z)-g(z)|$. Thus

$$|(\varphi_k(z;f),\ \varphi_l(z;f))-(\varphi_k(z;g),\ \varphi_l(z;g))|<2\alpha^{\frac{k+l}{2}}\Delta,$$

and so

$$|h_f-h_g|<2\alpha^n\left(\sum_{l=1}^{2n-1}\alpha^l\right)^{1/2}\Delta,$$

and

$$[\mathrm{tr}\,(A_f-A_g)^2]^{1/2}<2\Delta\sum_{k=1}^{2n}\alpha^k.$$

Thus

$$|\lambda(f)-\lambda(g)|<\alpha_0^n d_{n-1/2}^{-2}(f)\,\Delta. \tag{3.21}$$

We have assumed above that the vector $v=u_g$ exists. From the above calculations it is clear that this assumption is true if $|A_f^{-1}|\,|A_f-A_g|<1$, and this in turn holds if $|A_f^{-1}|\,[\mathrm{tr}(A_f-A_g)^2]^{1/2}<1$. Thus if we require that

$$2d_{n-1/2}^{-2}(f)\sum_{l=1}^{2n}\alpha^l\Delta<1,$$

then all the above conditions will be satisfied. If however

$$2d_{n-1/2}^{-2}(f)\sum_{l=1}^{2n}\alpha^l\Delta\geqslant 1,$$

then the inequality (3.21) can still be made to hold simply by increasing the constant α_0 if necessary. Thus (3.21) holds in all cases, and this completes the proof.

The following useful inequality is implied by the estimates given above. For any vector y, $|y| = 1$, we have

$$\left(\frac{1}{2\pi}\int_{\mathscr{C}}\left|\sum_{k=1}^{2n-1} y_k\varphi_k(z;f)\right|^2\left|\frac{dz}{z}\right|\right)^{1/2} \leqslant \left(\frac{1}{2\pi}\int_{\mathscr{C}}\left|\sum_{k=1}^{2n-1} y_k\varphi_k(z;g)\right|^2\left|\frac{dz}{z}\right|\right)^{1/2}$$
$$+\left(\frac{1}{2\pi}\int_{\mathscr{C}}\left|\sum_{k=1}^{2n-1} y_k[\varphi_k(z;f)-\varphi_k(z;g)]\right|^2\left|\frac{dz}{z}\right|\right)^{1/2}.$$

Recalling our previous estimates and the definition of $d_{n-1/2}$, we see that

$$d_{n-1/2}(g) \geqq d_{n-1/2}(f) - \alpha_1^n\Delta, \tag{3.22}$$

where α_1 is an absolute constant. Similarly we obtain

$$d_{n-1}(g) \geqq d_{n-1}(f) - \alpha_1^n\Delta. \tag{3.23}$$

We conclude this section with a technical lemma which will be useful in the future.

Lemma 3.4. *If $d_n(f) = 0$, $d_{n-1/2}(f) \neq 0$, and $t = (t_1, \cdots, t_n)$ is any vector such that $(\lambda, t) = 0$, then there exists a measure $\nu(dz)$ such that the conditions $\delta a_k = t_k$, $k = 1, \cdots, n$, are satisfied. The measure $\nu(dz)$ can be taken to be of the form $\nu(dz) = \zeta(z)\,dz/2\pi iz$, $z \in \mathscr{C}$, where*

$$\left(\frac{1}{2\pi}\int_{\mathscr{C}} |\zeta(z)|^2\left|\frac{dz}{z}\right|\right)^{1/2} \leqslant |t|\,d_{n-1/2}^{-1}. \tag{3.24}$$

Proof. Set

$$d_{n-1/2}^2(f) = \inf_{|y|=1}\frac{1}{2\pi}\int_{\mathscr{C}} |y_1\varphi_1 + \cdots + y_{m-1}\varphi_{m-1} + y_{m+1}\varphi_{m+1} + \cdots$$
$$\cdots + y_{2n}\varphi_{2n}|^2\left|\frac{dz}{z}\right|,$$

where $|y^2| = y_1^2 + \cdots + y_{m-1}^2 + y_{m+1}^2 + \cdots + y_{2n}^2$. If $m = 2l - 1$, then $\operatorname{Re}\lambda_l \neq 0$, and if $m = 2l$, then $\operatorname{Im}\lambda_l \neq 0$. We shall consider only the case $m = 2l - 1$, the case $m = 2l$ being very similar. Repeating the arguments of Lemma 3.1, we construct a measure $\nu(dz)$ satisfying (3.24) such that the relations $\delta a_k = t_k$ hold if $k < l$, or $\operatorname{Im}\delta a_l = \operatorname{Im} t_l$ for $1 < k \leqq n$. Clearly $\operatorname{Re}\sum_1^n \bar\lambda_k\delta a_k = 0$, and so $\operatorname{Re}\sum_1^n \bar\lambda_k(\delta a_k - t_k) = 0$. Thus $\operatorname{Re}\{\bar\lambda_l[\operatorname{Re}\delta a_l - \operatorname{Re} t_l]\} = 0$, and this means $\operatorname{Re}\delta a_l = \operatorname{Re} t_l$, so $\delta a_l = t_l$. Q.E.D.

§4. Usually the functionals whose extrema are studied with respect to the class S are in fact well-defined functionals with respect to the much wider class of functions regular in $|z| < 1$. It is clear that S can be embedded in various ways in the linear space of functions regular in $|z| < 1$. We shall assume that the functionals we are considering on S are really restrictions of functionals defined on some such linear space of normalized functions.

Thus one can consider concepts such as "linear functional" and "nonlinear functional", and apply them to the functionals considered on S. We shall also assume that our functionals have first and second differentials.

Let $f(z) \in S$, $f(z) = \sum_0^\infty a_k z^{k+1}$, and $a_0 = 1$. The vector $(a_1, \cdots, a_n)$ will be considered as an element of R^{2n} written in complex form, and will be denoted by x. We shall limit ourselves to consideration of functionals of the form

$$J[f] = F(x) \equiv F(a_1, \cdots, a_n),$$

where $F(x)$ is a twice continuously differentiable function on R^{2n} such that $\operatorname{grad} F \neq 0$ for $|x| \leqq \rho$, where ρ is sufficiently large. We shall denote by $dJ[f; \cdot]$ and $d^2J[f, \cdot]$ the first and second differentials of $J[f]$. We shall compute the change in the functional effected by the variation of the function $f(z)$ by an element of the class of admissible functions of the form (2.23). If

$$f_*(z, \varepsilon) = f(z) + \varepsilon \delta f(z) + \frac{\varepsilon^2}{2} \delta^2 f(z) + \dots,$$

then

$$J[f_*] - J[f] = \varepsilon dJ[f, \delta f] + \frac{\varepsilon^2}{2} (dJ[f, \delta^2 f] + d^2 J[f, \delta f]) + o(\varepsilon^2).$$

We shall call the functional $dJ[f, \delta f]$ the *first variation* of the functional $J[f]$, and we shall denote it by $\delta J[f]$, or $\delta J[f; \nu]$ if we wish to indicate the measure used to construct the variation of f. The expression

$$dJ[f, \delta^2 f] + d^2 J[f, \delta f] \tag{4.1}$$

will be called the *second variation* of the functional $J[f]$, and will be denoted by either $\delta^2 J[f]$ or $\delta^2 J[f; \nu]$. If at the function f the functional $J[f]$ has a local maximum, then

$$\delta J[f; \nu] = 0 \tag{4.2}$$

and

$$\delta^2 J[f; \nu] \leqq 0. \tag{4.3}$$

In the case of the function $J[f] = F(x)$ we have $dJ[f, \delta f] = (\operatorname{grad} F, y)$, where $y = (\delta a_1, \cdots, \delta a_n)$. If $\mathscr{A}$ is the Hessian matrix of F, then $d^2 J[f, \delta f] = (\mathscr{A} y, y)$. Thus at the extreme point of the functional $(\operatorname{grad} F, y) = 0$, and, recalling (2.13) and (2.14), we can write this formula as

$$\operatorname{Re} \left\{ \int_{\mathscr{C}} \sum_{k=1}^{n} [\bar{\lambda}_k \mathfrak{A}_k(z) + \lambda_k \overline{\mathfrak{B}_k(z)}] \, \nu(dz) + i c_1 \sum_{k=1}^{n} \bar{\lambda}_k k a_k \right\} = 0,$$

where

$$\lambda_k = \frac{\partial F}{\partial \xi_{2k-1}} + i \frac{\partial F}{\partial \xi_{2k}}, \quad \xi_{2k-1} + i \xi_{2k} = a_k.$$

Using standard notation, we write $\lambda_k = 2 \partial F / \partial \bar{a}_k$, and so

$$\lambda = \left(2\frac{\partial F}{\partial \bar{a}_1}, \ldots, 2\frac{\partial F}{\partial \bar{a}_n}\right),$$

where in general λ will not be normalized. From the arbitrary nature of ν and c_1 it follows that Euler's equation for the problem under consideration has the form

$$\sum_{k=1}^{n} [\bar{\lambda}_k \mathfrak{A}_k(z) + \lambda_k \mathfrak{B}_k(z)] = 0.$$

We also have the condition

$$\mathrm{Im} \sum_{k=1}^{n} \bar{\lambda}_k k a_k = 0. \tag{4.4}$$

Euler's equation is identical to (3.15), and so we can write it in the form

$$\left(\frac{Df(z)}{f(z)}\right)^2 Q[f(z)] + P(z) = 0. \tag{4.5}$$

From (3.19) and (4.4) it follows that for $|z| = 1$ the function $P(z)$ assumes only real values. Equation (4.5) was first considered by Schiffer for particular types of functionals, and so it is called the Schiffer equation. At a maximum point we have the necessary condition $(\operatorname{grad} F, \zeta) + (\mathscr{A}y, y) \leqq 0$, where $\zeta = (\delta^2 a_1, \cdots, \delta^2 a_n)$. This can be written as

$$\mathrm{Re} \sum_{k=1}^{n} \bar{\lambda}_k \delta^2 a_k + (\mathscr{H}y, y) \leqslant 0. \tag{4.6}$$

We shall change the first term of the left-hand side slightly after having rewritten it in terms of the function f and the measure $\nu(dz)$. We shall employ various results of §2, since we have already rewritten the desired expression there in the form of (2.21). We shall alter the three terms in brackets on the right-hand side of (2.21). From (2.9) it follows easily that

$$\frac{DF_1(z)}{Df(z)} = \int_{\mathscr{C}} \left[\frac{D^2 f(z)}{Df(z)} \frac{t}{t-z} - \mathscr{K}(z,t)\right] \nu(dt) + \int_{\mathscr{C}} \left[\frac{\bar{t}z}{(1-\bar{t}z)^2} + \frac{\bar{t}z}{1-\bar{t}z} \frac{D^2 f(z)}{Df(z)}\right] \overline{\nu(dt)} + ic_1 \frac{D^2 f(z)}{Df(z)},$$

where

$$\mathscr{K}(z,t) = \left(\frac{Df(t)}{f(z)-f(t)}\right)^2 - \left(\frac{Df(t)}{f(t)}\right)^2 - \frac{zt}{(z-t)^2}. \tag{4.7}$$

Thus the sum of the first three terms in brackets on the right-hand side of (2.21) is

$$\frac{1}{2}\iint_{\mathscr{C}\mathscr{C}} [\mathscr{K}(z,t)P(z) + \mathscr{K}(t,z)P(t)]\,\nu(dz)\,\nu(dt) - \frac{1}{2}\iint_{\mathscr{C}\mathscr{C}} \frac{\bar{t}z}{(1-\bar{t}z)^2} [P(z) + \overline{P(t)}]\,\nu(dz)\,\overline{\nu(dt)} + \int_{\mathscr{C}} P(z; F_1)\,\nu(dz),$$

and so

$$\operatorname{Re}\sum_{k=1}^{n}\bar{\lambda}_k\delta^2 a_k = 2\operatorname{Re}\Big\{\frac{1}{2}\iint_{\mathscr{C}\mathscr{C}}[\mathscr{K}(z,t)P(z)+\mathscr{K}(t,z)P(t)]\,\nu(dz)\,\nu(dt)$$

$$-\frac{1}{2}\iint_{\mathscr{C}\mathscr{C}}\frac{\bar{t}z}{(1-\bar{t}z)^2}[P(z)+\overline{P(t)}]\,\nu(dz)\overline{\nu(dt)}+\int_{\mathscr{C}}P(z;F_1)\,\nu(dz) \tag{4.8}$$

$$+ic_1\sum_{k=1}^{n}\bar{\lambda}_k k a_k(F_1)+\sum_{k=1}^{n}\bar{\lambda}_k\beta_k+ic_2\sum_{k=1}^{n}\bar{\lambda}_k k a_k\Big\}.$$

From (4.4) we see that the last term in the brackets in (4.8) is zero, and so we shall henceforth omit it. Next we evaluate

$$\int_{\mathscr{C}}P(z;F_1)\,\nu(dz)+ic_1\sum_{k=1}^{n}\bar{\lambda}_k k a_k(F_1)+\sum_{k=1}^{n}\bar{\lambda}_k\beta_k.$$

In order to simplify the calculations, we represent the sum in a different way. From (2.19) and (2.20) it follows that the sum equals

$$\operatorname{Re}\sum_{k=1}^{n}\bar{\lambda}_k\gamma_k, \tag{4.9}$$

where γ_k is the coefficient of z^{k+1} in the expansion of $\Gamma(z)=DF_1(z)\varphi_1(z)-F_1(z)\varphi_1(0)-\frac{1}{2}D^2f(z)\varphi_1^2(z)+\frac{1}{2}f(z)\varphi_1^2(0)$.

But $F_1(z)=Df(z)\varphi_1(z)+g(z)$, and so

$$\Gamma(z)=\frac{1}{2}D\,[Df(z)\,\varphi_1^2(z)]+Dg(z)\,\varphi_1(z)-g(z)\,\varphi_1(0)-Df(z)\,\varphi_1(z)\,\varphi_1(0)$$
$$+\frac{1}{2}f(z)\,\varphi_1^2(0).$$

The coefficients of $Df(z)\varphi_1^2(z)-f(z)\varphi_1^2(0)$ are easily calculated with the aid of (2.8). The coefficient of z^{k+1} is

$$\iint_{\mathscr{C}\mathscr{C}}\left[\frac{z\mathfrak{B}_k\left(\frac{1}{t}\right)-t\mathfrak{B}_k\left(\frac{1}{z}\right)}{z-t}+ka_k\right]\nu(dz)\,\nu(dt)+\iint_{\mathscr{C}\mathscr{C}}\frac{\bar{t}\mathfrak{B}_k(\bar{z})-\bar{z}\mathfrak{B}_k(\bar{t})}{\bar{z}-\bar{t}}\overline{\nu(dz)\,\nu(dt)}$$

$$+2\iint_{\mathscr{C}\mathscr{C}}\frac{z\bar{t}}{1-z\bar{t}}\left[\mathfrak{B}_k\left(\frac{1}{z}\right)-\mathfrak{B}_k(\bar{t})\right]\nu(dz)\overline{\nu(dt)}+2ic_1\int_{\mathscr{C}}\left[\mathfrak{B}_k\left(\frac{1}{z}\right)+ka_k\right]\nu(dz)$$

$$+2ic_1\int_{\mathscr{C}}\mathfrak{B}_k(\bar{z})\overline{\nu(dz)}-c_1^2ka_k.$$

The contribution of the function $\frac{1}{2}D[Df(z)\varphi_1^2(z)-f(z)\varphi_1^2(0)]$ to the sum (4.9) is

$$\frac{1}{2}\operatorname{Re}\Big\{\iint_{\mathscr{C}\mathscr{C}}\Big[\frac{zP^*(t)-tP^*(z)}{z-t}+\sum_{1}^{n}\bar{\lambda}_k k(k+1)a_k\Big]\nu(dz)\,\nu(dt)$$

$$+\iint_{\mathscr{C}\mathscr{C}}\frac{z\bar{t}}{1-z\bar{t}}[P^*(z)+\overline{P^*(t)}]\,\nu(dz)\overline{\nu(dt)}+2ic_1\int_{\mathscr{C}}\Big[P^*(z)+\sum_{1}^{n}\bar{\lambda}_k k(k+1)a_k\Big]\nu(dz)$$

$$-c_1^2\sum_{1}^{n}\bar{\lambda}_k k(k+1)a_k\Big\}, \tag{4.10}$$

where by $P^*(z)$ we denote

$$(4.11)\qquad P^*(z) = P^*(z; f) = \sum_1^n (k+1)\left[\bar{\lambda}_k \mathfrak{B}_k\left(\frac{1}{z}\right) - \lambda_k \overline{\mathfrak{B}}_k(z)\right].$$

Calculating the contribution to (4.9) of the function $Dg(z)\varphi_1(z) - g(z)\varphi_1(0)$, we first find that the coefficient of z^{k+1} is

$$\int_{\mathscr{C}}\left[\sum_{l=1}^{k-1}(k-l+1)z^{-l}\int_{\mathscr{C}}\left(\frac{Df(t)}{f(t)}\right)^2 q_{k-l}[f(t)]\,\nu(dt) + k\int_{\mathscr{C}}\left(\frac{Df(t)}{f(t)}\right)^2 q_k[f(t)]\,\nu(dt)\right]\nu(dz)$$
$$+\int_{\mathscr{C}}\sum_{l=1}^{k-1}(k-l+1)\bar{z}^l\int_{\mathscr{C}}\left(\frac{Df(t)}{f(t)}\right)^2 q_{k-l}[f(t)]\,\nu(dt)\overline{\nu(dz)} + ic_1k\int_{\mathscr{C}}\left(\frac{Df(t)}{f(t)}\right)^2 q_k[f(t)]\nu(dt).$$

We next set

$$(4.12)\qquad Q^*(w, z) = \overset{*}{Q}_n(w, z; f) = \sum_{k=2}^n \bar{\lambda}_k \sum_{l=1}^{k-1}(k-l+1)z^{-l}q_{k-l}(w)$$

and

$$(4.13)\qquad Q^*(w) = \overset{*}{Q}_n(w; f) = \sum_{k=1}^n k\bar{\lambda}_k q_k(w).$$

Then the contribution to (4.9) from $Dg(z)\varphi_1(z) - g(z)\varphi_1(0)$ is

$$\frac{1}{2}\operatorname{Re}\left\{\iint_{\mathscr{C}\mathscr{C}}\left[\left(\frac{Df(z)}{f(z)}\right)^2 Q^*[f(z), t] + \left(\frac{Df(t)}{f(t)}\right)^2 Q^*[f(t), z] + \left(\frac{Df(z)}{f(z)}\right)^2 Q^*[f(z)]\right.\right.$$
$$(4.14)\quad \left. + \left(\frac{Df(t)}{f(t)}\right)^2 Q^*[f(t)]\right]\nu(dz)\,\nu(dt) + \iint_{\mathscr{C}\mathscr{C}}\left[\left(\frac{Df(z)}{f(z)}\right)^2 Q^*\left[f(z), \frac{1}{\bar{t}}\right]\right.$$
$$\left.\left. + \overline{\left(\frac{Df(t)}{f(t)}\right)^2 Q^*\left[f(t), \frac{1}{\bar{z}}\right]}\right]\nu(dz)\,\nu(dt) + 2ic_1\int_{\mathscr{C}}\left(\frac{Df(z)}{f(z)}\right)^2 Q^*[f(z)]\,\nu(dz)\right\}.$$

Thus in order to finally determine (4.9), we must find the contribution by the function

$$\tfrac{1}{2}[Df(z) - f(z)]\varphi_1^2(0) - \varphi_1(0)[Df(z)\varphi_1(z) - f(z)\varphi_1(0)].$$

This contribution turns out to be

$$-\frac{1}{2}\operatorname{Re}\left\{\iint_{\mathscr{C}\mathscr{C}}\left[R(z) + R(t) + \sum_1^n \bar{\lambda}_k k a_k\right]\nu(dz)\,\nu(dt) + \iint_{\mathscr{C}\mathscr{C}}\overline{\left[R\left(\frac{1}{\bar{z}}\right) + R\left(\frac{1}{\bar{t}}\right)\right]}\right.$$
$$(4.15)\quad \left.\times\,\nu(dz)\overline{\nu(dt)} + 2ic_1\int_{\mathscr{C}}\left[R(z) - \overline{R\left(\frac{1}{\bar{z}}\right)} + \sum_1^n \bar{\lambda}_k k a_k\right]\nu(dz) - c_1^2\sum_1^n \bar{\lambda}_k k a_k\right\},$$

where

$$(4.16)\qquad R(z) = R_n(z; f) = \sum_1^n \bar{\lambda}_k \mathfrak{B}_k\left(\frac{1}{z}\right).$$

Thus (4.9) equals the sum of (4.10), (4.14), and (4.15), which turns out to be

$$\operatorname{Re}\sum_1^n \bar\lambda_k \gamma_k = \frac{1}{2}\operatorname{Re}\Big\{\iint_{\mathscr{C}\mathscr{C}}\Big[\frac{zP^*(t)-tP^*(z)}{z-t}+\Big(\frac{Df(z)}{f(z)}\Big)^2(Q^*[f(z),t]+Q^*[f(z)])$$

$$+\Big(\frac{Df(t)}{f(t)}\Big)^2(Q^*[f(t),z]+Q^*[f(t)])-R(z)-R(t)+\sum_1^n\bar\lambda_k k^2 a_k\Big]\nu(dz)\,\nu(dt)$$

$$+\iint_{\mathscr{C}\mathscr{C}}\Big[\frac{z\bar t}{1-z\bar t}(P^*(z)+\overline{P^*(t)})+\Big(\frac{Df(z)}{f(z)}\Big)^2 Q^*\Big[f(z),\frac{1}{\bar t}\Big]+\overline{\Big(\frac{Df(t)}{f(t)}\Big)^2 Q^*\Big[f(t),\frac{1}{\bar z}\Big]}$$

$$(4.17)\qquad \overline{-R\Big(\frac{1}{\bar z}\Big)-R\Big(\frac{1}{\bar t}\Big)}\Big]\nu(dz)\,\nu(dt)+2ic_1\int_{\mathscr{C}}\Big[\Big(\frac{Df(z)}{f(z)}\Big)^2 Q^*[f(z)]+P^*(z)-R(z)$$

$$+\overline{R\Big(\frac{1}{\bar z}\Big)}+\sum_1^n\bar\lambda_k k^2 a_k\Big]\nu(dz)-c_1^2\sum_1^n\bar\lambda_k k^2 a_k\Big\}.$$

Therefore from (4.8) and (4.17) it follows that

$$(4.18)\qquad \operatorname{Re}\sum_1^n\bar\lambda_k\delta^2 a_k=\operatorname{Re}\Big\{\iint_{\mathscr{C}\mathscr{C}}\mathfrak{A}(z,t)\,\nu(dz)\,\nu(dt)-\iint_{\mathscr{C}\mathscr{C}}\mathfrak{B}(z,\bar t)\,\nu(dz)\overline{\nu(dt)}$$

$$+2ic_1\int_{\mathscr{C}}\mathfrak{C}(z)\,\nu(dz)-c_1^2\sum_1^n\bar\lambda_k k^2 a_k\Big\},$$

where

$$\mathfrak{A}(z,t)=\mathfrak{A}_n(z,t;f)=\mathscr{K}(z,t)P(z)+\mathscr{K}(t,z)P(t)+\frac{zP^*(t)-tP^*(z)}{z-t}$$

$$(4.19)\qquad +\Big(\frac{Df(z)}{f(z)}\Big)^2[Q^*[f(z),t]+Q^*[f(z)]]+\Big(\frac{Df(t)}{f(t)}\Big)^2[Q^*[f(t),z]+Q^*[f(t)]]$$

$$-R(z)-R(t)+\sum_1^n\bar\lambda_k k^2 a_k,$$

$$\mathfrak{B}(z,\bar t)=\mathfrak{B}_n(z,\bar t;f)=\frac{z\bar t}{(1-z\bar t)^2}[P(z)+\overline{P(t)}]-\frac{z\bar t}{1-z\bar t}[P^*(z)+P^*(t)]$$

$$(4.20)\qquad -\Big(\frac{Df(z)}{f(z)}\Big)^2 Q^*\Big[f(z),\frac{1}{\bar t}\Big]-\overline{\Big(\frac{Df(t)}{f(t)}\Big)^2 Q^*\Big[f(t),\frac{1}{\bar z}\Big]}+\overline{R\Big(\frac{1}{\bar z}\Big)}+R\Big(\frac{1}{\bar t}\Big)$$

and

$$\mathfrak{C}(z)=\mathfrak{C}_n(z;f)=\Big(\frac{Df(z)}{f(z)}\Big)^2 Q^*[f(z)]+P^*(z)-R(z)+\overline{R\Big(\frac{1}{\bar z}\Big)}+\sum_1^n\bar\lambda_k k^2 a_k.$$

(4.20)

We shall not require this type of formulation for the second term in the formula $\delta^2J[f;\nu]=dJ[f,\delta^2 f]+d^2J[f,\delta f]$, since the quadratic form $d^2J[f,\delta f]=(\mathscr{A}y,y)$ is easily calculated from (2.13). Although difficult to substantiate at this point, let us merely make the statement that the most significant term in the formula for the second variation is the term $dJ[f,\delta^2 f]$; that is, the term arising from the nonlinearity of the class S. If J is a linear functional, then $d^2J[f,\delta f]=0$ and $\delta^2 J[f;\nu]=dJ[f,\delta^2 f]=J[\delta^2 f]$. In this case formulas (4.18) and (4.20) define the second variation of a linear functional.

§5. We shall now discuss some simple properties of the kernels $\mathfrak{A}(z,t)$ and $\mathfrak{B}(z,\bar{t})$. It is clear that $\mathfrak{A}(z,t)$ is a symmetric complex kernel, and that $\mathfrak{B}(z,\bar{t})$ is Hermitian.

Lemma 5.1. *If $f(z)\in S$, then $\mathfrak{A}(z,t)$ and $\mathfrak{B}(z,\bar{t})$ are analytic functions in the bicylinders $\{|z|<1,\ |t|<1\}$ and $\{|z|<1,\ |\bar{t}|<1\}$, respectively.*

The proof is given for the kernel $\mathfrak{A}(z,t)$, the details for $\mathfrak{B}(z,\bar{t})$ being very similar.

Since $f(z)$ is univalent, the kernel $\mathfrak{A}(z,t)$ for z fixed can have a singularity only at $t=0$. We shall show that at this point the kernel is in fact regular, and we shall compute $\mathfrak{A}(z,0)$. We first determine the Laurent expansions for the various terms of the kernel. By $\mathfrak{P}(t)$ we shall denote any power series in the variable t which converges in a neighborhood of $t=0$ and satisfies $\mathfrak{P}(0)=0$; in the following the series need not be the same at each occurence, even though we always use the notation $\mathfrak{P}(t)$. We first list some easily verified relations. We note that

$$\mathfrak{B}_k\left(\frac{1}{t}\right)=\sum_{l=1}^{k}(k-l+1)\,a_{k-l}t^{-l}=\frac{Df(t)}{t^{k+1}}-(k+1)\,a_k+\mathfrak{P}(t).$$

Thus, from (2.20),

$$P(t)=Df(t)\sum_1^n\bar{\lambda}_k t^{-k-1}-\sum_1^n\bar{\lambda}_k a_k+\mathfrak{P}(t), \tag{5.1}$$

and from (4.16) it follows that

$$R(t)=Df(t)\sum_1^n\bar{\lambda}_k t^{-k-1}-\sum_1^n(k+1)\,\bar{\lambda}_k a_k+\mathfrak{P}(t), \tag{5.2}$$

and upon using (4.11) we see that

$$P^*(t)=Df(t)\sum_1^n(k+1)\,\bar{\lambda}_k t^{-k-1}-\sum_1^n\bar{\lambda}_k\,(k+1)^2\,a_k+\mathfrak{P}(t). \tag{5.3}$$

From (2.12) we have that

$$\left(\frac{Df(t)}{f(t)}\right)^2 q_k\,[f(t)]=-\frac{Df(t)}{t^{k+1}}+a_k+\mathfrak{P}(t). \tag{5.4}$$

Since $\mathscr{K}(z,t)=-1+\mathfrak{P}(t)$,

$$P(z)\,\mathscr{K}(z,t)=-P(z)+\mathfrak{P}(t). \tag{5.5}$$

Now applying (5.1) we see that

$$P(t)\left[\left(\frac{Df(z)}{f(z)-f(t)}\right)^2-\left(\frac{Df(z)}{f(z)}\right)^2\right]=\left(\frac{Df(z)}{f(z)}\right)^2 P(t)\left[\left(\frac{f(z)}{f(z)-f(t)}\right)^2-1\right]$$
$$=\left(\frac{Df(z)}{f(z)}\right)^2 Df(t)\sum_1^n \bar{\lambda}_k t^{-k-1}\left[\left(\frac{f(z)}{f(z)-f(t)}\right)^2-1\right]+\mathfrak{P}(t).$$

But

$$Df(t)\left[\left(\frac{f(z)}{f(z)-f(t)}\right)^2-1\right]=t\frac{d}{dt}\left[\frac{f^2(z)}{f(z)-f(t)}-f(t)\right]$$
$$=t\frac{d}{dt}\left[\frac{f^2(t)}{f(z)-f(t)}\right]=-\sum_1^\infty (k+1)\,q_k\,[f(z)]\,t^{k+1},$$

and so

$$P(t)\left[\left(\frac{Df(z)}{f(z)-f(t)}\right)^2-\left(\frac{Df(z)}{f(z)}\right)^2\right]$$
$$=-\left(\frac{Df(z)}{f(z)}\right)^2\sum_{k=1}^n \bar{\lambda}_k t^{-k-1}\sum_{l=1}^\infty (l+1)\,q_l\,[f(z)]\,t^{l+1}+\mathfrak{P}(t).$$

From this relation and (4.13), (4.12) and (3.16) it follows that

$$P(t)\left[\left(\frac{Df(z)}{f(z)-f(t)}\right)^2-\left(\frac{Df(z)}{f(z)}\right)^2\right]=-\left(\frac{Df(z)}{f(z)}\right)^2[Q^*[f(z),t]$$
$$+Q^*[f(z)]+Q[f(z)]]+\mathfrak{P}(t). \tag{5.6}$$

From (5.5) and this latter equation we find that

$$A_1=P(z)\mathscr{K}(z,t)+P(t)\mathscr{K}(t,z)+\left(\frac{Df(z)}{f(z)}\right)^2[Q^*[f(z),t]+Q^*[f(z)]]$$
$$=-\left(\frac{Df(z)}{f(z)}\right)^2 Q[f(z)]-P(z)-\frac{zt}{(z-t)^2}P(t)+\mathfrak{P}(t).$$

With the aid of (5.1) we see that

$$P(t)\frac{zt}{(z-t)^2}=Df(t)\sum_{k=1}^n \bar{\lambda}_k t^{-k-1}\sum_{l=1}^n l\left(\frac{t}{z}\right)^l+\mathfrak{P}(t)$$

and from (5.3) it follows that

$$\frac{zP^*(t)-tP^*(z)}{z-t}=Df(t)\sum_{k=1}^n (k+1)\bar{\lambda}_k t^{-k-1}\sum_{l=0}^n\left(\frac{t}{z}\right)^l-\sum_{k=1}^n\bar{\lambda}_k(k+1)^2 a_k+\mathfrak{P}(t).$$

Using the above relations, we find that

$$\frac{zP^*(t)-tP^*(z)}{z-t}-P(t)\frac{zt}{(z-t)^2}=Df(t)\left[\sum_{k=0}^n (k+1)\bar{\lambda}_k t^{-k-1}\sum_{l=0}^n\left(\frac{t}{z}\right)^l\right.$$
$$\left.-\sum_{k=1}^n\bar{\lambda}_k t^{-k-1}\sum_{l=1}^n l\left(\frac{t}{z}\right)^l\right]-\sum_{k=1}^n\bar{\lambda}_k(k+1)^2 a_k+\mathfrak{P}(t)$$
$$=-Df(t)\,t\frac{d}{dt}\left\{\sum_{k=1}^n\bar{\lambda}_k t^{-k-1}\sum_{l=0}^n\left(\frac{t}{z}\right)^l\right\}-\sum_{k=1}^n\bar{\lambda}_k(k+1)^2 a_k+\mathfrak{P}(t)$$
$$=Df(t)\sum_{k=0}^n (k+1)\,t^{-k-1}\sum_{\substack{l=k\\ l\geqslant 1}}^n\bar{\lambda}_l z^{k-l}-\sum_{k=1}^n\bar{\lambda}_k(k+1)^2 a_k+\mathfrak{P}(t).$$

Now, using the above formula for A_1 and this latter equation, we see that

(5.7)
$$A_2 = A_1 + \frac{zP^*(t) - tP^*(z)}{z-t} = -\left(\frac{Df(z)}{f(z)}\right)^2 Q[f(z)] - P(z) + Df(t)\sum_{k=0}^{n}(k+1)t^{-k-1}\sum_{\substack{l=k\\ l\geqslant 1}}^{n}\bar{\lambda}_l z^{k-1} - \sum_{k=1}^{n}\bar{\lambda}_k(k+1)^2 a_k + \mathfrak{P}(t).$$

Thus it follows from (4.12), (4.13) and (2.12) that

$$\left(\frac{Df(t)}{f(t)}\right)^2 [Q^*[f(t), z] + Q^*[f(t)]] = \sum_{k=1}^{n}\bar{\lambda}_k \sum_{l=1}^{k-1}(k-l+1)z^{-l}\left[-\frac{Df(t)}{t^{k-l+1}} + a_{k-l}\right] + \sum_{k=1}^{n} k\bar{\lambda}_k\left[-\frac{Df(t)}{t^{k+1}} + a_k\right] + \mathfrak{P}(t).$$

Now we use (5.2) to deduce that

$$\left(\frac{Df(t)}{f(t)}\right)^2 [Q^*[f(t), z] + Q^*[f(t)]] = \sum_{k=1}^{n}\bar{\lambda}_k \sum_{l=0}^{k-1}(k-l+1)z^{-l}\left[-\frac{Df(t)}{t^{k-l+1}} + a_{k-l}\right]$$
$$+ R(t) + \sum_{k=1}^{n}\bar{\lambda}_k k a_k + \mathfrak{P}(t) = -Df(t)\sum_{k=1}^{n}\bar{\lambda}_k\sum_{l=0}^{k}(k-l+1)z^{-l}t^{l-k-1}$$
$$+ R(t) + R(z) + \sum_{k=1}^{n}\bar{\lambda}_k(2k+1)a_k + \mathfrak{P}(t)$$
$$= -Df(t)\sum_{k=0}^{n}(k+1)t^{-k-1}\sum_{\substack{l=k\\ l\geqslant 1}}^{n}\bar{\lambda}_l z^{k-l} + R(z) + R(t) + \sum_{k=1}^{n}\bar{\lambda}_k(2k+1)a_k + \mathfrak{P}(t),$$

and so from (5.7) we see that

$$A_3 = A_2 + \left(\frac{Df(t)}{f(t)}\right)^2 [Q^*[f(t), z] + Q^*[f(t)]]$$
$$= -\left(\frac{Df(z)}{f(z)}\right)^2 Q[f(z)] - P(z) + R(z) + R(t) - \sum_{k=1}^{n}\bar{\lambda}_k k^2 a_k + \mathfrak{P}(t).$$

Therefore

(5.8)
$$\mathfrak{A}(z, t) = A_3 - R(z) - R(t) + \sum_{k=1}^{n}\bar{\lambda}_k k^2 a_k = -\left(\frac{Df(z)}{f(z)}\right)^2 Q[f(z)] - P(z) + \mathfrak{P}(t).$$

Hence $\mathfrak{A}(z,t)$ is regular in $|t| < 1$ for $z \neq 0$, and so

$$\mathfrak{A}(z, t) = \frac{1}{2\pi i}\int_{|u|=\rho}\frac{\mathfrak{A}(z, u)}{u-t}\cdot du, \quad |t| < \rho < 1.$$

But $\mathfrak{A}(z,u) = \mathfrak{A}(u,z)$, and thus $\mathfrak{A}(u,z)$ is regular in $|z| < 1$; thus $\mathfrak{A}(z,t)$ is regular in the bicylinder $\{|z| < 1, |t| < 1\}$. From (5.8) we have

$$\mathfrak{A}(z, 0) = -\left(\frac{Df(z)}{f(z)}\right)^2 Q[f(z)] - P(z). \tag{5.9}$$

It is clear that $\mathfrak{A}(0,0) = 0$. The proof for $\mathfrak{B}(z,\bar{t})$ is analogous, and equally simple. It is important to note that

$$\mathfrak{B}(z,0) = 0. \tag{5.10}$$

This completes the proof of the lemma.

Lemma 5.2. *If $f(z) \in S$, then the function $\mathfrak{C}(z)$ in* (4.20) *is regular in $|z| < 1$, and $\mathfrak{C}(0) = 0$.*

Proof. It follows from (2.12), (5.2), and (5.3) that

$$\mathfrak{C}(z) = \sum_1^n \bar{\lambda}_k k[-z^{-k-1}Df(z) + a_k] + Df(z)\sum_1^n \bar{\lambda}_k (k+1) z^{-k-1} - \sum_1^n \bar{\lambda}_k (k+1)^2 a_k$$
$$- Df(z)\sum_1^n \bar{\lambda}_k a_k z^{-k-1} + \sum_1^n \bar{\lambda}_k (k+1) a_k + \sum_1^n \bar{\lambda}_k k^2 a_k + \mathfrak{P}(z) = \mathfrak{P}(z).$$

The next lemma will be very important in what follows. We consider the case when $J(f) = F(x)$.

Lemma 5.3. *If for a function $f(z) \in S$ the inequality $\delta^2 J \leqq 0$ holds for every measure $\nu(dz)$ and constant c_1, then $f(z)$ satisfies the differential equation* (4.5), *where* $\operatorname{Re} P(z) \geqq 0$ *for* $|z| = 1$.

Proof. Set $c_1 = 0$ and suppose that the mass of the measure is concentrated at the points z and z_1. Then from (4.18) we have

$$\operatorname{Re}\{\alpha^2\mathfrak{A}(z,z) + 2\alpha\beta\mathfrak{A}(z,z_1) + \beta^2\mathfrak{A}(z_1,z_1) - |\alpha|^2\mathfrak{B}(z,\bar{z})$$
$$- 2\alpha\bar{\beta}\mathfrak{B}(z,\bar{z}_1) - |\beta|^2\mathfrak{B}(z_1,\bar{z}_1)\} + (\mathcal{H}y, y) \leqslant 0,$$

where α and β are arbitrary. Letting $z_1 \to 0$, we see from Lemma 5.1 that

$$\operatorname{Re}\{\alpha^2\mathfrak{A}(z,z) + 2\alpha\beta\mathfrak{A}(z,0) - |\alpha|^2\mathfrak{B}(z,z)\} + (\mathcal{H}y, y) \leqslant 0. \tag{5.11}$$

But $y = (\delta a_1, \cdots, \delta a_n)$, and it is clear that δa_j is linear in α and β; i.e.

$$\delta a_k = \alpha\mathfrak{A}_k(z) + \beta\mathfrak{A}_k(0) + \bar{\alpha}\mathfrak{B}_k(\bar{z}) + \bar{\beta}\mathfrak{B}_k(0), \quad 1 \leqslant k \leqslant n.$$

From Lemma 2.2 we have $\mathfrak{A}_k(0) = \mathfrak{B}_k(0) = 0$ and $\delta a_k = \alpha\,\mathfrak{A}_k(z) + \bar{\alpha}\,\mathfrak{B}_k(\bar{z})$. Since β is arbitrary, it follows from (5.11) that $\mathfrak{A}(z,0) = 0$; recalling (5.9), we see that $f(z)$ satisfies equation (4.5). Any fixed singularities on $|z| = 1$ of this equation are algebraic critical points. According to a theorem of Painlevé, the equation (4.5) has no movable transcendental or essential singular points. Hence on $|z| = 1$ the function $f(z)$ has a finite number of algebraic critical points. It is easily seen that

$$(5.12)\quad \begin{aligned}\lim_{t\to z}\,[\mathcal{K}(z,t)\,P(z)+\mathcal{K}(t,z)\,P(z)] &= P(z)\left[\frac{1}{3}\frac{D^3 f(z)}{Df(z)}+\frac{1}{2}\left(\frac{D^2 f(z)}{Df(z)}\right)^2\right.\\ &\left. -2\left(\frac{Df(z)}{f(z)}\right)^2+\frac{1}{6}\right]-DP(z)\frac{D^2 f(z)}{Df(z)}.\end{aligned}$$

Therefore from (4.19) and (5.12) we conclude that on $|z|=1$ the function $\mathfrak{A}(z,z)$ will have singularities which can be the singularities of $f(z)$; and hence, with the exception of a finite number of points, it will be finite on $|z|=1$.

From (4.20) it follows that all the terms comprising $\mathfrak{A}(z,\bar{z})$, with the exception of $2|z|^2(1-|z|^2)^{-2}\operatorname{Re}P(z)$, are finite for $|z|=1$, except possibly at the singularities of $f(z)$.

In (5.11) set $\alpha=1$. Then $\delta a_k=\mathfrak{A}_k(z)+\mathfrak{B}_k(\bar{z})$, and it is obvious that this function is regular on $|z|=1$, with the exception of finitely many points. Thus, if in the inequality

$$\operatorname{Re}\{\mathfrak{A}(z,z)-\mathfrak{B}(z,\bar{z})\}+(\mathscr{A}y,y)\leqq 0$$

we let $|z|\to 1$, then with the exception of finitely many points on $|z|=1$ we have that $\operatorname{Re}P(z)\geqq 0$, and therefore this inequality holds for all z, $|z|=1$. This proves the lemma.

Remark 5.1. In the course of proving the lemma, we have shown that the only singularities on $|z|=1$ of a univalent function $f(z)$ satisfying (4.5) are a finite number of algebraic singularities. Thus $f(z)$ maps $|z|<1$ onto a domain whose boundary consists of finitely many analytic arcs.

If $f(z)$ gives rise to a local maximum of the functional $J(f)$, then, as shown above, the function $P(z)$ is real on $|z|=1$. Thus the previous lemma shows that $P(z)\geqq 0$ for $|z|=1$. This inequality is very well known, and it has been proved by various authors in various situations by the application of special variations. We can reformulate these results in the form of a theorem.

Theorem 3. *If $f(z)\in S$ gives rise to a maximum of the functional $J(f)$, then $f(z)$ satisfies the equation*

$$(5.13)\qquad \left(\frac{Df(z)}{f(z)}\right)^2 Q\,[f(z)]+P(z)=0,$$

in which the quasi-polynomial $P(z)$ is nonnegative on $|z|=1$.

We saw above that if $d_n(f)=0$, then the function $f(z)$ also satisfies equation (5.13); it must be stressed that in the case when $f(z)$ gives rise to a maximum, we can also conclude that $P(z)\geqq 0$ for $|z|=1$, a condition which cannot be satisfied if $d_n(f)=0$.

The concluding part of this section will be concerned with the proof of a technical proposition examining the dependence of the kernels $\mathfrak{A}(z,t)$, $\mathfrak{B}(z,\bar{t})$ and $\mathfrak{C}(z)$ on the function $f(z)$ and the vector λ. Let $f,g\in S$ be such that $d_n(f)=d_n(g)=0$, and denote by $\lambda(f)$ and $\lambda(g)$ their normalized associated vectors. Let $r_1 < r_2 < 1$ be fixed, and set

$$\Delta = \max_{|z|=r_2} |f(z) - g(z)| + |\lambda(f) - \lambda(g)|.$$

Proposition 5.1. *With the above notation,*

$$\max_{|z|\leqslant r_1 |t|\leqslant r_1} |\mathfrak{A}(r, t; f) - \mathfrak{A}(r, t; g)| < \alpha^n \Delta, \tag{5.14}$$

where the constant α *depends only on* r_1 *and* r_2. *Similar inequalities are true for the functions* $\mathfrak{B}$ *and* $\mathfrak{C}$.

Proof. Set $r_3 = \frac{1}{2}(r_1 + r_2)$. If $|z| \leqq r_1$ and $|t| \leqq r_1$, then

$$|\mathfrak{A}(z, t; f) - \mathfrak{A}(z, t; g)| \leqslant \max_{|\zeta|=r_3} |\mathfrak{A}(z, \zeta; f) - \mathfrak{A}(z, \zeta; g)|. \tag{5.15}$$

from the maximum principle. For $|z| = r_1$ and $|t| = r_3$ we shall investigate separately the terms comprising $\mathfrak{A}(z, t)$. Upon applying Littlewood's inequality for the coefficients of a univalent function, we see that $|P(z; f) - P(z; g)| < \alpha_0^n \Delta$ if $|z| = r_1$ or $|z| = r_3$. Similar inequalities hold for P^* and R. Using the theorem of Goluzin quoted in the proof of Proposition 3.1, we find that

$$|P(z; f)\mathscr{K}(z, t; f) - P(z; g)\mathscr{K}(z, t; g)| < \alpha_1^n \Delta,$$

if $|z| = r_1$, $|t| = r_3$. This inequality also holds if z and t are interchanged. Taking into account the inequality for large j (see the remarks on p. 36) which was proved in the course of the proof of Proposition 3.1, we can estimate the modulus of the difference of the terms defined by (4.12) and (4.13). After adding together the resulting ieequalities, we arrive at (5.14). The inequalities involving $\mathfrak{B}$ and $\mathfrak{C}$ are proved similarly, but more easily, since in these two cases the use of (5.15) is not necessary. This proves the proposition.

§6. **Definition** 6.1. A univalent function $f(z)$ satisfying an equation of the form

$$\left(\frac{Df(z)}{f(z)}\right)^2 Q[f(z)] + P(z) = 0, \tag{5.13}$$

where

$$Q(w) = \sum_1^n A_k w^{-k}, \quad P(z) = \sum_{-n}^n B_k z^{-k}, \quad B_k = \overline{B}_{-k}, \quad 0 \leqslant k \leqslant n,$$

and where $P(z) \geqq 0$ for $|z| = 1$, will be called an *extremal* univalent function.

Any univalent function $f(z) \in S$ which gives the maximum for some functional $J(f) = F(x)$ will be, in the above sense, an extremal function. The truth of the converse is not obvious. We shall show below that, generally speaking, the converse is true; but this is a deep result.

If $G_f = \{w : w = f(z), |z| < 1\}$ and if f is an extremal function, then ∂G_f consists of finitely many analytic arcs, as was shown in Remark 5.1. In a neighborhood of any point of regularity of $f(z)$ on $|z| = 1$ we have that

$$Q\,[f(z)]\left(\frac{df(z)}{f(z)}\right)^2 = -P(z)\left(\frac{dz}{z}\right)^2 \geqslant 0.$$

Consequently the boundary arcs of G_f are portions of the trajectory of the quadratic differential

$$Q(w)\,(dw/w)^2. \tag{6.1}$$

We shall mention here without proof some well-known facts concerning the structure of the entire set of trajectories of the quadratic differential (6.1). Proofs may be found in [**2**, **4**]. The zeros and poles of the quadratic differential are called its critical points. On the w-sphere the differential (6.1) will have the following critical points: the zeros of $Q(w)$, the point $w=0$, which is a pole of order $n+2$, and the point $w=\infty$, which is a pole of the first order if $A_1 \neq 0$, and a zero if $A_1 = A_2 = 0$. If $A_1 = 0$ and $A_2 \neq 0$, then $w=\infty$ is not a critical point. The local structure of the trajectories of the quadratic differential is very simple. Through each ordinary point there passes a unique trajectory of (6.1), consisting of an open arc. If w_0 is a zero of order κ of the quadratic differential, then there exist $\kappa+2$ trajectories beginning at w_0, with successive tangential limit directions each forming an angle of $2\pi/(\kappa+2)$. If w_0 is a first order pole, then there exists exactly one trajectory with end point w_0. The structure of the trajectories in a neighborhood of a pole of order $\kappa > 1$ is significantly more complicated.

LEMMA 6.1. *If $A_n \neq 0$, the point $w=0$ is a pole of order $n+2$. If some trajectory has $w=0$ as its end point, then it converges to $w=0$ along one of n directions, any successive pair of which is separated by an angle of $2\pi/n$. There exists an open neighborhood U of $w=0$ with the following properties:*

1. *Every trajectory passing through a point of U either converges to $w=0$ in both directions or leaves U entirely.*

2. *There exists an open neighborhood U_0 of $w=0$, $U_0 \subset U$, such that every trajectory passing through a point of U_0 converges to $w=0$ in at least one direction and remains in U_0.*

3. *If some trajectory lies entirely in U, and consequently converges to $w=0$ in both directions, then the tangent t to this trajectory as it approaches $w=0$ in the corresponding direction converges to one of two adjacent limit directions. The Jordan curve constructed by the adjunction of $w=0$ to this trajectory bounds a domain $\mathscr{D}$ which contains the angle formed by the two neighboring tangents. The tangent to any trajectory having points in common with $\mathscr{D}$ converges as $w \to 0$ to the line corresponding to these adjacent limit directions. The domain $\mathscr{D}$ is mapped by the appropriate branch of the function $\zeta = \int Q^{1/2}(w)\,w^{-1}dw$ to the half plane $\operatorname{Im}\zeta > \alpha$.*

4. *For each pair of adjacent limit directions there exists a trajectory possessing the properties listed in* 3.

The above remarks and Lemma 6.1 completely describe the local structure of the trajectories of the quadratic differential (6.1).

DEFINITION 6.2. A set $\mathfrak{M}$ on the w-sphere is called an *F-set* relative to the quadratic differential if each trajectory of the differential which intersects $\mathfrak{M}$ lies completely within $\mathfrak{M}$. Among the F-sets, the types described in the following definitions will play an important role in what follows.

DEFINITION 6.3. An *end domain* is a maximally connected open F-set possessing the following properties:

1) It does not contain critical points of the differential.

2) Through each point of it there passes a trajectory of the differential having the limit end points in each of the two possible directions at a given pole of the differential.

3) It is mapped by the function $\zeta = \int Q^{1/2}(w) w^{-1} dw$ to either the upper or lower half plane in the ζ plane, depending on which branch of the square root is chosen.

DEFINITION 6.4. A *strip-like domain* is a maximally connected open F-set possessing the following properties:

1) It contains no critical points of the differential.

2) Through each point of it there passes a trajectory of the differential having at some pole of the differential the end limit point in one direction, and at some pole of the differential the end limit point in the other direction.

3) It is mapped conformally onto the strip $\alpha < \operatorname{Im} \zeta < \beta$ by the function $\zeta = \int Q^{1/2}(w) w^{-1} dw$.

REMARK. The poles of the differential where the trajectories of the strip-like domains end must have order $\geqq 2$, and for the differential (6.1) these poles coincide with $w = 0$. For quadratic differentials not of this special form the classification of F-sets is richer, but we shall not require any information on this matter.

Let N be the set of all zeros and simple poles of the quadratic differential (6.1), including the point $w = \infty$ even if it is neither a simple pole nor a zero of (6.1). We denote by $\mathscr{G}$ the union of all those trajectories with end limit points in N, where we have also adjoined the point $w = 0$ to each such trajectory.

LEMMA 6.2. *The set $\mathscr{G}$ consists of a finite number of analytic arcs, each of which begins at some point of N. The set $\mathscr{G}$ is connected. The complement of $\mathscr{G}$ is an end domain, except for possibly finitely many strip domains. Each such domain is bounded by finitely many trajectories. Every boundary component of such a domain contains a point of N. If $w = 0$ is joined to two boundary components of the strip-like domain, then we get two Jordan curves, each containing a point of N.*

We thus see that the set $\mathscr{G}$ is a connected planar graph whose vertices are the points of the finite set N, together with the point $w = 0$. The edges of the graph are the trajectories of the differential (6.1) which include points of N.

We now return to the consideration of extremal functions. We have seen that ∂G_f consists of a portion of the trajectories of the differential. If some of these trajectories do not belong to $\mathscr{G}$, then from the connectedness of ∂G_f it follows that ∂G_f is a portion of some trajectory lying in one of the end domains or strip-like domains. But then $f(z)$ maps ∞ to the interior of $|z| < 1$, which is absurd. Therefore ∂G_f consists of a trajectory or part of a trajectory lying in $\mathscr{G}$ and containing the point $w = \infty$. Since any cycle of the graph $\mathscr{G}$ contains $w = 0$, it follows that ∂G_f does not contain a cycle. Therefore ∂G_f is a tree, which we denote by $\mathscr{T}_f$. The base, or root, of the tree will always be at $w = \infty$. All the above information can be formulated in the following theorem due to Schaeffer and Spencer [2].

Theorem 4 (Schaeffer and Spencer). *Any extremal function maps $|z| < 1$ onto a domain whose boundary is a tree $\mathscr{T}_f$. The base of the tree is at $w = \infty$, and $\mathscr{T}_f \subset \mathscr{G}$, where $\mathscr{G}$ is the graph of the differential* (6.1).

The converse of this theorem is also true, and is also due to Schaeffer and Spencer.

Theorem 5. *Let the differential* (6.1) *be given. Suppose that $\mathscr{G}$ is the graph of the differential and that a tree $\mathscr{T} \subset \mathscr{G}$ has its base at $w = \infty$ and also has the property that the conformal radius of the complement of $\mathscr{T}$ with respect to $w = 0$ is* 1. *Then there exists a function $f(z)$, mapping $|z| < 1$ onto the domain bounded by $\mathscr{T}$, which satisfies equation* (5.13), *where $P(z) \geqq 0$ for $|z| = 1$.*

The proof of this theorem is quite simple, and can be found in [2].

We now introduce a set of univalent functions which will play an important role in the future.

Definition 6.5. The set S_n is defined to be the set of all those extremal functions for which the function $Q(w)$ in (5.13) is a polynomial in $1/w$ of degree at most n.

We note that the set of all functions f for which $d_n(f) = 0$ is much larger than S_n. Thus it is very possible that in this set there are functions $f(z)$ such that the complement with respect to the w-sphere of G_f has nonempty interior.

Definition 6.6. The set S_n^* is defined to be the set of all those univalent functions f such that $d_n(f) = 0$ and such that the complement of G_f has an empty interior.

The following lemma characterizes S_n^*.

Lemma 6.3. *If $f \in S_n^*$, then for any associated vector λ the quasi-polynomial $P(z)$ of* (5.13) *assumes real values on $|z| = 1$. The boundary of the domain G_f is a finite tree whose branches are trajectories, or orthogonal trajectories, of the differential* (6.1).

PROOF. From (3.19), $\operatorname{Im} P(z) = \operatorname{Im} B_0$ if $|z| = 1$. We shall show that $\operatorname{Im} B_0 = 0$. We assume the opposite. Then $P(z) \neq 0$ if $|z| = 1$. According to the theorem of Painlevé, the function $f(z)$ solving equation (5.13) in $|z| < 1$ has on $|z| = 1$ a finite number of algebraic singularities. Let $z_1, \cdots, z_m$ be these singular points. For each particular point z_j there exists some disc $K_j = \{z : |z - z_j| \leqq \rho\}$ such that in this disc $f(z)$ has the local expansion

$$f(z) = \sum_{k \geqslant k_j} \alpha_{kj} (z - z_j)^{k/l_j}, \tag{6.2}$$

where l_j is an integer and where $k_j < 0$ if z_j is an algebraic pole. If z_j is not a pole, then $k_j = 0$ and $f(z_j) = w_j$, where w_j is a root of $Q(w)$. In this case it is easy to determine the boundary of L_j, where L_j is the image of the half-disc $K_j \cap \{z : |z| < 1\}$. This boundary will consist of an analytic Jordan arc (the image of the "semicircle" $|z - z_j| = \rho$, $|z| < 1$) and two analytic Jordan arcs intersecting at w_j and forming there some angle different from π. If z_j is a pole, then the image of K_j under the mapping $\zeta = 1/f(z)$ is as in the previous case, except that the point $\zeta = 0$ plays the role of w_j. The set $\{z : |z| = 1\} \backslash \bigcup K_j$ is composed of m arcs of the unit circle. On each such arc $f(z)$ is regular and satisfies $f'(z) \neq 0$. Thus the image of each such arc is an analytic Jordan arc γ. Each boundary point of γ lies on the boundary of some L_j, and so it is possible to analytically continue γ until we meet a zero of $Q(w)$. (Note that $Q(\infty) = 0$.) Hence ∂G_f is a finite graph whose vertices are the zeros of $Q(w)$. Each vertex of the graph, with the possible exception of $w = \infty$, meets at least two edges. On the other hand, this graph has no cycles. Thus it is a tree, and consequently has vertices which meet only one edge. This contradiction proves that $\operatorname{Im} B_0 = 0$, and so $\operatorname{Im} P(z) = 0$ for $|z| = 1$. It also follows from the above argument that ∂G_f is a finite tree. On each branch of the tree either $Q[f(z)](df(z)/f(z))^2 > 0$ or $Q[f(z)](df(z)/f(z))^2 < 0$, so each branch will either be a trajectory of the differential (6.2), or an orthogonal trajectory. This proves the lemma.

CHAPTER 2

THE COEFFICIENT REGION OF UNIVALENT FUNCTIONS

The functionals considered in the first chapter were essentially real-valued functionals defined on R^{2n}. Therefore the study of the extrema of these functionals amounts to the study of the extrema of a function defined on a closed set in R^{2n}. This set can be defined as follows.

Let $f(z) \in S$, $f(z) = \sum_0^\infty a_k z^{k+1}$, and $a_0 = 1$. We consider the mapping $F: S \to R^{2n}$, where $F(f) = x$, $x = (a_1, \cdots, a_n)$. Clearly F is continuous, and so the image of S under F is a compact set, which we shall denote by $\mathscr{V}_n$. This compact set is called the *2n-dimensional coefficient region*. The problem of determining the structure of this set was raised by Bieberbach. Clearly, the study of extremal problems is equivalent to the study of the boundary of $\mathscr{V}_n$, and in particular, to the study of its metric properties. A detailed knowledge of the structure of $\partial\mathscr{V}_n$ allows us to overcome those difficulties mentioned at the beginning of the first chapter. We shall discuss below some general properties of $\partial\mathscr{V}_n$.

§7. By S_x we shall denote the preimage of the point $x \in \mathscr{V}_n$. S_x is a closed set in S. Clearly, $x = 0$ is contained in $\mathscr{V}_n$. Since for any ρ, $0 < \rho \leqq 1$, the function $\rho^{-1}f(\rho z) \in S$ if $f \in S$, it follows that any point $x \in \mathscr{V}_n$, $x = (a_1, \cdots, a_n)$, can be joined to $x = 0$ by the path $(\rho a_1, \cdots, \rho^n a_n)$, $0 < \rho < 1$. The next lemma follows from this fact.

LEMMA 7.1 (SCHAEFFER AND SPENCER). *$\mathscr{V}_n$ is homeomorphic to the sphere $|x| \leqq 1$, and $\partial\mathscr{V}_n$ is homeomorphic to $|x| = 1$.*

The proof of this lemma is quite simple and may be found in [2]. From Theorem 2 it follows that for any $f \in S_x$, $x \in \partial\mathscr{V}_n$, we have $d_n(f) = 0$. It is well known that if for $f \in S$ the complement of G_f has nonempty interior, then from the Marty variation there exists a univalent function $g(x) = \sum_0^\infty b_k z^{k+1}$, $b_0 = 1$, where the vector $y = (b_1, \cdots, b_n)$ is any vector satisfying $|y - x| \leqq \epsilon$, provided ϵ is sufficiently small. Hence if $f \in S_x$, $x \in \partial\mathscr{V}_n$, then $f \in S_n^*$. In particular, all functions belonging to S_x are unbounded. The following lemma of Schaeffer and Spencer now follows easily.

LEMMA 7.2. *In order for $x \in \partial\mathscr{V}_n$ to be an interior point, it is necessary and sufficient that S_x contain a bounded function.*

The sufficiency follows from the remarks made above. To prove the necessity, we need note only that if $x = (a_1, \cdots, a_n)$ is an interior point, then so is $x_\rho = (\rho a_1, \cdots, \rho^n a_n)$, $\rho = 1 + \epsilon$ for some $\epsilon > 0$. But if $f \in S_{x_\rho}$, then $\rho f(z/\rho) \in S_x$, and $\rho f(z/\rho)$ is bounded in $|z| \leqq 1$.

It is easy to obtain a more precise characterization of S_x when $x \in \partial\mathscr{V}_n$. As was remarked by Schaeffer and Spencer, on $\partial\mathscr{V}_n$ it is possible to find an everywhere dense set $\mathscr{M}$ of points with the property that each point lies on the boundary of some sphere contained in the complement of $\mathscr{V}_n$. If $x^0 \in \mathscr{M}$, there exists a functional $J[f] = \mathscr{F}(x)$ which assumes its maximal value at every $f \in S_{x^0}$. In order to prove this, note that if y is the center of the corresponding sphere, then it suffices to take $\mathscr{F}(x) = |x-y|^{-1}$. Therefore S_{x^0} consists entirely of extremal functions, and in particular, S_{x^0} is finite dimensional. If $x \in \partial\mathscr{V}_n$, then one can find a sequence $\{x^k\} \subset \mathscr{M}$, $\lim_{k\to\infty} x^k = x$, and the sequence of extremal functions $f_k(z) \in S_{x^k}$ then converges to some $f \in S_x$. Clearly f itself is an extremal function. We have thus proved the following lemma, which was first proved by Schaeffer and Spencer [2].

LEMMA 7.3. *If $x \in \partial\mathscr{V}_n$, then $S_x \cap S_n \neq \emptyset$.*

§8. In this section we shall present an important theorem of Teichmüller. This theorem is in fact one of the strongest results known in the theory of extremal problems for univalent functions. Let $Q(w)(dw/w)^2$ be a quadratic differential, $\mathscr{G}$ its graph, and $\mathscr{T}$ a tree, not containing $w=0$, with base at $w = \infty$.

THEOREM 6 (TEICHMÜLLER [3]). *If the function $\varphi(w)$, $w = u + iv$, is regular and univalent outside $\mathscr{T}$, and if in a neighborhood of $w=0$ it has the expansion $\varphi(w) = w + \gamma w^{n+1} + \cdots$, then*

$$\frac{1}{2\pi}\iint\left(\left|\frac{Q^{1/2}[\varphi(w)]}{\varphi(w)}\varphi'(w)\right| - \left|\frac{Q^{1/2}(w)}{w}\right|\right)^2 du\,dv \leqslant \operatorname{Re}\gamma A_n, \tag{8.1}$$

where the range of integration is the entire w-sphere.

The proof of this striking and deep result is given in [2, 3, 4].

REMARK 8.1. In Teichmüller's original paper, as well as in other papers concerning this theorem, the conclusion of the theorem is given in the weaker form

$$\operatorname{Re}\gamma A_n \geqq 0 \tag{8.2}$$

and it is shown that in the case of equality we have $\varphi(w) \equiv w$.

We are not aware of any references in which the theorem is stated in the form given above, and so for the convenience of the reader we shall include a proof of inequality (8.1). Before so doing, however, we shall give some well-known corollaries of Theorem 6.

COROLLARY 8.1. *If $\operatorname{Re}\gamma A_n = 0$, then $\varphi(w) \equiv w$.*

To prove this, note that from (8.1) we have

$$\frac{w^2 Q[\varphi(w)]\varphi'^2(w)}{\varphi^2(w)Q(w)} = e^{i\theta(w)}, \qquad \operatorname{Im}\theta(w) = 0.$$

But the left-hand side of this latter relation is meromorphic in the complement of $\mathscr{T}$. Therefore $\theta(w) = \text{const}$, and since the left-hand side tends to 1 as $w \to 0$ we have $\theta(w) \equiv 0$. Thus $\varphi(w) = w$.

Let $f \in S_n$, $f(z) = \sum_0^\infty a_k z^{k+1}$ and $a_0 = 1$, and let $g(z) \in S$, $g(z) = \sum_0^{n-1} a_k z^{k+1} + \sum_n^\infty b_k z^{k+1}$.

Let $\mathscr{T}$ be a boundary tree for the function $f(z)$, and let $Q(w)(dw/w)^2$ be its corresponding differential. The function $\varphi(w) = g[f^{-1}(w)]$ is regular and univalent in the complement of $\mathscr{T}$. In a neighborhood of $w = 0$ we have $\varphi(w) = w + (b_n - a_n)w^{n+1} + \cdots$, and so from Theorem 6 it follows that

$$\operatorname{Re} A_n(b_n - a_n) > 0.$$

Also, if $\operatorname{Re} A_n(b_n - a_n) = 0$, then, according to Corollary 8.1, $f(z) \equiv g(z)$. Hence we have

Theorem 7 (**Teichmüller**). *Let $f(z) \in S_n$ and $f(z) \in S_x$. Then S_x contains exactly one function.*

The following theorem is a corollary of Teichmüller's theorem and Lemmas 7.2 and 7.3. It is contained in [**2**] in a slightly different form.

Theorem 8. *$\partial \mathscr{V}_n$ is homeomorphic to S_n.*

Proof. From Lemmas 7.2 and 7.3 and Theorem 7 it follows that there exists a one-to-one mapping of S_n onto $\partial \mathscr{V}_n$ which is clearly continuous, and is thus a homeomorphism.

We now state without proof some well-known facts which will be used in the proof of (8.1) as well as in the section following the proof.

Teichmüller's proof is based on the use of the metric

$$|d\zeta|^2 = |Q(w)(dw/w)^2|, \tag{8.3}$$

considered as a metric on the punctured w-sphere.

Definition 8.1. A *geodesic* is a curve each sufficiently short arc of which is, with respect to the metric (8.3), the shortest path joining the two ends of the arc.

Lemma 8.1. *A curve is a geodesic if and only if the following conditions hold:*

1) *Each of its points not belonging to the set N (the set of critical points of the differential) lies on an open arc on which*

$$\arg\{Q(w)(dw/w)^2\} = \text{const}. \tag{8.4}$$

2) *Each point of the curve which is a κ-fold zero of the differential is also a common point of two arcs such that* (8.4) *is satisfied on each of the arcs and such that the arcs form two angles, each of at least $2\pi/(\kappa + 2)$ radians.*

Lemma 8.2. *Any two points of the punctured w-sphere can be joined by a unique geodesic giving the shortest path in the metric* (8.3).

These lemmas are proved in [2, 4]. As we have already remarked, the graph $\mathscr{G}$ of the quadratic differential $Q(w)(dw/w)^2$ divides the w-sphere into a finite number of end domains or strip-like domains $\mathscr{D}_1, \cdots, \mathscr{D}_m$. Following Schaeffer and Spencer, for each domain $\mathscr{D}_j$ we choose a domain $\tilde{\mathscr{D}}_j \supset \mathscr{D}_j$ such that a branch of the function $\zeta = \int^w Q^{1/2}(w)w^{-1}dw$ is univalent in $\mathscr{D}_j$, can be analytically extended to $\tilde{\mathscr{D}}_j$, and maps $\tilde{\mathscr{D}}_j$ to the entire w-sphere slit by a finite number of vertical half-lines. The domain $\tilde{\mathscr{D}}_j$ is constructed as follows. We consider the differential $-Q(w)(dw/w)^2$ and its corresponding end domains and strip-like domains $\mathscr{D}_1^*, \cdots$ and its graph $\mathscr{G}^*$. The trajectories of this differential are orthogonal trajectories for the differential $Q(w)(dw/w)^2$. We consider those domains $\mathscr{D}_k^*$ which have nonempty intersection with $\mathscr{D}_j$, and we denote these by $\mathscr{D}_1^*, \cdots, \mathscr{D}_l^*$. We then take the union of these domains and those portions of the edges of the graph $\mathscr{G}^*$ which lie in $\mathscr{D}_j$. As a result we obtain a simply-connected domain, which we denote by $\tilde{\mathscr{D}}_j$.

In the case when $\mathscr{D}_j$ is an end domain and is mapped by $\int^w Q^{1/2}(w)w^{-1}dw$ onto the half-plane $\operatorname{Im}\zeta > 0$, the slits in the ζ-plane will be of the form

$$\operatorname{Re}\zeta = \xi_k, \quad \operatorname{Im}\zeta \leqq \eta_k \leqq 0, \tag{8.5}$$

and the point $\zeta_k = \xi_k + i\eta_k$ is the image of a point in the set N. If $\mathscr{D}_j$ is a strip-like domain and is mapped onto the strip $0 < \operatorname{Im}\zeta < d$, then the slits have the form

$$\begin{aligned} &\operatorname{Re}\zeta = \xi_k, \ -\infty < \operatorname{Im}\zeta \leqslant \eta_k \leqslant 0, \\ &\operatorname{Re}\zeta = \xi_k, \quad d \leqslant \eta_k^* \leqslant \operatorname{Im}\zeta, \end{aligned} \tag{8.6}$$

and both $\zeta_k = \xi_k + i\eta_k$ and $\zeta_k^* = \xi_k + i\eta_k^*$ are images of critical points. In some cases one of the half-lines defined by (8.6) might be absent. The proof of these facts can be found in [2].

Proof of Theorem 6. We shall use Teichmüller's proof. If $\mathscr{L}$ is a Jordan curve containing $w = 0$ in its interior, we shall denote by $G(\mathscr{L})$ that component of the complement of $\mathscr{L}$ (with respect to the w-sphere) which contains the point $w = \infty$. Let $\mathscr{L} = \{w : |w| = \rho\}$, where ρ is sufficiently small, and $\varphi\{\mathscr{L}\} = \mathscr{L}^*$. We shall compare the areas of the domains $G(\mathscr{L})$ and $\varphi\{G(\mathscr{L})\}$, calculated with respect to the metric (8.3). Set

$$J_\rho = \iint\limits_{G(\mathscr{L})} \left|\frac{Q(w)}{w^2}\right| du\, dv, \qquad J_\rho^* = \iint\limits_{\varphi\{G(\mathscr{L})\}} \left|\frac{Q(w)}{w^2}\right| du\, dv.$$

Clearly $J_\rho^* \leqq \iint_{G(\mathscr{L}^*)} |Q(w)/w^2|\, du dv$, and so

$$J_\rho^* - J_\rho \leqslant \iint\limits_{G(\mathscr{L}^*)} \left|\frac{Q(w)}{w^2}\right| du\, dv - \iint\limits_{G(\mathscr{L})} \left|\frac{Q(w)}{w^2}\right| du\, dv.$$

Let $r = r(\varphi)$ be the equation of the curve $\mathscr{L}^*$ in the polar coordinates $w = re^{i\varphi}$. Then

$$J_\rho^* - J_\rho \leqslant \int_0^{2\pi} d\varphi \int_\rho^{r(\varphi)} \left|\frac{Q(w)}{w^2}\right| r\,dr.$$

In a neighborhood of the point $w = 0$ we have

$$\varphi(w) = w + \gamma w^{n+1} + \cdots, \tag{8.7}$$

with $\gamma = |\gamma| e^{i\vartheta}$. Then $r(\varphi) = \rho + |\gamma| \rho^{n+1} \cos(n\varphi + \vartheta) + O(\rho^{n+2})$. Therefore

$$\int_0^{2\pi} d\varphi \int_\rho^{r(\varphi)} \left|\frac{Q(w)}{w^2}\right| r\,dr = |A_n| \rho^{-n-1} \int_0^{2\pi} [r(\varphi) - \rho]\, d\varphi + O(\rho^{-n}) \int_0^{2\pi} |r(\varphi) - \rho|\, d\varphi$$

$$= |A_n \gamma| \int_0^{2\pi} \cos(n\varphi + \vartheta)\, d\varphi + O(\rho) = O(\rho),$$

and consequently

$$J_\rho^* - J_\rho \leqslant O(\rho). \tag{8.8}$$

But

$$J_\rho^* = \iint_{\varphi\{G(\mathscr{C})\}} \left|\frac{Q(w)}{w^2}\right| du\,dv = \iint_{G(\mathscr{C})} \left|\frac{Q[\varphi(w)]\, \varphi'^2(w)}{\varphi^2(w)}\right| du\,dv$$

$$= \iint_{G(\mathscr{C})} \chi^2(w) \left|\frac{Q(w)}{w^2}\right| du\,dv,$$

where $\chi(w) = |w(\varphi'(w)/\varphi(w))\, Q^{1/2}[\varphi(w)\, Q^{-1/2}(w)]|$. We also have the identity

$$J_\rho^* = \iint_{G(\mathscr{C})} (\chi - 1)^2 \left|\frac{Q(w)}{w^2}\right| du\,dv + 2 \iint_{G(\mathscr{C})} \chi \left|\frac{Q(w)}{w^2}\right| du\,dv - \iint_{G(\mathscr{C})} \left|\frac{Q(w)}{w^2}\right| du\,dv.$$

From the above we obtain

$$J_\rho^* - J_\rho = \iint_{G(\mathscr{C})} (\chi - 1)^2 \left|\frac{Q(w)}{w^2}\right| du\,dv + 2 \iint_{G(\mathscr{C})} (\chi - 1) \left|\frac{Q(w)}{w^2}\right| du\,dv. \tag{8.9}$$

We shall find a lower estimate for the second integral on the right-hand side of (8.9). Let $\mathscr{D}$ be one of the end domains of the quadratic differential, and let $\widetilde{\mathscr{D}}$ be its extension as previously defined. We take a branch of the function

$$\zeta = \int^w Q^{1/2}(w) \frac{dw}{w} \tag{8.10}$$

such that $\mathscr{D}$ is mapped onto the upper half plane. We set $\mathscr{D}_\rho = \mathscr{D} \cap G(\mathscr{L})$, and we consider

$$j_\rho = \iint_{\mathscr{D}_\rho} (\chi(w) - 1) \left|\frac{Q(w)}{w^2}\right| du\,dv. \tag{8.11}$$

In the ζ-plane we consider the circle $|\zeta| = (2/n) |A_n^{1/2}| \rho^{-n/2} = R$, and we let $\mathscr{D}_\rho^*$ be the preimage of the half-disc $|\zeta| \leqq R$, $\operatorname{Im} \zeta > 0$ under the mapping

$\zeta=\zeta(w)$. If in the integral (8.11) we change the domain of integration from $\mathscr{D}_\rho$ to $\mathscr{D}_\rho^*$, then the resulting integral has the growth rate $O(\rho)$. From (8.7) we have

$$\frac{w^2Q\,[\varphi\,(w)]\,\varphi'^2\,(w)}{\varphi^2\,(w)\,Q\,(w)}=1+O\,(w^n). \tag{8.12}$$

Therefore $\chi(w)=1+O(w^n)$, and so

$$(\chi-1)\,|\,Q(w)/w^2|=O(w^{-2}). \tag{8.13}$$

We note that

$$\zeta=\pm(2/n)A_n^{1/2}w^{-n/2}[1+\alpha w+\cdots]. \tag{8.14}$$

For $|\zeta|=R$ and $w=re^{i\varphi}$ it follows from the latter formula that the preimage of the semicircle $|\zeta|=R$, $\operatorname{Im}\zeta>0$ satisfies $r=\rho+O(\rho^2)$. Thus the change in area caused by the change from $\mathscr{D}_\rho$ to $\mathscr{D}_\rho^*$ is of the order $O(\rho^3)$, and in view of (8.13) we see that

$$j_\rho=\iint\limits_{\mathscr{D}_\rho^*}(\chi-1)\left|\frac{Q\,(w)}{w^2}\right|du\,dv+O\,(\rho).$$

Set $L_\zeta=\{\zeta:|\zeta|=R,\ \operatorname{Im}\zeta>0\}$, and let L_w be the preimage of L_ζ. Intersecting L_ζ with the straight line $\operatorname{Im}\zeta=\eta$, $\eta<R$, we get two points $\zeta_1=-\xi+i\eta$, $\zeta_2=\xi+i\eta$, and we denote their preimages, which lie on L_w, by w_1 and w_2. The geodesic λ_η connecting w_1 and w_2 is mapped into the line segment connecting ζ_1 and ζ_2, and its length is equal to 2ξ. Let l_η be the length of the geodesic connecting $\varphi(w_1)$ and $\varphi(w_2)$. Then

$$\int\limits_{\varphi\{\lambda_\eta\}}\left|\frac{Q^{1/2}\,(w)}{w}\,dw\right|=\int\limits_{\lambda_\eta}\left|\frac{Q^{1/2}\,[\varphi\,(w)]}{\varphi\,(w)}\,\varphi'\,(w)\,dw\right|=\int\limits_{\lambda_\eta}\chi\,(w)\left|\frac{Q^{1/2}\,(w)}{w}\,dw\right|\geqslant l_\eta.$$

For sufficiently small ρ the points $\varphi(w_1)$ and $\varphi(w_2)$ will lie in $\tilde{\mathscr{D}}$. Thus under the mapping $\zeta=\zeta(w)$ the geodesic connecting these points will be mapped to some polygonal line in the punctured ζ plane. (Here we have used Lemma 8.1.) The length of this polygonal line is clearly not less than the length of its projection onto the real axis, and so

$$l_\eta\geqslant\operatorname{Re}\{\zeta\,(\varphi\,(w_2))-\zeta\,(\varphi\,(w_1))\}=2\xi+\operatorname{Re}\{\zeta\,(\varphi\,(w_2))-\zeta\,(w_2)-\zeta\,(\varphi\,(w_1))+\zeta\,(w_1)\}.$$

We obtain from (8.14) that for small $|w|$

$$\zeta\,(\varphi\,(w))-\zeta\,(w)=(\varphi\,(w)-w)\frac{d\zeta}{dw}(1+O\,(w))$$

$$=\gamma\,\frac{w^{n+1}}{2\zeta}\,\frac{d\zeta^2}{dw}\,(1+O\,(w))=-\frac{2}{n\zeta}\,\gamma A_n\,(1+O\,(w)).$$

Hence

$$l_\eta \geqslant 2\xi + \operatorname{Re}\left\{\frac{2\gamma}{n} A_n\left(\frac{1}{\zeta_1} - \frac{1}{\zeta_2}\right) + O\left(\frac{\rho}{R}\right)\right\} = 2\xi - \frac{4\xi}{nR^2}\operatorname{Re}\gamma A_n + O\left(\frac{\rho}{R}\right),$$

and so

$$\int_{\lambda_\eta} \chi(w)\left|Q^{1/2}(w)\frac{dw}{w}\right| > 2\xi - \frac{4\xi}{nR^2}\operatorname{Re}\gamma A_n + O\left(\frac{\rho}{R}\right). \tag{8.15}$$

Integrating this inequality with respect to η from 0 to R, we find that

$$\int_0^R d\eta \int_{\lambda_\eta} \chi(w)\left|Q^{1/2}(w)\frac{dw}{w}\right| > \frac{\pi}{2}R^2 - \frac{\pi}{n}\operatorname{Re}\gamma A_n + O(\rho).$$

But

$$\frac{\pi}{2}R^2 = \iint_{\mathscr{D}_\rho^*}\left|\frac{Q(w)}{w^2}\right| du\,dv,$$

and

$$\int_0^R d\eta \int_{\lambda_\eta} \chi(w)\left|\frac{Q^{1/2}(w)}{w}dw\right| = \iint_{\mathscr{D}_\rho^*}\chi(w)\left|\frac{Q(w)}{w^2}\right| du\,dv.$$

Hence

$$j_\rho \geqslant -\frac{\pi}{n}\operatorname{Re}\gamma A_n + O(\rho). \tag{8.16}$$

In the case of a strip-like domain we must integrate (8.15) with respect to η from 0 to $d < \infty$. As the result of this integration, the second and third terms on the right-hand side of (8.15) will contribute $O(d/R) + O(d\rho/R) = O(\rho)$, and we find that

$$j_\rho \geqq O(\rho). \tag{8.17}$$

Summing (8.16) over the end domains and (8.17) over the strip-like domains, we conclude that

$$\iint_{G(\mathscr{C})} (\chi(w) - 1)\left|\frac{Q(w)}{w}\right| du\,dv \geqslant -\pi\operatorname{Re}\gamma A_n + O(\rho),$$

and upon recalling (8.8) and (8.9), and letting $\rho \to 0$, we deduce (8.1). Q.E.D.

§9. From inequality (8.1) one can deduce an estimate for $|\varphi(w) - w|$ when w lies in the complement of $\mathscr{T}$. Some important corollaries will follow from this estimate. In accordance with the properties of an extremal function, we shall assume that 1 is the conformal radius of the complement of the tree $\mathscr{T}$.

The proof will be reasonably simple, but a bit lengthy. In the course of the proof we shall denote absolute constants by the letter α, and constants depending only on n will be denoted by β. As usual, these constants need not be the same at each occurrence.

We shall first prove some simple propositions.

Proposition 9.1. *Let $f(\zeta)$ be regular in $|\zeta| < r_0$, and $\zeta = \xi + i\eta$. Assume that $f(\zeta)$ satisfies*

$$\frac{1}{2\pi} \iint\limits_{|\zeta|<r_0} [|f(\zeta)|^{1/2} - 1]^2 \, d\xi \, d\eta \leqslant \Delta^2 < \infty. \tag{9.1}$$

Then

$$|f(0)| \leqslant \left(1 + 2\sqrt{\frac{2}{3}} \frac{\Delta}{r_0}\right)^2. \tag{9.2}$$

Proof. From Jensen's formula,

$$\log |f(0)|^{1/2} \leqslant \frac{1}{2\pi} \int_0^{2\pi} \overset{+}{\log} |f(re^{i\varphi})|^{1/2} \, d\varphi,$$

where $r < r_0$. It then follows from well-known inequalities that

$$|f(0)|^{1/2} \leqslant \frac{1}{2\pi} \int_0^{2\pi} \max [1, |f(re^{i\varphi})|^{1/2}] \, d\varphi \leqslant 1 + \frac{1}{2\pi} \int_0^{2\pi} ||f(ze^{i\varphi})|^{1/2} - 1| \, d\varphi.$$

Multiplying this inequality by r and integrating from r_1 to $2r_1 < r_0$, we find after simplification that

$$|f(0)|^{1/2} \leqslant 1 + \frac{1}{3\pi r_1^2} \int_{r_1}^{2r_1} r\,dr \int_0^{2\pi} ||f(re^{i\varphi})|^{1/2} - 1| \, d\varphi.$$

Applying Schwarz' inequality to the second integral and using (9.1), we see that $|f(0)|^{1/2} \leqq 1 + \sqrt{2/3}\Delta r_1^{-1}$. Squaring this inequality and letting $r_1 \to r_0/2$, we see that (9.2) holds. Q.E.D.

Proposition 9.2. *If the domain $\mathscr{D}$ is a half plane, a disc, or a strip, and if $f(\zeta)$ is regular in $\mathscr{D}$ and satisfies*

$$\frac{1}{2\pi} \iint\limits_{\mathscr{D}} [|f(\zeta)|^{1/2} - 1]^2 \, d\xi \, d\eta \leqslant \Delta^2 < \infty,$$

then $|f(\zeta)| \geqq 1/8$ if ζ is at at least a distance $\lambda_0 \Delta$ from $\partial\mathscr{D}$, where λ_0 is some absolute constant.

Proof. We shall consider the case when $\mathscr{D}$ is the disc $|\zeta| \leqq r_0$, the proofs in the other cases being similar. Let $\mathscr{E} = \{\zeta : |f(\zeta)| \leqq 1/4\}$. Clearly

$$\iint\limits_{\mathscr{E}} d\xi \, d\eta \leqslant 2 \cdot \iint\limits_{\mathscr{E}} [1 - |f(\zeta)|^{1/2}] \, d\xi \, d\eta.$$

Schwarz' inequality gives that meas $\mathscr{E} \leqq 8\pi\Delta^2$. From Proposition 9.1 it follows that

$$|f(\zeta)| \leqslant \left(1 + 2\sqrt{\frac{2}{3}} \frac{\Delta}{r_0 - |\zeta|}\right)^2.$$

From Cauchy's formula we have that

$$f'(\zeta)=\frac{1}{2\pi i}\int\limits_{|t|=\frac{1}{2}(r_0+|\zeta|)}\frac{f(t)}{(t-\zeta)^2}\,dt.$$

Hence

$$|f'(\zeta)|\leqslant\frac{2}{r_0-|\zeta|}\left(1+4\sqrt{\frac{2}{3}}\frac{\Delta}{r_0-|\zeta|}\right)^2.$$

Suppose that at some point ζ_0 we have $|f(\zeta_0)|\leqq 1/8$. Then in the domain $\mathscr{D}_0$ formed by the intersection of the discs $|\zeta|\leqq\frac{1}{2}(|\zeta_0|+r_0)$ and $|\zeta-\zeta_0|\leqq\rho$ we have

$$|f(\zeta)-f(\zeta_0)|=\left|\int\limits_{\zeta_0}^{\zeta}f'(\zeta)\,d\zeta\right|\leqslant\frac{4\rho}{r_0-|\zeta_0|}\left(1+8\sqrt{\frac{2}{3}}\frac{\Delta}{r_0-|\zeta_0|}\right)^2.$$

Assuming that ρ satisfies

$$\frac{4\rho}{r_0-|\zeta_0|}\left(1+8\sqrt{\frac{2}{3}}\frac{\Delta}{r_0-|\zeta_0|}\right)^2=\frac{1}{8},$$

we see that in $\mathscr{D}_0$ we have $|f(\zeta)|\leqq 1/4$, so $\mathscr{D}_0\subseteq\mathscr{E}$. Thus meas $\mathscr{D}_0\leqq$ meas $\mathscr{E}$, and since meas $\mathscr{D}_0\geqq\pi\rho^2/3$, we have $\rho\leqq\sqrt{24}\Delta$, and this in turn gives

$$32\sqrt{24}\frac{\Delta}{r_0-|\zeta_0|}\left(1+8\sqrt{\frac{2}{3}}\frac{\Delta}{r_0-|\zeta_0|}\right)^2\geqslant 1,$$

Therefore $\Delta/(r_0-|\zeta_0|)\geqq 1/\lambda_0$, where λ_0 is an absolute constant. Q.E.D.

Proposition 9.3. *Assume that $\mathscr{D}$ is either a half plane or a disc, and that the function $f(\zeta)$ satisfies the hypotheses of Proposition 9.2. Then*

$$|f(\zeta)-e^{i\theta}|<\alpha\frac{\Delta}{h}\left(1+\frac{\Delta}{h}\right),$$

where h is the distance from ζ to $\partial\mathscr{D}$. If $\mathscr{D}$ is the disc $|\zeta|<r_0$, then $\theta=\arg f(0)$.

Proof. Since the proofs are similar if $\mathscr{D}$ is either a half plane or a disc, we shall assume that $\mathscr{D}$ is the half plane $\eta=\operatorname{Im}\zeta>0$. From Propositions 9.1 and 9.2 it follows that for $\eta\geqq\lambda_0\Delta=\eta_0$, $1/8\leqq|f(\zeta)|\leqq(1+2\sqrt{2/3}\lambda_0^{-1})^2$. Hence in the half plane $\eta\geqq\eta_0$ the function $\log f(\zeta)$ is regular and bounded. From the inequality $|\log x|\leqq|x-1|\max(1/x,1)$, it follows that for $\eta\geqq\eta_0$

$$\log^2|f(\xi+i\eta)|\leqq\alpha(|f(\xi+i\eta)|^{1/2}-1)^2. \tag{9.3}$$

The integral

$$\int\limits_{-\infty}^{\infty}(|f(\xi+i\eta)|^{1/2}-1)^2\,d\xi$$

exists for almost all $\eta>0$, and so this integral is finite for $\eta_1=\eta_0+\epsilon$, where $\epsilon>0$ can be chosen arbitrarily small. In the half plane $\eta\geqq\eta_1$ we have $\log f(\zeta)=u(\zeta)+i\bar{u}(\zeta)+i\theta$, and so $u(\xi+i\eta_1)$ and $\bar{u}(\xi+i\eta_1)$ belong to $L^2(-\infty,\infty)$. For $\eta>\tau>\eta_1$ we have from Schwarz' formula that

$$\log f(\xi + i\eta) - i\theta = \frac{1}{\pi i} \int_{-\infty}^{\infty} u(t + i\tau) \frac{dt}{t - \xi - i(\eta - \tau)} = a + ib.$$

It is easily verified that $|e^{a+ib} - 1| \leqq |a + ib| e^{|a|}$. In our case,

$$a = \frac{1}{\pi} \int_{-\infty}^{\infty} u(t + i\tau) \frac{(\eta - \tau)\, dt}{(t - \xi)^2 + (\eta - \tau)^2}.$$

Thus $|a| \leqq \max\{\log 8, 2\log(1 + 2\lambda_0^{-1}\sqrt{2/3})\}$. Using Schwarz' inequality, we see that

$$|a + ib| \leqslant \sqrt{\frac{2}{\eta - \tau}} \left(\frac{1}{2\pi} \int_{-\infty}^{\infty} u^2(t + i\tau)\, dt \right)^{1/2}.$$

and so

$$|f(\zeta) - e^{i\theta}| < \frac{\alpha}{(\eta - \tau)^{1/2}} \left(\frac{1}{2\pi} \int_{-\infty}^{\infty} u^2(t + i\tau)\, dt \right)^{1/2}.$$

Upon using inequality (9.3), we see that

$$|f(\zeta) - e^{i\theta}| < \frac{\alpha}{(\eta - \tau)^{1/2}} \left(\frac{1}{2\pi} \int_{-\infty}^{\infty} (|f(t + i\tau)|^{1/2} - 1)^2\, dt \right)^{1/2}.$$

Integrating this inequality with respect to τ from η_1 to $\eta/2$, assuming that $\eta > 2\eta_1$, and applying Schwarz' inequality, we see after simplifying that

$$|f(\zeta) - e^{i\theta}| < \frac{\alpha}{\eta - 2\eta_1} \left(\log \frac{2(\eta - \eta_1)}{\eta} \right)^{1/2} \Delta.$$

We now let $\eta_1 \to \eta_0$ and we assume that $\eta \geqq 4\eta_0$. Then we conclude that

$$|f(\zeta) - e^{i\theta}| < \frac{\alpha}{\eta} \Delta.$$

Let $\eta < 4\eta_0 = 4\lambda_0\Delta$. From (9.1) we see that

$$|f(\zeta) - e^{i\theta}| \leqslant 1 + \left(1 + 2\sqrt{\frac{2}{3}} \frac{\Delta}{\eta}\right)^2 < \frac{2}{\eta}\left(\eta + 2\sqrt{\frac{2}{3}}\Delta\right)\left(1 + 2\sqrt{\frac{2}{3}}\frac{\Delta}{\eta}\right)$$

$$< 2\left(4\lambda_0 + 2\sqrt{\frac{2}{3}}\right)\frac{\Delta}{\eta}\left(1 + 2\sqrt{\frac{2}{3}}\frac{\Delta}{\eta}\right) < \alpha\frac{\Delta}{\eta}\left(1 + \frac{\Delta}{\eta}\right)$$

as required. If we had considered the case of a disc, we would use Schwarz' formula for the disc, and we would also find that $\theta = \arg f(0)$. Q.E.D.

In the following we shall denote the integral on the left-hand side of (8.1) by Δ^2. We set

$$g(w) = \frac{w^2 Q[\varphi(w)]\, \varphi'^2(w)}{\varphi^2(w)\, Q(w)}.$$

If $\mathscr{D}$ is any end domain or strip-like domain, then

$$\frac{1}{2\pi}\iint_{\mathscr{D}}(|g(w)|^{1/2}-1)^2\left|\frac{Q(w)}{w^2}\right|du\,dv<\Delta^2.$$

Let $w(\zeta)$ be the inverse to the function in (8.10). We set $f(\zeta)=g[w(\zeta)]$.

Lemma 9.1. *In any end domain $\mathscr{D}$ we have*

$$|g(w)-1|\leqslant\alpha\frac{\Delta}{\rho(w)}\left(1+\frac{\Delta}{\rho(w)}\right),$$

where $\rho(w)$ is the shortest distance (with respect to the metric (8.3)) from the point w to the set consisting of all the critical points of the differential which lie on the boundary of the domain $\widetilde{\mathscr{D}}$ and the portions of the tree $\mathscr{T}$ lying in $\widetilde{\mathscr{D}}$.

Proof. In the domain $\widetilde{\mathscr{D}}$, portions of the branches of the tree $\mathscr{T}$ can lie either on the boundary of $\mathscr{D}$ or in some of the domains $\mathscr{D}_1^*,\cdots,\mathscr{D}_m^*$. Under the mapping of $\widetilde{\mathscr{D}}$ by the function (8.10), the image of these branches will be a finite number of line segments I_j parallel to the real axis and lying either on or below the real axis. Each such segment begins at the image of a critical point. The ζ-plane is slit by these segments. These segments together with the vertical segments defined earlier give us a complete system of segments. Let $\mathscr{E}$ be the maximal connected component in the upper half plane of the complement of these segments. $\mathscr{E}$ is a simply connected domain, and $f(\zeta)$ is regular in $\mathscr{E}$. It is clear that

$$\frac{1}{2\pi}\iint_{\mathscr{E}}(|f(\zeta)|^{1/2}-1)^2\,d\xi\,d\eta<\Delta^2. \tag{9.4}$$

Let ζ_0 be any point in the upper half plane, and set $d=\inf_{\zeta\in\partial\mathscr{E}}|\zeta-\zeta_0|=|\zeta^*-\zeta_0|$. It is obvious that ζ^* is either the image of a critical point, or is a point of some slit I_j; i.e. by definition $d=\rho(w_0)$, $w_0=w(\zeta_0)$. In the disc $|\zeta-\zeta_0|<d$ the function $f(\zeta)$ is regular, and so, from Proposition 9.3,

$$|f(\zeta)-e^{i\theta}|<\alpha\frac{\Delta}{d-|\zeta-\zeta_0|}\left(1+\frac{\Delta}{d-|\zeta-\zeta_0|}\right). \tag{9.5}$$

The expansion (8.12) is valid in a neighborhood of the point $w=0$, and so $f(\zeta)\to1$ as $|\zeta|\to\infty$. Using this remark and applying Proposition 9.3 (in the case of a half plane), we find that

$$|f(\zeta)-1|<\alpha\frac{\Delta}{\eta}\left(1+\frac{\Delta}{\eta}\right). \tag{9.6}$$

In (9.5) and (9.6) we set $\zeta=\zeta_0+id/2=\zeta_1$, and we conclude that

$$|f(\zeta_1)-e^{i\theta}|<2\alpha\frac{\Delta}{d}\left(1+\frac{2\Delta}{d}\right)\quad\text{and}\quad|f(\zeta_1)-1|<2\alpha\frac{\Delta}{d}\left(1+\frac{2\Delta}{d}\right).$$

From these two inequalities and inequality (9.5), applied at the point $\zeta=\zeta_0$, we find that $|f(\zeta_0)-1|<\alpha(\Delta/d)(1+\Delta/d)$. Q.E.D.

LEMMA 9.2. *In any strip-like domain* $\mathscr{D}$

$$|g(w)-1|<\alpha\frac{\Delta}{\rho(w)}\left(1+\frac{\Delta}{\rho(w)}\right)\left(n+\frac{l}{d}\right),$$

where $\rho(w)$ *is the shortest distance from* w *to* $\partial\mathscr{D}$ *(with respect to the metric* (8.3)), *where* d *is the width of the strip onto which* $\mathscr{D}$ *is mapped, and where* l *is the sum of the lengths of the branches of the tree* $\mathscr{T}$, *the lengths being computed via the metric* (8.3).

PROOF. Suppose that $\mathscr{D}$ is mapped onto the strip $-\infty<\xi<\infty$, $0<\eta<d$ in the plane $\zeta=\xi+i\eta$. We consider the domain $\widetilde{\mathscr{D}}$ and the slits referred to in the proof of the previous lemma, and we also define $\mathscr{G}$ to be the domain which contains our strip and is the maximal connected component of the complement of the set of slits. In $\mathscr{G}$ the function $f(\zeta)$ is regular, and (9.4) clearly holds. The domain $\mathscr{G}$ contains two half planes $\xi<\xi_-$ and $\xi>\xi_+$. From (8.12) and Proposition 9.3 it follows that in these half planes we have

$$|f(\zeta)-1|<\alpha\frac{\Delta}{\xi_--\xi}\left(1+\frac{\Delta}{\xi_--\xi}\right),\quad |f(\zeta)-1|<\alpha\frac{\Delta}{\xi-\xi_+}\left(1+\frac{\Delta}{\xi-\xi_+}\right). \tag{9.7}$$

We shall first estimate the quantity $|f(\xi+id/2)-1|$. The domain $\mathscr{G}$ contains some strips which correspond to the strip-like domains $\mathscr{D}_j^*\subset\widetilde{\mathscr{D}}$. Let k_0 be the number of these strips. Clearly k_0 is not greater than the number k of strip-like domains of the differential $-Q(w)(dw/w)^2$. Some of these strips will contain straight line slits, parallel to the ξ axis, which are images of the branches of the tree $\mathscr{T}$. We shall project all horizontal slits onto the line $\eta=d/2$. After so doing, we get a set j_0 containing at most $2k+2$ segments. Let j_1 be the complement with respect to the line $\eta=d/2$ of the set j_0. Each interval of j_1 defines either some strip $a<\xi<b$ in which $f(\zeta)$ is regular, or one of the half planes $\xi<\xi_-$ or $\xi>\xi_+$. For each such strip $a<\xi<b$ it follows from Proposition 9.3 that

$$|f(\zeta)-e^{i\theta}|<\frac{\alpha\Delta}{\dfrac{d_0}{2}-\left|\zeta-\dfrac{a+b+id}{2}\right|}\left(1+\frac{\Delta}{\dfrac{d_0}{2}-\left|\zeta-\dfrac{a+b+id}{2}\right|}\right), \tag{9.8}$$
$$d_0=\sqrt{d^2+(b-a)^2},$$

where $|\zeta-(a+b+id)/2|<\frac{1}{2}d_0$. In any disc $|\zeta-\zeta^*|<d/2$, $\zeta^*=\xi^*+id/2$, we have

$$|f(\zeta)-e^{i\vartheta^*}|<\frac{\alpha\Delta}{\dfrac{d}{2}-|\zeta-\zeta^*|}\left(1+\frac{\Delta}{\dfrac{d}{2}-|\zeta-\zeta^*|}\right). \tag{9.9}$$

For any interval of the set j_0 we write down the system of inequalities (9.9) derived by taking the centers of the discs to be the end points of the interval, and then covering the interval by discs with centers a distance $d/2$ apart. Finally, we arrive at (9.9) with the centers of the discs at the points $\xi_--d/4+id/2$ and $\xi_++d/4+id/2$. As a result of this construction, we get a system of m overlapping discs, which we index from left to right. The disc with index p will correspond to either θ or θ^*, which we denote by θ_p. Let K_p and K_{p+1} be two overlapping discs. We apply to the segment $\{\zeta: \zeta = \xi + id/2\} \cap K_p \cap K_{p+1}$ the inequalities for the discs K_p and K_{p+1}. As a result, we find that $|e^{i\theta_p} - e^{i\theta_{p+1}}| < (\alpha\Delta/d)(1+\Delta/d)$. In addition, applying inequality (9.7) at the points $\zeta = \xi_- - d/4 + id/2$ and $\zeta = \xi_+ + d/4 + id/2$, we deduce inequalities for $|e^{i\vartheta_1} - 1|$ and $|e^{i\vartheta_m} - 1|$ in which the two moduli cannot exceed $(\alpha\Delta/d)(1+\Delta/d)$. Hence for any p we have

$$|e^{i\vartheta_p} - 1| < \alpha m \frac{\Delta}{d}\left(1 + \frac{\Delta}{d}\right).$$

But m is at most $\alpha(l_0/d + k)$, where l_0 is the length of the horizontal slits, and where $l_0/d \leq l/d$ and $k < \alpha n$. Hence for any p

$$|e^{i\vartheta_p} - 1| < \alpha \frac{\Delta}{d}\left(1 + \frac{\Delta}{d}\right)\left(n + \frac{l}{2}\right).$$

From (9.7), (9.8), and (9.9) it follows that

$$\left|f\left(\xi + i\frac{d}{2}\right) - 1\right| < \alpha \frac{\Delta}{d}\left(1 + \frac{\Delta}{d}\right)\left(n + \frac{l}{d}\right), \quad -\infty < \xi < \infty.$$

If $\zeta_0 = \xi_0 + i\eta_0$ is any point of the strip, then from the latter inequality and from (9.9), applied with $\zeta^* = \xi_0 Qid/2$, we deduce the statement of the lemma.

Using these results, we shall establish for $g(w) - 1$ an estimate independent of the type of domain to which w belongs. Suppose that the disc $K_0 = \{w: |w - w_0| \leq r_0\}$ is free from critical points of the differential and from points of the tree $\mathscr{T}$. Set $\mu = \min_{w \in K_0} |Q^{1/2}(w) w^{-1}|$. From the definition of the quantity Δ it follows that

$$\frac{1}{2\pi} \iint_{K_0} (|g(w)|^{1/2} - 1)^2 \, du \, dv < \left(\frac{\Delta}{\mu}\right)^2.$$

Suppose $\mathscr{D}_1, \cdots, \mathscr{D}_\nu$ are those end and strip-like domains which intersect K_0, and suppose that among these there are no end domains. We draw an orthogonal trajectory through the center of the disc. By the definition of an end point, this orthogonal trajectory contains either a point of the set N or the point $w = 0$. Thus this trajectory must intersect the circumference ∂K_0. Let λ be that maximal arc of the orthogonal trajectory which has no interior points on the circumference ∂K_0. Then the end points of λ lie on the circumference. The length of λ in the metric (8.3) satisfies the inequality

$$l_\lambda = \int_\lambda \left| Q^{1/2}(w) \frac{dw}{w} \right| > 2r_0\mu.$$

Let d_p be the width of the strip domain $\mathscr{D}_p$ in the metric (8.3). Then

$$\sum_{p=1}^{\nu} d_p \geqslant l_\lambda > 2r_0\mu. \tag{9.10}$$

We denote by m_n the maximal number of strip-like domains in the complement of the graph $\mathscr{G}$ of the differential. It is easy to see that $m_n < \alpha n$. Thus from (9.10) it follows that $\max_p d_p > \alpha r_0\mu/n$. We have thus proved the following proposition.

PROPOSITION 9.4. *If* $\min_{w\in K_0}|Q^{1/2}(w)w^{-1}| = \mu > 0$, *then the disc* K_0 *intersects either an end domain, or a strip-like domain whose width is at least* $\alpha r_0\mu/n$.

LEMMA 9.3. *If under the hypothesis of Proposition* 9.4 *the disc* K_0 *does not intersect* $\mathscr{T}$, *then*

$$|g(w)-1| < \alpha n\frac{\Delta_0}{r_0-|w-w_0|}\left(1+\frac{\Delta_0}{r_0-|w-w_0|}\right)\left(1+\frac{l}{\mu r_0}\right), \tag{9.11}$$

where $\Delta_0 = \Delta/\mu$.

PROOF. Consider the disc $K_1 = \{w : |w-w_0| \leqq r_1\}$, $r_1 = r_0/2$. This disc intersects either an end domain $\mathscr{D}_q$ or a strip-like domain $\mathscr{D}_p$ of width at least $\alpha r_0\mu/n$. In the first case we choose $w^* \in \mathscr{D}_q \cap K_1$ and apply Lemma 9.1. Since K_0 does not intersect $\mathscr{T}$ or N, we have $\rho(w^*) > \alpha\mu r_0$, and consequently

$$|g(w^*)-1| < \alpha\frac{\Delta_0}{r_0}\left(1+\frac{\Delta_0}{r_0}\right). \tag{9.12}$$

In the second case, choose a point w^*, lying in the intersection of $\mathscr{D}_p$ and $K_2 = \{w : |w-w_0| \leqq r_2\}$, $r_2 = 3r_0/4$, whose distance from $\partial\mathscr{D}_p$ is at least $\alpha r_0\mu/n$. In particular, if $K_1 \subset \mathscr{D}_p$, then we set $w^* = w_0$, and it is obvious that the distance from w^* to $\partial\mathscr{D}_p$ is at least $\alpha r_0\mu$. If we cannot choose a point as specified above, then choose $w_1 \in \partial K_1 \cap \partial\mathscr{D}_p$. From w_1 we construct an orthogonal trajectory inside $\mathscr{D}_p$, extending the trajectory until it intersects a second component of $\partial\mathscr{D}_p$. We have thus constructed an arc λ of length $l_\lambda > \alpha r_0\mu/n$. If the point bisecting λ lies in K_2, then we choose it as the point w^*. If the bisecting point lies outside K_2, then λ intersects the circumference ∂K_2 in a point w^*, and the distance from w_1 to w^* is at least $r_0\mu/4$. Hence, in view of Lemma 9.2, we have

$$|g(w^*)-1| < \alpha n\frac{\Delta_0}{r_0}\left(1+\frac{\Delta_0}{r_0}\right)\left(1+\frac{l}{\mu r_0}\right). \tag{9.13}$$

By Proposition 9.3 we have in K_0 that

$$|g(w)-e^{i\theta}| < \alpha\frac{\Delta_0}{r_0-|w-w_0|}\left(1+\frac{\Delta_0}{r_0-|w-w_0|}\right).$$

Applying this inequality and either (9.12) or (9.13), we arrive at (9.11). Q.E.D.

PROPOSITION 9.5. *Set* $p(w) = \prod_{\nu=1}^m (w-w_\nu)$, $w_\nu \neq 0$, $1 \leqq \nu \leqq m$. *There exists in the plane a sector* $L_\epsilon = \{w : |\arg w - \theta| \leqq \epsilon\}$ *of opening* $2\epsilon \geqq \pi/4m$ *such that* $|p(w)| > (|w|/\alpha)^m$, $w \in L_\epsilon$.

PROOF. We construct as follows a sequence of sectors $L_1, L_2, \cdots$, and of open sets $L_1', L_2', \cdots$, whose structure will be described below. Divide the plane into two half planes by a straight line through $w=0$. That half plane which contains in its closure $n_1 \geqq m/2$ roots of $p(w)$ we denote by L_1. In the second half plane L_1'' we construct two open sectors, each of opening $\epsilon_1 = \pi 2^{-3}$, which border on L_1. These two sectors form a set L_1', and the remaining part of L_1'' will be denoted by L_1^*. In L_1' lie $n_1' \leqq m/2$ roots. We bisect the sector L_1^*, and we denote by L_2 the sector so formed whose closure contains at least half the roots in L_1^*. The number n_2 of roots lying in L_2 is clearly $\leqq m/2$. In the remaining sector L_2'' we construct two open sectors, of opening $\epsilon_2 = \pi 2^{-5}$ each, which border on L_2 and L_1'. These two sectors form a set L_2', and the complementary section will be called L_2^*. The number n_2' of roots lying in L_2' satisfies $n_2' \leqq m/2^2$. In the sector L_2^* there lie at most $m/2^2$ roots. We continue this process, and we see that at the kth stage the sectors forming the set L_k' have an opening of $\epsilon_k = \pi 2^{-2k-1}$, and the sector L_k^* contains at most $m2^{-k}$ roots. Let q denote the first stage at which the sector L_q^* contains at most one root. Then the sector L_{q+1}^* will be determined. Note that the set L_{q+1}' will not contain any roots. From the construction it is clear that the opening of the sector L_{q+1}^* is

$$2\varepsilon = \frac{\pi}{2^{q+1}} - \frac{\varepsilon_1}{2^q} - \frac{\varepsilon_2}{2^{q-1}} - \cdots - 2\varepsilon_{q+1} = \frac{\pi}{2^{q+1}}\left(1 - \frac{1}{2^2} - \cdots - \frac{1}{2^{q+1}}\right) > \frac{\pi}{2^{q+2}}.$$

If s is defined by the condition $2^s \leqq m < 2^{s+1}$, then $q \leqq s$. Hence $2\epsilon > \pi/4m$. We now give a lower estimate for the modulus of the polynomial $p(w)$. Clearly

$$p(w) = \prod_{j=1}^{q+1} \prod_{w_\nu \in L_j} (w - w_\nu) \prod_{j=1}^{q} \prod_{w_\nu \in L_j'} (w - w_\nu),$$

where by convention we set $\prod_{w_\nu \in L_j'} (w - w_\nu) = 1$ in the case when the set in question contains no roots. We shall use the elementary estimates $|w - w_\nu| \geqq |w|\,|\sin \varphi_\nu|$, if $|\varphi_\nu| \leqq \pi/2$, where $\varphi_\nu = \arg(w/w_\nu)$, and the inequality $|w - w_\nu| \geqq |w|$ if $\pi/2 \leqq |\varphi_\nu| \leqq \pi$. If $w \in L_{q+1}^*$, it follows that

$$\prod_{w_\nu \in L_j} |w - w_\nu| \geqslant \prod_{w_\nu \in L_j} |w| \sin \varepsilon_j = (|w| \sin \varepsilon_j)^{n_j} \geqslant \left(\frac{|w|}{2^{2j}}\right)^{n_j},$$

where n_j is the number of roots in L_j. By construction $n_j \leqq m/2^{j-1}$. Similarly

$$\prod_{w_\nu \in L_j'} |w - w_\nu| \geqslant \prod_{w_\nu \in L_j'} |w| \sin \varepsilon_{j+1} = (|w| \sin \varepsilon_{j+1})^{n_j'} \geqslant \left(\frac{|w|}{2^{2j+2}}\right)^{n_j'},$$

where n_j' is the number of roots in L_j'. By construction $n_j' \leqq m/2^j$. If L_j' contains no roots, we set $\prod_{w_\nu \in L_j'} |w| \sin \epsilon_{j+1} = 1$. From these inequalities

$$|P(w)| \geqslant |w|^m 2^{-\sum_1^{q+1} 2jn_j - \sum_1^{q} (2j+2)n_j'}$$

But

$$\sum_1^{q+1} 2jn_j + \sum_1^{q} (2j+2)n_j' < m \sum_1^{\infty} \frac{2j}{2^{j-1}} + m \sum_1^{\infty} \frac{j+1}{2^{j-1}} = m \sum_1^{\infty} \frac{3j+1}{2^{j-1}} = 14m.$$

Thus $|p(w)| > 2^{-14m}|w|^m$. Q.E.D.

REMARK 9.1. Set $p(w) = p_1(w)p_2(w)$, where $p_k(w)$ is a polynomial of degree m_k; $k = 1, 2$, $m = m_1 + m_2$. If $w \in L_\epsilon$, then $|p_1(w)| > |w|^{m_1} 2^{-14m}$.

PROPOSITION 9.6. *Let* $p(w) = \prod_1^m (w - w_\nu)$. *In the annulus* $0 < a \leqq |w| \leqq b$ *it is possible to find an annulus* $K_0 = \{w : r_0 - h \leqq |w| \leqq r_0 + h\}$ *of width* $2h \geqq (b-a)/4m$, *where* $r_0 > a + (b-a)/8$, *such that*

$$(\max_{w \in K_0} |p(w)|)(\min_{w \in K_0} |p(w)|)^{-1} \leqslant \left(\alpha \frac{b}{b-a}\right)^m.$$

PROOF. The proof is similar to the proof of Proposition 9.5, except that in place of sectors we choose annuli. Hence we omit the details.

In accordance with Proposition 9.6, we construct within the annulus $1/8 \leqq |w| \leqq 1/4$ an annulus K_0 of width $2h_0 \geqq (32(n-1))^{-1}$ in which the estimate

$$\left(\max_{w \in K_0} \left| \frac{Q^{1/2}(w)}{w} \right|\right)\left(\min_{w \in K_0} \left| \frac{Q^{1/2}(w)}{w} \right|\right)^{-1} \leqslant \alpha^n$$

holds. We set

$$\mu = \min_{w \in K_0} \left| \frac{Q^{1/2}(w)}{w} \right|, \qquad \Delta_0 = \frac{\Delta}{\mu}.$$

The annulus K_0 does not intersect N, and from Koebe's 1/4 theorem and the hypothesis on K_0 we also see that K_0 does not intersect $\mathscr{T}$. Hence it follows from Lemma 9.3 that

$$g(w) - 1 | < \alpha n \frac{\Delta_0}{\delta}\left(1 + \frac{\Delta_0}{\delta}\right)\left(1 + \frac{l}{\mu h_0}\right),$$

for $w \in K_0$, where δ is the distance from w to the boundary of K_0.

We next estimate l. Let γ be any branch of the tree $\mathscr{T}$; let $d(\gamma)$ be its length with respect to the metric (8.3). If w_1 and w_2 are the end points of γ, then

$$d(\gamma) \leqslant \int_{\gamma_0} \left| \frac{Q^{1/2}(w)}{w} dw \right|,$$

where γ_0 is any path joining w_1 and w_2, and also lying in the domain $|w| \geqq 1/4$. Therefore

$$d(\gamma) < \max_{|w| = 1/4} \left| \frac{Q^{1/2}(w)}{w} \right| \int_{\gamma_0} \left| \frac{dw}{w^{3/2}} \right|.$$

It is clear that no matter what points w_1 and w_2 we choose, the path γ_0 can be chosen so that the latter integral is bounded above by some absolute constant α_0. Therefore

$$l = \sum_{\gamma \subset \mathscr{T}} d(\gamma) < \alpha_0 \max_{|w| = 1/4} \left| \frac{Q^{1/2}(w)}{w} \right| \sum_{\gamma \subset \mathscr{T}} 1 < \alpha_1 n \max_{|w| = 1/4} \left| \frac{Q^{1/2}(w)}{w} \right|.$$

Taking the inequality $h_0 \geqq (64(n-1))^{-1}$ into account, we find that

$$\frac{l}{\mu h_0} < \alpha_2 n^2 \left(\max_{w \in K_0} \left| \frac{Q^{1/2}(w)}{w} \right| \right) \left(\min_{w \in K_0} \left| \frac{Q^{1/2}(w)}{w} \right| \right)^{-1} < \alpha^n$$

and so in K_0

(9.14)
$$|g(w) - 1| < \beta \frac{\Delta_0}{\delta}\left(1 + \frac{\Delta_0}{\delta}\right), \qquad \beta = \alpha^n.$$

We shall denote below by β constants of the type α^n, where α is an absolute constant; similarly we shall write $\beta_k = \alpha_k^n$. We shall also assume that Δ_0 is sufficiently small, and in particular $\Delta_0 \leqq \beta_0$. We also introduce the notation

$$h = 1/64(n-1).$$

Set

(9.15)
$$g(w) = 1 + \chi(w).$$

In view of our assumptions on Δ_0, and of the inequality $h_0 \geqq h$, it follows from (9.14) that for $|w| = r_0$

(9.16)
$$|\chi(w)| < \beta_1 \Delta_0.$$

Equation (9.15) can be written as follows:

(9.17)
$$\frac{w^2 Q[\varphi(w)]}{\varphi^2(w) Q(w)} \varphi'^2(w) = 1 + \chi(w).$$

Set $\psi(w) = \varphi(w) - w$ and $\xi(w) = \psi(w)/w$. If w_ν, $1 \leqq \nu \leqq n-1$, are the finite zeros of the function $Q(w)$, then it is easy to verify that

$$\frac{w^2 Q[\varphi(w)]}{\varphi^2(w) Q(w)} = (1+\xi)^{-n-2} \prod_1^{n-1} \left(1 + \frac{w}{w - w_\nu} \xi\right).$$

We now consider the function

$$\mathscr{F}(w, \xi) = (1+\xi)^{\frac{n+2}{2}} \left[\prod_1^{n-1} \left(1 + \frac{w}{w - w_\nu} \xi\right) \right]^{-1/2},$$

where the square root is chosen so that $\mathscr{F}(w, 0) = 1$. If w is such that

(9.18)
$$\left| \frac{w}{w - w_\nu} \right| < \frac{1}{4h}, \qquad 1 \leqslant \nu \leqslant n-1,$$

then $\mathscr{F}(w, \xi)$ is a regular function of ξ in the disc $|\xi| < 4h$. We define the functions $\mathscr{F}_1(w, \xi)$ and $\mathscr{F}_2(w, \xi)$ by the relations

$$\mathscr{F}(w, \xi) = 1 + \xi \mathscr{F}_1(w, \xi) \text{ and } \mathscr{F}(w, \xi) = 1 + \xi a_1(w) + \xi^2 \mathscr{F}_2(w, \xi),$$

where

$$a_1(w) = \left. \frac{\partial \mathscr{F}(w, \xi)}{\partial \xi} \right|_{\xi = 0} = \frac{w}{2} \left[\frac{n+2}{w} - \sum_1^{n-1} \frac{1}{w - w_\nu} \right].$$

We shall now estimate $\mathscr{F}_1(w,\xi)$ and $\mathscr{F}_2(w,\xi)$ under the assumptions that $|\xi| \leqq h$ and that (9.18) is satisfied. Cauchy's theorem gives

$$\mathscr{F}_1(w,\xi) = \frac{1}{2\pi i} \int\limits_{|t|=2h} \mathscr{F}(w,t) \frac{dt}{t(t-\xi)}$$

and so

$$|\mathscr{F}_1(w,\xi)| < h^{-1}(1+h)^{\frac{n+2}{2}} 2^{\frac{n-1}{2}} = A. \tag{9.19}$$

Similarly we find that

$$|\mathscr{F}_2(w,\xi)| < h^{-1}A. \tag{9.20}$$

Set

$$p(w) = \prod_1^{n-1} (w - w_\nu), \qquad q(w) = p(w)\Big(\prod_{|w_\nu|>1/2} w_\nu\Big)^{-1},$$

where again by convention $\prod_{|w_\nu|>1/2} w_\nu = 1$ if there are no w_ν with $|w_\nu| > 1/2$. From (8.12) it follows that the function

$$\eta(w) = q(w)w^{-n}[g(w) - 1]$$

is regular in the disc $|w| \leqq r_0$, and so from (9.16) we see that for $|w| \leqq r_0$

$$|\eta(w)| < \beta_1(2 + 1/r_0)^{n-1}2^{-j}r_0^{-1}\Delta_0,$$

where j is the number of zeros of $Q(w)$ lying in the disc $|w| \leqq 1/2$. Hence

$$|\eta(w)| \leqq \beta_2\Delta_0,$$

if $|w| \leqq r_0$.

Lemma 9.4. *There exists a ray* $\arg w = \theta$ *on which* $|\varphi(w) - w| < \beta_3\Delta_0$, *provided* $\Delta_0 \leqq \beta_0$ *and* $|w| \leqq r_0$.

Proof. From Proposition 9.5 there exists a sector $L_\epsilon = \{w : |\arg w - \theta| \leqq \epsilon\}$ of opening $2\epsilon \geqq \pi/4(n-1)$, in which the inequality $|w^{n-1}/p(w)| < \alpha_*^n$ is satisfied. We now write the relation $g(w) = 1 + (w^n/q(w))\eta(w)$ in the form of the differential equation

$$\psi'(w) + 1 = \mathscr{F}(w,\xi)\left[1 + \frac{w^n}{q(w)}\eta(w)\right]^{1/2}, \tag{9.21}$$

which will be examined on the ray $\arg w = \theta$. For $|w| \leqq r_0$, $\arg w = \theta$, we see from Remark 9.1 that

$$\left|\frac{w^{n-1}}{q(w)}\right| \leqslant \alpha_*^n \prod_{|w_\nu|>\frac{1}{2}} \frac{|w|}{1-\left|\frac{w}{w_\nu}\right|} \leqslant \alpha_*^n \prod_{|w_\nu|>\frac{1}{2}} \frac{r_0}{1-2r_0} \leqslant \alpha_*^n. \tag{9.22}$$

Hence from the previous estimate for $|\eta(w)|$ it follows that

$$\left|\frac{w^n}{q'(w)}\eta(w)\right| < \frac{1}{4}\alpha_*^n\beta_2\Delta_0,$$

if $\arg w = \theta$, $|w| \leqq r_0$. Therefore, assuming that $\beta_2 \alpha_*^n \Delta_0 \leqq 1$, we find that $|(w^n/q(w))\eta(w)| < 1/4$. Set

$$\left[1 + \frac{w^n}{q(w)} \eta(w)\right]^{1/2} = 1 + \frac{w^n}{q(w)} \eta_1(w).$$

From the above inequalities it follows that

$$|\eta_1(w)| < |\eta(w)| \leqq \beta_2 \Delta_0.$$

Equation (9.21) can be written in the form

$$\psi'(w) = \frac{1}{2}\left(\frac{n+2}{w} - \sum_1^{n-1} \frac{1}{w - w_\nu}\right)\psi(w) + \frac{w^n}{q(w)} \eta_1(w) + \xi \Phi(w, \xi),$$

where

$$\Phi(w, \xi) = \frac{w^n}{q(w)} \eta_1(w) \mathscr{F}_1(w, \xi) + \xi \mathscr{F}_2(w, \xi).$$

On the ray $\arg w = \theta$ the inequality (9.18) is satisfied, and since $\xi(w) = O(w^n)$,

$$\psi(w) = \left[\frac{w^{n+2}}{q(w)}\right]^{1/2} \int_0^w \left[\frac{q(t)}{t^{n+2}}\right]^{1/2} \left[\frac{t^n}{q(t)} \eta_1(t) + \xi \Phi(t, \xi)\right] dt, \quad \arg t = \theta,$$

for small $|w|$, $\arg w = \theta$. Set

$$M(w_0) = \max\left|\xi(w)\left[\frac{q(w)}{w^{n+1}}\right]^{1/2}\right|,$$

where the maximum is taken subject to the conditions $\arg w = \theta$ and $|w| \leqq |w_0|$. Assuming that on the segment $\arg t = \theta$, $0 \leqq |t| \leqq |w|$ we have $|\xi| \leqq h$ (and for sufficiently small $|w|$ this is true), we see from the integral representation for $\psi(w)$ that

$$M(w) \leqslant \max_{0 \leqslant |u| \leqslant |w|} |u|^{-1/2} \int_0^{|u|} \left[\left|\frac{t^{n-2}}{q(t)}\right|^{-1/2} \beta_2 \Delta_0 + M(t)|t|^{-1/2} |\Phi(t, \xi)|\right] |dt|,$$

and, moreover, from (9.19) and (9.20)

$$|\Phi(t, \xi)| < A\left[\left|\frac{t^n}{q(t)}\right| \beta_2 \Delta_0 + h^{-1} M(t) \left|\frac{t^{n+1}}{q(t)}\right|^{1/2}\right].$$

Using the estimate (9.22) we have that

$$|\Phi(t, \xi)| < [\alpha_*^{n/2} \beta_2 \Delta_0 + h^{-1} M(t)]\, \alpha_*^{n/2} A |t|.$$

Hence

$$M(w) < 2\alpha_*^{n/2} \beta_2 \Delta_0 + \alpha_*^{n/2} A \int_0^{|w|} M(t)\, [\alpha_*^{n/2} \beta_2 \Delta_0 + h^{-1} M(t)]\, |dt|.$$

We next set $M(w) = \alpha_*^{n/2} \beta_2 \Delta_0 N(w)$ and $\kappa = (1/4h)\alpha_*^n \beta_2 \Delta_0 A$. Then from the latter inequality it follows by the monotonicity of $M(w)$ that for $|w| \leqq r_0 < 1/4$

$$N(w) \leqq 2 + \kappa h N(w) + \kappa N^2(w).$$

Restricting Δ_0 by the condition $\kappa \leqq 1/32$, we see that

$$\Delta_0 \leqslant \frac{h}{8} (\alpha_*^n \beta_2 A)^{-1}.$$

Then for small $|w|$ we have that $N(w) < 4$, i.e.

$$M(w) < 4\alpha_*^{n/2} \beta_2 \Delta_0.$$

From this latter inequality we deduce $|\xi(w)| \leqq h$ by noting that

$$|\xi(w)| = \left| \xi(w) \left[\frac{q(w)}{w^{n+1}} \right]^{1/2} \right| \left| \frac{w^{n+1}}{q(w)} \right|^{1/2} \leqslant M(w) \left| \frac{w^{n+1}}{q(w)} \right|^{1/2}$$

and, in view of (9.22),

$$|\xi(w)| \leqslant M(w)\, \alpha_*^{n/2}. \tag{9.23}$$

Since $M(w)$ is continuous on the segment $\arg w = \theta$, $0 \leqq |w| \leqq r_0$, the inequality $M(w) < 4\alpha_*^{n/2}\beta_2\Delta_0$ will be satisfied everywhere on this segment, and from (9.23) we get

$$|\varphi(w) - w| < \alpha_*^n \beta_2 \Delta_0,$$

provided that $\Delta_0 \leqq \beta_0 = \alpha_0^n$. This proves the lemma.

Lemma 9.5. *In the disc $|w| \leqq r_0$ the inequality*

$$|\varphi(w) - w| < \beta_4 \Delta_0$$

holds provided $\Delta_0 \leqq \beta_0$.

Proof. We write (9.17) as the differential equation

$$\psi'(w) + 1 = \mathscr{F}(w, \xi)\,[1 + \chi(w)]^{1/2}.$$

We set $[1 + \chi(w)]^{1/2} = 1 + \chi_1(w)$, and $\Phi(w, \xi) = \chi_1(w)\, \mathscr{F}_1(w, \xi) + \xi \mathscr{F}_2(w, \xi)$. Then

$$\psi'(w) = \frac{1}{2} \left(\frac{n+2}{w} - \sum_{1}^{n-1} \frac{1}{w - w_\nu} \right) \psi(w) + \chi_1(w) + \xi \Phi(w, \xi).$$

From (9.16) with $|w| = r_0$, and from $\Delta_0 \leqq \beta_0$, it follows that $|\chi_1(w)| < \beta_1 \Delta_0$. The above equation will be considered as a linear differential equation for $\psi(w)$. On the circle $|w| = r_0$ we solve the Cauchy problem for this equation, with initial values $w = w_*$, where $\arg w_* = \theta$. From Lemma 9.4 we have $|\psi(w_*)| < \beta_4 \Delta_0$, and, assuming that inequalities (9.18) hold for $|w| = r_0$, in a sufficiently small neighborhood of w_* we find that

$$\psi(w) = \psi(w_*) \left[\frac{w^{n+2} p(w_*)}{w_*^{n+2} p(w)} \right]^{1/2} + \left[\frac{w^{n+2}}{p(w)} \right]^{1/2} \int_{w_*}^{w} \left[\frac{p(t)}{t^{n+2}} \right]^{1/2} [\chi_1(t) + \xi \Phi(t, \xi)]\, dt,$$

$$|t| = r_0.$$

Set $M(w) = \max |\psi(t)|$, where the maximum is taken subject to the conditions $|t| = r_0$ and $\theta \leqq \arg t \leqq \arg w$. Noting that for $|t| = |w| = r_0$

$$\left| \frac{w^{n+2} p(t)}{t^{n+2} p(w)} \right|^{1/2} = \left| \frac{w^2 Q(t)}{t^2 Q(t)} \right|^{1/2},$$

we see, in view of the construction of the annulus K_0, that

$$M(w) < \alpha_*^n \left[|\psi(w_*)| + \int_{w_*}^{w} [|\chi_1(t)| + M(t)|t|^{-1}|\Phi(t,\xi)|]|dt| \right].$$

From this point on the arguments used are very similar to those of the previous lemma, and so we omit them. The lemma is now proved.

An immediate corollary of Lemma 9.5 is

Lemma 9.6. *There exist absolute constants α_0, α_1 and a disc $|w| \leqq r_0$ of radius $r_0 > 9/64$ such that*

$$|\varphi(w) - w| \leqq \alpha_0^n \Delta,$$

provided $\Delta \leqq \alpha_1^n$ and the function $Q(w)$ is normalized by the condition $\sum_1^n |A_k^2| = 1$.

Proof. The existence of the desired disc follows from Lemma 9.5, and the estimate

$$|\varphi(w) - w| \leqq \alpha^n \Delta_0$$

holds if $\Delta_0 < \alpha_2^{-n}$. But $\Delta_0 = \Delta\mu^{-1}$, where $\mu = \min_{w \in K_0} |Q^{1/2}(w)/w|$. On the other hand, in K_0 we have

$$\mu > \alpha_*^{-n} \max_{w \in K_0} \left| \frac{Q^{1/2}(w)}{w} \right|.$$

Noting that

$$\max_{w \in K_0} \left| \frac{Q^{1/2}(w)}{w} \right| \geqslant \left(\frac{1}{2\pi} \int_{|w| = \frac{1}{4}} \left| \frac{Q(w)}{w^2} \right|^2 \left| \frac{dw}{w} \right| \right)^{1/4} = \left(\sum_{k=1}^{n} |A_k|^2 4^{2k+4} \right)^{1/4} > 1,$$

we see that $\alpha^n \Delta_0 < (\alpha\alpha_*)^n \Delta$. If the above lower estimate for μ holds, it is easily seen that the inequality $|\varphi(w) - w| < \alpha_0^n \Delta$ holds if we require that $\Delta < \alpha_1^{-n}$. Here $\alpha_0 = \alpha\alpha^*$. Q.E.D.

This result is quite sufficient for our applications. However, for the sake of generality, we shall show that it is easy to extend it to a much wider domain.

Let $\mathscr{C}_r$ be the level line of the function $|f^{-1}(w)|$ defined by

$$\mathscr{C}_r = \{w : |f^{-1}(w)| = r < 1\}.$$

It is easy to see that the following result is true.

Lemma 9.7. *If $\mathscr{D}_r$ is a domain bounded by a level line $\mathscr{C}_r$, then for $w \in \mathscr{D}_r$ the inequality*

$$|\varphi(w) - w| < \beta(n,r)\Delta$$

holds provided $\Delta \leqq \Delta(n,r)$, where $\beta(n,r)$ and $\Delta(n,r)$ are constants depending only on n and r, and provided $Q(w)$ is normalized as in Lemma 9.6.

We shall not include the proof, since it is essentially the same as the proof of Lemma 9.4, the only difference being that in Lemma 9.4 the construction based on the existence of the annulus K_0 is replaced by a construction depending on Proposition 9.6. In this case we must consider the doubly connected domain bounded by the level lines $\mathscr{L}_r$ and $\mathscr{L}_{(1+r)/2}$ (which is analogous to annulus $1/8 \leqq |w| \leqq 1/4$), and construct in it a doubly connected domain analogous to K_0. This is easily done if we transfer ourselves to the z-plane via the mapping $z = f^{-1}(w)$, and then in place of the polynomial $p(w)$ considered in Proposition 9.6 we consider $p[f(z)]$, which obviously amounts to the consideration of some polynomial in the annulus $r \leqq |z| \leqq (1+r)/2$. An application of the distortion theorems for the class S then easily leads to an analogue of Proposition 9.6 for doubly connected domains bounded by $\mathscr{L}_r$ and $\mathscr{L}_{(1+r)/2}$. The calculations from this point on are essentially a repetition of the previous argument.

Theorem 9. *Let* $f(z) \in S_n$, $f(z) = \sum_0^\infty a_k z^{k+1}$, $a_0 = 1$, *be an extremal function. Let* $g(z) \in S$, $g(z) = \sum_0^{n-1} a_k z^{k+1} + \sum_n^\infty b_k z^{k+1}$. *Then in the disc* $|z| \leqq r_*$, $r_* > 1/9$, *the inequality*

$$|f(z) - g(z)|^2 < \alpha^n \left\{ \sum_1^n |A_k|^2 \right\}^{-\frac{1}{2}} \operatorname{Re} A_n (b_n - a_n)$$

holds, where α *is an absolute constant.*

Proof. Set $\varphi(w) = g(f^{-1}(w))$. Then $\varphi(w) = w + (b_n - a_n) w^{n+1} + \cdots$. On the basis of (8.1) we can write

$$\Delta^2 = \left\{ \sum_1^n |A_k|^2 \right\}^{-\frac{1}{2}} \operatorname{Re} A_n (b_n - a_n),$$

where $Q(w)$ has been normalized as in Lemma 9.6. From this lemma we conclude that there exists a disc $|w| \leqq r_0$ of radius $r_0 > 9/64$ in which the inequality $|\varphi(w) - w|^2 \leqq \alpha_0^{2n} \Delta^2$ holds, provided $\Delta \leqq \alpha_1^{-n}$. Hence in the domain $\mathscr{D}_0$ consisting of the image of this disc under the mapping $w = f(z)$ we have

$$|f(z) - g(z)|^2 \leqq \alpha_0^{2n} \Delta^2, \tag{9.24}$$

provided

$$\Delta \leqq \alpha_1^{-n}. \tag{9.25}$$

From a classical theorem of Bieberbach it follows that $\mathscr{D}_0 \supset \{z : |z| \leqq r_*\}$, where it is easily seen that $r_* > 1/9$. The quantity r_* can be taken to be independent of the function $f(z)$. Inequality (9.24), which holds in $\mathscr{D}_0$, clearly holds in the disc $|z| \leqq r_*$. If however (9.25) is not satisfied, it follows from

$$|f(z) - g(z)|^2 < 2|f(z)|^2 + 2|g(z)|^2 < \frac{4r_*^2}{(1-r_*)^4}$$

that (9.24) holds, but with some other absolute constant α_2 replacing α_0. Then (9.24) holds everywhere, with $\alpha = \max(\alpha_0, \alpha_2)$.

§10. We shall generalize Theorem 9 to a form which applies to functions $g(z) \in S$ whose first $n-1$ Taylor coefficients do not satisfy the hypothesis of Theorem 9. We first must prove some technical propositions.

An extremal function $f(z) \in S_n$ satisfies equation (5.13), which is defined in a unique way by the normalized associated vector λ. We recall that

$$\sum_1^n [\bar{\lambda}_k \mathfrak{A}_k(z) + \lambda_k \mathfrak{B}_k(z)] \equiv \left(\frac{Df(z)}{f(z)}\right)^2 Q[f(z)] + P(z), \tag{10.1}$$

where

$$Q(w) = \sum_1^n \bar{\lambda}_k q_k(w) = \sum_1^n A_k w^{-k}.$$

Set $f(z) = \sum_0^\infty a_k z^{k+1}$, and for any natural number s write

$$[f(z)]^s = \sum_{k=s-1}^\infty a_k^{(s)} z^{k+1}, \qquad a_{s-1}^{(s)} = 1. \tag{10.2}$$

Since

$$q_k(w) = \frac{1}{2\pi i} \int_{|z|=\rho} \frac{f^2(z)}{f(z)-w} \frac{dz}{z^{k+2}},$$

where ρ is chosen such that the only singularity of the integrand in the disc $|z| \leq \rho$ is at $z=0$, it is easily seen that

$$q_k(w) = -\sum_{l=1}^k a_k^{(l+1)} w^{-l}, \qquad k \geqslant 1.$$

Hence

$$A_l = -\sum_{k=l}^n \bar{\lambda}_k a_k^{(l+1)}, \qquad 1 \leqslant k \leqslant n. \tag{10.3}$$

and so

$$A_n = -\bar{\lambda}_n. \tag{10.4}$$

We now prove a simple proposition which will be helpful in the future.

Proposition 10.1. *If $|\lambda| = 1$ and A_l is defined by* (10.3), *then*

$$\alpha^{2n} \leqslant \sum_1^n |A_l|^2 \leqslant \alpha_1^{2n},$$

where α and α_1 are absolute constants.

Proof. It is easy to solve (10.3) for $\bar{\lambda}_1, \cdots, \bar{\lambda}_n$. To do so, set $z = f^{-1}(w) = \sum_0^\infty b_k w^{k+1}$, $b_0 = 1$, and for any natural number s write

$$z^s = \sum_{k=s-1}^\infty b_k^{(s)} w^{k+1}.$$

Then

$$\sum_{k=s}^n a_k^{(s+1)} b_n^{(k+1)} = \delta_{ns},$$

where δ_{ns} is the Kronecker delta. Therefore

$$\bar{\lambda}_l = -\sum_{s=l}^{n} A_s b_s^{(l+1)}, \qquad l = 1, \ldots, n. \tag{10.5}$$

From this we obtain

$$1 = \sum_1^n |\lambda_l|^2 = \sum_{l=1}^n \left|\sum_{s=l}^n A_s b_s^{(l+1)}\right|^2 \leqslant \sum_{l=1}^n \sum_{s=l}^n |A_s|^2 \sum_{k=l}^n |b_k^{(l+1)}|^2$$

$$\leqslant \sum_{s=1}^n |A_s|^2 \sum_{l=1}^n \sum_{k=l}^n |b_k^{(l+1)}|^2.$$

From the well-known Loewner inequality we have

$$|b_k| \leqslant 2^{k+1} \frac{(2k+1)!!}{(k+2)!} < \frac{4^{k+1}}{k+1},$$

and so

$$z \ll \sum_1^\infty \frac{(4w)^k}{k} = \log \frac{1}{1-4w}.$$

In the same way

$$z^l \ll \left(\log \frac{1}{1-4w}\right)^l,$$

and so

$$|b_j^{(l)}| < \frac{1}{2\pi i} \int_{|w|=\rho} \left(\log \frac{1}{1-4w}\right)^l \frac{dw}{w^{j+2}} < \rho^{-j-1} \frac{1}{2\pi} \int_{|w|=\rho} |\log(1-4w)|^l \left|\frac{dw}{w}\right|.$$

Set $\rho = 1/8$. Then $|\log(1-4w)|$ is bounded by some absolute constant α^*. Set $\alpha_0 = \max(8, \alpha^*)$. Then

$$|b_j^{(l)}| < \alpha_0^{j+l+1}, \quad j = l, l+1, \cdots.$$

Therefore

$$\sum_{l=1}^n \sum_{k=l}^n |b_k^{(l+1)}|^2 < \sum_{l=1}^n \sum_{k=l}^n \alpha_0^{2(l+s+2)},$$

and so $\alpha^{2n} \leqq \sum_1^n |A_k|^2$. The other inequality of the proposition can be proved similarly. From (10.3) we have

$$\sum_{l=1}^n |A_l|^2 \leqslant \sum_{s=1}^n |\lambda_s|^2 \sum_{l=1}^n \sum_{k=l}^n |a_k^{(l+1)}|^2 = \sum_{l=1}^n \sum_{k=l}^n |a_k^{(l+1)}|^2.$$

By the well-known inequality $|a_k| \leqq e(k+1)$ of Littlewood we have $f(z) \ll ez/(1-z)^2$, and so $(f(z))^l \ll e^l z^l (1-z)^{-2l}$. Thus $|a_k^{(l)}| < e^l 2^{k+l+1}$. Since the remaining calculations are extremely elementary, we omit them. This completes the proof.

THEOREM 10. *Let $f(z) \in S_n$, $f(z) = \sum_0^\infty a_k z^{k+1}$, $a_0 = 1$, be an extremal function. Let $g(z) \in S$, $g(z) = \sum_0^\infty b_k z^{k+1}$. Assume that $d_{n-1}(g) > 0$. Then for $|z| \leqq r_*$*

$$|f(z)-g(z)|^2<\alpha^n \operatorname{Re}\sum_1^n \bar{\lambda}_k(a_k-b_k)+\alpha_1^n d_{n-1}^{-2}(g)(1+d_{n-1}^{-2}(g))\sum_1^{n-1}|a_k-b_k|^2,$$

where α and α_1 are absolute constants.

Proof. We first assume that $g(z)$ satisfies the conditions of Theorem 2; that is,

$$\left(\sum_1^{n-1}|a_k-b_k|^2\right)^{1/2}<\alpha \min(n^{-\beta}d_{n-1}^2(g),\, n^{-\beta_0}d_{n-1}(g)), \tag{10.6}$$

where α, β and β_0 are the constants of Theorem 2, and $d_{n-1}(g)$, in accord with §3, is computed on the circumference $\mathscr{C}$ of the disc of radius $r^*=\frac{1}{2}r_*$.

As in Theorem 2 we construct the variation $g_*(z)$ of the function $g(z)$ such that

$$g_*(z)=\sum_0^{n-1}a_k z^{k+1}+\sum_n^{\infty}b_k^* z^{k+1}.$$

From Theorem 10 we conclude that in the disc $|z|\leq r_*$

$$|f(z)-g_*(z)|^2<\alpha^n\left(\sum_1^n|A_k|^2\right)^{-1/2}\operatorname{Re}A_n(b_n^*-a_n).$$

Taking into account (10.4) and Proposition 10.1, we write this latter inequality in the form

$$|f(z)-g_*(z)|^2<\frac{1}{4}\alpha_1^n \operatorname{Re}\bar{\lambda}_n(a_n-b_n^*). \tag{10.7}$$

According to Lemma 2.1,

$$g_*(z)=g(z)+\sum_1^{\infty}\varepsilon^k g_k(z),\qquad \text{where } \varepsilon=2\left(\sum_1^{n-1}|a_k-b_k|^2\right)^{1/2}$$

As was shown in the proof of Theorem 2,

$$b_l^*=b_l+\varepsilon\delta b_l+\sum_{k=2}^{\infty}\varepsilon^k b_{l,k}=b_l+\varepsilon\delta b_l+\varepsilon^2 B_l,\qquad l\geqslant 1. \tag{10.8}$$

Denote the variation of the coefficients of $f(z)$ by $\delta a_k=\delta a_k(f)$, where these are defined in terms of the measure $\nu(dz)=\zeta(z)dz/2\pi iz$ used to construct the function $g_*(z)$. Since $f(z)$ is extremal, $\operatorname{Re}\sum_1^n\bar{\lambda}_k\delta a_k=0$. Using this relation, we deduce the identity

$$\begin{aligned} j=\operatorname{Re}\bar{\lambda}_n(a_n-b_n^*)&=\operatorname{Re}\sum_1^n\bar{\lambda}_k(a_k-b_k)+\operatorname{Re}\sum_1^{n-1}\bar{\lambda}_k(\varepsilon\delta b_k+b_k-a_k)\\ &\quad+\varepsilon\operatorname{Re}\sum_1^n\bar{\lambda}_k(\delta a_k-\delta b_k)+\operatorname{Re}\bar{\lambda}_n(\varepsilon\delta b_n-b_n^*+b_n). \end{aligned}$$

From (10.8) and the equalities $b_l^*=a_l$, $1\leq l\leq n-1$, we get

$$j = \operatorname{Re}\sum_1^n \bar{\lambda}_k (a_k - b_k) - \varepsilon^2 \operatorname{Re}\sum_1^n \bar{\lambda}_k B_k + \varepsilon \operatorname{Re}\sum_1^n \bar{\lambda}_k (\delta a_k - \delta b_k).$$

From (3.9) and the fact that $\epsilon AB\Omega \leqq \frac{1}{2}$ we find that

$$|B_k| < \alpha_0 n^2 A \sum_{k=2}^{\infty} \varepsilon^{k-2} B^{k-1} \Omega^k k^{-2} < \frac{\alpha_0}{2} n^2 AB\Omega^2,$$

for $1 \leqq l \leqq n$. Thus

$$\left| \operatorname{Re}\sum_1^n \bar{\lambda}_k B_k \right| < \frac{\alpha_0}{2} n^{5/2} AB\Omega^2. \tag{10.9}$$

Recalling inequality (3.5) and the significance of the constants A and B in the proof of Theorem 2, we deduce from (10.9) the inequality

$$\left| \operatorname{Re}\sum_1^n \bar{\lambda}_k B_k \right| < \alpha_2 n^{\beta_1} d_{n-1}^{-2}, \tag{10.10}$$

where α_2 and β_1 are absolute constants. On the basis of formula (2.13) we find that

$$\delta a_k - \delta b_k = \int_{\mathscr{C}} [\mathfrak{A}_k(z; f) - \mathfrak{A}_k(z; g)]\, \zeta(z) \frac{dz}{2\pi iz}$$
$$+ \int_{\mathscr{C}} [\mathfrak{B}_k(\bar{z}; f) - \mathfrak{B}_k(\bar{z}; g)]\, \overline{\zeta(z) \frac{dz}{2\pi iz}}.$$

Using Proposition 3.1, we get

$$|\delta a_k - \delta b_k| < \alpha_3^k \Delta \frac{1}{2\pi} \int_{\mathscr{C}} |\zeta(z)| \left| \frac{dz}{z} \right| < \alpha_3^k \sqrt{2n}\, d_{n-1}^{-1} \Delta,$$

where $\Delta = \max_{|z|=r_*} |f(z) - g(z)|$. Thus

$$\left| \operatorname{Re}\sum_1^n \bar{\lambda}_k (\delta a_k - \delta b_k) \right| < \alpha_4^n d_{n-1}^{-1} \Delta. \tag{10.11}$$

From (10.10) and (10.11) we find that

$$j \leqslant \operatorname{Re}\sum_1^n \bar{\lambda}_k (a_k - b_k) + \varepsilon^2 \alpha_2 n^{\beta_1} d_{n-1}^{-2} + \varepsilon \alpha_4^n d_{n-1}^{-1} \Delta. \tag{10.12}$$

It is easily seen that

$$|f(z) - g_*(z)|^2 \geqslant \frac{1}{2} |f(z) - g(z)|^2 - |g(z) - g_*(z)|^2.$$

But $g_*(z) = g(z) + \epsilon \sum_1^\infty \epsilon^{k-1} g_k(z)$, and so in accordance with (3.8)

$$|g_*(z) - g(z)| < A\Omega\varepsilon\, [|g(z)| + 2|Dg(z)|] \sum_1^\infty (\varepsilon B\Omega)^{k-1} k^{-2}.$$

Bieberbach's inequality shows that $|g(z)|+2|Dg(z)|<5r_*(1-r_*)^{-3}$ for $|z|\leqq r_*$. Recalling the significance of the constant A in (3.5), we find that

$$|g_*(z)-g(z)|<\alpha_5 n^{\beta_2}d_{n-1}^{-1}\varepsilon,$$

where α_5 and β_2 are absolute constants. Thus

$$|f(z)-g_*(z)|^2\geqslant\frac{1}{2}|f(z)-g(z)|^2-\alpha_5^2 n^{2\beta_2}d_{n-1}^{-2}\varepsilon^2. \tag{10.13}$$

From (10.7), (10.12) and (10.13) we get that

$$|f(z)-g(z)|^2\leqslant\frac{1}{2}\alpha_1^n\left\{\mathrm{Re}\sum_1^n\bar{\lambda}_k(a_k-b_k)+\varepsilon\alpha_4^n d_{n-1}^{-1}\Delta\right\}+\alpha_6^n d_{n-1}^{-2}\varepsilon^2.$$

We now apply this inequality at the point on $|z|=r_*$ which maximizes the function $|f(z)-g(z)|$. Since

$$\varepsilon\alpha_4^n d_{n-1}^{-1}\Delta\leqslant\alpha_1^{-n}\Delta^2+\frac{1}{4}(\alpha_1\alpha_4^2)^n d_{n-1}^{-2}\varepsilon^2,$$

we have

$$\max|f(z)-g(z)|^2<\alpha_1^n\,\mathrm{Re}\sum_1^n\bar{\lambda}_k(a_k-b_k)+\frac{1}{4}\alpha_7^n d_{n-1}^{-2}\varepsilon^2,$$

where α_7 is an absolute constant. Taking into account the value of ϵ, we have for $|z|\leqq r_*$ that

$$|f(z)-g(z)|^2<\alpha_1^n\,\mathrm{Re}\sum_1^n\bar{\lambda}_k(a_k-b_k)+\alpha_7^n d_{n-1}^{-2}\sum_1^{n-1}|a_k-b_k|^2. \tag{10.14}$$

This inequality holds if (10.6) is true. If (10.6) is not true, then with a suitable choice of a constant α_8 it is clear that

$$|f(z)-g(z)|^2\leqslant\alpha_1^n\,\mathrm{Re}\sum_1^n\bar{\lambda}_k(a_k-b_k)+\alpha_8^n d_{n-1}^{-2}(1+d_{n-1}^{-2})\sum_1^{n-1}|a_k-b_k|^2. \tag{10.15}$$

Combining (10.14) and (10.15), we see that the theorem is proved.

§11. From Theorem 10 we can deduce some very interesting corollaries. Assume that $g(z)$ is also an extremal function, and that $g(z)\in S_n$ and $d_{n-1}(f)>0$. Then from Theorem 10 we get two inequalities, first by applying the theorem to the pair $f(z)$, $g(z)$, and then to the pair $g(z)$, $f(z)$. Adding these two inequalities, we find that

$$\begin{aligned}|f(z)-g(z)|^2<\frac{1}{2}\alpha^n\,\mathrm{Re}\sum_1^n(\bar{\lambda}_k(f)-\bar{\lambda}_k(g))(a_k-b_k)\\+\frac{1}{2}\alpha_1^n\mathcal{M}(f,g)\sum_1^{n-1}|a_k-b_k|^2,\end{aligned} \tag{11.1}$$

where

$$\mathcal{M}(f,g)=d_{n-1}^{-2}(f)+d_{n-1}^{-2}(g)+d_{n-1}^{-4}(f)+d_{n-1}^{-4}(g). \tag{11.2}$$

If $\mathcal{M}(f,g) < \infty$, then in accord with Remark 3.3 we have $d_{n-1/2}(f) > 0$ and $d_{n-1/2}(g) > 0$. We set

$$d_{n-1/2}(f, g) = \max\{d_{n-1/2}(f), d_{n-1/2}(g)\}. \tag{11.3}$$

Lemma 11.1. *If $f,g \in S_n$ and $\mathcal{M}(f,g) < \infty$, then*

$$|\lambda(f) - \lambda(g)| < \alpha^n d_{n-1/2}^{-2}(f, g)\,[d_{n-1/2}^{-4}(f, g) + \mathcal{M}(f, g)]^{1/2}\left(\sum_1^n |a_k - b_k|^2\right)^{1/2}, \tag{11.4}$$

where α is an absolute constant.

Proof. From Schwarz' inequality we see that

$$\operatorname{Re}\sum_1^n (\bar\lambda_k(f) - \bar\lambda_k(g))(a_k - b_k) \leqslant |\lambda(f) - \lambda(g)|\left(\sum_1^n |a_k - b_k|^2\right)^{1/2}.$$

Set

$$\Delta^2 = \sum_1^n |a_k - b_k|^2.$$

Inequality (11.1) gives

$$|f(z) - g(z)|^2 < \frac{1}{2}\alpha^n |\lambda(f) - \lambda(g)|\,\Delta + \frac{1}{2}\alpha_1^n \mathcal{M}(f, g)\,\Delta^2.$$

From Lemma 3.3 with $2r^* = r_*$ we have

$$|\lambda(f) - \lambda(g)| < \alpha_0^n \max_{|z|=r_*} |f(z) - g(z)|\, d_{n-1/2}^{-2}(f, g).$$

The latter two inequalities give

$$|\lambda(f) - \lambda(g)|^2 < \alpha_2^n d_{n-1/2}^{-4}(f, g)\,[|\lambda(f) - \lambda(g)|\,\Delta + \mathcal{M}(f, g)\,\Delta^2].$$

But

$$\alpha_2^n d_{n-1/2}^{-4}(f, g)\,|\lambda(f) - \lambda(g)|\,\Delta < \frac{1}{2}|\lambda(f) - \lambda(g)|^2 + \frac{1}{2}\alpha_2^{2n} d_{n-1/2}^{-8}(f, g)\,\Delta^2.$$

Therefore

$$|\lambda(f) - \lambda(g)|^2 < [\alpha_2^{2n} d_{n-1/2}^{-4}(f, g) + 2\alpha_2^n \mathcal{M}(f, g)]\, d_{n-1/2}^{-4}(f, g)\,\Delta^2$$

and (11.4) follows from this.

Inequality (11.4) is very important, since it enables us to progress toward a solution to the problem of examining the smoothness of the boundary of the coefficient region.

§12. We established earlier that the sets S_n and $\partial\mathscr{V}_n$ are homeomorphic. Thus on the set $\{z: |z| < 1\} \times \partial\mathscr{V}_n$ there is defined a complex valued function $f(z,x)$ which is regular and univalent in $|z| < 1$ for fixed x, and which is continuous on $\partial\mathscr{V}_n$ for fixed z. If $x \in \partial\mathscr{V}_n$ is fixed, then $f(z,x)$ is an extremal function which depends on the point x. For the quantities connected with $f(z,x)$ we shall employ the notation $d_k(x)$, $d_{n-1/2}(x)$, and so forth.

PROPOSITION 12.1. *The functions* $d_{n-1/2}(x)$ *and* $d_k(x)$, $k < n$, *are continuous on* $\partial\mathscr{V}_n$.

PROOF. The proposition follows easily from inequalities (3.22) and (3.23), and the continuity of $f(z,x)$. If $d_{n-1/2}(x) > 0$, then the normalization condition defines a unique vector $\lambda(x)$, associated with the function $f(z,x)$. If $d_{n-1/2}(x) = 0$, then there are an infinite number of associated vectors.

The metric properties of $\partial\mathscr{V}_n$ in a neighborhood of x depend in an essential way on whether or not $d_{n-1/2}(x) > 0$. We introduce the set

$$\mathscr{R} = \{x : x \in \mathscr{V}_n, d_{n-1/2}(x) > 0\}$$

which will be very important in what follows. It is clear that $\mathscr{R}$ is an open set. We set

$$A_1 = \partial\mathscr{V}_n \setminus \mathscr{R}.$$

From Lemma 3.3 it follows that $\lambda(x)$ is continuous on $\mathscr{R}$. We now explain the geometric interpretation of the vector $\lambda(x)$, but first we must introduce some auxiliary technical concepts to be used in the course of the various proofs. In the (u,v) plane take a concave curve lying in the first quadrant of the disc $\{u, v : u^2 + v^2 \leq 1\}$, and defined by the equation $v = \varphi_\theta(u)$, $0 \leq u \leq 1$. For $0 \leq u \leq \frac{1}{2}\cot\theta$, $\pi/4 \leq \theta < \pi/2$, we set $\varphi_\theta(u) = \tan\theta u$, and for $3/4 \leq u \leq 1$ we set $\varphi_\theta(u) = (1-u^2)^{1/2}$. For $\frac{1}{2}\cot\theta < u < 3/4$ we define $\varphi_\theta(u)$ arbitrarily, subject to the conditions that it be three times continuously differentiable for $0 \leq u < 1$, that it be concave, and that it lie in the first quadrant of the unit circle. The curve so constructed will be denoted by $\mathscr{L}_\theta$. We denote by ν the unit normal to $\mathscr{L}_\theta$ pointing into the domain bounded by $\mathscr{L}_\theta$ and the segment $0 \leq u \leq 1$ of the u axis. If ν_0 is the unit vector of the u axis, then in view of the construction we have

$$|\nu - \nu_0| \geq 2\sin(\pi/4 - \theta/2). \tag{12.1}$$

In R^{2n} we form the hypersurface of revolution defined by the equation $(\xi_1^2 + \cdots + \xi_{2n-1}^2)^{1/2} - \epsilon\varphi_\theta(\xi_{2n}/\epsilon) = 0$. This hypersurface bounds a convex solid which we call the *test solid*. A cone-like point of the boundary of the test solid will be called a *cusp*, and the axis of revolution will be called the *axis* of the test solid. The quantity 2θ will be called the *opening* of the test solid. It is clear that the diameter of the test solid is ϵ. We will denote the test solid by $T_{\theta,\epsilon} = T$. For any point $x \in \partial T$ which is not a cusp, we can find a sphere contained in T and tangent to ∂T at x.

DEFINITION 12.1. We say that *it is possible to touch the set* $\partial\mathscr{V}_n$ with the test solid T at the point x if the following conditions are satisfied:

1. The point x is a cusp of T, and the axis of rotation is parallel to the vector $\lambda(x)$.
2. The set $T\setminus\{x\}$ lies in the complement of $\mathscr{V}_n$.

DEFINITION 12.2. The point $x \in \partial\mathscr{V}_n$ is called *accessible from without* if for any $\theta < \pi/2$ it is possible to find $\epsilon > 0$ such that the solid $T_{\theta,\epsilon}$ touches $\partial\mathscr{V}_n$ at the point x.

DEFINITION 12.3. We say that the sphere J is *tangent* to the set $\partial\mathscr{V}_n$ at the point x if the following conditions are satisfied:

1. The center of the sphere lies on a straight line through x parallel to the vector $\lambda(x)$.
2. $\partial J \cap \partial\mathscr{V}_n = \{x\}$.

PROPOSITION 12.2. *For any point $x^0 \in \mathscr{R}$ there exists a sphere $J \subset \mathscr{V}_n$ tangent to $\partial\mathscr{V}_n$ at the point x^0. In a sufficiently small neighborhood of the point x^0, the radius of such a sphere can be taken to be the same for all points of the neighborhood.*

PROOF. Let $t = (t_1 + it_2, \cdots, t_{2n-1} + it_{2n})$ be any vector orthogonal to the vector $\lambda(x^0)$. To ease the notation we write $f(z) = f(z, x^0)$. According to Lemma 3.4, we can construct a measure $\nu(dz) = \zeta(z)dz/2\pi iz$ with support $\mathscr{L} = \{z: |z| = r_0\}$ such that $\delta a_l(f; \nu) = t_{2l-1} + it_{2l}$, $l = 1, \cdots, n$. Corresponding to this measure we construct an admissible function of the form (2.2) which, in turn, defines a variation $f_*(z)$ of $f(z)$. From inequality (3.24) it follows that if $|t| d_{n-1/2}^{-1}(x^0) < \alpha$, where α is an absolute constant, then Lemma 2.1 is applicable. We shall take all the arbitrary constants $c_1, c_2, \cdots$ equal to zero. The varied function then has the form

$$f_*(z) = \sum_{k=0}^{\infty} f_k(z, t), \qquad f_0(z, t) = f(z).$$

As was shown in the proof of Theorem 2, the functions $f_k(z, t)$ are homogeneous polynomials in $t_1, \cdots, t_{2n}$ of degree k.

Set $f_*(z) = \sum_0^\infty a_l(t) z^{l+1}$ and $a_0 = 1$. Then

$$a_l(t) = a_l + t_{2l-1} + it_{2l} + \sum_{k=2}^{\infty} P_{lk}(t), \qquad l = 1, \ldots, n,$$

where $P_{lk}(t)$ is a homogeneous polynomial and $(a_1, \cdots, a_n) = x^0$. From inequality (2.7) it follows that

$$|P_{lk}(t)| < (C|t|)^k, \quad k \geqq 2, \quad l = 1, \cdots, n,$$

where C is a constant depending only on $d_{n-1/2}(x^0)$ and n. From this estimate we see that for $|t| \leqq \rho$, where ρ depends only on C, the function $P_l(t) = \sum_2^\infty P_{lk}(t)$ is an analytic function of the variables $t_1, \cdots, t_{2n}$. Hence the equations

$$x_l = a_l + t_{2l-1} + it_{2l} + P_l(t), \qquad l = 1, \cdots, n, \tag{12.2}$$

where $|t| \leqq \rho$ and $(\lambda, t) = 0$, are the parametric equations of some analytic hypersurface contained in $\mathscr{V}_n$, since the equations (12.2) define the homeomorphic image of the disc $|t| \leqq \rho$, $(\lambda, t) = 0$, provided $\rho \leqq \rho_0$. The constant ρ_0 depends only on C. This hyperplane will be denoted by σ. Note that the hyperplane tangent to σ at the point $t = 0$ is orthogonal to the vector $\lambda(x^0)$. We join each point $x = (x_1, \cdots, x_n) \in \sigma$ to the point $(0, \cdots, 0)$ by a line $(ux_1, \cdots, u^n x_n)$, $0 \leqq u \leqq 1$. As a result we get a domain contained in $\mathscr{V}_n$. Note that the tangent vector to this line at the point x is not orthogonal to $\lambda(x^0)$. In fact, this vector is $\tau = (x_1, 2x_2, \cdots, nx_n)$, and

$$(\lambda(x^0), \tau) = \operatorname{Re} \sum_{l=1}^{n} l\overline{\lambda_l(x^0)}\, a_l + O(|t|).$$

The function $f(z)$ satisfies equation (5.13), and from (3.19) we get

$$\sum_{l=1}^{n} l\overline{\lambda_l(x^0)}\, a_l = \frac{1}{2\pi} \int_0^{2\pi} P(e^{i\varphi})\, d\varphi > 0.$$

Since the latter integral is bounded away from zero by a constant depending only on n, we see that for sufficiently small $|t|$ we have $(\lambda(x^0), \tau) > \vartheta > 0$. We may without loss of generality assume that our choice of the quantity ρ insures the truth of this inequality. In this way we see that the domain we have constructed contains some segment normal to σ at the point $t = 0$. Thus there exists some neighborhood of $t = 0$ such that at any point of this neighborhood the normal to σ intersects the above domain in a segment. The existence of a sphere J tangent to $\partial\mathscr{V}_n$ at the point x^0 now follows immediately. From the construction it is also clear that for two sufficiently close points x^0 and x^1 the radii of the spheres can be chosen to be near to each other in size, and this in turn implies the existence of a neighborhood of the point x^0 such that for all points in this neighborhood the radii of the corresponding spheres can be chosen to be equal. This completes the proof of the proposition.

Proposition 12.3. *Any point of the set $\mathscr{R}$ is accessible from without.*

Proof. Let x^0 be any point in $\mathscr{R}$. Then $d_{n-1/2}(x^0) > 0$. By $\mathscr{W}_0$ we denote the set $\partial\mathscr{V}_n \cap \{x : |x - x^0| \leqq r_0\}$, and by $\mathscr{W}$ the set $\partial\mathscr{V}_n \cap \{x : |x - x^0| \leqq r\}$, $r < r_0$. If r_0 is sufficiently small, then $\inf_{y \in \mathscr{W}_0} d_{n-1/2}(y) > 0$. Therefore the vector-valued function $\lambda(x)$ is continuous on $\mathscr{W}_0$. We choose the quantity $\theta < \pi/2$ sufficiently close to $\pi/2$ so that $\delta_0 = \sin(\pi/4 - \theta/2)$ will be sufficiently small. We then define $\epsilon_0 > 0$ depending on δ_0 so that $|\lambda(x) - \lambda(y)| \leqq \delta_0$ whenever $|x - y| \leqq \epsilon_0$, $x, y \in \mathscr{W}_0$. We also assume that our choice of δ_0 implies $\epsilon_0 \leqq r_0 - r$. This point will be very important in our arguments. We assume that for a point $x^1 \in \mathscr{W}$ we can find a solid $T_{\theta\epsilon}$, of diameter $\epsilon \leqq \epsilon_0$, which is positioned so that condition 1 of Definition 12.1 is satisfied. We intend to prove that the set of interior points of the solid $T_{\theta\epsilon}$ does not intersect $\mathscr{V}_n$. Then $\partial T_{\theta\epsilon} \cap \partial\mathscr{V}_n = \{x^1\}$; that is, at the point x^1 it is possible to touch $\partial\mathscr{V}_n$ by a test solid of opening 2θ and diameter ϵ. To prove this, suppose it is false, so there exists $y \in \partial T_{\theta\epsilon} \cap \partial \mathscr{V}_n$, $y \neq x^1$. We construct a sphere J lying in $T_{\theta\epsilon}$ and tangent to $\partial T_{\theta\epsilon}$ at y, so that $\partial J \cap \partial \mathscr{V}_n = \{y\}$. Let y_0 be the center of the sphere. The functional $\mathscr{F}[f] = |x - y_0|^{-1}$ attains its upper bound at the function $f(z, y)$. Consequently $\lambda(y) = (y - y_0)/|y - y_0|$. From (12.1) we have $|\lambda(x^1) - \lambda(y)| \geqq 2\sin(\pi/4 - \theta/2) = 2\delta_0$. On the other hand, $|x^1 - y| \leqq \epsilon \leqq \epsilon_0$, and $|y - x^0| \leqq |y - x^1| + |x^1 - x^0| \leqq \epsilon_0 + r \leqq r_0$. Thus $y \in \mathscr{W}_0$, and so $|\lambda(x^1) - \lambda(y)| \leqq \delta_0$, which is a contradiction. Therefore we have shown that at x^1 it is possible to touch the set $\partial \mathscr{V}_n$ by a test solid of opening 2θ and diameter ϵ. Denote by A_θ the set of all points $x \in \mathscr{W}$ at which it is possible to touch $\partial \mathscr{V}_n$ by a test solid of opening 2θ and diameter ϵ_0. We shall show that A_θ is not empty. In §7 we noted that on $\partial \mathscr{V}_n$ there exists an everywhere dense set $\mathscr{M}$ consisting of those points which can be touched by a sphere lying in the complement of $\mathscr{V}_n$. Let $x^* \in \mathscr{M}$, and let y be the center of the corresponding sphere. As we have already seen, $\lambda(x^*) = (y - x^*)/|y - x^*|$; that is, the center of the sphere lies on the straight line through the point x^* having the direction of the vector $\lambda(x^*)$. From this it follows that at x^* it is possible to touch the set $\partial \mathscr{V}_n$ by a test solid of opening 2θ and of some diameter ϵ_1. We shall show that $x^* \in A_\theta$. We consider the set $\mathscr{E}$ of positive real numbers ϵ having the property that there exists a test solid of opening 2θ and diameter ϵ touching $\partial \mathscr{V}_n$ at x^*. Since $\epsilon_1 \in \mathscr{E}$, $\mathscr{E}$ is not empty. Also, if $\epsilon' \in \mathscr{E}$ and $\epsilon < \epsilon'$, then $\epsilon \in \mathscr{E}$, since $T_{\theta\epsilon} \subset T_{\theta\epsilon'}$, and these test solids both have cusps at x^* and the same axis of revolution. Set $\epsilon^* = \sup_{\epsilon \in \mathscr{E}} \epsilon$, and suppose that $\epsilon^* \leqq \epsilon_0$. Then there exists a sequence $\{\epsilon_k\}$, $k = 1, 2, \cdots$, such that $\epsilon_1 < \epsilon_2 < \cdots$ and $\lim_{k\to\infty} \epsilon_k = \epsilon^*$. Let $T_1, T_2, \cdots$ be the corresponding sequence of test solids. Since $T_1 \subset T_2 \subset \cdots$, this sequence of convex solids has a limit, which we denote by T^*. Since the set $T^* \backslash \partial T^*$ does not intersect $\mathscr{V}_n$, we may, in view of the remarks made above, touch the set $\partial \mathscr{V}_n$ at x^* by T^*, since $\epsilon^* \leqq \epsilon_0$. Thus $\epsilon^* \in \mathscr{E}$. Note that $\{x : x \in T^*, |x - x^*| \geqq h\} \cap \partial \mathscr{V}_n = \emptyset$ if h is sufficiently small. Thus there exists $\epsilon' > \epsilon^*$ such that the test solid $T_{\theta\epsilon'}$ touches $\partial \mathscr{V}_n$ at the point x^*, i.e. $\epsilon' \in \mathscr{E}$, which is a contradiction. Therefore $\epsilon^* > \epsilon_0$, and hence $\epsilon_0 \in \mathscr{E}$, or, in other words, $x^* \in A_\theta$. Hence $\mathscr{M} \cap \mathscr{W} \subset A_\theta$. We next show that A_θ is closed. Let the sequence $\{x^k\}$, $x^k \in A_\theta$, $k = 1, 2, \cdots$, converge to some point $x \in \mathscr{W}$. Each point x^k corresponds to some test body T_k of opening 2θ and diameter ϵ_0. From this sequence of convex bodies we select a convergent subsequence, and we denote its limit by T. The solid T is oriented so that it satisfies condition 1 of Definition 12.2. Clearly the set of interior points of T does not intersect $\partial \mathscr{V}_n$, and so $x \in A_\theta$. Thus A_θ is closed, and consequently $A_\theta = \mathscr{W}$. Since θ is arbitrary, the points of the set $\mathscr{W}$ are accessible from without. Q.E.D.

Proposition 12.3 can be used to examine the structure of the boundary of the set $\mathscr{V}_n$ in a neighborhood of the points of $\mathscr{R}$. Let $x^0 \in \mathscr{R}$. Denote by Π_{x^0} the hyperplane passing through the point x^0 and oriented so that $\lambda(x^0)$ is its direction vector. We shall consider the orthogonal projection of the set $\mathscr{W}$

onto Π_x, and we shall denote it by $\mathfrak{U}$. We shall show that if r is sufficiently small, then the projection operation determines a one-to-one relationship between $\mathscr{W}$ and $\mathfrak{U}$. We take r so small that $|\lambda(x)-\lambda(x^0)|<1$, $x\in\mathscr{W}$, and we take a fixed quantity θ near to $\pi/2$. Then we have $\theta>2\arcsin\frac{1}{2}=\pi/3$. It is clear from Propositions 12.2 and 12.3 that at each point $x\in\mathscr{W}$ there exists a sphere I_x and a test solid T_x of diameter ϵ and opening 2θ which touches $\partial\mathscr{V}_n$ at x. We choose the radius of the sphere I_x to be independent of x. Through x we construct a straight line $\mathscr{I}_x$ normal to the hyperplane Π_{x^0}, and by x^* we denote the point of their intersection. Note that since $\theta>\pi/3$, the line $\mathscr{I}_x$ intersects the sphere I_x and the solid T_x in segments, the sum of whose lengths is $\geqq c_0>0$, where c_0 is independent of x. Therefore the line $\mathscr{I}_x$ can intersect $\mathscr{W}$ in at most a finite number of points. Assume that $\mathscr{I}_x$ has more than one point of intersection with $\mathscr{W}$, and let x^1 be the nearest of these points to x. Then the segment (x^1,x) does not intersect $\partial\mathscr{V}_n$, since for any point y on this segment we have $|x^0-y|\leqq r$; also we have $|x-x^0|\leqq r$ and $|x^1-x^0|\leqq r$, and so if $y\in\partial\mathscr{V}_n$, then $y\in\mathscr{W}$, which is a contradiction. Therefore the segment (x^1,x) lies either in the complement of $\mathscr{V}_n$ or in $\mathscr{V}_n\backslash\partial\mathscr{V}_n$. Orienting the line $\mathscr{I}_x$ to have the direction of the vector $\lambda(x^0)$, we have (in view of our choice of r) that $(\lambda(x^0),\lambda(x))\geqq\frac{1}{2}$ and $(\lambda(x^0),\lambda(x^1))\geqq\frac{1}{2}$. From these inequalities it follows that the segment (x^1,x) must intersect one of the test bodies, and also one of the spheres, corresponding to the points x and x^1. From this contradiction it follows that the line $\mathscr{I}_x$ has only one common point with $\mathscr{W}$. In this way we have established a one-to-one correspondence between $\mathscr{W}$ and $\mathfrak{U}$, and moreover the mapping $\mathscr{W}\to\mathfrak{U}$ is continuous. Thus $\mathfrak{U}$ is homeomorphic to $\mathscr{W}$. We now write down the analytic details of the projection. At x we introduce a local coordinate system $(y_1,\cdots,y_{2n})$, where we have taken x^0 as our origin and the y_{2n}-axis to be normal to Π_{x^0}, and where we have taken the other axes to be arbitrary, subject to the restriction that they lie in Π_{x^0}. For ease of notation we denote the vector $(y_1,\cdots,y_{2n-1})$ by y, we denote y_{2n} by τ, and the vector $\mathfrak{y}=(y_1,\cdots,y_{2n})$ we shall write in the form (y,τ). The connection between the old and new coordinate systems is given by

$$x=x^0+U\mathfrak{y}, \tag{12.3}$$

where U is an orthogonal matrix. The set $\mathscr{W}$ can be considered as the graph of some continuous function defined on $\mathfrak{U}$. In this way the equation of the boundary of $\mathscr{V}_n$ in a neighborhood of the point x^0 will be of the form

$$\tau=\psi(y) \tag{12.4}$$

in the local coordinate system.

We now consider the structure of the set $\mathfrak{U}$. Let $x^1\in\mathscr{W}$, $|x^1-x^0|<r$, and let x^1_* be the projection of x^1. On the boundary of the sphere I_x we consider the set $\mathscr{M}_\epsilon=\{x:x\in\partial I_{x^1},|x-x^1|<\epsilon\}$, where $\epsilon>0$ is small. $\mathscr{M}_\epsilon$ is homeomorphic to a $(2n-1)$-dimensional open sphere. We now project $\mathcal{M}_\epsilon$ orthogonally onto the hyperplane Π_{x^0}. For $\epsilon > 0$ sufficiently small, this projection operation is a homeomorphism. Set $\mathcal{M}_\epsilon^*$ equal to the projection of $\mathcal{M}_\epsilon$. We shall show that $\mathcal{M}_\epsilon^* \subset \mathfrak{U}$. Take θ sufficiently near to $\pi/2$, and let the test solid T of opening 2θ and diameter δ be tangent to $\partial\mathcal{V}_n$ at x^1. We assume that we have taken ϵ so small that any straight line $\mathcal{I}_x$, $x \in \mathcal{M}_\epsilon$, which is orthogonal to Π_{x^0} will intersect T. If x^2 is the point of intersection of $\mathcal{I}_x$ and ∂T which is nearest to x, then the segment (x, x^2) contains a point $x^3 \in \partial\mathcal{V}_n$. Since $|x^3 - x^0| \leq |x^3 - x| + |x - x^1| + |x^1 - x^0| < \epsilon + |x^3 - x| + |x^1 - x^0|$, we see that $|x^3 - x^0| < r$, provided only that ϵ is sufficiently small, since $|x^3 - x|$ becomes small together with ϵ. Thus $x^3 \in \mathcal{W}$, and it follows that $\mathcal{M}_\epsilon^* \subset \mathfrak{U}$, as required. Let $\mathfrak{U}^0$ be the projection of the set $\mathcal{W}^0 = \{x : x \in \partial\mathcal{V}_n, |x - x^0| < r_0\}$. From the above it follows that $\mathfrak{U}^0$, considered as a subset of Π_{x^0}, is open. Consider the component containing the point $x^0_* = x^0$. This component will contain some $(2n-1)$-dimensional sphere centered at x^0. Let R_0 be the radius of the largest such sphere. In what follows we shall consider the function (12.4) as having been restricted to the set $\{\mathfrak{Y} : \mathfrak{Y} = (y, 0), |y| < R, R \leq R_0\}$. Trusting that no misunderstanding will arise, we denote this set by $\mathfrak{U}$ or $\mathfrak{U}_R$, and its image by $\mathcal{W}$ or $\mathcal{W}_R$.

Lemma 12.1. *At any point x of the set $\mathcal{R}$ there exists a hyperplane tangent to $\partial\mathcal{V}_n$ which is also orthogonal to the vector $\lambda(x)$.*

Proof. Let x^0 be an arbitrary point of $\mathcal{R}$. From Propositions 12.2 and 12.3 it follows that the set of points touching $\partial\mathcal{V}_n$ at x^0 lies in a hyperplane Π_{x^0}. We must show only that this "tangent plane" contains all rays lying in Π_{x^0} and emanating from x^0. But this follows since the projection of $\mathcal{W}$ contains a $(2n-1)$-dimensional sphere lying in Π_{x^0} and centered at x^0. Thus the lemma is proved.

Lemma 12.2. *The function $\psi(y)$ is continuously differentiable on* $\mathfrak{U}$.

Proof. Let $x \in \partial\mathcal{V}_n$ correspond to $(y, 0)$. In the local coordinate system the normal vector to $\partial\mathcal{V}_n$ at $(y, \psi(y))$ will be $\mu(y) = U'\lambda(x)$. Since $\lambda(x)$ is a continuous vector-valued function on $\mathcal{W}$, $\mu(y)$ is also continuous on $\mathfrak{U}$. Set $\mu(y) = (\mu_1(y), \cdots, \mu_{2n}(y))$, and note that $\mu(0) = (0, \cdots, 0, 1)$. Since $|\mu(y) - \mu(0)| = |\lambda(x) - \lambda(x^0)| < 1$, we have that $\mu_{2n}(y) \geq c_0 > 0$. Let $(y^1, 0) \in \mathfrak{U}$ and take $|y^1 - y|$ sufficiently small. From Propositions 12.2 and 12.3 it follows that

$$(\psi(y^1) - \psi(y))\mu_{2n}(y) + \sum_{k=1}^{2n-1} (y_k^1 - y_k)\mu_k(y)$$
$$= o([(\psi(y') - \psi(y))^2 + |y' - y|^2]^{1/2}).$$

From this, in view of our lower bound for $\mu_{2n}(y)$, we have

$$(\psi(y') - \psi(y))\mu_{2n}(y) + \sum_{k=1}^{2n-1} (y_k^1 - y_k)\mu_k(y) = o(|y^1 - y|).$$

Hence $\psi(y)$ is differentiable on $\mathfrak{U}$, and $\partial\psi(y)/\partial y_k = -\mu_k(y)/\mu_{2n}(y)$, $k = 1, \cdots, 2n-1$. Also note that we have shown that the partial derivatives of $\psi(y)$ are continuous on $\mathfrak{U}$, and so the lemma is proved.

CHAPTER 3

CONSTRUCTION OF THE VARIATION OF UNIVALENT FUNCTIONS AND OF FUNCTIONALS FOR ADMISSIBLE FUNCTIONS OF A GENERAL FORM

The investigation of variational problems for functionals of the form

$$J[f] = F(x),\ f(z) = \sum_{k=0}^{\infty} a_k z^{k+1},\ a_0 = 1,\ x = (a_1, \ldots, a_n),$$

leads to the study of the extrema of the function $F(x)$ on the compact set $\mathscr{V}_n$, and, to be more precise, on its boundary $\partial\mathscr{V}_n$. Since $\partial\mathscr{V}_n$ is homeomorphic to S_n, it is clear that we need not study an entire neighborhood $U(f)$ of an extremal function $f(z)$, but that we may limit ourselves to a study of the subset $U \cap S_n$. The subset $U \cap S_n$ is considerably less complicated than is the entire neighborhood U, and, as a rule, this subset can be characterized by a constructive process. If $f(z)$ does not belong to a certain exceptional set, then any function $g(z) \in U \cap S_n$ can be represented as a variation of $f(z)$. However, the class of admissible functions of the form (2.2) is not sufficiently wide to meet this goal, and we must consider a class of admissible functions of a more general type. After a detailed study of the structure of the boundary of the coefficient region $\mathscr{V}_n$, we shall investigate those extremal functions belonging to the above-mentioned exceptional set.

§13. We first present some technical facts which we shall need later on.

LEMMA 13.1. *Let the function $t = \Phi(x)$ map the segment $(0, l)$, $l \leq \infty$, homeomorphically onto itself. Assume that for $0 \leq x \leq l$*

$$\alpha^{-1} \leq \Phi'(x) \leq 2, \qquad \alpha > \tfrac{1}{2}. \tag{13.1}$$

Let $g(x), h(x) \in L^2(0, l)$ be connected by the relation

$$\frac{1}{2}(h(x) + h(\Phi(x))) + \frac{1}{2\pi i}\int_0^l \frac{h(t) - h(\Phi(t))}{t - x}\, dt = g(x). \tag{13.2}$$

Set

$$\begin{aligned}\beta = \frac{1}{\pi}\Bigg(\int_0^l\int_0^l \left|\frac{1}{x - t} - \frac{\Phi'(t)}{\Phi(x) - \Phi(t)}\right|^2 \frac{\Phi'(x)}{\Phi'(t)}\, dx\, dt\Bigg)^{1/2} \\ + \frac{1}{2\pi^2}\int_0^l \frac{dt}{\Phi'(t)}\left(\int_{-\infty}^{0} + \int_l^{\infty}\right)\left|\frac{1}{t - x} - \frac{\Phi'(t)}{\Phi(t) - x}\right|^2 dx.\end{aligned} \tag{13.3}$$

Then, if $\beta < 1$,

$$\left(\int_0^l |h(x)|^2\,dx\right)^{1/2} < \frac{4(1+\sqrt{\alpha})}{1-\beta}\left(\int_0^l |g(x)|^2\,dx\right)^{1/2}.$$

PROOF. As usual, the L^2 norm will be denoted by $\|h\|$, $h \in L^2(0,l)$. Note that in view of (13.1), if $h(x) \in L^2(0,l)$, then $h[\Phi(x)] \in L^2(0,l)$. Consider the following two functions, which are regular for $z \notin (0,l)$:

$$\xi(z) = \frac{1}{2\pi i}\int_0^l \frac{h(t)-g(t)}{t-z}\,dt \text{ and } \eta(z) = \frac{1}{2\pi i}\int_0^l \frac{h[\Phi(t)]-g(t)}{t-z}\,dt.$$

By a classical theorem, the limits $\xi(x+i0)$ and $\xi(x-i0)$ exist almost everywhere on $(0,l)$, and these functions also belong to $L^2(0,l)$. Similar results also hold for $\eta(z)$. We shall show that the function $\xi(z)$, considered as a function defined for $\operatorname{Im} z > 0$, is analytically continuable across the segment $(0,l)$, and its analytic continuation into the half plane $\operatorname{Im} z < 0$ is the function $\eta(z)$. Before beginning, however, we note that by Schwarz' inequality

$$|\xi(z)| < \frac{1}{2\sqrt{\pi y}}\|h-g\|, \tag{13.4}$$

where $\operatorname{Im} z = y > 0$. A similar inequality holds in the lower half plane for $\eta(z)$. Let $\mathscr{D}_{1\epsilon}$ be the rectangle with vertices at $i\epsilon$, $1+i\epsilon$, $l+1$, i, and let $\mathscr{D}_{2\epsilon}$ be the rectangle symmetric to $\mathscr{D}_{1\epsilon}$ with respect to the real axis. We shall orient $\partial\mathscr{D}_{1\epsilon}$ and $\partial\mathscr{D}_{2\epsilon}$ in the usual way. Take $z \in \mathscr{D}_{1\epsilon}$. Then

$$\xi(z) = \frac{1}{2\pi i}\int_{\partial\mathscr{D}_{1\varepsilon}} \frac{\xi(\zeta)}{\zeta-z}\,d\zeta \qquad \text{and} \qquad 0 = \frac{1}{2\pi i}\int_{\partial\mathscr{D}_{2\varepsilon}} \frac{\eta(\zeta)}{\zeta-z}\,d\zeta.$$

Letting $\epsilon \to 0$, we see that the function $\xi(x+i\epsilon)$ converges to

$$\xi(x+i0) = \frac{1}{2}(h(x)-g(x)) + \frac{1}{2\pi i}\int_0^l \frac{h(t)-g(t)}{t-x}\,dt,$$

and the function $\eta(x-i\epsilon)$ to

$$\eta(x-i0) = -\frac{1}{2}(h[\Phi(x)]-g(x)) + \frac{1}{2\pi i}\int_0^l \frac{h[\Phi(t)]-g(t)}{t-x}\,dt.$$

In view of inequality (13.4) and the corresponding inequality for $\eta(z)$, the integrals over the vertical sides of the rectangles will be convergent. Thus we see that

$$\begin{aligned}\xi(z) = \frac{1}{2\pi i}\int_0^l \frac{dx}{x-z}\left[\frac{1}{2}(h(x)-g(x)) + \frac{1}{2\pi i}\int_0^l \frac{h(t)-g(t)}{t-x}\,dx\right]& \\ + \frac{1}{2\pi i}\int_{\partial\mathscr{D}^*_{10}} \frac{\xi(\zeta)}{\zeta-z}\,d\zeta&,\end{aligned} \tag{13.5}$$

where $\partial\mathcal{D}_{10}^* = \partial\mathcal{D}_{10} \setminus \{z : z = x, 0 \leq x \leq l\}$. Similarly

$$0 = -\frac{1}{2\pi i}\int_0^l \frac{dx}{x-z}\left[-\frac{1}{2}(h[\Phi(x)] - g(x)) + \frac{1}{2\pi i}\int_0^l \frac{h[\Phi(t)] - g(t)}{t-x}\,dt\right] + \frac{1}{2\pi i}\int_{\partial\mathcal{D}_{20}^*} \frac{\eta(\zeta)}{\zeta - z}\,d\zeta, \tag{13.6}$$

where $\partial\mathcal{D}_{20}^*$ has the same orientation as does $\partial\mathcal{D}_{10}^*$. Combining (13.5) and (13.6), and recalling (13.2), we find that

$$\xi(z) = \frac{1}{2\pi i}\int_{\partial\mathcal{D}_{10}^*} \frac{\xi(\zeta)}{\zeta - z}\,d\zeta + \frac{1}{2\pi i}\int_{\partial\mathcal{D}_{20}^*} \frac{\eta(\zeta)}{\zeta - z}\,d\zeta.$$

In this way $\xi(z)$ can be continued across the segment $(0, l)$ into the lower half plane, and it is clear that its continuation there is the function $\eta(z)$. The function constructed in this way will be denoted by $\chi(z)$. It is clear that $\chi(z)$ is regular outside the rays $(-\infty < \operatorname{Re} z \leq 0)$ and $(l \leq \operatorname{Re} z < \infty)$. Estimate (13.14) and the corresponding estimate for $\eta(z)$ allow us to apply Cauchy's theorem to the function $\chi(z)$, where as a contour of integration we use the two sides of the above rays and a neighborhood of infinity of very large radius. In the limit we find that

$$\chi(z) = \frac{1}{2\pi i}\left(\int_{-\infty}^0 + \int_l^\infty\right)\frac{\xi(t+i0) - \eta(t-i0)}{t-z}\,dt,$$

and so

$$\chi(z) = -\frac{1}{4\pi^2}\left(\int_{-\infty}^0 + \int_l^\infty\right)\frac{dt}{t-z}\int_0^l \frac{h(x) - h[\Phi(x)]}{x-t}\,dx.$$

Set $h_1(x) = h(x) - h[\Phi(x)]$. We use the following notation: if $f(x) \in L^2(-\infty, \infty)$, then $\tilde{f}(x)$ will denote the conjugate function

$$\tilde{f}(x) = -\frac{1}{\pi}\int_{-\infty}^\infty \frac{f(t)}{t-x}\,dt.$$

It is a classical result that

$$\int_{-\infty}^\infty |\tilde{f}(x)|^2\,dx = \int_{-\infty}^\infty |f(x)|^2\,dx. \tag{13.7}$$

We thus have that

$$\chi(z) = \frac{1}{4\pi}\left(\int_{-\infty}^0 + \int_l^\infty\right)\frac{\tilde{h}_1(t)}{t-z}\,dt.$$

Set $z = x + i\epsilon$, $\epsilon > 0$. Then it is easily seen that

$$\chi(x + i\varepsilon) = \frac{i}{4\pi}\int_0^\infty e^{u(ix-\varepsilon)}\,du\left(\int_{-\infty}^0 + \int_l^\infty\right)e^{-iut}\tilde{h}_1(t)\,dt,$$

for $-\infty < x < \infty$. By Plancherel's theorem,

$$\int_{-\infty}^{\infty} |\chi(x+i\varepsilon)|^2\,dx = \frac{1}{8\pi}\int_0^{\infty} e^{-2\varepsilon u}\left|\left(\int_{-\infty}^{0}+\int_l^{\infty}\right) e^{-iut}\tilde{h}_1(t)\,dt\right|^2 du.$$

As $\epsilon\to 0$, $\chi(x+i\epsilon)$ converges to $\chi(x+i0)$. Since $h_1(t)\in L^2(-\infty,\infty)$, we may, on the right-hand side, take the limit inside the integral, and we find that

$$\int_{-\infty}^{\infty} |\chi(x+i0)|^2\,dx = \frac{1}{8\pi}\int_0^{\infty}\left|\left(\int_{-\infty}^{0}+\int_l^{\infty}\right) e^{iut}\tilde{h}_1(t)\,dt\right|^2 du. \tag{13.8}$$

In a completely analogous way, we get that

$$\int_{-\infty}^{\infty} |\chi(x-i0)|^2\,dx = \frac{1}{8\pi}\int_{-\infty}^{0}\left|\left(\int_{-\infty}^{0}+\int_l^{\infty}\right) e^{-iut}\tilde{h}_1(t)\,dt\right|^2 du. \tag{13.9}$$

We combine (13.8) and (13.9), and then apply Plancherel's theorem again. We deduce that

$$J=\int_{-\infty}^{\infty}(|\chi(x+i0)|^2+|\chi(x-i0)|^2)\,dx = \frac{1}{4}\left(\int_{-\infty}^{0}+\int_0^{\infty}\right)|\tilde{h}_1(t)|^2\,dt.$$

But

$$\tilde{h}_1(x) = -\frac{1}{\pi}\int_0^l h(t)\left(\frac{1}{t-x}-\frac{\psi'(t)}{\psi(t)-x}\right)dt,$$

where $\psi(t)=\Phi^{-1}(t)$. Thus, if $x\notin(0,l)$, Schwarz' inequality gives that

$$|\tilde{h}_1(x)|^2\leqslant \pi^{-2}\|h\|^2\int_0^l\left|\frac{1}{t-x}-\frac{\psi'(t)}{\psi(t)-x}\right|^2 dt,$$

and so

$$J\leqslant\frac{1}{4\pi^2}\|h\|^2\left(\int_{-\infty}^{0}+\int_l^{\infty}\right)dx\int_0^l\left|\frac{1}{t-x}-\frac{\Phi'(t)}{\Phi(t)-x}\right|^2\frac{dt}{\Phi'(t)}=\frac{1}{2}A\|h\|^2. \tag{13.10}$$

Now we evaluate the integral J. We have $\chi(x+i0)=\xi(x+i0)$, and so

$$\int_{-\infty}^{\infty}|\chi(x+i0)|^2\,dx=\frac{1}{4}\left(\int_{-\infty}^{0}+\int_l^{\infty}\right)|\tilde{h}(x)-\tilde{g}(x)|^2\,dx+\frac{1}{4}\int_0^l|h(x)-g(x)$$

$$+\,i(\tilde{h}(x)-\tilde{g}(x))|^2\,dx=\frac{1}{2}\|h-g\|^2-\frac{1}{2}\,\mathrm{Im}\left\{\int_0^l\overline{(h(x)-g(x))}\,(\tilde{h}(x)-\tilde{g}(x))\,dx\right\}$$

where we have used (13.7). Setting $h_0(x)=h[\Phi(x)]$, and going through an argument analogous to the one above, we see that

$$\int_{-\infty}^{\infty}|\chi(x-i0)|^2\,dx=\frac{1}{2}\|h_0-g\|^2+\frac{1}{2}\,\mathrm{Im}\left\{\int_0^l\overline{(h_0(x)-g(x))}\,(\tilde{h}_0(x)-\tilde{g}(x))\,dx\right\}.$$

Therefore

$$J=\frac{1}{2}\|h-g\|^2+\frac{1}{2}\|h_0-g\|^2-\frac{1}{2}J_0,$$

where

$$J_0=\operatorname{Im}\left\{\int_0^l \overline{(h_0(x)-g(x))}\,(\tilde h_0(x)-\tilde g(x))\,dx-\int_0^l\overline{(h(x)-g(x))}\,(\tilde h(x)-\tilde g(x))\,dx\right\}.$$

Using the obvious equation

$$\int_{-\infty}^{\infty}\varphi(x)\,\tilde\psi(x)\,dx=-\int_{-\infty}^{\infty}\tilde\varphi(x)\,\psi(x)\,dx,$$

which holds for any two L^2 functions, we write J_0 as

$$J_0=\operatorname{Im}\left\{\int_0^l[\overline{h_0(x)}\,\tilde h_0(x)-\overline{h(x)}\,\tilde h(x)]\,dx+2\int_0^l\overline{g(x)}\,(\tilde h(x)-\tilde h_0(x))\,dx\right\}.$$

Note that

$$\int_0^l\overline{h(x)}\,\tilde h(x)\,dx=\int_0^l\overline{h_0(x)}\,\tilde h_0(x)\,\Phi'(x)\,dx+\frac{1}{\pi}\int_0^l\overline{h_0(x)}\,\Phi'(x)\,dx$$
$$\times\int_0^l h(t)\left[\frac{\psi'(t)}{\psi(t)-x}-\frac{1}{t-\Phi(t)}\right]dt.$$

Applying Schwarz' inequality, we thus find that

$$\left|\int_0^l[\overline{h_0(x)}\,\tilde h_0(x)-\overline{h(x)}\,\tilde h(x)]\,dx\right|\leqslant\|h_0(\Phi'-1)\|\|\tilde h_0\|$$
$$+\|h\|^2\left(\pi^{-2}\int_0^l\int_0^l\left|\frac{\psi'(t)}{\psi(t)-x}-\frac{1}{t-\Phi(x)}\right|^2\Phi'(x)\,dx\,dt\right)^{1/2}.$$

Set

$$B=\left(\pi^{-2}\int_0^l\int_0^l\left|\frac{\Phi'(t)}{\Phi(t)-\Phi(x)}-\frac{1}{t-x}\right|^2\frac{\Phi'(x)}{\Phi'(t)}\,dx\,dt\right)^{1/2}.$$

The latter inequality can be strengthened by using the fact that $|\Phi'(x)-1|\leq 1$ and $\|\tilde h_0(x)\|<\|h_0\|$. Then

$$\left|\int_0^l[\overline{h_0(x)}\,\tilde h_0(x)-\overline{h(x)}\,\tilde h(x)]\,dx\right|<\|h_0\|^2+B\|h\|^2.$$

Furthermore, Schwarz' inequality gives

$$\int_0^l\overline{g(x)}\,(\tilde h(x)-\tilde h_0(x))\,dx\,\Big|\leqslant\|g\|\|\tilde h-\tilde h_0\|\leqslant\|g_0\|(\|\tilde h\|+\|\tilde h_0\|)<\|g\|(\|h\|+\|h_0\|).$$

Thus we have

$$J_0\leqslant\|h_0\|^2+B\|h\|^2+2\|g\|(\|h\|+\|h_0\|),$$

and so

$$J \geqslant \frac{1}{2}(\|h-g\|^2+\|h_0-g\|^2)-\frac{1}{2}[\|h_0\|^2+B\|h\|^2+2\|g\|(\|h|+\|h_0\|)].$$

But

$$\|h-g\|^2+\|h_0-g\|^2 \geqslant \|h\|^2+\|h_0\|^2+2\|g\|^2-2\|g\|(\|h\|+\|h_0\|)$$

so

$$> \|h\|^2+\|h_0\|^2-2\|g\|(\|h\|+\|h_0\|),$$

(13.11) $$J > \frac{1}{2}[(1-B)\|h\|^2-4\|g\|(\|h\|+\|h_0\|)].$$

Inequalities (13.10) and (13.11) imply the inequalities

$$(1-A-B)\|h\|^2 < 4\|g\|(\|h\|+\|h_0\|) < 4(1+\sqrt{\alpha})\|g\|\cdot\|h\|,$$

as required. Q.E.D.

We now extend the above lemma to the case when the contour of integration is a Jordan curve γ. We shall always assume that the following conditions hold with respect to γ: there exists a parameterization

$$w=w(\sigma), \quad w \in \gamma, \qquad 0 \leqq \sigma \leqq l \leqq \infty,$$

of γ, where $w(\sigma)$ is a three times continuously differentiable function of σ satisfying

(13.12) $$|w^{(\nu)}(\sigma)| \leqq M, \quad \nu=1,2,3, \qquad 0 \leqq \sigma \leqq l.$$

In addition, we assume that

(13.13) $$|w(\sigma_1)-w(\sigma_2)| \geqq M^{-1}|\sigma_1-\sigma_2|,$$

and, in particular, this condition must be satisfied when σ is the arc length parameter. Note that the parameterization defines an orientation on γ, and if γ happened to be oriented already, we assume that our parameterization preserves the orientation.

Let

$$t=\Phi(w), \qquad w,t \in \gamma,$$

be a homeomorphism of γ onto itself having the following properties: $\Phi(w)$ is three times continuously differentiable as a function defined on γ, and its derivatives satisfy

(13.14) $$|[\Phi(w)-w]^{(\nu)}| \leqslant \Delta\{|w(0)|+|w-w(0)|+1\}^{-1}, \qquad \nu=0,\dots,3,$$

where Δ is a sufficiently small quantity. The end conditions on γ are assumed to be as follows:

$$\Phi(w(0))=w(0), \quad \Phi(w(l))=w(l), \quad \Phi'(w(0))=\Phi'(w(l))=1,$$

(13.15) $$\Phi''(w(0))=\Phi''(w(l))=0.$$

The homeomorphism $t=\Phi(w)$ gives rise to a homeomorphism of $(0,l)$ onto itself. To see this, set

$$\Phi(w(\sigma)) = w(\Phi_0(\sigma)).$$

Then $\Phi_0(\sigma)$ is a homeomorphism. From (13.14) and (13.15) it follows, upon using (13.12) and (13.13), that

$$(13.16)\qquad |[\Phi_0(\sigma) - \sigma]^{(\nu)}| < M_0\Delta(\sigma+1)^{-1}, \quad \nu = 0, \dots 3,$$

where M_0 is a constant depending only on M and $|w(0)|$. From (13.15) we have

$$(13.17)\qquad \Phi_0(0) = 0, \quad \Phi_0(l) = l, \quad \Phi_0'(0) = \Phi_0'(l) = 1.$$

Verification of (13.16) is easy, and we do not carry it out.

Now on γ we consider the space of square integrable functions L_γ^2, and we set

$$\|f\|^2 = \int_0^l |f(w(\sigma))|^2 |w'(\sigma)| d\sigma = \int_\gamma |f(w)|^2 |dw|.$$

Singular integrals of the type

$$\frac{1}{2\pi i}\int_\gamma \frac{f(t)}{t-w}\,dt, \qquad w \in \gamma,$$

will be taken in the sense of their Cauchy principal value, and if $f \in L_\gamma^2$, then this integral will exist almost everywhere on γ and is an element of L_γ^2.

Lemma 13.2. *Assume given an arc γ and a homeomorphism $\Phi(w)$ satisfying (13.12) – (13.15). If on γ*

$$(13.18)\quad \frac{1}{2}(h(w) + h(\Phi(w))) + \frac{1}{2\pi i}\int_\gamma h(t)\left(\frac{1}{t-w} - \frac{1}{t-\Phi(w)}\right)dt = g(w),$$

where $h, g \in L_\gamma^2$, then

$$\|h\| \leq A\|g\|$$

provided $\Delta \leq \Delta_0$. A is a constant depending on M, $w(0)$, and Δ.

Proof. Set $h_0(\sigma) = h(w(\sigma))$ and $g_0(\sigma) = g(w(\sigma))$. Clearly $h(\Phi(w)) = h_0(\Phi_0(\sigma))$. Consider the integral

$$j = \frac{1}{2\pi i}\int_\gamma h(t)\left(\frac{1}{t-w} - \frac{1}{t-\Phi(w)}\right)dt$$
$$= \frac{1}{2\pi i}\int_0^l h_0(\tau)\left(\frac{1}{w(\tau)-w(\sigma)} - \frac{1}{w(\tau)-w(\Phi_0(\sigma))}\right)w'(\tau)\,d\tau.$$

Setting

$$K(\tau,\sigma) = \frac{w'(\tau)}{w(\tau)-w(\sigma)} - \frac{1}{\tau-\sigma},$$

and using (13.12) and (13.15), we easily deduce the inequality

$$(13.19)\ K(\tau,\sigma)| < M_1(1+|\tau-\sigma|)^{-1}, \quad \left|\frac{\partial}{\partial\sigma}K(\tau,\sigma)\right| \leq M_1(1+|\tau-\sigma|)^{-1},$$

where M_1 is a constant depending on M.

Noting the identity

$$j = \frac{1}{2\pi i}\int_0^l h_0(\tau)\left(\frac{1}{\tau-\sigma} - \frac{1}{\tau-\Phi_0(\sigma)}\right)d\tau + \frac{1}{2\pi i}\int_0^l h_0(\tau)\,[K(\tau,\sigma) - K(\tau,\Phi_0(\sigma))]\,d\tau = \frac{1}{2\pi i}\int_0^l \frac{h_0(\tau)-h_0(\Phi_0(\tau))}{\tau-\sigma}\,d\tau + h_1(\sigma),$$

we see that

$$h_1(\sigma) = \frac{1}{2\pi i}\int_0^l h_0(\Phi_0(\tau))\left(\frac{1}{\tau-\sigma} - \frac{\Phi_0'(\tau)}{\Phi_0(\tau)-\Phi_0(\sigma)}\right)d\tau + \frac{1}{2\pi i}\int_0^l h_0(\tau)\,[K(\tau,\sigma) - K(\tau,\Phi_0(\sigma))]\,d\tau = h_{11}(\sigma) + h_{12}(\sigma),$$

where the meaning of h_{11} and h_{12} is clear. Thus (13.18) implies

$$\frac{1}{2}(h_0(\sigma) + h_0(\Phi_0(\sigma))) + \frac{1}{2\pi i}\int_0^l \frac{h_0(\tau)-h_0(\Phi_0(\tau))}{\tau-\sigma}\,d\tau = g_0(\sigma) - h_1(\sigma).$$

From (13.16) and (13.17) it follows that for $\Delta \leqq \Delta_0$ the conditions of the previous lemma will hold, and for small Δ_0 we have $\beta \leqq \frac{1}{2}$ and $\alpha \leqq 2$. Then

$$\|h_0\| < B(1+\sqrt{2})\,\|g_0 - h_1\|,$$

where the norm is the $L^2(0,l)$ norm. But

$$\|g_0 - h_1\| \leqslant \|g_0\| + \|h_1\| \leqslant \|g_0\| + \|h_{11}\| + \|h_{12}\|.$$

In view of (13.19),

$$|h_{12}(\sigma)| < \frac{M_1}{2\pi}\,|\Phi_0(\sigma) - \sigma|\int_0^l |h_0(\tau)|\,\frac{d\tau}{1+|\tau-\sigma|} < \frac{M_1}{\pi}\,\|h_0\|\,|\Phi_0(\sigma) - \sigma|.$$

Thus from (13.16)

$$\|h_{12}(\sigma)\| < \frac{M_2}{\pi}\,\Delta\,\|h_0\|.$$

According to a known inequality,

$$\|h_{11}\| \leqslant \|h_0\|\left(\frac{1}{4\pi^2}\int_0^l\int_0^l \left|\frac{1}{\tau-\sigma} - \frac{\Phi_0'(\tau)}{\Phi_0(\tau)-\Phi_0(\sigma)}\right|^2 \frac{1}{\Phi_0'(\sigma)}\,d\tau\,d\sigma\right)^{1/2},$$

and from (13.16) we get

$$\|h_{11}\| < M_3\Delta\,\|h_0\|.$$

Thus

$$\left[1 - 8(1+\sqrt{2})\left(\frac{1}{\pi}M_2 + M_3\right)\Delta\right]\|h_0\| < 8(1+\sqrt{2})\,\|g_0\|.$$

For $\Delta < \Delta_0$ we may take the term in brackets to be at least $\frac{1}{2}$, and so

$$\|h_0\| < 16(1+\sqrt{2})\,\|g_0\|.$$

Upon using (13.12) and (13.13), we arrive at the desired conclusion. Q.E.D.

REMARK 13.1. Lemma 13.2 continues to hold if a minus sign is placed before the integral in (13.18). This case can in fact be reduced to the previous one by taking conjugates in (13.2).

Let $\mathscr{T}$ be an oriented tree with finitely many branches, where the branches are arcs satisfying (13.12) and (13.13). We assume that at each branch point of the tree, each pair of one-sided tangents forms an angle different from zero. If $w=\infty$ is a branch point, then we assume that the tree formed by the transformation $\zeta=1/w$ satisfies the previous conditions at $\zeta=0$. Below we shall only be considering trees of the above type. By defining the integral over $\mathscr{T}$ to be the sum of the integrals over the oriented branches, we extend the concepts of a singular integral and of an $L^2_{\mathscr{T}}$ space to the case when we integrate over the tree $\mathscr{T}$. The norm is taken to be

$$\|f\|^2_{\mathscr{T}}=\sum_r\int_{\gamma_r}|f(w)|^2\,|dw|,$$

where the sum is taken over all branches of the tree $\mathscr{T}$.

REMARK 13.2. Consider the singular integral

$$\tilde{f}(w)=-\frac{1}{\pi}\int_{\mathscr{T}}\frac{f(t)}{t-w}\,dt,\qquad w\in\mathscr{T}.$$

If $f\in L^2_{\mathscr{T}}$, then this integral exists almost everywhere on $\mathscr{T}$ and is an element of $L^2_{\mathscr{T}}$. The inequality

$$\|\tilde{f}\|_{\mathscr{T}}<A_{\mathscr{T}}\|f\|_{\mathscr{T}}\tag{13.20}$$

holds, where $A_{\mathscr{T}}$ depends only on the tree $\mathscr{T}$. This inequality does not follow automatically from the classical case because of the difficulties encountered at the branch points of the tree. These difficulties can be overcome quite easily, however, and for the sake of brevity we shall omit the details.

Suppose we are given a homeomorphism $t=\Phi(w)$ of the tree $\mathscr{T}$ onto itself such that on each branch $\gamma_r\subset\mathscr{T}$ the restricted mapping satisfies (13.14) and (13.15). The following lemma is a trivial corollary of Lemma 13.2.

LEMMA 13.3. *Suppose that the homeomorphism $t=\Phi(w)$, $t,w\in\mathscr{T}$, satisfies the above hypothesis. If $h,g\in L^2_{\mathscr{T}}$ and*

$$\frac{1}{2}(h(w)+h(\Phi(w)))+\frac{1}{2\pi i}\int_{\mathscr{T}}h(t)\left(\frac{1}{t-w}-\frac{1}{t-\Phi(w)}\right)dt=g(w),\tag{13.21}$$

then

$$\|h\|_{\mathscr{T}}<B\|g\|_{\mathscr{T}},$$

provided $\Delta\leqq\Delta_0$.

PROOF. We first make a remark. If any of the branches begin at $w=\infty$, the parameterization of this branch is usually taken so that the value of the parameter $\sigma=0$ corresponds to the finite end of the branch. However, such a parameterization induces an orientation opposite to the one assumed earlier. Thus we change the orientation, which amounts to changing the sign in the integral. However, due to Remark 13.1, this will cause no difficulties.

Let $w \in \gamma_r$. Then $\Phi(w) \in \gamma_r$. From (13.21) we have

$$\frac{1}{2}(h(w) + h(\Phi(w))) + \frac{1}{2\pi i}\int_{\gamma_r} h(t)\left(\frac{1}{t-w} - \frac{1}{t-\Phi(w)}\right) dt = g^*(w),$$

where

$$g^*(w) = g(w) - \frac{1}{2\pi i}\int_{\mathscr{T}/\gamma_r} h(t)\left(\frac{1}{t-w} - \frac{1}{t-\Phi(w)}\right) dt = g(w) - g_r(w).$$

According to Lemma 13.2,

$$\| h \|^2_{\gamma_r} < A^2 \| g^* \|^2_{\gamma_r} < 2A^2 (\| g \|^2_{\gamma_r} + \| g_r \|^2_{\gamma_r}),$$

and using Schwarz' inequality we find that

$$\| g_r(w) \|^2_{\gamma_r} \leqslant \frac{1}{4\pi^2} \int_{\mathscr{T}\setminus\gamma_r} | h(t) |^2 | dt | \int_{\gamma_r} | dw | \int_{\mathscr{T}\setminus\gamma_r} \left| \frac{1}{t-w} - \frac{1}{t-\Phi(w)} \right|^2 | dt |.$$

If $w \in \gamma_r$, $t \in \gamma_l$, $r \neq l$, and if the branch γ_l does not begin at the end of the branch γ_r, then from (13.14) we see that

$$\int_{\gamma_r} | dw | \int_{\gamma_l} \left| \frac{1}{t-w} - \frac{1}{t-\Phi(w)} \right|^2 | dt | < C\Delta^2.$$

Suppose that γ_r and γ_l begin at the same point w_j. Then in a neighborhood of w_j we find from (13.14) and (13.15) that

$$\frac{1}{t-w} - \frac{1}{t-\Phi(w)} = O\left(\Delta\left(\frac{w-w_j}{w-t}\right)^2\right).$$

In view of our earlier hypothesis on the tree $\mathscr{T}$, the quantity $(w-w_j)/(w-t)$ will remain bounded for any $t \in \gamma_l$ and $w \in \gamma_r$ lying in some neighborhood of w_j. The case $w_j = \infty$ causes no problems here due to the properties of $\mathscr{T}$ in a neighborhood of $w_j = \infty$. Thus

$$\frac{1}{4\pi^2}\int_{\gamma_r} | dw | \int_{\mathscr{T}\setminus\gamma_r} \left| \frac{1}{t-w} - \frac{1}{t-\Phi(w)} \right|^2 | dt | < C_0\Delta^2,$$

where the constant C_0 is independent of z. Hence

$$(1 + 2A^2C_0\Delta^2) \| h \|^2_{\gamma_r} < 2A^2 \left(\| g \|^2_{\gamma_r} + C_0\Delta^2 \int_{\mathscr{T}} | h(w) |^2 | dw |\right).$$

Summing these inequalities over all the branches of $\mathscr{T}$, we find that

$$(1 + 2A^2C_0\Delta^2) \| h \|^2_{\mathscr{T}} < 2A^2 (\| g \|^2_{\mathscr{T}} + C_0 N\Delta^2 \| h \|^2_{\mathscr{T}}),$$

where N is the number of branches. If $\Delta \leqq \Delta_0$, then the inequalities stated in the lemma follow. Q.E.D.

§14. We now consider the structure of the solution of the equation (13.21). Assume that on each branch of the tree $\mathcal{T}$ the function $g(w)$ is twice continuously differentiable, with

(14.1) $$|g^{(\nu)}(w)| < K, \qquad \nu = 1, 2,$$

where K is a constant independent of the particular branch under consideration.

LEMMA 14.1. *Let $h, g \in L^2_{\mathcal{T}}$ be connected by* (13.21), *and assume that $g(w)$ satisfies* (14.1). *Then*

(14.2) $$\begin{aligned} h(w) &= g(w) + \frac{1}{2}(g(w) - g(\Phi(w))) \\ &\quad - \frac{1}{2\pi i}\int_{\mathcal{T}} g(t)\left(\frac{1}{t-w} - \frac{1}{t-\Phi(w)}\right)dt + h_1(w), \end{aligned}$$

where $h_1(w)$ satisfies the inequality

$$\|h_1\|_{\mathcal{T}} < B_0(K + \|g\|)\Delta^2.$$

PROOF. If we define $h_1(w)$ by (14.2), it is easily verified that $h_1(w)$ satisfies the equation

$$\frac{1}{2}(h_1(w) + h_1(\Phi(w))) + \frac{1}{2\pi i}\int_{\mathcal{T}} h_1(t)\left(\frac{1}{t-w} - \frac{1}{t-\Phi(w)}\right)dt = g_1(w),$$

where

$$\begin{aligned} g_1(w) &= \frac{1}{4}(g(w) - 2g(\Phi(w)) + g(\Phi(\Phi(w)))) \\ &\quad - \frac{1}{4\pi i}\int_{\mathcal{T}} g(t)\left(\frac{1}{t-w} - \frac{2}{t-\Phi(w)} + \frac{1}{t-\Phi(\Phi(w))}\right)dt \\ &\quad - \frac{1}{4\pi i}\int_{\mathcal{T}} (g(t) - g(\Phi(t)))\left(\frac{1}{t-w} - \frac{1}{t-\Phi(w)}\right)dt \\ &\quad - \frac{1}{4\pi^2}\int_{\mathcal{T}}\left(\frac{1}{t-w} - \frac{1}{t-\Phi(w)}\right)dt\int_{\mathcal{T}} g(s)\left(\frac{1}{s-t} - \frac{1}{s-\Phi(t)}\right)ds. \end{aligned}$$

Since $g \in L^2_{\mathcal{T}}$, each of the terms in the right-hand side of this latter formula will also belong to $L^2_{\mathcal{T}}$ (see Remark 13.2). Hence $g_1 \in L^2_{\mathcal{T}}$. Similar remarks hold for h_1. Thus from Lemma 13.3 we have

$$\|h_1\|_{\mathcal{T}} < B\|g_1\|_{\mathcal{T}}.$$

We now estimate $\|g_1\|_{\mathcal{T}}$. Let $G_1(w) = g(w) - g(\Phi(w))$ and $G_2(w) = G_1(w) - G_1(\Phi(w))$. Also put

$$k(t, w) = \frac{\Phi'(t)}{\Phi(t) - \Phi(w)} - \frac{1}{t-w}.$$

Simple computations give

$$j_0(w) = \frac{1}{4\pi i}\int_{\mathcal{T}} g(t)\left(\frac{1}{t-w} - \frac{2}{t-\Phi(w)} + \frac{1}{t-\Phi(\Phi(w))}\right)dt$$

$$(14.3) \qquad = \frac{1}{4\pi i}\int_{\mathcal{T}} \frac{G_2(t)}{t-w}\,dt - \frac{1}{4\pi i}\int_{\mathcal{T}} G_1(\Phi(t))\,k(t,w)\,dt$$

$$- \frac{1}{4\pi i}\int_{\mathcal{T}} g(\Phi(t))\,[k(t,w) - k(t,\Phi(w))]\,dt.$$

Furthermore,

$$\frac{1}{4\pi i}\int_{\mathcal{T}} G_1(t)\left(\frac{1}{t-w} - \frac{1}{t-\Phi(w)}\right)dt = \frac{1}{4\pi i}\int_{\mathcal{T}} G_2(t)\,\frac{dt}{t-w}$$

$$- \frac{1}{4\pi i}\int_{\mathcal{T}} G_1(\Phi(t))\,k(t,w)\,dt.$$

Therefore

$$j_1(w) = j_0(w) + \frac{1}{4\pi i}\int_{\mathcal{T}} G_1(t)\left(\frac{1}{t-w} - \frac{1}{t-\Phi(w)}\right)dt = \frac{1}{2\pi i}\int_{\mathcal{T}} \frac{G_2(t)}{t-w}\,dt$$

$$(14.4) \qquad - \frac{1}{2\pi i}\int_{\mathcal{T}} G_1(\Phi(t))\,k(t,w)\,dt - \frac{1}{4\pi i}\int_{\mathcal{T}} g(\Phi(t))\,[k(t,w) - k(t,\Phi(w))]\,dt,$$

and in addition we note that

$$j_2(w) = \frac{1}{4\pi^2}\int_{\mathcal{T}}\left(\frac{1}{t-w} - \frac{1}{t-\Phi(w)}\right)dt\int_{\mathcal{T}} g(s)\left(\frac{1}{s-t} - \frac{1}{s-\Phi(t)}\right)ds$$

$$= \frac{i}{\pi}\int_{\mathcal{T}} \frac{j_0(t)}{t-w}\,dt - \frac{1}{4\pi^2}\int_{\mathcal{T}} k(t,w)\,dt\int_{\mathcal{T}} g(s)\left(\frac{1}{s-\Phi(t)} - \frac{1}{s-\Phi(\Phi(t))}\right)ds.$$

From this we find that

$$j_2(w) = \frac{i}{\pi}\int_{\mathcal{T}} \frac{j_0(t)}{t-w}\,dt - \frac{1}{4\pi^2}\int_{\mathcal{T}} k(t,w)\,dt\int_{\mathcal{T}} \frac{G_1(s)}{s-\Phi(t)}\,ds$$

$$(14.5) \qquad + \frac{1}{4\pi^2}\int_{\mathcal{T}} k(t,w)\,dt\int_{\mathcal{T}} g(\Phi(s))\,k(s,\Phi(t))\,ds.$$

We now estimate the integral

$$I_0^2 = \iint_{\mathcal{T}\mathcal{T}} |k(t,w)|^2\,|dt|\,|dw| \quad \text{и} \quad I_1^2 = \iint_{\mathcal{T}\mathcal{T}} |k(t,w) - k(t,\Phi(w))|^2\,|dt|\,|dw|.$$

For any $t, w \in \mathcal{T}$ we have from (13.14) that

$$k(t,w) = \frac{\Phi(w) - \Phi(t) - (w-t)\,\Phi'(t)}{(\Phi(w) - \Phi(t))\,(w-t)} = O\left(\frac{|t|^{-1} + |w|^{-1}}{|w-t|^2+1}\,\Delta\right) + O\left(\frac{\Delta}{t\,(|w-t|+1)}\right),$$

if $|t| \geqq C$ and $|w| \geqq C$. Therefore $I_0^2 = O(\Delta^2)$. We note that

$$\frac{\partial}{\partial w}\,k(t,w) = \frac{\Phi'(t)\,\Phi'(w)}{[\Phi(t) - \Phi(w)]^2} - \frac{1}{(t-w)^2}\,.$$

If $|t-w| \leqq \rho$ where ρ is sufficiently small, it follows from (13.13) and (13.15) that $\partial k(t,w)/\partial w = O(\Delta)$. If however $|t-w| > \rho$, then $\partial k(t,w)/\partial w = O(\Delta(t-w)^{-2})$. Thus, for all $t, w \in \mathcal{T}$,

$$\frac{\partial}{\partial w} k(t, w) = O\left(\frac{\Delta}{|t-w|^2+1}\right).$$

It now follows from (13.14) and (13.12) that

$$k(t, w) - k(t, \Phi(w)) = \int_{\Phi(w)}^{w} \frac{\partial}{\partial \tau} k(t, \tau)\, d\tau = O\left(\Delta^2 w^{-1} \frac{1}{|w-t|^2+1}\right).$$

Thus $I_1^2 = O(\Delta^4)$.

Now we estimate the norm of $j_k(w)$, $k = 0, 1, 2$. Each of these functions is a sum of integrals. Using the triangle inequality, we see that we must estimate the norms of the various terms on the right-hand sides of equations (14.3) – (14.5). Some of these have the form

$$f(w) = \int_{\mathcal{T}} \varphi(t)\, \mathcal{H}(t, w)\, dt,$$

and we shall now derive an estimate for integrals of this type. We have, after applying Schwarz' inequality three times, that

$$\int_{\mathcal{T}} |f(w)|^2 |dw| = \int_{\mathcal{T}} |\varphi(t_1)|\, |dt_1| \int_{\mathcal{T}} |\varphi(t_2)|\, |dt_2| \int_{\mathcal{T}} |\mathcal{H}(t_1, w)\, \mathcal{H}(t_2, w)|\, |dw|$$

$$< \iint_{\mathcal{T}\mathcal{T}} |\varphi(t_1)\varphi(t_2)|\, |dt_1|\, |dt_2| \left(\int_{\mathcal{T}} |\mathcal{H}(t_1, w)|^2 |dw|\right)^{1/2} \left(\int_{\mathcal{T}} |\mathcal{H}(t_2, w)|^2 |dw|\right)^{1/2}$$

$$< \|\varphi\|_{\mathcal{T}}^2 \iint_{\mathcal{T}\mathcal{T}} |\mathcal{H}(t, w)|^2 |dt|\, |dw|.$$

The norms of the singular integrals can be estimated by using Remark 13.2. Using the fact that for $f \in L^2_{\mathcal{T}}$

$$\|f(\Phi(w))\|_{\mathcal{T}} < C \|f(w)\|_{\mathcal{T}},$$

and using our estimates for I_0 and I_1, we find that

(14.6) $$\|j_k\|_{\mathcal{T}} < C_0 (\|G_2\|_{\mathcal{T}} + \Delta \|G_1\|_{\mathcal{T}} + \Delta^2 \|g\|_{\mathcal{T}}), \quad k = 0, 1.$$

With regard to the estimate for $\|j_2\|_{\mathcal{T}}$, the only difficult term is the last term in the right-hand side of (14.5). We denote this term by $j_3(w)$. As a result of a two-fold application of Schwarz' inequality we see that

$$|j_3(w)| < \frac{C}{4\pi^2} \|g\|_{\mathcal{T}} \int_{\mathcal{T}} |k(t, w)|\, |dt| \left(\int_{\mathcal{T}} |k(s, \Phi(t))|^2 |ds|\right)^{1/2}$$

$$< \frac{C^2}{4\pi^2} \|g\|_{\mathcal{T}} I_0 \left(\int_{\mathcal{T}} |k(t, w)|^2 |dt|\right)^{1/2}.$$

From this we get

$$\| j_3 \|_{\mathscr{T}} < \frac{C^2}{4\pi^2} I_0^2 \| g \|_{\mathscr{T}}.$$

Thus, using (14.6) for $k=0$ and the last inequality, we have

$$\| j_2 \|_{\mathscr{T}} < C_1 (\| G_2 \|_{\mathscr{T}} + \Delta \| G_1 \|_{\mathscr{T}} + \Delta^2 \| g \|_{\mathscr{T}}). \tag{14.7}$$

From (14.6) and (14.17) we find that

$$\| g_1 \|_{\mathscr{T}} < C_2 (\| G_2 \|_{\mathscr{T}} + \Delta \| G_1 \|_{\mathscr{T}} + \Delta^2 \| g \|_{\mathscr{T}}). \tag{14.8}$$

From (14.1) we have $|G_1(w)| < K|\Phi(w) - w|$, and so $\| G_1 \|_{\mathscr{T}} < C_4 K \Delta$ by (13.14). Furthermore,

$$G_1 (w) = g' (w) - g' (\Phi (w)) + [1 - \Phi' (w)] g' (\Phi (w)).$$

Thus from (13.14) and (14.1) it follows that $|G_1(w)| < C_5 K\Delta$, and so $|G_2(w)| = |G_1(w) - G_1(\Phi(w))| < C_6 K |\Phi(w) - w| \Delta$ and $\| G_2 \|_{\mathscr{T}} < C_7 K \Delta^2$. Taking these estimates into account, we find from (14.8) that $\| g_1 \|_{\mathscr{T}} < C_8 \Delta^2 (K + \| g \|_{\mathscr{T}})$, which implies that

$$\| h_1 \|_{\mathscr{T}} < B_0 \Delta^2 (K + \| g \|_{\mathscr{T}}), \quad \text{Q.E.D.}$$

Lemma 14.2. *Let* $g \in L^2_{\mathscr{T}}$ *satisfy condition* (14.1), *and suppose* $g' \in L^2_{\mathscr{T}}$. *Suppose the homeomorphism* $\Phi(w)$ *satisfies conditions* (13.14) *and* (13.15). *If the equation*

$$\frac{1}{2} (h (w) + h (\Phi (w))) + \frac{1}{2\pi i} \int_{\mathscr{T}} h (t) \left(\frac{1}{t - w} - \frac{1}{t - \Phi (w)} \right) dt = g (w) \tag{13.21}$$

has a solution $h \in L^2_{\mathscr{T}}$, *then for* $\Delta \leqq \Delta_0$

$$h (w) = g (w) - \frac{1}{2} (\Phi (w) - w) g' (w) + \frac{\Phi (w) - w}{2\pi i} \int_{\mathscr{T}} \frac{g' (t)}{t - w} dt + \sigma (w) + h_0 (w), \tag{14.9}$$

where

$$\sigma (w) = \sum \alpha_r \log \frac{w_r - \Phi (w)}{w_r - w},$$

with the sum being taken over all finite branch points of the tree, and, moreover,

$$\| h_0 \|_{\mathscr{T}} < B \Delta^2 (K + \| g \|_{\mathscr{T}} + \| g' \|_{\mathscr{T}}).$$

The constants α_r *depend linearly on the values of the function* g *at the branch points of the tree* $\mathscr{T}$.

Proof. On the basis of Lemma 14.1, in order to obtain (14.9) it is sufficient to reformulate the expression $g(w) - g(\Phi(w))$ and the integral

$$j (w) = \frac{1}{2\pi i} \int_{\mathscr{T}} g (t) \left(\frac{1}{t - w} - \frac{1}{t - \Phi (w)} \right) dt.$$

Set $g(w) - g(\Phi(w)) = (w - \Phi(w)) g'(w) + G(w)$. As in the proof of Lemma

14.1, we make some obvious transformations. We have

$$j(w)=\frac{1}{2\pi i}\int_{\mathscr{T}}\frac{(t-\Phi(t))\,g'(t)}{t-w}\,dt-\frac{1}{2\pi i}\int_{\mathscr{T}}g(t)\,k(t,w)\,dt+j_1(w)=j_0(w)+j_1(w),$$

where

$$j_1(w)=\frac{1}{2\pi i}\int_{\mathscr{T}}G_1(t)\,k(t,w)\,dt+\frac{1}{2\pi i}\int_{\mathscr{T}}\frac{G(t)}{t-w}\,dt.$$

We note that for $t, w\in\mathscr{T}$ the quantity $\zeta=(\Phi(t)-\Phi(w))/(t-w)$ lies in the disc $|\zeta-1|<C\Delta$, as follows from (13.14) and (13.15). Thus for $\Delta\leqq\Delta_0$ we can define a single-valued branch of $\log\zeta$, $\zeta\in\mathscr{T}$, where we choose the branch satisfying $\log 1=0$. Since $\partial\log\zeta/\partial t=k(t,w)$, we find upon integrating by parts and making the obvious transformations that

$$j_0(w)=\frac{w-\Phi(w)}{2\pi i}\int_{\mathscr{T}}\frac{g'(t)}{t-w}\,dt+\frac{1}{2\pi i}\int_{\mathscr{T}}g'(t)\left(1-\frac{\Phi(t)-\Phi(w)}{t-w}\right)dt$$
$$+\frac{1}{2\pi i}\int_{\mathscr{T}}g'(t)\log\zeta\,dt-\sum_r\alpha_r\log\frac{w_r-\Phi(w)}{w_r-w}.$$

Therefore

$$j(w)=-\frac{\Phi(w)-w}{2\pi i}\int_{\mathscr{T}}\frac{g'(t)}{t-w}\,dt-\sum_r\alpha_r\log\frac{w_r-\Phi(w)}{w_r-w}+j_2(w),$$

where

$$j_2(w)=j_1(w)+\frac{1}{2\pi i}\int_{\mathscr{T}}g'(t)\left[\log\zeta+1-\frac{\Phi(t)-\Phi(w)}{t-w}\right]dt.$$

If we set

$$h_0(w)=\frac{1}{2}G(w)+h_1(w)-j_2(w),$$

then formula (14.9) will follow from (14.2), and all that remains is to estimate $\|h_0\|_{\mathscr{T}}$. Recalling the inequalities derived in the proof of the previous lemma, we have

$$\|j_1\|_{\mathscr{T}}<C_0(\|G\|_{\mathscr{T}}+K\Delta^2).$$

It is clear that

$$\left|\log\zeta+1-\frac{\Phi(t)-\Phi(w)}{t-w}\right|<C_1\left|1-\frac{\Phi(t)-\Phi(w)}{t-w}\right|^2.$$

Thus the norm of the integral

$$\frac{1}{2\pi i}\int_{\mathscr{T}}g'(t)\left[\log\zeta+1-\frac{\Phi(t)-\Phi(w)}{t-w}\right]dt$$

is at most

$$C_2\|g'\|_{\mathscr{T}}\Big(\iint\limits_{\mathscr{T}\mathscr{T}}\Big|1-\frac{\Phi(t)-\Phi(w)}{t-w}\Big|^4|dt||dw|\Big)^{1/2}<C_3\Delta^2\|g'\|_{\mathscr{T}},$$

since

$$1-\frac{\Phi(t)-\Phi(w)}{t-w}=O\Big(\Delta\frac{|t|^{-1}+|w|^{-1}}{|t-w|+1}\Big)$$

on the strength of (13.14). We have thus shown that

$$\|j_2\|_{\mathscr{T}}\leqq C_3(\|G\|_{\mathscr{T}}+(K+\|g'\|_{\mathscr{T}})\Delta^2),$$

and so

$$\|h_0\|_{\mathscr{T}}<C_4(\|G\|_{\mathscr{T}}+\|h_1\|_{\mathscr{T}}+(K+\|g'\|_{\mathscr{T}})\Delta^2).$$

But it is easily seen that $g(w)-g(\Phi(w))=(w-\Phi(w))g'(w)+O[K(\Phi(w)-w)^2]$. Thus $\|G\|_{\mathscr{T}}<CK\Delta^2$. Hence from Lemma 14.1 we get that

$$\|h_0\|_{\mathscr{T}}<B\Delta^2(K+\|g\|_{\mathscr{T}}+\|g'\|_{\mathscr{T}}).\quad \text{Q.E.D.}$$

REMARK 14.1. Lemma 14.2 gives the structure of the solution of the integral equation (13.21); this will prove useful in the solution of some of our problems. The first two terms in our expression of the solution were determined as asymptotic series in terms of Δ. We should point out that nowhere did our arguments make use of the fact that $\mathscr{T}$ was a tree (that is, a graph without cycles and multiple branches). All our arguments continue to hold if we assume only that $\mathscr{T}$ is a planar graph satisfying the conditions imposed on the geometrical positioning of the branches of the graph. Since we shall require the result of Lemma 14.2 only in the case when $\mathscr{T}$ is a tree, we have formulated it in this way.

§15. We shall assume below that $w=0$ lies in the complement of the tree $\mathscr{T}$. Let the function $z=\psi(w)$ map the complement of $\mathscr{T}$ onto the disc $|z|<1$, with $\psi(0)=0$ and $\psi'(0)>0$. Denote by $\mathscr{L}_r$ the level curve $\mathscr{L}_r=\{w:|\psi(w)|=r, r<1\}$. The doubly connected domain bounded by $\mathscr{L}_r$ and the tree $\mathscr{T}$ will be denoted by $\mathscr{G}(\mathscr{L}_r,\mathscr{T})$. The domain bounded by $\mathscr{L}_r$ and containing $w=0$ will be denoted by $\mathscr{G}(\mathscr{L}_r)$, and the complement of $\mathscr{T}$ will be denoted by $\mathscr{G}(\mathscr{T})$. We orient $\mathscr{T}$ in an arbitrary manner. The orientation of each branch $\gamma\subset\mathscr{T}$ defines a left and right side of the branch. In connection with this we shall employ the usual notation of "+" and "−" to denote one-sided limits.

We now construct a class of univalent functions which will play an important role in what follows. In the final analysis this class will really be defined by the tree $\mathscr{T}$, and we denote it by $U_{\mathscr{T}}$.

To this class belong functions univalent and regular in some domain $\mathscr{G}(\mathscr{L}_r,\mathscr{T})$, and which depend on a positive parameter ϵ. If $\varphi(w,\epsilon)\in U_{\mathscr{T}}$, then the following conditions will be satisfied.

1. For any branch $\gamma\subset\mathscr{T}$ there exist continuous limits $\varphi_+(w,\epsilon)$ and $\varphi_-(w,\epsilon)$,

$w \in \gamma$. These functions are four times continuously differentiable on γ, and their derivatives are bounded by a constant independent of the branch and of ϵ, if $\epsilon \leqq \epsilon_0$.

2. For any branch γ

$$\varphi_+\{\gamma, \epsilon\} = \varphi_-\{\gamma, \epsilon\}.$$

3. Let the branch γ have end points w_0 and w_1. For $\epsilon \leqq \epsilon_0$ the inequality

$$|\varphi_+^{(\nu)}(w, \varepsilon) - \varphi_-^{(\nu)}(w, \varepsilon)| < C_0 \varepsilon \min\{1, |w - w_0|, |w - w_1|\},$$
$$\nu = 0, 1, 2, 3,$$

is satisfied on γ, where the constant C_0 is independent of ϵ. If $w_0 = \infty$, the right-hand side above is replaced by $C_0\epsilon \min\{1, |w|^{-1}, |w - w_1|\}$.

4. For any branch $\gamma \subset \mathscr{T}$,

$$|\varphi_-(w, \epsilon) - \varphi_-(t, \epsilon)| > C_0^{-1}|w - t|, \quad w, t \in \gamma.$$

5. In the domain $\mathscr{G}(\mathscr{L}_r, \mathscr{T})$,

$$|\varphi(w, \epsilon) - w| < C_0\epsilon|w|.$$

From 1, 2 and 4 we see by setting

$$\varphi_-(\Phi(w, \epsilon), \epsilon) = \varphi_+(w, \epsilon) \tag{15.1}$$

that on any branch γ there is defined a homeomorphism $\Phi(w, \epsilon)$ mapping γ onto itself. For ease of notation we shall henceforth omit ϵ from our symbols. From 3 and 4 it follows that on any branch we have

$$|[\Phi(w) - w]_-^{(\nu)}| < C_1 \varepsilon \min\{1, |w - w_0|, |w - w_1|\}, \quad \nu = 0, 1, 2, 3. \tag{15.2}$$

To see this, note that the identity

$$\varphi_-(\Phi(w)) - \varphi_-(w) = \varphi_+(w) - \varphi_-(w)$$

is valid, and so from 3 and 4 it follows that

$$C_0^{-1} |\Phi(w) - w| < C_0 \varepsilon \min\{1, |w - w_0|, |w - w_1|\}. \tag{15.3}$$

Now $\Phi(w)$ is differentiable since φ_+ and φ_- are, so we have

$$\varphi'_-(\Phi(w))\Phi'(w) = \varphi'_+(w),$$

and so

$$\Phi'(w) - 1 = \frac{\varphi'_+(w) - \varphi'_-(\Phi(w))}{\varphi'_-(\Phi(w))} = \frac{\varphi'_+(w) - \varphi'_-(w) + \varphi'_-(w) - \varphi'_-(\Phi(w))}{\varphi'_-(\Phi(w))}.$$

From this we find that

$$|\Phi'(w) - 1| < C_2(\varepsilon\theta_w + |\varphi'_-(w) - \varphi'_-(\Phi(w))|),$$
$$\theta_w = \min\{1, |w - w_0|, |w - w_1|\}.$$

But

$$\varphi_-'(\Phi(w)) - \varphi_-'(w) = \int_w^{\Phi(w)} \varphi_-''(\tau)\, d\tau,$$

and on the basis of condition 1 above and (13.13) we deduce that

$$|\varphi_-'(\Phi(w)) - \varphi_-'(w)| < C_3 |\Phi(w) - w|.$$

Then from (15.3) we see that

$$|\Phi'(w) - 1| < C_1 \varepsilon \min\{1, |w - w_0|, |w - w_1|\}.$$

The verification of inequality (15.2) for $\nu = 2, 3$ is analogous. In particular, the right-hand side of (15.2) takes the previously-mentioned form when $w_0 = \infty$, since in this case we introduce the variable ζ before performing our calculations as above.

In this way we get a homeomorphism $\Phi(w)$ of the tree $\mathscr{T}$ onto itself, and this homeomorphism satisfies conditions (13.14) and (13.15) if $\epsilon \leqq \epsilon_0$. Set $K_R = \{w : |w| \leqq R\}$, $\mathscr{T}_R = \mathscr{T} \cap K_R$ and

$$\varphi(w, \epsilon) = w + \epsilon g(w, \epsilon).$$

We take R large enough so that $\mathscr{C}_r \subset K_R$. If $w \in \mathscr{G}(\mathscr{C}_r, \mathscr{T})$, then Cauchy's theorem gives

$$\varphi(w) - w = \frac{\varepsilon}{2\pi i} \int_{\mathscr{L}} \frac{w^2 g(t, \varepsilon)}{t^2 (t - w)}\, dt + \frac{\varepsilon}{2\pi i} \int_{\mathscr{C}_r} \frac{w^2 g(t, \varepsilon)}{t^2 (w - t)}\, dt,$$

where $\mathscr{L}$ is a Jordan curve lying in $\mathscr{G}(\mathscr{C}_r, \mathscr{T})$ and containing w in its interior. After changing the path of integration we get

$$\varphi(w) - w = \frac{w^2}{2\pi i} \int_{\mathscr{T}_R} \frac{\varphi_+(t) - \varphi_-(t)}{t^2(t - w)}\, dt + \frac{\varepsilon w^2}{2\pi i} \int_{\partial K_R} \frac{g(t)\, dt}{t^2 (t - w)} + \frac{\varepsilon}{2\pi i} \int_{\mathscr{C}_r} \frac{w^2 g(t, \varepsilon)}{t^2 (w - t)}\, dt.$$

Now let $R \to \infty$. The integral over $\mathscr{T}_R$ will approach a limit, due to condition 3. The integral over ∂K_R tends to zero, due to condition 5. Thus

$$\varphi(w) - w = \frac{w^2}{2\pi i} \int_{\mathscr{T}} \frac{\varphi_+(t) - \varphi_-(t)}{t^2 (t - w)}\, dt + \frac{\varepsilon}{2\pi i} \int_{\mathscr{C}_r} \frac{w^2 g(t, \varepsilon)}{(w - t)\, t^2}\, dt. \tag{15.4}$$

The components of the right-hand side of (15.4) have definite interpretations. Both terms are sums of admissible functions with respect to w. The second term is regular on $\mathscr{T}$ and has the same form as an admissible function of the class defined in (2.1). The first term is a function regular outside of $\mathscr{T}$. We introduce the notation

$$g(w, \varepsilon; \infty) = g(w; \infty) = \frac{1}{2\pi i} \int_{\mathscr{C}_r} \frac{w^2 g(t, \varepsilon)}{(w - t)\, t^2}\, dt.$$

Set $\varphi_+(w) - \varphi_-(w) = \epsilon\omega(w)$ and

$$g(w, \varepsilon; 0) = g(w; 0) = \frac{1}{2\pi i} \int_{\mathscr{T}} \frac{w^2 \omega(t)}{(t - w)\, t^2}\, dt. \tag{15.5}$$

It is clear that if $r < r' < 1$, then

$$g(w;\ \infty) = \frac{1}{2\pi i} \int_{\mathscr{C}_{r'}} \frac{w^2 g(t;\ \infty)}{(w-t)\, t^2}\, dt. \tag{15.6}$$

In what follows we may assume that the integration is over the level curve $\mathscr{C}_r$, since at the beginning we could have started with a level curve $\mathscr{C}_{r-\delta}$, where $\delta > 0$ is sufficiently small. We can now state a formula for the function $\omega(w)$ which will give us valuable information concerning the properties of $\omega(w)$. Using the Sohockiĭ-Plemelj formulas we have

$$\varphi_+(w) = w + \frac{1}{2}\varepsilon\omega(w) + \frac{\varepsilon w^2}{2\pi i}\int_{\mathscr{T}} \frac{\omega(t)}{t^2(t-w)}\, dt + \varepsilon g(w;\ \infty),$$

$$\varphi_-(w) = w - \frac{1}{2}\varepsilon\omega(w) + \frac{\varepsilon w^2}{2\pi i}\int_{\mathscr{T}} \frac{\omega(t)}{t^2(t-w)}\, dt + \varepsilon g(w;\ \infty).$$

and from these relations and formula (15.1) we get that

$$\begin{aligned} &\frac{\varepsilon}{2}(\omega(w) + \omega(\Phi(w))) + \frac{\varepsilon}{2\pi i}\int_{\mathscr{T}} \omega(t)\left(\frac{1}{t-w} - \frac{1}{t-\Phi(w)}\right) dt \\ &\quad = \varepsilon\,(g(\Phi(w);\ \infty) - g(w;\ \infty)) + (\Phi(w) - w)\left(1 - \frac{\varepsilon}{2\pi i}\int_{\mathscr{T}} \omega(t)\,\frac{dt}{t^2}\right). \end{aligned} \tag{15.7}$$

It is clear that $\omega(w) \in L^2_{\mathscr{T}}$, and the right-hand side of the above equation also belongs to $L^2_{\mathscr{T}}$. We next apply Lemma 14.2. Note that if the right-hand side of (15.7) is twice continuously differentiable on each branch $\gamma \in \mathscr{T}$, then (15.2) insures that the hypotheses of Lemma 14.2 will be satisfied. But this differentiability condition is trivially satisfied, as can be seen from (15.2) and the analyticity of $g(w;\ \infty)$ on $\mathscr{T}$. Note that $K = C\epsilon$ and also that the norm of the right-hand side above, and the norm of its derivative, are no larger than $C\epsilon$. We can also take the constant Δ in Lemma 14.2 to be $C\epsilon$. Denoting the right-hand side of (15.7) by $G(w)$, we see easily that

$$G(w) = (\Phi(w) - w)\left(\varepsilon g'(w;\ \infty) + 1 - \frac{\varepsilon}{2\pi i}\int_{\mathscr{T}} \omega(t)\,\frac{dt}{t^2}\right) + G_1(w),$$

where

$$\|G_1(w)\| < C\epsilon^3, \qquad \|G_1'(w)\| < C\epsilon^3. \tag{15.8}$$

Therefore

$$G'(w) = \Phi'(w) - 1 + G_2(w), \tag{15.9}$$

where

$$\|G_2(w)\| < C\epsilon^2, \tag{15.10}$$

and so

$$G(w)-\frac{1}{2}(\Phi(w)-w)G'(w)+\frac{\Phi(w)-w}{2\pi i}\int_{\mathcal{T}}\frac{G'(t)\,dt}{t-w}$$
$$=(\Phi(w)-w)\Big(1+\varepsilon g'(w;\infty)-\frac{\varepsilon}{2\pi i}\int_{\mathcal{T}}\omega(t)\frac{dt}{t^2}-\frac{1}{2}(\Phi'(w)-1)$$
$$+\frac{1}{2\pi i}\int_{\mathcal{T}}\frac{\Phi'(t)-1}{t-w}dt\Big)+G_3(w),$$

where

$$G_3(w)=G_1(w)-\frac{1}{2}(\Phi(w)-w)G_2(w)+\frac{\Phi(w)-w}{2\pi i}\int_{\mathcal{T}}\frac{G_2(t)}{t-w}dt.$$

Now combining (15.8), (15.10) and (13.20), we deduce that

$$\|G_3(w)\|<C_1\epsilon^3.$$

Finally we note that $\sigma(w)\equiv 0$, since at the ends of the tree $\mathcal{T}$ the function $G(w)$ is zero. Thus, applying (14.9) to the function $\epsilon\omega(w)$, we find that

$$\omega(w)=\frac{1}{\varepsilon}(\Phi(w)-w)\Big[1+\varepsilon g'(w;\infty)-\frac{\varepsilon}{2\pi i}\int_{\mathcal{T}}\omega(t)\frac{dt}{t^2}$$
$$-\frac{1}{2}(\Phi'(w)-1)+\frac{1}{2\pi i}\int_{\mathcal{T}}\frac{\Phi'(t)-1}{t-w}dt\Big]+\omega_1(w),$$

where $\|\omega_1\|<C\epsilon^2$. If in place of $\omega(t)$ in the integral $\int_{\mathcal{T}}\omega(t)t^{-2}dt$ we put the dominant term of the latter formula, we find that

$$\omega(w)=\frac{1}{\varepsilon}(\Phi(w)-w)\Big(1+\varepsilon g'(w;\infty)-\frac{1}{2\pi i}\int_{\mathcal{T}}\frac{\Phi(t)-t}{t^2}dt$$
$$-\frac{1}{2}(\Phi'(w)-1)+\frac{1}{2\pi i}\int_{\mathcal{T}}\frac{\Phi'(t)-1}{t-w}dt\Big)+\omega_0(w),$$

where

$$\|\omega_0(w)\|<C\epsilon^2. \tag{15.11}$$

Integrating by parts, we see that

$$\frac{1}{2\pi i}\int_{\mathcal{T}}\frac{\Phi(t)-t}{t^2}dt=\frac{1}{2\pi i}\int_{\mathcal{T}}\frac{\Phi'(t)-1}{t}dt,$$

since $\Phi(w)-w$ vanishes at the end points of the tree. Thus we finally conclude that

$$\omega(w)=\frac{1}{2}(\Phi(w)-w)\Big(1+\varepsilon g'(w;\infty)-\frac{1}{2}(\Phi'(w)-1)$$
$$+\frac{w}{2\pi i}\int_{\mathcal{T}}\frac{\Phi'(t)-1}{t(t-w)}dt\Big)+\omega_0(w). \tag{15.12}$$

These results are reformulated in the following lemma.

Lemma 15.1. *Let* $\varphi(w,\epsilon)\in U_{\mathscr{T}}$. *Then*

$$\varphi(w,\epsilon)=w+\epsilon g(w,\epsilon;0)+\epsilon g(w,\epsilon;\infty), \tag{15.13}$$

where

$$g(w,\varepsilon;0)=\frac{w^2}{2\pi i}\int_{\mathscr{T}}\frac{\omega(t,\varepsilon)}{t^2(t-w)}\,dt,$$

and where $g(w,\epsilon;\infty)$ *is defined by* (15.6). *The function* $\omega(w)$, $\omega\in\mathscr{T}$, *satisfies* (15.12) *and* (15.11).

If the tree $\mathscr{T}$ *is the "boundary tree" of a function* $f(z)\in S_n$, *then by substituting* $w=f(z)$ *in* (5.13) *we get a function of the form*

$$\epsilon g(z,\epsilon)=\epsilon g(f(z);0)+\epsilon g(f(z),\epsilon;\infty)$$

which is admissible with respect to $f(z)$.

§16. For $x^0\in\partial\mathscr{V}_n$, denote by $\mathscr{T}_{x^0}$ the boundary tree of the function $f(z,x^0)$. In this section we shall examine the neighborhood $\mathfrak{W}=\{x:x\in\partial\mathscr{V}_n, |x-x^0|<\epsilon_0\}$ of the point x^0.

The results we shall give below will be valid when ϵ_0 is a sufficiently small number, and for the sake of brevity we shall not repeat this assumption on ϵ_0 each time we state a result. We shall show that under certain conditions on the trees $\mathscr{T}_{x^0}$ and $\mathscr{T}_x$ it is possible to construct a conformal mapping of the domain $\mathscr{G}(\mathscr{L}_r,\mathscr{T}_{x^0})$ onto some neighborhood of the tree $\mathscr{T}_x$ in such a manner that we get a function of class $U_{\mathscr{T}_{x^0}}$. We assume that $x^0\in\mathscr{R}$ and $d_{n-1}(x^0)\neq 0$. Then $d_{n-1}(x)\neq 0$ if $x\in\mathfrak{W}$. According to Lemma 11.1 the vector $\lambda(x)$ satisfies the Lipschitz condition $|\lambda(x^1)-\lambda(x^2)|<C|x^1-x^2|$ on $\mathfrak{W}$. As before, we set

$$Q(w,x)=\sum_{1}^{n}A_k(x)w^{-k}\text{ and } P(z,x)=\sum_{-n}^{n}B_k(x)z^{-k}.$$

From (10.3) it follows that the functions $A_k(x)$ satisfy a Lipschitz condition on $\mathfrak{W}$. Also it is clear that the functions $B_k(x)$ depend linearly on the coordinates of the vectors $\lambda(x)$ and x. Set m equal to the number of branches of the tree $\mathscr{T}_{x^0}$; also, in the following we assume that the trees of extremal functions have been oriented. Under the mapping of the domain $\mathscr{G}(\mathscr{T}_{x^0})$ onto the disc $|z|<1$ by the function $z=f^{-1}(w,x^0)$, each branch $\gamma_j\subset\mathscr{T}_{x^0}$ corresponds to two arcs of the unit circle, namely

$$I_j=f_+^{-1}\{\gamma_j,x^0\}\quad\text{and}\quad I_j'=f_-^{-1}\{\gamma_j,x^0\}.$$

Thus the circle $|z|=1$ is divided into $2m$ arcs

$$I_1,\cdots,I_m,I_1',\cdots,I_m' \tag{16.1}$$

and between any two arcs I_j and I_j' we have a natural one-to-one relationship, defined by

$$f(z,x^0)=w,\quad f(z',x^0)=w,\quad z\in I_j,\quad z'\in I_j',\quad w\in\gamma_j.$$

It is clear that the end points of the arcs under consideration are images of end points of branches of the tree. If the branch γ_j is a terminal branch, then the arcs I_j and I_j' have a common boundary point—the image of the end point of the tree. Otherwise the arcs I_j and I_j' lie a positive distance apart.

Let $w_1(x^0), w_2(x^0), \cdots$ be the end points (vertices) of the tree $\mathscr{T}_{x^0}$, and denote by $K_{j\rho}$ the open disc $K_{j\rho} = \{w : |w - w_j(x^0)| < \rho\}$. At the base of the tree we take the disc $\{w : |w| > \rho^{-1}\}$. This entire system of discs will be denoted by $\mathscr{K}_\rho$. To this system $\mathscr{K}_\rho$ we add those discs of radius ρ centered at the zeros of $Q(w, x^0)$ which do not belong to $\mathscr{T}_{x^0}$ and which lie in the domain $|w| \geqq 1/16$. We denote this supplemented system of discs by $\mathscr{K}_{\rho^*}$. For $\rho \leqq \rho^*$ any two discs of the system $\mathscr{K}_{\rho^*}$ will be separated by a distance not less than ρ^*, and the distance from $\mathscr{T}_{x^0}$ to any disc $K_{j\rho} \subset \mathscr{K}_{\rho^*} \setminus \mathscr{K}_\rho$ will be at least ρ^*. In what follows, we shall always assume $\rho \leqq \rho^*$. Let $\mathscr{G}_\rho$ be the domain formed by removing from the annulus $1/12 \leqq |w| \leqq \rho^{-1}$ the tree $\mathscr{T}_{x^0}$ and the system of discs $\mathscr{K}_{\rho^*}$. Denote by G_ρ the image of $\mathscr{G}_\rho$ under the mapping $z = f^{-1}(w, x^0)$. If $K_{j\rho} \subset \mathscr{K}_{\rho^*}$ but $w_j(x^0) \notin \mathscr{T}_{x^0}$, then the image of $w_j(x^0)$ will be a zero $z_j(x^0)$ of the function $P(z, x^0)$. It is clear that $f^{-1}\{K_{j\rho}, x^0\} \subset \{z : |z| < 1\}$. If however $w(x^0) \in \mathscr{T}_{x^0}$, then to this point there correspond points lying on $|z| = 1$ which are boundary points of the arcs (16.1). The disc $K_{j\rho}$ divides the branches of $\mathscr{T}_{x^0}$ into curvilinear sectors which we denote by $\sigma_{jl\rho}$, $l = 1, \cdots, p$. It is clear that p is the degree of the vertex $w_j(x^0)$. Thus $w_j(x^0)$ corresponds to p points $z_{jl}(x^0)$, $l = 1, \cdots, p$. We denote the image of the sector $\sigma_{jl\rho}$ by $\tau_{jl\rho}$. All the above remarks also apply to the disc $\{w : |w| > \rho^{-1}\}$. We note that all the zeros of $P(z, x^0)$ which lie on $|z| = 1$ belong to the set of points $z_{jl}(x^0)$. Set $k_{j\delta} = \{z : |z - z_j(x^0)| < \delta\}$, $k_{jl\delta} = \{z : |z - z_{jl}(x^0)| < \delta\}$, and $k_{jl\delta}^* = k_{jl\delta} \cap \{z : |z| \leqq 1\}$. For a given ρ we can find a largest $\delta = \delta(\rho)$ such that

$$k_{j\delta} \subset f^{-1}\{K_{j\rho}, x^0\} \quad \text{and} \quad k_{jl\delta}^* \subset \tau_{jl\rho}. \tag{16.2}$$

Denote by D_ρ the domain formed from $\{z : |z| < 1\}$ by the removal of those domains $\tau_{jl\rho}$, $l = 1, \cdots, p$, which correspond to the vertices of the tree $\mathscr{T}_{x^0}$. We now consider the question of whether the functions $f(z, x)$ converge as $x \to x^0$ to the function $f(z, x^0)$. We first make the following remark.

Remark 16.1. There exists an absolute constant $r_0 > 1/9$ such that $\max_{|z| \in r_0} |f(r, x^0) - f(z, x)| < A_0 |x^0 - x|$ if $x \in \mathfrak{W}$. The constant A_0 is defined in terms of the given neighborhood $\mathfrak{W}$. The proof of these statements is quite easy, and follows from the application of inequality (11.1) and Lemma 11.1 to the functions $f(z, x^0)$ and $f(z, x)$. We note at this point that henceforth various constants denoted by the letters C, M, L, and so forth, will depend on ρ and on the given neighborhood $\mathfrak{W}$. We shall suppress this dependence for the sake of notational ease.

Proposition 16.1. *If* $x \in \mathfrak{W}$, *then the inequality* $|f(z, x^0) - f(z, x)| < c|x^0 - x|$ *holds in* D_ρ.

PROOF. Set $G_\rho^* = G_\rho \cap \{z : r_0 \leqq |z| \leqq 1\}$, and note that if $z \in G_\rho^*$, then $f(z, x^0) \in \mathscr{G}_\rho$, and in addition $|f(z, x^0)| \geqq 1/12 + \vartheta$, where ϑ is an absolute constant. Hence, if $|t| \leqq \rho/2$ and $z \in G_\rho^*$, then

$$f(z, x^0) + t \in \mathscr{G}_{\rho/2}. \tag{16.3}$$

Since we can assume that, with $x \in \mathfrak{W}$, the zeros of $Q(w, x)$ which lie in $|w| \geqq 1/16$ will in fact lie in the discs $K_{j\rho/4}$ and in the disc $\{w : |w| > 4\rho^{-1}\}$, we see that $\min_{w \in \mathscr{G}_{\rho/2}} |Q(w, x)| \geqq C_0 > 0$, where C_0 is independent of x. We then see that for $w \in \mathscr{G}_{\rho/2}$ and $x \in \mathfrak{W}$,

$$| Q^{-1/2}(w, x^0) - Q^{-1/2}(w, x) | < C_1 | x^0 - x |. \tag{16.4}$$

Similar remarks are true for the function $P(z, x)$. To see this, we can assume that all the zeros of $P(z, x)$ which lie in the annulus $r_0 \leqq |z| \leqq 1$ are actually located in the discs $K_{j\delta/2}$ and $K_{jl\delta/2}^*$. From (16.2) it then follows that $\min_{z \in G_\rho^*} | P(z, x) | \geqq C_0^*$, where C_0^* is independent of x. Thus for $x \in G_\rho^*$ and $x \in \mathfrak{W}$,

$$| P^{1/2}(z, x^0) - P^{1/2}(z, x) | < C_2 | x^0 - x |. \tag{16.5}$$

Let z_1 be any point of the boundary of D_ρ. If $z_1 r_0 / |z_1| \in G_\rho^*$, we set $z_0 = z_1 r_0 / |z_1|$. If $z_1 r_0 / |z_1| \notin G_\rho^*$, then we take z_0 to be the nearest point, with $|z_0| = r$, to the domain G_ρ^*. We join z_0 and z_1 by a polygonal line lying in G_ρ^*. It is clear that such a curve can be chosen so that its length does not exceed some constant L which is independent of z_1. We set $t(z) = f(z, x) - f(z, x^0)$, and we shall use the equation satisfied by $f(z, x)$ to write down a differential equation for $t(z)$. It is clear from Remark 16.1 that with an appropriate definition of the image of the neighborhood $\mathfrak{W}$ we have $|t(z_0)| < \rho/2$. The following transformations will be defined on that subset $E \subset \lambda$ for which we have $|t(z)| \leqq \rho/2$. Recalling (16.3) and the above remarks, we see that

$$Dt(z) = -Df(z, x^0) + (f(z, x^0) + t(z)) i P^{1/2}(z, x) Q^{-1/2}(f + t, x).$$

Set

$$R(z) = i(f(z, x^0) + t(z)) [P^{1/2}(z, x) Q^{-1/2}(f + t, x) - P^{1/2}(z, x^0) \times Q^{-1/2}(f + t, x^0)].$$

Then if $z \in E$, it follows from (16.4) and (16.5) that

$$|R(z)| < M_0 |x^0 - x|.$$

Set

$$Q^{-1/2}(w + t, x^0) = Q^{-1/2}(w, x^0) + tQ_1(w, t)$$

and

$$Q^{-1/2}(w + t, x^0) = Q^{-1/2}(w, x^0) - \frac{t}{2} \frac{Q'(w, x^0)}{Q^{3/2}(w, x^0)} + t^2 Q_2(w, t).$$

If $w \in \mathscr{G}_\rho$ and $|t| \leqq \rho/2$, then

(16.6) $$|Q_1| \leqq M, \qquad |Q_2| \leqq M_2.$$

Recalling the relation

$$\frac{Df(z, x^0)}{f(z, x^0)} = iP^{1/2}(z, x^0)\,Q^{-1/2}(f(z, x^0), x^0)$$

and suppressing the variable x^0 in the notation, we have the following equation for $t(z)$:

(16.7) $$Dt(z) = \left[\frac{Df(z)}{f(z)} - \frac{1}{2}Df(z)\frac{Q'(f(z))}{Q^{3/2}(f(z))}\right]t(z) + S(t, z)t^2 + R(z),$$

where $S(t,z) = iP^{1/2}(z)[Q_1 + f(z)Q_2]$. From the estimate (16.6) it follows that for $z \in E$

$$|S(t,z)| \leqq M.$$

Solving (16.7) as a nonhomogeneous linear equation, we get

(16.8) $$t(z) = \frac{f(z)}{Q^{1/2}(f(z))}\left\{\int_{z_0}^{z}\frac{Q^{1/2}(f(z))}{f(z)}[S(t, z)t^2(z) + R(z)]\frac{dz}{z} + t(z_0)\frac{Q^{1/2}(f(z_0))}{f(z_0)}\right\}.$$

If $w \in \mathscr{G}_\rho$, then

$$\mu \leqslant \left|\frac{Q^{1/2}(w, x^0)}{w}\right| \leqslant \mu^{-1}, \qquad \mu \leqslant 1.$$

Set $A = (A_0 + M_0Lr_0^{-1})\mu^{-2}$, $A > A_0$ and $B = MLr_0^{-1}\mu^2$. We assume $|x^0 - x| \leqq \min\{\rho/4A, 1/8AB\}$. Also assume that on the arc $(z_0, z_*) \subset \lambda$ the inequality $|t(z)| \leqq 2A|x^0 - x|$ is satisfied if $z \neq z_*$. Setting $z = z_*$ in (16.8), we deduce that

$$|t(z_*)| \leqslant 4A^2B|x^0 - x|^2 + A|x^0 - x| \leqslant \frac{1}{2}A|x^0 - x| + A|x^0 - x|$$
$$= \frac{3}{2}A|x^0 - x|.$$

Since $|t(z_0)| < 2A|x^0 - x|$ and $|t(z)|$ is continuous on λ, we conclude that $|t(z)| < 2A|x^0 - x|$ for $z \in \lambda$, and so $|t(z)| < 2A|x^0 - x|$, as desired. Q.E.D.

We now estimate $|f(z, x^0) - f(z, x)|$ in a neighborhood of a terminal vertex of the tree $\mathscr{T}_{x^0}$. Let $w_j(x^0)$ be such a vertex, with $z_j(x^0)$ its image; we assume that $Q(w_j(x^0), x^0) \neq 0$. Then $z_j(x^0)$ is a double zero of $P(z, x^0)$. Let $\partial_0\tau_{j\rho} = \partial\tau_{j\rho} \cap \{z : |z| < 1\}$. The arc $\partial_0\tau_{j\rho}$ meets $|z| = 1$ at two points z_j^1 and z_j^2. Let l be the image of $\partial_0\tau_{j\rho}$ under the mapping $w = f(z, x)$. From Proposition 16.1 it follows that the Jordan curve l lies in an annulus bounded by the circles $|w - w_j| = \rho - C|x^0 - x|$ and $|w - w_1| = \rho + C|x^0 - x|$. If $|x^0 - x|$ is sufficiently small, it is clear that $w_j^1 = f(z_j^1, x) \in \mathscr{T}_x$, $w_j^2 = f(z_j^2, x) \in \mathscr{T}_x$, and

$$|w_j^1 - w_j^2| < 2C\epsilon.$$

Let γ_{0j} be a branch of $\mathscr{T}_{x^0}$ beginning at w_j, and let γ_j be a branch of $\mathscr{T}_x$ containing the point w_j^1. Let Γ_{0j} and Γ_j be the trajectories of the differentials $Q(w, x^0)(dw/w)^2$ and $Q(w, x)(dw/w)^2$, respectively, where $\gamma_{0j} \subset \Gamma_{0j}$ and $\gamma_j \subseteq$

Γ_j. Since $Q(w,x)\neq 0$ in $K_{j2\rho}$, we see that for $|x^0-x|$ small the arcs $K_{j2\rho}\cap \Gamma_{0j}$ and $K_{j2\rho}\cap\Gamma_j$ will be as near to each other as desired. This is so since the solutions of an equation of the form $dw/ds=F(w)$ depend continuously on the initial conditions and on the right-hand side. Since Γ_{0j} intersects $\partial K_{j\rho}$ in at most two points, $w_j^0=f(z_j^1,x^0)=f(z_j^2,x^0)$ and w^*, also Γ_j intersects $\partial K_{j\rho}$ in two points, one of which is very close to w^*. Thus Γ_j intersects the arc l at a point $\widetilde{w}$, near to w^*. Since $|f^{-1}(\widetilde{w},x)|<1$, we have $\widetilde{w}\notin\mathscr{T}_x$. Thus Γ_j contains a terminal vertex $w_j(x)$ of $\mathscr{T}_x$ which lies between the points $\widetilde{w}$ and w_j^1. The image $z_j(x)$ of the vertex $w_j(x)$ is a double zero of $P(z,x)$, and it is clear that $z_j(x)\in\tau_{j\rho}$. Thus we conclude that $P(z,x)$ has a double zero $z_j(x)$ such that $z_j(x)\to z_j(x^0)$ as $x\to x^0$.

Proposition 16.2. *If $w_j(x^0)$ is a terminal vertex, and if $Q(w_j(x^0),x^0)\neq 0$, then the inequality $|f(z,x^0)-f(z,x)|<C|x^0-x|$ holds on $\tau_{j\rho}$.*

Proof. It is easy to see that there exists a disc $|z-z_j(x^0)|\leq\delta_0$ in which the function $f(z,x)$ is regular, provided $\epsilon=|x^0-x|$ is small. Assuming ρ sufficiently small, we consider in this disc the line $\lambda=\{z:|f(z,x^0)-f(z_j(x^0),x^0)|=\rho\}$. It is clear that, for small ρ, λ is homeomorphic to a circle. Let λ_1 be that part of λ lying in the disc $|z|<1$, and set $\lambda_2=\lambda\backslash\lambda_1$. Note that for small ϵ the distance from $z_j(x)$ to λ is bounded away from zero. Applying the remarks of Proposition 16.1 directly to the function $f(z,x)$ and the path λ_2, we see that the desired inequality is proved.

Now we shall consider not arbitrary points $x\in\mathfrak{W}$, but only those for which the tree $\mathscr{T}_x$ is isomorphic to the tree $\mathscr{T}_{x^0}$. If $|x^0-x|$ is small, the multiplicity of the zero $w_j(x^0)$ will equal the number of zeros of $Q(w,x)$ in each disc $K_{j\rho}$. For simplicity these zeros will be called the zeros corresponding to $w_j(x^0)$. Set

$$Q(w,x^0)=\sum_{k=k_0}^{n}A_k(x^0)w^{-k},\qquad A_1(x^0)=\dots=A_{k_0-1}(x^0)=0,\qquad A_{k_0}(x^0)\neq 0.$$

Definition 16.1. The set $\mathfrak{D}_{x^0}\subset\mathfrak{W}$ consists of precisely those points $x\in\mathfrak{W}$ satisfying the following conditions:

1. $A_1(x)=\cdots=A_{k_0-1}(x)=0$, $A_{k_0}(x)\neq 0$, and the multiplicity of the base of $\mathscr{T}_x$ is the same as the multiplicity of the base of $\mathscr{T}_{x^0}$.
2. If $w_j(x^0)$ is a vertex of $\mathscr{T}_{x^0}$ which is also a zero of $Q(w,x^0)$, then the zeros corresponding to $w_j(x)$ have the same multiplicity as $w_j(x^0)$ and they are vertices of the tree $\mathscr{T}_x$. The multiplicities of $w_j(x^0)$ and $w_j(x)$ are the same.
3. If the terminal vertex $w_j(x^0)$ is a zero of $Q(w,x^0)$, then the zeros corresponding to $w_j(x)$ have the same multiplicity and are terminal vertices of $\mathscr{T}_x$.

If $x\in\mathfrak{D}_{x^0}$, then a natural one-to-one correspondence exists between the nonterminal vertices of the trees $\mathscr{T}_x$ and $\mathscr{T}_{x^0}$. Definition 16.1 and Proposition 16.2 establish a correspondence between the terminal vertices.

Proposition 16.3. *If $w_j(x)$ and $w_j(x^0)$ are corresponding vertices of the trees $\mathscr{T}_x$, $x \in \mathfrak{D}_{x^0}$, and $\mathscr{T}_{x^0}$, then*

$$|w_j(x) - w_j(x^0)| < C|x - x^0|, \tag{16.9}$$

where C is independent of x.

Proof. Let the vertex $w_j(x^0)$ be a zero of $Q(w, x^0)$ of multiplicity k. Then

$$w_j(x^0) - w_j(x) = \frac{1}{2\pi i k} \int\limits_{|w - w_j(x^0)| = \rho} \left[\frac{Q'(w, x^0)}{Q(w, x^0)} - \frac{Q'(w, x)}{Q(w, x)} \right] w \, dw.$$

From this we see that $|w_j(x^0) - w_j(x)| < C|x^0 - x|$. Now suppose that $w_j(x^0)$ is not a zero of $Q(w, x^0)$. Then $w_j(x^0)$ is a terminal vertex, and its image $z_{j1}(x^0)$ is a double zero of $P(z, x^0)$. According to Proposition 16.2, the vertex $w_j(x^0)$ is associated with a terminal vertex $w_j(x)$. Note that $w_j(x^0) = f(z_{j1}(x^0), x^0)$ and $w_j(x) = f(z_{j1}(x), x)$.

Thus, in accordance with Proposition 16.2,

$$\begin{aligned} |w_j(x^0) - w_j(x)| &< |f(z_{j1}(x^0), x^0) - f(z_{j1}(x), x^0)| + C|x^0 - x| \\ &< C_1 |z_{j1}(x^0) - z_{j1}(x)| + C|x^0 - x|. \end{aligned}$$

But $z_{j1}(x)$ is a double zero, and, as above, it is easy to show that $|z_{j1}(x^0) - z_{j1}(x)| < C_2|x^0 - x|$. From this (16.9) follows in the case of terminal vertices.

We shall denote the branches of the tree $\mathscr{T}_{x^0}$ by $\gamma_j(x^0)$. If $x \in \mathfrak{D}_{x^0}$, then the corresponding branch of $\mathscr{T}_x$ will be denoted by $\gamma_j(x)$. Set $l_j(x)$ equal to the length of the branch $\gamma_j(x)$, computed with respect to the metric $|d\zeta|^2 = |Q(w, x)(dw/w)^2|$.

Proposition 16.4. *If $x \in \mathfrak{D}_{x^0}$, then $|l_j(x^0) - l_j(x)| < C_0|x^0 - x|$.*

Proof. Suppose that the branch $\gamma_j(x)$ has as its end points the vertices $w_j(x)$ and $w_{j+1}(x)$, and that $\gamma_j(x^0)$ has as its end points $w_j(x^0)$ and $w_{j+1}(x^0)$. By definition,

$$l_j(x) = \int\limits_{\gamma_j(x)} \left| Q^{1/2}(w, x) \frac{dw}{w} \right|.$$

We make the change of variable

$$w = \frac{w_{j+1}(x) - w_j(x)}{w_{j+1}(x^0) - w_j(x^0)} (t - w_j(x^0)) + w_j(x) = \alpha t + \beta.$$

From Proposition 16.3 we have that $|\alpha - 1| < C|x^0 - x|$ and $|\beta| < C|x^0 - x|$. If $\tilde{\gamma}_j(x)$ is the image of $\gamma_j(x)$ on the t-sphere, then

$$l_j(x) = \int\limits_{\tilde{\gamma}_j(x)} \left| Q^{1/2}(\alpha t + \beta, x) \frac{\alpha \, dt}{\alpha t + \beta} \right|.$$

The arc $\tilde{\gamma}_j(x)$ joins the points $w_j(x^0)$ and $w_{j+1}(x^0)$. If in the latter integral the path of integration is changed to $\gamma_j(x^0)$, we see that

$$l_j(x) < \int_{\gamma_j(x^0)} \left| Q^{1/2}(\alpha t+\beta, x)\frac{\alpha\, dt}{\alpha t+\beta}\right|.$$

Therefore

$$l_j(x) - l_j(x^0) < \int_{\gamma_j(x^0)} \left| \left[Q^{1/2}(\alpha t+\beta, x)\frac{\alpha}{\alpha t+\beta} - Q^{1/2}(t, x^0)\frac{1}{t}\right] dt\right|.$$

Let $w_j(x^0)$ be a zero with multiplicity $k \geqq 0$, and $w_{j+1}(x^0)$ a zero with multiplicity l. It is easily seen that on $\gamma_j(x^0)$

$$\left| Q(t, x^0) - Q(\alpha t+\beta, x)\left(\frac{\alpha t}{\alpha t+\beta}\right)^2\right| < C|x^0 - x|\,|t - w_j(x^0)|^k\,|t - w_{j+1}(x^0)|^l. \tag{16.10}$$

It follows from this that

$$\left| Q^{1/2}(t, x^0) - Q^{1/2}(\alpha t+\beta, x)\frac{\alpha t}{\alpha t+\beta}\right| < C_1|x^0 - x|.$$

Thus we conclude that $l_j(x) - l_j(x^0) < C_2|x^0 - x|$; reversing the roles of $\gamma_j(x)$ and $\gamma_j(x^0)$, we also have $l_j(x^0) - l_j(x) < C_3|x^0 - x|$. If the branch $\gamma_j(x)$ happens to begin at the base of the tree, the proof proceeds along similar lines, the only change being that the change of variable has the form $w = t+\beta$, $\gamma = w_j(x) - w_j(x^0)$, and inequality (16.10) then becomes

$$\left| Q(t, x^0) - Q(\alpha t+\beta, x)\left(\frac{t}{t+\beta}\right)^2\right| < C|x^0 - x|\,|t - w_j(x^0)|^k\,|t|^{-l}.$$

This completes the proof of the proposition.

We now proceed to the construction of a conformal mapping of a neighborhood of the tree $\mathscr{T}_{x^0}$ onto a neighborhood of the tree $\mathscr{T}_x$, provided $x \in \mathfrak{O}_{x^0}$. Suppose that there are no zeros of $P(z, x^0)$ in the annulus $r_1 \leqq |z| < 1$, and set $r = (1+r_1)/2$. All our construction will take place in the domain $\mathscr{G}(\mathscr{C}_r, \mathscr{T}_{x^0})$. We set $\epsilon = |x^0 - x|$ and

$$l_j(x) - l_j(x^0) = \mu_j\epsilon, \qquad j = 1, \cdots, m.$$

From Proposition 16.4 we have $|\mu_j| \leqq C_0$, $j = 1, \cdots, m$, where C_0 is independent of x, provided $|x^0 - x| \leqq \epsilon_0$. We shall orient the trees $\mathscr{T}_{x^0}$ and $\mathscr{T}_x$ so that the orientation of isomorphic branches coincides. The orientation of each arc (16.1) will be the same as the orientation induced by $|z| = 1$. It is clear that

$$l_j(x^0) = \int_{\gamma_j(x^0)} \left| Q^{1/2}(w, x^0)\frac{dw}{w}\right| = \frac{1}{i}\int_{I_j} P^{1/2}(z, x^0)\frac{dz}{z} = \frac{1}{i}\int_{I_j'} P^{1/2}(z, x^0)\frac{dz}{z},$$

if the branch of the function $P^{1/2}(z, x^0)$ is defined on I_j and I_j' in an appropriate way.

Proposition 16.5. *There exists a function* $\chi(z)$, *regular in the annulus* $r_1 \leqq |z| \leqq r_1^{-1}$ *and assuming real values on* $|z| = 1$, *having a tenth order zero at the end points of the intervals* (16.1), *and satisfying*

$$\frac{1}{i}\int_{I_j} P^{1/2}(z, x^0)\chi(z)\frac{dz}{z} = \frac{1}{i}\int_{I_j'} P^{1/2}(z, x^0)\chi(z)\frac{dz}{z} = \mu_j, \qquad j = 1, \ldots, m.$$

Also, if $r_1 \leqq |z| \leqq r_1^{-1}$, *then* $|\chi(z)| \leqq A$, *where A is a constant depending only on* x^0.

Proof. Set $q(x) = \prod_j [(z - z_j)^2/zz_j]^5$, where the product is taken over the set of all end points of the intervals I_j, I_j'. On the circle $|z| = 1$ we consider the system of functions

$$\varphi_j(z) = \alpha_j \begin{cases} P^{1/2}(z, x^0)\, q(z), & \text{if } z \in I_j \\ 0, & \text{if } z \notin I_j \end{cases},$$

$$\varphi_{j+m}(z) = \alpha_{j+m} \begin{cases} P^{1/2}(z, x^0)\, q(z), & \text{if } z \in I_j', \\ 0 \quad , & \text{if } z \notin I_j', \end{cases}$$

where α_j is chosen so that the functions $\varphi_j(z)$ form an orthonormal system (their orthogonality is clear). Let $(f, g) = \int_0^{2\pi} f(e^{i\theta})g(e^{i\theta})\,d\theta$ be the scalar product in the real Hilbert space of functions square-integrable on $|z| = 1$. We construct real trigonometric polynomials $t_j(z)$, $j = 1, \cdots, m$, so that

$$\|\varphi_j(z) - t_j(z)\| \leqq 1/4m, \qquad j = 1, 2, \cdots, 2m.$$

It is clear that the degrees of these polynomials will be bounded by some number M, and M can be easily estimated by Jensen's theorem and the fact that $\varphi_j(z)$ has a bounded tenth derivative. It is also clear that

$$(\varphi_j, t_k) = \delta_{jk} + \beta_{jk},$$

where $|\beta_{jk}| \leqq 1/4m$. Set $\chi_0(z) = \sum_1^{2m} u_j t_j(z)$ and define u_j so that

$$(\varphi_j, \chi_0) = \alpha_j\mu_j, \qquad (\varphi_{j+m}, \chi_0) = \alpha_{j+m}\mu_j, \qquad j = 1, 2, \ldots, m,$$

or, equivalently, so that

$$\sum_{k=1}^{2m} u_k(\delta_{jk} + \beta_{jk}) = \alpha_j\mu_j, \quad \sum_{k=1}^{2m} u_k(\delta_{m+jk} + \beta_{m+jk}) = \alpha_{j+m}\mu_j,$$

where $j = 1, \cdots, m$. The inverse of the matrix of this system has norm at most 2, as follows from the inequalities satisfied by the β_{jk}. Thus

$$\left(\sum_1^{2m} u_k^2\right)^{1/2} < 2\left(\sum_1^{m} \mu_i^2(\alpha_j^2 + \alpha_{j+m}^2)\right)^{1/2} < C_1.$$

Set $\chi(z) = q(z)\chi_0(z)$. Clearly $\chi(z)$ satisfies all the conditions of the lemma, since $|\chi(z)| \leqq C_2$ for $|z| = 1$, which implies

$$|\chi(z)| \leqq C_2 r_1^{-(M+10m)},$$

for $r_1 \leqq |z| \leqq r_1^{-1}$. Q.E.D.

We now set $\xi(w) = \chi[f^{-1}(w, x^0)]$. The function $\xi(w)$ is regular and bounded in the domain $\mathscr{G}(\mathscr{L}_r, \mathscr{T}_{x^0})$, and by construction it is real-valued on the tree $\mathscr{T}_{x^0}$. Since $\chi(z)$ is regular on the circle $|z| = 1$, and since $f^{-1}(w, x^0)$ can be continued analytically across each branch of the tree from both the positive and negative sides of the branch, it follows that $\xi(w)$ can be analytically continued in the same way. The singular points of $\xi(w)$ will be the vertices of the tree. We shall examine in some detail the behavior of $\xi(w)$ in a neighborhood of a vertex of $\mathscr{T}_{x^0}$. For the remainder of this section we shall fix the quantity ρ, and it will not be displayed explicitly in each index. Let $K_j \subset \mathscr{K}$, and let $\sigma_{j1}, \cdots, \sigma_{jp}$ be the curvilinear sectors into which the branches of $\mathscr{T}_{x^0}$ divide K_j. Consider the function $\xi(w)$ in the sector σ_{jl}. Let τ_{jl} be the image of σ_{jl}, and $z_{jl}(x^0)$ the image of $w_j(x^0)$. The point $z_{jl}(x^0)$ is an algebraic critical point of the function $f(z, x^0)$, and in some neighborhood of $z_{jl}(x^0)$ we have the expansion

$$f(z) = w_j(x^0) + b_{l0}(z - z_{jl})^{\alpha_{jl}} + \dots,$$

where α_{jl} satisfies the condition $-2 \leqq \alpha_{jl} \leqq 2$. In σ_{jl} we have

$$z = z_{jl} = C_{l0}(w - w_j(x^0))^{\beta_{jl}} + \dots,$$

where $\beta_{jl} \geqq \frac{1}{2}$ or $\beta_{jl} \leqq -\frac{1}{2}$, and $z \in \tau_{jl}$. Since z_{jl} is a tenth order zero of $\chi(z)$, we see that for $\beta_{jl} \geqq \frac{1}{2}$

$$|\xi^{(s)}(w)| \leqslant C|w - w_j(x)|^{5-s}, \quad s = 0, \dots, 4, \quad w \in \sigma_{jl}. \tag{16.11}$$

If $\beta_{jl} \leqq -\frac{1}{2}$, then $w_j(x^0) = \infty$, and in this case we have in place of (16.11) that

$$|\xi^{(s)}(w)| \leqslant C|w|^{-5}, \quad s = 0, \dots, 4. \tag{16.11'}$$

The constant C in these inequalities can be chosen to be independent of j and l.

We now have need of a fairly simple technical proposition.

Proposition 16.6. *Let the function $\varphi(u) = u + \cdots$ be regular in $|u| < R$, and suppose $|\varphi(u)| < M$ for $|u| < R$. Then $\varphi(u)$ is univalent in the disc $|u| < R^2/4M$. The equation $\varphi(u) = a$ has a unique solution if $|a| < R^2/16M$, and moreover the solution u satisfies $|u| < 4|a|$.*

Proof. Let $\varphi(u) = \sum_0^\infty b_k u^{k+1}$, $b_0 = 1$. Then $|b_k| < MR^{-k-1}$, $k = 1, 2, \cdots$. For $|u_1|, |u_2| \leqq r$ we have

$$|\varphi(u_1) - \varphi(u_2)| \geqslant |u_1 - u_2| \left(1 - MR^{-1} \sum_1^\infty (k+1)(rR^{-1})^k\right).$$

If $r \leqq R^2/4M$ then $r/R \leqq R/4M < 1/4$, since $1 < MR^{-1}$ by Schwarz' Lemma. Therefore

$$|\varphi(u_1)-\varphi(u_2)|\geqslant|u_1-u_2|\left(1-\frac{1}{2}\left(1-\frac{r}{2R}\right)\left(1-\frac{r}{R}\right)^{-2}\right)\geqslant\frac{2}{9}|u_1-u_2|.$$

Set $\vartheta=R^2/4M$, $u=\vartheta z$ and $\psi(z)=\vartheta^{-1}\varphi(\vartheta z)$. The function $\psi(z)\in S$, and so if $|w|<1/4$, Koebe's 1/4 theorem insures that the equation $\psi(z)=w$ has a unique solution. Also, $|z|<4|w|$. Setting $w=a\vartheta^{-1}$, we arrive at the conclusion of the proposition. Q.E.D.

We shall attempt to find a function mapping a neighborhood of $\mathscr{T}_{x^0}$ conformally onto a neighborhood of $\mathscr{T}_x$ by examining the solutions of the equation

$$Q(t,x)\left(\frac{dt}{t}\right)^2=Q(w,x^0)\,(1+\varepsilon\xi(w))^2\left(\frac{dw}{w}\right)^2. \tag{16.12}$$

All our constructions and analysis will take place in the domain $\mathscr{G}(\mathscr{L}_{r_0},\mathscr{T}_{x^0})$. We take the set of discs $\mathscr{K}_\rho$ and we fix ρ; below we set $\rho_0=\rho/2$. By the definition of r_1, the function $Q(w,x^0)$ does not vanish in $\mathscr{G}(\mathscr{L}_{r_0},\mathscr{T}_{x^0})\setminus\mathscr{K}_{\rho_0}$. Thus, if $\epsilon=|x-x^0|\leqq\epsilon_0$, then

$$\left|\frac{Q(w,x)}{w^2}\right|\geqslant C_0,\quad w\in\mathscr{G}(\mathscr{C}_{r_0},\mathscr{T}_{x^0})\setminus\mathscr{K}_{\rho_0}. \tag{16.13}$$

In particular, (16.13) holds if $x=x^0$. Since the inequality

$$\left|\frac{Q(w,x)}{w^2}-\frac{Q(w,x^0)}{w^2}\right|<C\varepsilon$$

is also true, we see that

$$\left|\sqrt{\frac{Q(w,x)}{w^2}}-\sqrt{\frac{Q(w,x^0)}{w^2}}\right|<C_1\varepsilon,\quad w\in\mathscr{G}(\mathscr{C}_{r_0},\mathscr{T}_{x^0})\setminus\mathscr{K}_{\rho_0}, \tag{16.14}$$

provided we take the same branch of the square root. Making the a priori statement that the solution of equation (16.12) satisfies the condition $|t-w|\leqq\delta$, where δ is sufficiently small, we are led to the equation

$$\left[\frac{Q(t,x)}{t^2}\right]^{1/2}dt=\left[\frac{Q(w,x^0)}{w^2}\right]^{1/2}(1+\varepsilon\xi(w))\,dw, \tag{16.15}$$

which we shall investigate in the domain $\mathscr{G}(\mathscr{L}_{r_0},\mathscr{T}_{x^0})$.

Lemma 16.1. *Suppose $x^0\in\mathscr{R}$ and $d_{n-1}(x^0)>0$. If $\epsilon\leqq\epsilon_0$ and $x\in\mathfrak{O}_{x^0}$, then in the domain $\mathscr{G}(\mathscr{L}_r,\mathscr{T}_{x^0})$, where r is independent of x, there exists a univalent function $\varphi(w,\epsilon)\in U_{\mathscr{T}_{x^0}}$ which maps $\mathscr{G}(\mathscr{L}_r,\mathscr{T}_{x^0})$ onto a doubly connected domain, one of whose boundary components is the tree $\mathscr{T}_x$. Under this mapping the branches of $\mathscr{T}_{x^0}$ will be isomorphic to the branches of $\mathscr{T}_x$, and the vertices of $\mathscr{T}_{x^0}$ will be isomorphic to the vertices of $\mathscr{T}_x$.*

The proof will be divided into four steps.

Step 1. We consider the solution of Cauchy's problem for equation (16.15) in an arbitrary neighborhood of a point $w\in\mathscr{G}(\mathscr{L}_{r_0},\mathscr{T}_{x^0})\setminus\mathscr{K}_\rho$. By construction, the function $\xi(w)$ can be analytically continued across any branch of $\mathscr{T}_{x^0}$, and the continuation can be made from either side of the branch.

Thus for any point w_0 in the closure of $\mathscr{G}(\mathscr{L}_{r_0}, \mathscr{T}_{x^0})\setminus\mathscr{K}_\rho$ we can find a disc of radius $2\rho^*$ centered at w_0 such that $\xi(w)$ is regular in this disc. Moreover, if w_0 lies on $\mathscr{T}_{x^0}$, then we distinguish two cases: when it "lies" on the left, the continuation of $\xi(w)$ is carried out from the left side to the right, and when it "lies" on the right side, the continuation is made in the opposite direction. We also require that $2\rho^*$ be chosen so that any disc as above lies in the complement of $\mathscr{K}_{\rho_0}$. Such a choice of ρ^* can be made since the closure of $\mathscr{G}(\mathscr{L}_{r_0}, \mathscr{T}_{x^0})\setminus\mathscr{K}_\rho$ is compact.

We shall construct a solution in the disc $|w-w_0|\leqq\rho^*$. Let the initial conditions satisfy $|t(w_0)-w_0|\leqq\delta$, where δ is sufficiently small. Set $t=w+\eta(w)$; then we change (16.15) into the integral equation

$$\int_{t_0}^{w+\eta}\left[\frac{Q(v,x)}{v^2}\right]^{1/2}dv=\int_{w_0}^{w}\left[\frac{Q(u,x^0)}{u^2}\right]^{1/2}(1+\varepsilon\xi(u))\,du,$$

where $t_0=t(w_0)$, and the integration is along a line segment. (For this it is necessary that the points t_0 and $w+\eta$ lie in the disc $|w-w_0|<2\rho^*$, and this is satisfied ex post facto.) The latter equation can now be written in the form

$$\int_{w}^{w+\eta}\left[\frac{Q(v,x)\,w^2}{Q(w,x)\,v^2}\right]^{1/2}dv=G(w),$$

where

$$G(w)=\left[\frac{Q(w,x)}{w^2}\right]^{-1/2}\left\{\int_{w_0}^{w}\left(\left[\frac{Q(u,x^0)}{u^2}\right]^{1/2}-\left[\frac{Q(u,x)}{u^2}\right]^{1/2}\right)du\right.$$
$$\left.+\varepsilon\int_{w_0}^{w}\left[\frac{Q(u,x^0)}{u^2}\right]^{1/2}\xi(u)\,du-\int_{w_0}^{t_0}\left[\frac{Q(u,x)}{u^2}\right]^{1/2}du\right\}.$$

Taking into account (16.13), (16.14), and the estimate

$$|Q(w,x)/w^2|\leqq C_2,$$

which holds for $|w|\geqq 1/16$ (since all our work is taking place in a neighborhood of the tree $\mathscr{T}_{x^0}$, the domain $\mathscr{G}(\mathscr{L}_{r_0}, \mathscr{T}_{x^0})$ is obviously contained in the above-mentioned domain), we see that

$$|G(w)|<C_0^{-\frac{1}{2}}\left[\varepsilon\rho^*\left(C_1+AC_2^{\frac{1}{2}}\right)+C_2^{\frac{1}{2}}\,|t_0-w_0|\right]=C_3.$$

The function

$$F(w,\eta)=\int_{w}^{w+\eta}\left[\frac{Q(v,x)\,w^2}{Q(w,x)\,v^2}\right]^{1/2}dv,$$

satisfies $|F(w,\eta)|<(C_2/C_0)^{1/2}\rho^*$ for $|w-w_0|\leqq\rho^*$ and $|\eta|\leqq\rho^*$. Since $F_\eta(w,0)=1$, Proposition 16.6 implies that this function is univalent as a function of η in the disc $|\eta|<(\rho^*/4)(C_0/C_2)^{1/2}$, and as a result we have that for $C_3<(\rho^*/16)(C_0/C_2)^{1/2}$, our equation has a solution which satisfies

$$|\eta(w)| < 4C_3.$$

It is clear that $\eta(w)$ and $t(w)$ are regular functions in the disc $|w - w_0| \leqq \rho^*$. Assuming that $|t_0 - w_0| \leqq C\epsilon$, we set $C_3 = C_4\epsilon$; it follows that for $\epsilon \leqq \epsilon_0$ the previous conditions will be satisfied, and if $|w - w_0| \leqq \rho^*$ we also have

$$|t(w) - w| \leqq 4C_4\epsilon.$$

Upon writing (16.15) in the form

$$\frac{dt(w)}{dw} - 1 = \left[\frac{Q(w, x^0)\, t^2}{Q(t, x)\, w^2}\right]^{1/2} (1 + \varepsilon\xi(w)) - 1,$$

we see from the last inequality that $|dt(w)/dw - 1| < C_5\epsilon$. Also, by repeated differentiation of the above equation, we can derive the following estimates for higher derivatives:

$$|(d^s/dw^s)^s[t(w) - w]| < C_6\epsilon, \qquad 0 \leqq s \leqq 4, \tag{16.16}$$

provided $|w - w_0| \leqq \rho^*$.

Let l be any smooth arc lying in $\mathscr{G}(\mathscr{L}_{r_0}, \mathscr{T}_{x^0}) \setminus \mathscr{K}_0$ which does not pass through any branch of $\mathscr{T}_{x^0}$; let w_0 and w_1 be the end points of l. The results we have derived so far allow us to construct a solution of Cauchy's problem for the equation (16.15) in some neighborhood of l. We assume that $|t(w_0) - w_0| \leqq C\epsilon$. On l we choose points $w_0, w_1, \cdots, w_N = w_1$ so that

$$|w_j - w_{j-1}| = \frac{1}{2}\rho^*, \ 1 \leqslant j \leqslant N-1, \text{ and } |w_1 - w_{N-1}| \leqslant \frac{1}{2}\rho^*.$$

We now construct a system of discs of radius ρ^* with centers at the points w_j. Having defined the solution in the jth disc, we can continue it analytically into the $(j+1)$th disc (since the discs overlap), and we can then use the value of the solution at the center of the $(j+1)$th disc as our initial condition for the construction of a solution in the $(j+1)$th disc. By the theorem on the uniqueness of a solution of a differential equation, the solution we construct in the $(j+1)$th disc will be the analytic continuation of the solution in the jth disc. If ϵ is sufficiently small, we get as a result a function analytic in the union of all the discs. We apply this construction to the branches of the tree $\mathscr{T}_{x^0}$, keeping in mind that in this case we have two cases, corresponding to the two sides of the branches. Since $\mathscr{T}_{x^0}$ has only finitely many branches, the number N of discs involved in the construction will be bounded, and so a similar number of analytic continuations will be required, if $\epsilon \leqq \epsilon_0$. It is clear that inequality (16.16) continues to hold for the solutions constructed as described above.

Finally, we make the following important remark. The solution $t(w)$ which we have constructed will be a univalent function; this follows from (16.16), provided ϵ is sufficiently small. Therefore, if $t(w_0) \in \mathscr{T}_x$ and $w_0 \in \mathscr{T}_{x^0}$, in view of equation (16.15) we see that the function $t(w)$ will map an arc of a branch of $\mathscr{T}_{x^0}$ to an arc of some branch of $\mathscr{T}_x$.

Step 2. We next construct the solution of (16.12) in the curvilinear sectors σ_{jl} formed by the disc $K_{j\rho}$ and the branches of the tree $\mathcal{T}_{x^0}$. Let $w_j(x^0)$ be any vertex of $\mathcal{T}_{x^0}$, and let $w_j(x)$ be the corresponding vertex of $\mathcal{T}_x$. Let k be the multiplicity of the zero $w_j(x^0)$ of the function $Q(w,x^0)$. Set $u = w - w_j(x^0)$ and $v = t - w_j(x)$; if $w_j(x^0) = \infty$, set $u = w^{-1}$ and $v = t^{-1}$. Consider the functions

$$u^k q_0(u) = \frac{Q(w_j(x^0)+u,\, x^0)}{(w_j(x^0)+u)^2}, \qquad v^k q(v) = \frac{Q(w_j(x)+v,\, x)}{(w_j(x)+v)^2}$$

if $w_j(x^0) \neq \infty$, and

$$u^k q_0(u) = Q\left(\frac{1}{u},\, x^0\right), \qquad v^k q(v) = Q\left(\frac{1}{v},\, x\right)$$

if $w_j(x^0) = \infty$. Set $\xi_0(u) = \xi[w_j(x^0)+u]$. Writing (16.12) in terms of our new variable, we find that

$$v^k q(v)\,dv^2 = u^k q_0(u)(1-\epsilon\xi_0(u))^2 du^2,$$

if $w_j(x^0) \neq \infty$; in the other case the exponent k is changed to $k-2$. We shall complete the construction only in the case $w_j(x^0) \neq \infty$, since the details are very similar if $w_j(x^0) = \infty$.

We first note that without loss of generality we may assume that ρ has been chosen so that every disc $K_{j\rho_1}$, $\rho_1 = 2\rho$, contains no zeros or poles of $Q(w,x^0)$ except $w_j(x^0)$. Thus for $\epsilon \leqq \epsilon_0$ there will lie in the disc $|w - w_j(x)| \leqq 2\rho$ only one zero of $Q(w,x)$, namely $w_j(x)$, if $k \geqq 1$, and the pole at $w = 0$ will not be in the disc. Therefore for $|u| \leqq 2\rho$ we have

$$\begin{aligned} |q_0(u)\,u^k - q(u)\,u^k| &= \Big|\sum_{\nu=1}^{n} \left(A_\nu(x^0)(w_j(x^0)+u)^{-\nu-2} - A_\nu(x)(w_j(x)+u)^{-\nu-2}\right)\Big| \\ &\leqslant \sum_{\nu=1}^{n} |A_\nu(x^0) - A_\nu(x)|\,|w_j(x)+u|^{-\nu-2} \\ &\quad + \sum_{\nu=1}^{n} |A_\nu(x^0)|\,|(w_j(x^0)+u)^{-\nu-2} - (w_j(x)+u)^{-\nu-2}| \leqslant C_7\varepsilon. \end{aligned}$$

Hence $|q_0(u) - q(u)| \leqq C_8\epsilon$ if $|u| \leqq 2\rho$, and, since $|q_0(u)| \geqq C_9$, while $|q(u)| \geqq C_9$ for $|u| \leqq 2\rho$, we know that

$$|q_0^{1/2}(u) - q^{1/2}(u)| \leqq C_{10}\epsilon, \tag{16.17}$$

provided we choose the same branch of the square root. We set

$$q_0^{1/2}(u) = \sum_0^\infty \alpha_\nu u^\nu, \qquad q^{1/2}(u) = \sum_0^\infty \beta_\nu u^\nu.$$

In what follows, the transformations will be made in a purely formal manner. Taking square roots in (16.12) and changing to our new variables, we find that

$$v^{k/2} q^{1/2}(v)\,dv = u^{k/2} q_0^{1/2}(u)(1+\epsilon\xi(u))\,du. \tag{16.18}$$

We first construct a solution to the equation

$$v^{k/2}q(v)\,dv = u^{k/2}q_0(u)\,du$$

and upon integrating we find that

$$v^{\frac{k}{2}+1}\sum_0^\infty \frac{\beta_\nu v^\nu}{\frac{k}{2}+\nu+1} = u^{\frac{k}{2}+1}\sum_0^\infty \frac{\alpha_\nu u^\nu}{\frac{k}{2}+\nu+1}.$$

We shall seek a solution of the form $v = u(1+\chi(u))$. Substituting this into the above equation, we are led to the equation

$$(1+\chi)^{\frac{k}{2}+1}\sum_0^\infty \frac{\beta_\nu u^\nu(1+\chi)^\nu}{\frac{k}{2}+\nu+1} = \sum_0^\infty \frac{\alpha_\nu u^\nu}{\frac{k}{2}+\nu+1},$$

for χ, which can be written in the form $F(u,\chi) = G(u)$, where

$$F(u,\chi) = q^{-\frac{1}{2}}(u)\left\{(1+\chi)^{\frac{k}{2}+1}\sum_0^\infty \frac{\beta_\nu u^\nu(1+\chi)^\nu}{\frac{k}{2}+\nu+1} - \sum_0^\infty \frac{\beta_\nu u^\nu}{\frac{k}{2}+\nu+1}\right\},$$

and

$$G(u) = q^{-\frac{1}{2}}(u)\sum_0^\infty \frac{(\alpha_\nu-\beta_\nu)u^\nu}{\frac{k}{2}+\nu+1}.$$

We see by applying Bessel's inequality to (16.17) that

$$\sum_0^\infty |\alpha_\nu-\beta_\nu|^2(2\rho)^{2\nu} \leqslant C_{10}^2\varepsilon^2.$$

Thus for $|u| \leqq 2\rho$ Schwarz' inequality yields that

$$|G(u)| < C_9^{-\frac{1}{2}}C_{10}\left(\sum_0^\infty\left(\frac{k}{2}+\nu+1\right)^{-2}\right)^{\frac{1}{2}}\varepsilon.$$

It is easily seen that $F(u,0) = 0$ and $F_\chi(u,0) = 1$ for $|u| \leqq 2\rho$. Furthermore, if $|\chi| \leqq 1/2$ and $|u| \leqq 4\rho/3$, then we find in the same manner as above that

$$|F(u,\chi)| < 2^{\frac{k}{2}+2}C_9^{-\frac{1}{2}}\left(\sum_0^\infty|\beta_\nu 2\rho|^{2\nu}\right)^{1/2}\left(\sum_0^\infty\left(\frac{k}{2}+\nu+1\right)^{-2}\right)^{1/2} = C_{11}.$$

It thus follows in accordance with the assumptions (16.6) that $F(u,\chi)$ is a univalent function of χ, provided $|\chi| \leqq 1/16C_{11}$ and $|u| \leqq 4\rho/3$. This means that there exists a solution of the equation under consideration, and moreover the solution satisfies

$$|\chi(u)| \leqq C_{12}\epsilon$$

for $|u| \leqq 4\rho/3$. It is easily verified that for $|u| \leqq 5\rho/4$ the derivatives of $\chi(u)$ satisfy

(16.19) $$|\chi^{(s)}(u)| \leqq C_{13}\epsilon, \qquad 0 \leqq s \leqq 4.$$

As we have already remarked, the discs $K_{j\rho_1}$, $\rho_1 = 2\rho$, are divided by the branches emanating from the vertex $w_j(x^0)$ into p curvilinear sectors σ_{jl}, $1 \leqq l \leqq p$, $p \geqq 1$. We may assume that ρ has been chosen so small that each branch emanating from $w_j(x^0)$ intersects the circle $\partial K_{j\rho_1}$ in only one point, and we may also assume that those branches of $\mathscr{T}_{x^0}$ not emanating from $w_j(x^0)$ do not intersect $\partial K_{j\rho_1}$ at all. The existence of such a ρ is insured by our knowledge of the structure of the trajectories of the differential $Q(w, x^0)(dw/w)^2$. Under the mapping $u = w - w_j(x^0)$ (or $u = w^{-1}$) the branches of $\mathscr{T}_{x^0}$ are mapped to trajectories of the differential $u^k q_0(u)du^2$ which begin at the point $u = 0$, and the images of the curvilinear sectors σ_{jl} will be curvilinear sectors which we denote by σ_l, $1 \leqq l \leqq p$, $p \geqq 1$. It is clear that $\xi_0(u)$ is a regular function in each sector σ_l. The solution constructed above will be denoted by $v_0(u)$, $v_0(u) = u(1 + \chi(u))$. We seek a solution of (16.18) of the form $v = v_0(u) + v_1(u)$ in the sector σ_l. In view of the definition of $v_0(u)$, we have

$$\int_{v_0(u)}^{v_0(u)+v_1(u)} \tau^{\frac{k}{2}} q^{\frac{1}{2}}(\tau)\,d\tau = \varepsilon \int_0^u \xi_0(\tau)\,\tau^{\frac{k}{2}} q_0^{\frac{1}{2}}(\tau)\,d\tau.$$

The integral on the right will be taken over a path lying in σ_l, and the integral on the left over the line segment joining $v_0(u)$ and $v_0(u) + v_1(u)$. The following calculations will be purely formal in nature, although they can be completely justified. In the integral on the left we make the variable change $\tau = v_0(u)(1 + t)$ and we set $v_1 = v_0 y$. Then

$$\int_0^y (1+t)^{\frac{k}{2}+1} q^{\frac{1}{2}}[v_0(1+t)]\,dt = \frac{\varepsilon}{[v_0(u)]^{\frac{k}{2}+1}} \int_0^u \xi_0(\tau)\,\tau^{\frac{k}{2}} q_0^{\frac{1}{2}}(\tau)\,d\tau.$$

Dividing this equation by $q^{1/2}[v_0(u)]$ and recalling the differential equation for $v_0(u)$, we find that

$$\int_0^y (1+t)^{\frac{k}{2}+1} \left(\frac{q[v_0(1+t)]}{q(v_0)}\right)^{1/2} dt = \varepsilon \frac{v_0'(u)}{v_0(u)} \int_0^u \xi_0(\tau) \left[\frac{\tau^k q_0(\tau)}{u^k q_0(u)}\right]^{1/2} d\tau.$$

The trajectories of differential $u^k q_0(u)du^2$ emanating from $u = 0$ are analytic arcs, and moreover the end point $u = 0$ is a point of analyticity of the arc (this will be proved after we prove this lemma). For small ρ these arcs will not differ appreciably in $|u| \leqq 2\rho$ from their tangents at the point $u = 0$. Therefore the path of integration in the integral on the right-hand side of the last formula can always be taken so that $|\tau/u|$ remains bounded. Denoting by $G_1(u)$ the right-hand side of the last equation, we see from (16.11) or (16.11′) that

$$|G_1^{(s)}(u)| \leqq C_{14}|u|^{5-s}\epsilon, \qquad 0 \leqq s \leqq 4,$$

if $u \in \sigma_l$, $|u| \leqq 5\rho/4$. Applying, as before, Proposition 16.6 to the last equation, we conclude that there exists a solution of this equation which is regular in the domain $\sigma_l \cap \{u : |u| \leqq 5\rho/4\}$ and which satisfies the inequality $|y(u)| \leqq C_{15}|u|^5\epsilon$. Then by differentiating the equation and using the estimates for the derivatives of $G_1(u)$ we find that $|y^{(s)}(u)| \leqq C_{16}|u|^{5-s}\epsilon$, $0 \leqq s \leqq 4$. Thus, if we set $v_1(u) = v_0(u)y(u)$, we see that

$$|v_1^{(s)}(u)| < C_{17}|u|^{6-s}\epsilon, \qquad 0 \leqq s \leqq 4, \tag{16.20}$$

provided $u \in \sigma_l \cap \{u : |u| \leqq 5\rho/4\}$. Thus we have a solution of equation (16.18) in the form $v(u) = v_0(u) + v_1(u) = u + u\chi(u) + v_1(u)$. Since the function $\xi_0(u)$ can be continued analytically across those boundary components of σ_l belonging to the trajectories of the differential, it is easily seen that $v_1(u)$, and hence $v(u)$, can be continued analytically across the desired arcs.

Changing back to the variables w and t, we conclude that in each sector σ_{jl} of the disc $K_{j\rho_2}$, $\rho_2 = 5\rho/4$, there exists a solution $t_{jl}(w)$ of equation (16.15) which can be represented in the form

$$\begin{aligned} t_{jl}(w) &= t_j(w) + \tau_{jl}(w), \qquad t_j(w) = v_0[w - w_j(x^0)], \\ \tau_{jl}(w) &= v_1[w - w_j(x^0)], \end{aligned} \tag{16.21}$$

where $t_j(w)$ is regular in the disc $K_{j\rho_2}$ and $\tau_{jl}(w)$ satisfies (in view of (16.20)) the inequality

$$|\tau_{jl}^{(s)}(w)| \leqslant C_{17}|w - w_j(x^0)|\,\varepsilon, \qquad 0 \leqslant s \leqslant 4. \tag{16.22}$$

If $w_j(x^0) = \infty$, it is easily seen that

$$|\tau_{jl}^{(s)}(w)| \leqslant C_{17}|w|^{-1}\varepsilon, \qquad 0 \leqslant s \leqslant 4. \tag{16.23}$$

By construction, $t_{jl}[w_j(x^0)] = w_j(x)$ and

$$|t_{jl}(w) - w| \leqq C_{18}|w|\epsilon. \tag{16.24}$$

In addition we note that from (16.19) it follows that

$$|[t_j(w) - w]^{(s)}| \leqq C_{19}\epsilon, \qquad 1 \leqq s \leqq 4. \tag{16.25}$$

In view of (16.15), the function $t_{jl}(w)$ maps those boundary components of the sector σ_{jl} belonging to branches of $\mathscr{T}_{x^0}$ onto arcs lying on the trajectories of the differential $Q(w,x)(dw/w)^2$ which emanate from the vertex $w_j(x)$. If $|w_0 - w_j(x^0)| = \rho$ and w_0 lies on a branch $\gamma \in \mathscr{T}_{x^0}$ emanating from $w_j(x^0)$, then by Proposition 16.1 we can find a point w_0^*, $|w_0^* - w_j(x^0)| = \rho$, which lies on the corresponding branch of $\mathscr{T}_x$, for which $|w_0^* - w_0| < C\epsilon$. On the other hand, in view of inequality (16.24), $t_{jl}(w_0)$ lies on a trajectory of the differential $Q(w,x)(dw/w)^2$, and $|t_{jl}(w_0) - w_0| \leqq C_{18}|w_0|\epsilon$. Thus $t_{jl}(w_0)$ lies on the branch of $\mathscr{T}_x$ corresponding to the branch $\gamma \subset \mathscr{T}_{x^0}$, and so $t_{jl}(w)$ maps those boundary components of σ_{jl} belonging to the branches of $\mathscr{T}_{x^0}$ to the corresponding branches of $\mathscr{T}_x$. It is clear that $t_{jl}(w)$ is defined in a larger domain than σ_{jl}, since $t_{jl}(w)$ is analytically continuable across the above mentioned boundary components of σ_{jl}.

Step 3. Take one of the curvilinear sectors of the disc $K_{j\rho_2}$, say σ_{jl}, and let γ_j^0 be that branch of $\mathscr{T}_{x^0}$ having end points at $w_j(x^0)$ and $w_p(x^0)$, say, which contains one of the boundary components of σ_{jl}. Suppose that σ_{jl} lies to the left of the branch; let σ_{pk} be a curvilinear sector of $K_{p\rho_2}$ such that one of its boundary components lies on γ_j^0 and such that it lies to the left of the branch. Denote by w_j^* and w_p^* the points of intersection of $\partial K_{j\rho}$ and $\partial K_{p\rho}$ with γ_j^0. As was shown in step 1, it is possible to analytically continue the solution $t_{jl}(w)$ constructed in σ_{jl} along the branch from w_j^*. As a result we obtain a solution $\varphi_j(w)$ of equation (16.15) which is in particular defined in a neighborhood of w_p^*. Since $t_{jl}(w_j^*) \in \mathscr{T}_x$, the solution $\varphi_j(w)$ will map the arc $(w_j(x^0), w_p^*)$ of the branch γ_j^0 to a corresponding branch of $\mathscr{T}_x$. On the other hand, in a neighborhood of w_p^* we have defined a solution $t_{pk}(w)$, which we constructed in the sector σ_{pk} of the disc $K_{p\rho_2}$. Set $t_{pk}(w_p^*) = t_p^*$ and $\varphi_j(w_p^*) = t_p$. If $t_p^* = t_p$, then by the uniqueness theorem we have in a neighborhood of w_p^* that

$$t_{pk}(w) \equiv \varphi_j(w),$$

and so the process of analytic continuation leads us from the local solution $t_{jl}(w)$ in the sector σ_{jl} to the local solution $t_{pk}(w)$ in σ_{pk}. The equality of t_p^* and t_p follows from properties of the function $\xi(w)$. As a matter of fact, if γ_j is a branch of $\mathscr{T}_x$ different from γ_j^0, it follows from (16.15) that

$$\int_{w_j(x)}^{t_p} Q^{1/2}(t, x)\frac{dt}{t} = \int_{w_j(x^0)}^{w_p^*} Q^{1/2}(w, x^0)(1 + \varepsilon\xi_+(w))\frac{dw}{w},$$

where the integration on the left is along γ_j, and on the right along γ_j^0; the plus sign on $\xi(w)$ indicates that the analytic continuation takes place on the "left side" of γ_j^0. By the construction of $t_{pk}(w)$ we have

$$\int_{t_p^*}^{w_p(x)} Q^{1/2}(w, x)\frac{dt}{t} = \int_{w_p^*}^{w_p(x^0)} Q^{1/2}(w, x^0)(1 + \varepsilon\xi_+(w))\frac{dw}{w}.$$

and so

$$\left|\left(\int_{w_j(x)}^{t_p} + \int_{t_p^*}^{w_p(x)}\right) Q^{1/2}(t, x)\frac{dt}{t}\right| = \left|\int_{w_j(x_0)}^{w_p(x^0)} Q^{1/2}(w, x^0)(1 + \varepsilon\xi_+(w))\frac{dw}{w}\right|$$

$$= \lambda_j(x^0) + \varepsilon\mu_j = \lambda_j(x) = \left|\int_{w_j(x)}^{w_p(x)} Q^{1/2}(t, x)\frac{dt}{t}\right|,$$

in view of the construction of $\xi(w)$. The fact that $t_p = t_p^*$ follows immediately from these equalities, since the differential $Q^{1/2}(t,x)dt/t$ is real-valued and of constant sign on γ_i.

Continuing this process of analytic continuation, we eventually return to the original sector σ_{jl}, and thus we have constructed a regular univalent solution in some neighborhood of the tree $\mathcal{T}_{x^0}$. This solution will be denoted by $\varphi(w,\epsilon)$. (The remarks in the preceding paragraph insure that $\varphi(w,\epsilon)$ is single-valued.) It is clear that $\varphi(w,\epsilon)$ is regular in some domain $\mathcal{G}(\mathcal{L}_r, \mathcal{T}_{x^0})$, provided we choose r in a suitable way.

Step 4. We shall now show that $\varphi(w,\epsilon) \in U_{\mathcal{T}_{x^0}}$. Clearly $\varphi(w,\epsilon)$ is a univalent function. To see this, note that it follows from (16.16), (16.23) and (16.25) that $\varphi(w,\epsilon)$ maps $\mathcal{L}_r$ univalently to a Jordan curve, and furthermore $\varphi_+(w,\epsilon)$ and $\varphi_-(w,\epsilon)$ map each branch of the tree $\mathcal{T}_{x^0}$ univalently to a corresponding branch of $\mathcal{T}_x$. Thus $\varphi(w,\epsilon)$ is univalent in $\mathcal{G}(\mathcal{L}_r, \mathcal{T}_{x^0})$. The above remarks also make it obvious that conditions 1, 2, 4 and 5 of §15 are satisfied. We must therefore verify only condition 3. Let σ_{jl} and σ_{jl+1} be two consecutive curvilinear sectors having a common boundary component—an arc of the branch γ_j^0. Suppose σ_{jl} lies to the left and σ_{jl+1} to the right of γ_j^0. Then, according to (16.21),

$$[t_{jl}(w)]_+ - [t_{jl+1}(w)]_- = [\tau_{jl}(w)]_+ - [\tau_{jl+1}(w)]_-$$

and condition 3 follows from (16.22) and (16.23). Thus $\varphi(w,\epsilon) \in U_{\mathcal{T}_{x^0}}$, as required.

Remark 16.2. In order to use the results of §15 for $\varphi(w,\epsilon)$, we must verify that the branches of the tree $\mathcal{T}_{x^0}$ are analytic curves, especially in a neighborhood of a vertex. For this purpose we consider the equation

$$Q(w,x^0)(dw/w)^2 = Au^k du^2$$

in a neighborhood of a zero $w_j(x^0)$ of order k. We choose the constant A so that one of the solutions of this equation has the form $w - w_j(x^0) = u + a_1u^2 + \cdots = \psi(u)$. The trajectories of the differential $Au^k du^2$ emanating from $u=0$ will be the rays $\arg u = \theta$, where θ is defined by the equation

$$\exp\{i[\arg A + (k+2)\theta]\} = 1.$$

In this way we see that the equations of the trajectories of our differential will have the form $w - w_j(x^0) = \psi(ye^{i\theta})$, $y > 0$. But the function $\psi(ye^{i\theta})$ is regular in a neighborhood of $y=0$. Thus each of the trajectories lies on an analytic curve $w - w_j(x^0) = \psi(ye^{i\theta})$, $\operatorname{Im} y = 0$, and in particular this implies that $w_j(x^0)$ is a regular point of the trajectory, as required.

Later on we shall require a slightly modified version of Proposition 16.1. We shall not assume $d_{n-1}(x^0) \neq 0$, but instead that $d_{n-1/2}(x^0) \neq 0$.

Proposition 16.7. *Given any $\epsilon > 0$, the inequality $|f(z,x^0) - f(z,x)| \leq \epsilon$ holds in the domain D_ρ if $|x - x^0| \leq h$.*

Proof. We used the condition $d_{n-1}(x^0) \neq 0$ only to establish the inequality

$|\lambda(x)-\lambda(x^0)|<C|x-x^0|$. Since $\lambda(x)\to\lambda(x^0)$ as $x\to x^0$, and since in any disc $|z|\leqq r<1$ we have $\lim_{x\to x^0} f(z,x)=f(z,x^0)$, it follows that $\lim_{x\to x^0} Q(w,x)=Q(w,x^0)$, $w\neq 0$, and $\lim_{x\to x^0} P(z,x)=P(z,x^0)$, $z\neq 0,\infty$. The remainder of the proof consists of a repetition of the proof of Proposition 16.1, and hence will be omitted.

§17. We shall now reconsider the class of functions $U_{\mathscr{T}}$ defined in §15. We assume that $\mathscr{T}$ is the boundary tree of some extremal function $f(z)\in S_n$. Each function $\varphi(w,\epsilon)\in U_{\mathscr{T}}$ defines some function admissible with respect to a given $f(z)$. To see this, note that from (15.13) we have $\varphi(w,\epsilon)=w+\epsilon g(w,\epsilon)$, where $g(w,\epsilon)=g(w,\epsilon;0)+g(w,\epsilon;\infty)$. The function $g[f(z),\epsilon]$ will be admissible with respect to $f(z)$. We shall denote the class of functions $g[f(z),\epsilon]$ by $\mathscr{D}_f$, and for convenience we denote by $\mathscr{D}_f^0$ the class of admissible functions of the form (2.2). It is clear that $\mathscr{D}_f^0\subset\mathscr{D}_f$. In particular, we should note that the class $\mathscr{D}_f$ is nonlinear, while $\mathscr{D}_f^0$ is linear over the reals. By using Theorem 1 we can use admissible functions of class $\mathscr{D}_f$ to construct a variation of $f(z)$. In order to do this, note that for any $g(z,\epsilon)\in\mathscr{D}_f$ we can find an annulus $r_1\leqq|z|\leqq r_2$, $r_2<1$, in which

$$|g(z,\epsilon)|<C,$$

where C is independent of ϵ. We then set

$$F(z,\epsilon)=f(z)+\epsilon g(z,\epsilon).$$

For sufficiently small ϵ the conditions of Theorem 1 are satisfied, and so we can construct a regular change of variable $z=\psi(\zeta)$ such that $F_*(z,\epsilon)=F[\psi(z),\epsilon]$ is regular and univalent in $|z|<1$. The choice of the arbitrary constants will be made in accordance with the fairly strict conditions of Lemma 1.4. Then from this lemma we have

$$F'_*(0,\epsilon)>0.$$

Denote by $\mathscr{T}_\epsilon$ the image of the tree $\mathscr{T}$. The function $F_*(z,\epsilon)$ maps the disc $|z|<1$ onto a domain $\mathscr{G}(\mathscr{T}_\epsilon)$. In the case when the conformal radius at $w=0$ of $\mathscr{G}(\mathscr{T}_\epsilon)$ is 1 we have $F_*(z,\epsilon)\in S$. It is clear from Theorem 1 that there exists a disc $|z|\leqq r^*$ in which

$$F_*(z,\varepsilon)=\sum_0^\infty \varepsilon^k F_k(z,\varepsilon),\qquad F_0(z,\varepsilon)=f(z). \tag{17.1}$$

We note that this is not a power series, since $F_k(z,\epsilon)$ is a function of ϵ. It is easy to see that the kth term of this series is of the order $O(\epsilon^k)$. However, it is impossible to expand the above series in powers of ϵ, since $F_k(z,\epsilon)$ is not a regular function of ϵ in a neighborhood of $\epsilon=0$. To simplify the notation, we shall not indicate below the dependence of F_k on ϵ. After normalization, we set $f_*(z,\epsilon)=F_*(z,\epsilon)\,[F'_*(0,\epsilon)]^{-1}$, and

$$(17.2)\qquad f_*(z,\varepsilon)=\sum_0^\infty \varepsilon^k f_k(z),\quad f_0(z)=f(z).$$

If $F_*'(0,\epsilon)=1$ (i.e. the conformal radius of $\mathscr{D}(\mathscr{T}_\epsilon)$ at $w=0$ is 1), the normalization is redundant, but in either case it will be technically more convenient to work with formula (17.2) than with (17.1). From the above remarks and Lemma 16.1 we deduce

Lemma 17.1. *Let $x^0\in\mathscr{R}$, $d_{n-1}(x^0)\neq 0$ and $x\in\mathfrak{Q}_{x^0}$. If $\epsilon=|x-x^0|$ is sufficiently small, then $f(z,x)$ can be represented as a variation of $f(z,x^0)$ by*

$$f(z,x)=\sum_{k=0}^\infty \varepsilon^k f_k(z),\qquad f_0(z)=f_0(z,x^0).$$

Some of the results of §§2 and 4 carry over to the class $\mathscr{D}_f$. Before exploring this area, however, we first reformulate (15.12). Let γ be an arbitrary branch of the tree $\mathscr{T}$. If γ does not emanate from the base of the tree, then corresponding to the given orientation we introduce on γ a parameter such that the arc γ is given by an equation

$$w=w(\sigma),\qquad 0\leqq\sigma\leqq l,$$

where $w(\sigma)$ is an analytic function of σ on the interval $0\leqq\sigma\leqq l$. Note that the function $w(\sigma)$ satisfies

$$(17.3)\qquad \begin{gathered} m|\sigma_1-\sigma_2|\leqq|w(\sigma_1)-w(\sigma_2)|,\qquad 0\leqq\sigma_1,\sigma_2\leqq l,\\ |w^{(k)}(\sigma)|\leqq M,\qquad k=1,\cdots,4,\end{gathered}$$

where the constants m and M are defined in terms of the tree $\mathscr{T}$. If γ emanates from the base of the tree, we introduce a new parameterization of the tree (even if it does not have the same orientation) so that the value $\sigma=0$ will correspond to a vertex different from the base. Inequality (17.3) continues to hold in this case.

Let $\Phi(w)=\Phi(w,\epsilon)$ be the homeomorphism of the branch γ onto itself defined by formula (15.1). We set

$$\Phi[w(\sigma)]=w[\Phi_0(\sigma)].$$

Clearly $\Phi_0(\sigma)$ depends on ϵ, but we shall suppress this dependence in the notation. It is convenient to set $\Phi_0(\sigma)=\sigma+\epsilon\chi(\sigma)$ and $\chi(\sigma)=\chi(\sigma,\epsilon)$. From (16.2) and (17.3) it follows that

$$(17.4)\qquad |\chi^{(k)}(\sigma)|<C\min\{1,\sigma,l-\sigma\},\quad k=0,\dots,3,\quad 0\leqslant\sigma\leqslant l.$$

If $l=\infty$, then in the above inequality we must change $l-\sigma$ to $1/\sigma$. The constant C is independent of ϵ. Set

$$\Phi(w)-w=\varepsilon w'(\sigma)\chi(\sigma)+\frac{\varepsilon^2}{2}w''(\sigma)\chi^2(\sigma)+\varepsilon^3\chi^3(\sigma)\Phi_1(w,\varepsilon).$$

It is not difficult to see that

$$|\Phi_1(w,\epsilon)| < A_1,$$

where A_1 is independent of ϵ. Below $A_2, A_3, \cdots$ will denote constants not depending on ϵ. The function

$$\beta(w) = \beta(w,\epsilon) = w'(\sigma)\chi(\sigma), \quad w \in \mathscr{T}, \tag{17.5}$$

is three times continuously differentiable, and from inequality (17.4) it follows that both the function and its derivatives vanish at the vertices of $\mathscr{T}$. It is easily seen that

$$\Phi'(w) - 1 = \varepsilon \frac{d\beta(w)}{dw} + \varepsilon^2 \chi(\sigma)\Phi_2(w, \varepsilon),$$

where

$$|\Phi_2(w, \varepsilon)| < A_2, \left|\frac{d\Phi_2(w, \varepsilon)}{dw}\right| < A_2.$$

Thus we have

$$\frac{1}{\varepsilon}[\Phi(w) - w]\frac{1}{2\pi i}\int_{\mathscr{T}} \frac{(\Phi'(t) - 1)\, w}{t(t-w)} dt = \varepsilon\beta(w)\frac{1}{2\pi i}\int_{\mathscr{T}} \frac{w\beta'(t)}{t(t-w)} dt + \varepsilon^2\chi(\sigma)\Phi_3(w, \varepsilon).$$

It is clear that $|\Phi_3(w,\epsilon)| < A_3$, and it is not difficult to verify that

$$\frac{1}{2}[\Phi(w) - w]\left[1 - \frac{1}{2}(\Phi'(w) - 1)\right] = \beta(w) - \frac{\varepsilon}{2} w'(\sigma)\chi(\sigma)\chi'(\sigma) + \varepsilon^2\chi(\sigma)\Phi_4(w, \varepsilon),$$

where $|\Phi_4(w,\epsilon)| < A_4$. Keeping in mind the above formulas, we see from (15.12) that

$$\begin{aligned} \omega(w) = \beta(w) &- \frac{\varepsilon}{2} w'(\sigma)\chi(\sigma)\chi'(\sigma) + \varepsilon\beta(w) g'(w, \varepsilon; \infty) \\ &+ \varepsilon \frac{\beta(w)}{2\pi i}\int_{\mathscr{T}} \frac{w\beta'(t)}{t(t-w)} dt + \omega_0(w) + \varepsilon^3\chi(\sigma)\Phi_5(w, \varepsilon), \end{aligned} \tag{17.6}$$

where $|\Phi_5(w,\epsilon)| < A_5$.

We shall now give a slightly changed formula for $g(w,\epsilon;\infty)$. Again the dependence on ϵ will not be explicitly indicated. From (15.6) we have

$$g(w;\infty) = \frac{1}{2\pi i}\int_{\mathscr{C}_r} \frac{w^2}{w-t} g(t;\infty)\frac{dt}{t^2}.$$

Introducing the change of variable $t = f(\zeta)$, we have

$$g(w;\infty) = \int_{|\zeta|=r} \frac{w^2}{w - f(\zeta)}\left(\frac{Df(\zeta)}{f(\zeta)}\right)^2 \frac{g[f(\zeta);\infty]}{Df(\zeta)}\frac{d\zeta}{2\pi i\zeta}.$$

The function $g[f(\zeta);\infty]$ is regular in $r < |\zeta| < 1$. We set

$$\frac{1}{Df(\zeta)} g[f(\zeta); \infty] = \sum_{-\infty}^{\infty} h_k \zeta^{-k}.$$

If $w \in \mathcal{G}(\mathcal{C}_r, \mathcal{T})$, the last integral remains unchanged if we add to the function $(1/Df(\zeta))g[f(\zeta); \infty]$ any function regular in $|\zeta| < 1$ and vanishing at $\zeta = 0$. We set

$$h_0(\zeta) = \sum_{1}^{\infty} (h_k \zeta^{-k} - \overline{h}_k \zeta^k) + h_0.$$

Then

$$g(w; \infty) = \int_{|\zeta|=r} \frac{w^2}{w - f(\zeta)} \left(\frac{Df(\zeta)}{f(\zeta)}\right)^2 h_0(\zeta) \frac{d\zeta}{2\pi i \zeta}.$$

For convenience of notation we introduce the measure

(17.7) $$\nu(d\zeta) = h_0(\zeta) \frac{d\zeta}{2\pi i \zeta}$$

on the circle $\mathcal{C} = \{\zeta : |\zeta| = r\}$. Then we get for $g(w; \infty)$ a formula having the generic form of an admissible function of the type (2.2), namely

(17.8) $$g(w; \infty) = \int_{\mathcal{C}} \frac{w^2}{w - f(\zeta)} \left(\frac{Df(\zeta)}{f(\zeta)}\right)^2 \nu(d\zeta).$$

Consider the function

(17.9) $$h(\zeta) = h_0(\zeta) - \operatorname{Re} h_0$$

which is regular in $r < |\zeta| < r^{-1}$, and which satisfies

(17.10) $$\overline{h(\zeta)} = -h(\zeta)$$

for $|\zeta| = 1$. We next proceed to define the functions $F_1(z, \epsilon)$ and $F_2(z, \epsilon)$ appearing in (17.1). The admissible function has the form

$$g[f(z); 0] + g[f(z); \infty] = g(z, \epsilon).$$

Simultaneously with the variation defined by this admissible function, we consider the variation defined by the admissible function

(17.11) $$g[f(z); \infty].$$

In accord with (17.8), this function has the standard form of the variation (2.2). The variation of $f(z)$ by the admissible function (17.11) will be denoted by $\tilde{F}_*(z, \epsilon)$, and the normalized function by $\tilde{f}_*(z, \epsilon)$. According to Theorem 1,

$$\tilde{F}_*(z, \varepsilon) = \sum_{0}^{\infty} \varepsilon^k \tilde{F}_k(z, \varepsilon), \qquad \tilde{F}_0(z, \varepsilon) = f(z),$$

and

$$\tilde{f}_*(z, \varepsilon) = \sum_{0}^{\infty} \varepsilon^k \tilde{f}_k(z, \varepsilon), \qquad \tilde{f}_0(z, \varepsilon) = f(z).$$

The functions $\widetilde{F}_1(z,\epsilon)$ and $\widetilde{F}_2(z,\epsilon)$ are defined by (2.9) and (2.16). Since $g[f(z);0]$ is regular in the disc $|z|<1$, we have

(17.12) $$F_1(z,\epsilon) = \widetilde{F}_1(z,\epsilon) + g[f(z);0].$$

Since $g'(0,\epsilon;0)=0$, the constant c_1 must be chosen so that

$$\operatorname{Im} \widetilde{F}_1'(0,\varepsilon) = \operatorname{Im} \varphi_1(0) = \operatorname{Im}\left(\int_{\mathscr{C}} \nu(dz) + ic_1\right) = 0.$$

Recalling (17.7), we set $c_1 = -\operatorname{Im} h_0$, and

(17.13) $$\varphi_1(0) = \int_{\mathscr{C}} \nu(dz) + ic_1 = \operatorname{Re} h_0.$$

It is easy to see that

(17.14) $$f_1(z,\epsilon) = \widetilde{f}_1(z,\epsilon) + g[f(z);0].$$

We now define $F_2(z,\epsilon)$. Applying the same transformations as in §2, we get

$$F_2(z,\varepsilon) = -\frac{1}{2} D^2 f(z)\, \varphi_1^2(z) + D\widetilde{F}_1(z)\, \varphi_1(z) + Df(z)\, \psi(z) + Df(z)\, g'[f(z);\ 0]\, \varphi_1(z).$$

Set $\psi(z) = \psi_0(z) + \psi_1(z)$, where $\psi_0(z)$ is defined so that

$$\widetilde{F}_2(z) = -\frac{1}{2} D^2 f\, \varphi_1^2 + D\widetilde{F}_1 \varphi_1 + Df\psi_0$$

will be regular. Then $\psi_0(z)$ is in fact defined by the formulas of §2. We set $F_2^*(z,\epsilon) = Df\psi_1 + Dfg'[f(z);0]\varphi_1$. We define ψ_1 so that this function will be regular in $|z|<1$, and we see that

$$\psi_1(z) = -\int_{\mathscr{C}} g'[f(\zeta);\ 0]\, \frac{\zeta}{\zeta - z}\, \nu(d\zeta) - \int_{\mathscr{C}} \overline{g'[f(\zeta);\ 0]}\, \frac{\bar{\zeta} z}{1 - \bar{\zeta} z}\, \overline{\nu(d\zeta)} + ic_2^*.$$

The arbitrary constants will be chosen so that

$$\operatorname{Im} \widetilde{F}_2'(0,\epsilon) = 0 \quad \text{and} \quad \operatorname{Im} F_2^{*\prime}(0,\epsilon) = 0.$$

It is easily seen that $f_2(z,\epsilon)$ is defined by the formula

(17.15) $$f_2(z,\epsilon) = \widetilde{f}_2(z,\epsilon) + f_2^*(z,\epsilon),$$

where

(17.16) $$f_2^*(z,\varepsilon) = Df(z)\, g'[f(z);\ 0]\, \varphi_1(z) - \varphi_1(0)\, g[f(z);\ 0] + Df(z)\, \psi_1(z) - f(z)\, \psi_1(0).$$

We shall reformulate the expressions we have obtained for $f_1(z,\epsilon)$ and $f_2(z,\epsilon)$. We will pick out the dominant terms (with respect to ϵ) and the terms of higher orders. It is true that this division into dominant terms and terms of higher order is somewhat arbitrary, since the dominant terms depend on

ϵ. The division will be accomplished as follows: set

(17.17) $$\eta(w, \varepsilon) = \frac{1}{2\pi i} \int_{\mathscr{T}} \frac{w^2 \beta(t)}{t^2 (t-w)} dt.$$

We note that $\beta(t)$ is a function of ϵ, but we shall not indicate this dependence in the notation. Consider the function

$$\omega_1(w) = \beta(w) g'(w, \varepsilon; \infty) + \beta(w) \frac{1}{2\pi i} \int_{\mathscr{T}} \frac{w\beta'(t)}{t(t-w)} dt.$$

From (15.5), (17.6) and (15.11) we see that in the disc $|z| \leqq r < 1$

$$g[f(z); 0] = \eta[f(z), \varepsilon] + \frac{\varepsilon}{2\pi i} \int_{\mathscr{T}} \frac{\left[\omega_1(w) - \frac{1}{2} w'(\sigma) \chi(\sigma) \chi'(\sigma)\right] f^2(z)}{w^2 (w - f(z))} dw + O(\varepsilon^2),$$

where the constant in the remainder term is independent of z and ϵ. With

(17.18) $$\eta_1[f(z), \varepsilon] = \frac{1}{2\pi i} \int_{\mathscr{T}} \frac{\left[\omega_1(w) - \frac{1}{2} w'(\sigma) \chi\chi'\right] f^2(z)}{w^2 (w - f(z))} dw$$

we have

$$g[f(z); 0] = \eta[f(z), \epsilon] + \epsilon\eta_1[f(z), \epsilon] + O(\epsilon^2).$$

We now reformulate the expression defining $\tilde{f}_1(z)$. Consider the formula (2.11). Since

$$\frac{f^2(z)}{f(z) - f(\zeta)} \left(\frac{Df(\zeta)}{f(\zeta)}\right)^2 + \frac{Df(z)\zeta}{\zeta - z} - f(z)$$

is a regular function of ζ in the disc $|\zeta| < 1$ which vanishes at $\zeta = 0$, we can replace the measure $\nu(d\zeta) = h_0(\zeta) d\zeta / 2\pi i \zeta$ in (2.11) by the measure $h(\zeta) d\zeta / 2\pi i \zeta$. Thus

$$\tilde{f}_1(z, \varepsilon) = \int_{|\zeta|=r} \left[\frac{f^2(z)}{f(z) - f(\zeta)} \left(\frac{Df(\zeta)}{f(\zeta)}\right)^2 + \frac{Df(z)\zeta}{\zeta - z} - f(z)\right] h(\zeta) \frac{d\zeta}{2\pi i \zeta}$$
$$+ Df(z) \int_{|\zeta|=r} \overline{\frac{\bar{\zeta} z}{1 - \bar{\zeta} z} h(\zeta) \frac{d\zeta}{2\pi i \zeta}} + iC_1 (Df(z) - f(z)).$$

If $|z| < r$, this relation can be written as

$$\tilde{f}_1(z, \varepsilon) = \int_{|\zeta|=r} \frac{f^2(z)}{f(z) - f(\zeta)} \left(\frac{Df(\zeta)}{f(\zeta)}\right)^2 h(\zeta) \frac{d\zeta}{2\pi i \zeta} + \int_{|\zeta|=r} \left[\left(\frac{Df(z)\zeta}{\zeta - z} - f(z)\right)\right.$$
$$\left.\times h(\zeta) \frac{d\zeta}{2\pi i \zeta} + \overline{\frac{Df(z)\bar{\zeta} z}{1 - \bar{\zeta} z} h(\zeta) \frac{d\zeta}{2\pi i \zeta}}\right] + ic_1 (Df(z) - f(z)).$$

In the second integral we change the path of integration to the circle $|\zeta| = 1$, and recalling (17.10) we see that

$$\tilde{f}_1(z, \zeta) = \int\limits_{|\zeta|=r} \frac{f^2(z)}{f(z) - f(\zeta)} \left(\frac{Df(\zeta)}{f(\zeta)}\right)^2 h(\zeta) \frac{d\zeta}{2\pi i \zeta}$$
$$+ (Df(z) - f(z))\left(\frac{1}{2\pi i} \int\limits_{|\zeta|=1} h(\zeta) \frac{d\zeta}{\zeta} + ic_1\right).$$

Since $c_1 = -\operatorname{Im} h_0$, it follows that

$$\tilde{f}_1(z, \varepsilon) = \int\limits_{|\zeta|=r} \frac{f^2(z)}{f(z) - f(\zeta)} \left(\frac{Df(\zeta)}{f(\zeta)}\right)^2 h(\zeta) \frac{d\zeta}{2\pi i \zeta}.$$

Consider the function

$$\alpha(w) = Df(z)h(z), \quad z = f^{-1}(w), \tag{17.19}$$

which is regular in $\mathscr{G}(\mathscr{L}_r, \mathscr{T})$. Assuming that $w \in \mathscr{G}(\mathscr{L}_r)$, we also consider the function

$$\xi(w, \varepsilon) = \frac{1}{2\pi i} \int\limits_{\mathscr{E}_r} \frac{w^2 \alpha(t)}{(w - t)\, t^2}\, dt. \tag{17.20}$$

It is clear that $\xi(w, \epsilon)$ is regular in the domain $\mathscr{G}(\mathscr{T})$. This follows since in view of the regularity of $\alpha(w)$ in $\mathscr{G}(\mathscr{L}_r, \mathscr{T})$ we can change the path of integration in (17.20) to any level curve $\mathscr{L}_\rho$, provided $\rho < 1$ is sufficiently close to 1. Thus

$$\tilde{f}_1(z, \epsilon) = \xi[f(z), \epsilon].$$

Recalling the above relations, we find that

$$f_1(z, \varepsilon) = \xi[f(z), \varepsilon] + \eta[f(z), \varepsilon] + \varepsilon \eta_1[f(z), \varepsilon] + O(\varepsilon^2).$$

We would like to define the first variation of $f(z)$ with respect to the class $\mathscr{D}_f$ of admissible functions to be

$$\lim_{\varepsilon \to 0} \varepsilon^{-1} [f_*(z, \varepsilon) - f(z)] = \delta f(z)$$

provided the limit exists in the disc $|z| < 1$. However, it will prove expedient to broaden the above definition by taking the limit along some subsequence $\{\epsilon_k\}$. This is especially useful since for variations of the class $\mathscr{D}_f$ we can always find such a subsequence. It is in this sense that we shall define the first variation.

Lemma 17.2. *For any function $g(z, \epsilon) \in \mathscr{D}_f$ there exists a subsequence $\{\epsilon_k\}$ such that the functions $\alpha(w, \epsilon_k)$ converge uniformly on any annulus bounded by level curves $\mathscr{L}_{r_1}$ and $\mathscr{L}_{r_2}$, $r < r_1 < r_2 < 1$, to a function $\alpha(w)$ regular in the domain $\mathscr{G}(\mathscr{L}_r, \mathscr{T})$. If $\mathscr{K}_\delta$ is a set of discs of radius $\delta > 0$ centered at the vertices of the tree, then the subsequence $\{\xi(w, \epsilon_k)\}$ converges uniformly in the complement of $\mathscr{K}_\delta$. The subsequence $\{\beta(w, \epsilon_k)\}$ converges uniformly on $\mathscr{T}$ to a twice continuously differentiable function $\beta(w)$ which, together with its derivatives, vanishes at the vertices of $\mathscr{T}$. The functions $\beta'(w, \epsilon_k)$ and $\beta''(w, \epsilon_k)$ converge uniformly on*

$\mathscr{T}$ to $\beta'(w)$ and $\beta''(w)$. The subsequence $\{\eta(w,\epsilon_k)\}$ converges uniformly in $\mathscr{G}(\mathscr{T})$ to the function

$$\eta(w)=\frac{1}{2\pi i}\int_{\mathscr{T}}\frac{w^2\beta(t)\,dt}{t^2(t-w)},$$

and the functions $w^{-1}\eta_\pm(w,\epsilon_k)$ and $\eta'_\pm(w,\epsilon_k)$ tend uniformly on $\mathscr{T}$ to the functions $w^{-1}\eta_\pm(w)$ and $\eta'_\pm(w)$, respectively.

Proof. According to condition 5 of §15 the functions $w^{-1}g(w,\epsilon;\infty)$ are uniformly bounded in $\mathscr{G}(\mathscr{L}_{r_2},\mathscr{T})$. Thus these functions form a normal family, and so there exists a subsequence $\{w^{-1}g(w,\epsilon_k;\infty)\}$ converging in $\mathscr{G}(\mathscr{L}_{r_2},\mathscr{T})$ to a regular function. Thus the functions $(1/Df(z))g[f(r),\epsilon_k;\infty]$, $k=1,2,\cdots$, will converge in any annulus $r_1\leqq|z|\leqq r_2$, $r<r_1<r_2<1$.

We set

$$\frac{1}{Df(z)}g[f(z),\varepsilon_k;\infty]=\sum_{-\infty}^{\infty}h_{kl}z^{-l}.$$

For $|z|>r_1$ we have

$$\sum_{l=1}^{\infty}h_{kl}z^{-l}=\frac{1}{2\pi i}\int_{|\zeta|=r_1}\frac{1}{Df(\zeta)}g[f(\zeta),\varepsilon_k;\infty]\frac{d\zeta}{z-\zeta}$$

and so the series $\sum_{l=1}^{\infty}h_{kl}z^{-l}$ converges for $|z|>r$. Thus the functions

$$h_k(z)=\sum_{l=1}^{\infty}(h_{kl}z^{-l}-\bar h_{kl}z^{l})+i\operatorname{Im}h_{k0}$$

converge in the annulus $r<|z|<r^{-1}$ to a function $h(z)$. In the same way the functions $\alpha(w,\epsilon_k)$ converge to the function $\alpha(w)=Df(z)h(z)$, $z=f^{-1}(w)$, in any annulus bounded by the level curves $\mathscr{L}_{r_1}$ and $\mathscr{L}_{r_2}$, $r<r_1<r_2<1$.

We set

$$\xi(w)=\frac{1}{2\pi i}\int_{\mathscr{C}_{r'}}\frac{w^2\alpha(t)}{t^2(w-t)}dt,\qquad r'>r,$$

for $w\in\mathscr{G}(\mathscr{L}_{r'})$. This function is regular in $\mathscr{G}(\mathscr{T})$. If $w\in\mathscr{G}(\mathscr{L}_{r'},\mathscr{T})$, then

$$\xi(w)=-\alpha(w)+\frac{1}{2\pi i}\int_{\mathscr{C}_{r'}}\frac{w^2\alpha(t)}{t^2(w-t)}dt. \tag{17.21}$$

An analogous formula also holds for the functions $\xi(w,\epsilon_k)$. Since $\{h_k(z)\}$ converges to $h(z)$ in the annulus $r<|z|<r^{-1}$, it follows that $\xi(w,\epsilon_k)$ converges to $\xi(w)$ in the complement of $\mathscr{K}_\delta$ for any $\delta>0$. The function $\beta(w,\epsilon)$, $w\in\mathscr{T}$, is three times continuously differentiable, and by (17.4) it vanishes together with its derivatives at the vertices of $\mathscr{T}$. Thus we can find a subsequence $\{\epsilon'_k\}$ such that $\{\beta(w,\epsilon'_k)\}$ converges, together with its first two derivatives, to a twice continuously differentiable function $\beta(w)$. Clearly we may take

$\epsilon_k' = \epsilon_k$, $k = 1, 2, \cdots$, and it is also clear that $\beta(w) = w'(\sigma)\chi(\sigma)$, where $\chi(\sigma)$ satisfies (17.4) for $k \leq 2$.

We set

$$\eta(w) = \frac{1}{2\pi i} \int_{\mathscr{T}} \frac{w^2 \beta(t)}{t^2 (t - w)} dt.$$

Clearly on each compact subset of $\mathscr{G}(\mathscr{T})$ the subsequence $\{\eta(w, \epsilon_k)\}$ converges uniformly to $\eta(z)$. Let $w \in \mathscr{T}$, and set

$$\mathscr{I}(w) = \frac{1}{2\pi i} \int_{\mathscr{T}} \frac{w\beta(t, \varepsilon_k)}{t^2 (t - w)} dt.$$

Since $\beta(t, \epsilon_k)$ is continuously differentiable and vanishes at the vertices of $\mathscr{T}$, it follows from well-known properties of Cauchy-type integrals that $\mathscr{I}_k(w)$ is bounded on $\mathscr{T}$. Recalling (17.4), we find that the functions $\mathscr{I}_k(w)$ are uniformly bounded, which implies that

$$\left| \frac{1}{w} \eta_{\pm}(w, \varepsilon_k) \right| < A, \qquad w \in \mathscr{T}. \tag{17.22}$$

By Privalov's theorem, the family $\{\mathscr{I}_k(w)\}$ is equicontinuous. Since the functions $w^{-1}\mathscr{I}_k(w)$ converge in $L^2_{\mathscr{T}}$ to the function

$$\frac{1}{w}\mathscr{I}(w) = \frac{1}{2\pi i} \int_{\mathscr{T}} \frac{\beta(t)}{t^2 (t - w)} dt,$$

it follows that the subsequence $\{\mathscr{I}_k(w)\}$ converges uniformly on $\mathscr{T}$ to a function $\mathscr{I}(w)$. From this we deduce the uniform convergence on $\mathscr{T}$ of the functions $w^{-1}\eta_{\pm}(w, \epsilon_k)$ to the functions $w^{-1}\eta_{\pm}(w)$, respectively. Observing that

$$\eta'(w, \varepsilon) = \frac{1}{2\pi i} \int_{\mathscr{T}} \frac{w}{t(t - w)} \beta'(t, \varepsilon)\, dt,$$

and recalling (17.4), we can establish the convergence of $\eta'_{\pm}(w, \epsilon_k)$ to $\eta'_{\pm}(w)$, respectively. In passing we note that

$$|\eta'_{\pm}(w, \epsilon_k)| < A_1, \quad w \in \mathscr{T}. \tag{17.23}$$

This completes the proof of the lemma.

In the above manner we have defined for each admissible function $g(z, \epsilon) \in \mathscr{D}_f$ a first variation (possibly not unique)

$$\delta f(z) = \xi[f(x)] + \eta[f(z)].$$

We shall give one characteristic property of the first variation. It will prove convenient to consider the first variation as a function of $w \in \mathscr{G}(\mathscr{T})$. Then

$$\delta f[f^{-1}(w)] = \xi(w) + \eta(w).$$

Note that in accordance with (17.21) we have

$$\xi_+(w) - \xi_-(w) = -[\alpha_+(w) - \alpha_-(w)]$$

for $w \in \mathscr{T}$. But for $w \in \mathscr{T}$

$$\alpha_+(w)\left[\frac{Q(w)}{w^2}\right]^{1/2} = \left[\frac{Q(w)}{w^2}\right]^{1/2}[Df(z)\,h(z)]_+ = i\,[\sqrt{P(z)}\,h(z)]_\pm.$$

Applying (17.10), we see that

$$\text{(17.24)} \qquad \operatorname{Im}\left\{\alpha_+(w)\left[\frac{Q(w)}{w^2}\right]^{1/2}\right\} = 0, \qquad w \in \mathscr{T},$$

and so

$$\text{(17.25)} \qquad \operatorname{Im}\left\{\left[\frac{Q(w)}{w^2}\right]^{1/2}(\xi_+(w) - \xi_-(w))\right\} = 0.$$

According to the formula for jump discontinuities,

$$\eta_+(w) - \eta_-(w) = \beta(w) = \chi(\sigma)\,w'(\sigma).$$

Thus

$$\text{(17.26)} \quad \operatorname{Im}\left\{\left[\frac{Q(w)}{w^2}\right]^{1/2}(\eta_+(w) - \eta_-(w))\right\} = \chi(\sigma)\operatorname{Im}\left\{\left[\frac{Q(w)}{w^2}\right]^{1/2} w'(\sigma)\right\} = 0.$$

In this way we have shown that

$$\text{(17.27)} \qquad \operatorname{Im}\left\{\left[\frac{Q(w)}{w^2}\right]^{1/2}[(\delta f(f^{-1}(w)))_+ - (\delta f(f^{-1}(w)))_-]\right\} = 0.$$

This is the above-mentioned characteristic property of the first variation of an extremal function.

§18. To each extremal function $f(z) \in S_n$ there corresponds a normalized associated vector λ, defined by (5.13). The vector λ defines a linear functional

$$L(g) = \operatorname{Re}\sum_{k=1}^{n}\bar{\lambda}_k b_k, \qquad g \in S, \quad g(z) = \sum_{0}^{\infty} b_k z^{k+1}, \qquad b_0 = 1,$$

on the class S. This linear functional plays a major role in questions concerning the smoothness of the boundary of the coefficient region. We shall calculate the second variation of this functional for admissible functions of class $\mathscr{D}_f$. We set

$$f_1(z,\epsilon) = \xi[f(z),\epsilon] + \eta[f(z),\epsilon].$$

Using the formulas for $f_1(z,\epsilon)$ established in the previous section, we see that for $|z| \leq r$

$$f_*(z,\epsilon) = f(z) + \epsilon\hat{f}_1(z,\epsilon) + \epsilon^2\hat{f}_2(z,\epsilon) + O(\epsilon^3),$$

where

$$\text{(18.1)} \qquad \hat{f}_2(z,\epsilon) = f_2(z,\epsilon) + \eta_1[f(z),\epsilon].$$

It is clear that at the function $f(z)$ the functional L is stationary, and therefore

$$L(f_* - f) = o(\epsilon).$$

To prove this, we evaluate $L(\hat{f}_1)$. Since $\xi[f(z),\epsilon]$ is the first variation of $f(z)$ with respect to admissible functions of class $\mathscr{D}_f^0$, it follows that $L(\xi) = 0$. Thus $L(\hat{f}_1) = L(\eta)$. Recalling the definition of $q_k(w)$, we get

$$L(\eta) = -\operatorname{Re}\left\{\frac{1}{2\pi i}\int_{\mathscr{T}} \beta(w,\varepsilon)\frac{Q(w)}{w^2}\,dw\right\}. \tag{18.2}$$

But $Q(w)(\beta(w,\epsilon)/w^2)\,dw = \chi Q(w)w'(\sigma)\,dw/w^2$, and so this expression is real on $\mathscr{T}$. Therefore $L(\eta) = 0$, and so $L(\hat{f}_1) = 0$. Therefore

$$L(f_*) = L(f) + \epsilon^2 L(\hat{f}_2) + O(\epsilon^3).$$

On the basis of (17.15) and (18.1), $L(\hat{f}_2) = L(\tilde{f}_2) + L(f_2^*) + L(\eta_1)$. But $2\tilde{f}_2$ is the second variation of $f(z)$ with respect to the admissible function (2.2), with the measure $\nu(dz)$ defined by (17.7). Thus we can calculate $L(2\tilde{f}_2)$ via formula (4.18), and all that remains is to evaluate $L(f_2^*)$ and $L(\eta_1)$. We note that

$$g[f(z),\epsilon;0] = \eta[f(z),\epsilon] + O(\epsilon).$$

Using this relation, we can identify the dominant term of $f_2^*(z,\epsilon)$. Using (17.16) and the formula for $\psi_1(z)$ established in the previous section, we have

$$f_2^*(z,\varepsilon) = Df(z)\eta'[f(z),\varepsilon]\varphi_1(z) - \varphi_1(0)\eta[f(z),\varepsilon] + Df(z)\psi_1^*(z) - f(z)\psi_1^*(0) + O(\varepsilon),$$

where

$$\psi_1^*(z) = -\int_{\mathscr{C}} \eta'[f(\zeta),\varepsilon]\frac{\zeta}{\zeta - z}\nu(d\zeta) - \int_{\mathscr{C}} \overline{\eta'[f(\zeta),\varepsilon]}\frac{\bar{\zeta}z}{1-\bar{\zeta}z}\overline{\nu(d\zeta)} + ic_2^*.$$

For the sake of computational ease we put

$$f_{21}^*(z,\varepsilon) = Df(z)\eta'[f(z),\varepsilon]\varphi_1(z) - \eta[f(z),\varepsilon]\varphi_1(0) \tag{18.3}$$

and

$$f_{22}^*(z,\epsilon) = Df(z)\psi_1^*(z) - f(z)\psi_1^*(0). \tag{18.4}$$

Set

$$\eta[f(z),\varepsilon] = \sum_0^\infty b_k(\varepsilon)z^{k+1}.$$

Then from (17.17) it follows that

$$b_k(\varepsilon) = -\frac{1}{2\pi i}\int_{\mathscr{T}} q_k(w)\frac{\beta(w)}{w^2}\,dw, \qquad k = 1, 2, \ldots. \tag{18.5}$$

We note that $Df(z)\eta'[f(z),\epsilon] = \sum_0^\infty (k+1)b_k(\epsilon)z^{k+1}$.

From (18.3) we have

$$L(f_{21}^*) = \operatorname{Re}\left\{\int_{\mathscr{C}} P(z;\eta)\,\nu(dz) + ic_1\sum_1^n k\bar{\lambda}_k b_k(\varepsilon)\right\},$$

where $P(z;\eta)$, in agreement with (2.20), has the form

$$P(z;\eta) = \sum_{l=1}^n \bar{\lambda}_l \sum_{k=1}^{l-1}(l-k+1)\,b_{l-k}(\varepsilon)\,z^{-l}$$
$$+\sum_{k=1}^n \lambda_l \sum_{k=1}^{l-1}(l-k+1)\,\overline{b_{l-k}(\varepsilon)}\,z^l + \sum_{k=1}^n k\bar{\lambda}_k b_k(\varepsilon).$$

If we use (18.5), (4.12) and (4.13), we find that

$$P(z;\eta) = -\frac{1}{2\pi i}\int_{\mathscr{T}}[Q^*(w,z)+Q^*(w)]\frac{\beta(w)}{w^2}\,dw - \overline{\frac{1}{2\pi i}\int_{\mathscr{T}} Q^*\left(w,\frac{1}{\bar{z}}\right)\frac{\beta(w)}{w^2}\,dw}.$$

Thus we conclude that

$$L(f_{21}^*) = -\operatorname{Re}\left\{\frac{1}{2\pi i}\int_{\mathscr{T}}\frac{\beta(w)}{w^2}\,dw\left[\int_{\mathscr{C}}[Q^*(w,z)+Q^*(w)]\,\nu(dz)\right.\right.$$
$$\left.\left.+\frac{1}{2\pi i}\int_{\mathscr{C}} Q^*\left(w,\frac{1}{\bar{z}}\right)\overline{\nu(dz)} + ic_1 Q^*(w)\right]\right\}.$$

Applying (17.13), we write this relation as

$$L(f_{21}^*) = -\operatorname{Re}\left\{\frac{1}{2\pi i}\int_{\mathscr{T}}\frac{\beta(w)}{w^2}\,dw\left[\int_{\mathscr{C}}\left(Q^*(w,z)\,\nu(dz)\right.\right.\right.$$
$$\left.\left.\left.+Q^*\left(w,\frac{1}{\bar{z}}\right)\overline{\nu(dz)}\right) + \operatorname{Re} h_0 Q^*(w)\right]\right\}.$$

We now use (17.7) and (17.9). Since $\overline{h_0(z)} = 2\operatorname{Re} h_0 - h_0(z)$ for $|z|=1$, we have

$$\int_{\mathscr{C}}\left[Q^*(w,z)\,\nu(dz) + Q^*\left(w,\frac{1}{\bar{z}}\right)\overline{\nu(dz)}\right] = \int_{|z|=1}[Q^*(w,z)\,h_0(z)$$
$$+Q^*(w,z)\,\overline{h_0(z)}]\frac{dz}{2\pi iz} = \frac{\operatorname{Re} h_0}{\pi i}\int_{|z|=1} Q^*(w,z)\frac{dz}{z} = 0.$$

It follows that

$$L(f_{21}^*) = -\operatorname{Re} h_0 \operatorname{Re}\left\{\frac{1}{2\pi i}\int_{\mathscr{T}} Q^*(w)\frac{\beta(w)}{w^2}\,dw\right\}. \tag{18.6}$$

Using (18.4) and (17.7), we find that

$$L(f_{22}^*) = -\operatorname{Re}\left\{\int_{|z|=r} P(z)\,\eta'[f(z),\varepsilon]\,h_0(z)\frac{dz}{2\pi iz}\right\}.$$

In this formula we replace $h_0(z)$ by $h(z)$, where $h_0(z) = h(z) + \operatorname{Re} h_0$.

We now evaluate the integral

$$\mathcal{J} = \frac{1}{2\pi i} \int\limits_{|z|=r} P(z)\, \eta'[f(z), \varepsilon] \frac{dz}{z}.$$

Using formula (5.1), we find that

$$\mathcal{J} = \frac{1}{2\pi i} \int\limits_{|z|=r} \sum_1^n \bar{\lambda}_k z^{-k-2} Df(z)\, \eta'[f(z), \varepsilon]\, dz = \sum_1^n (k+1) \bar{\lambda}_k b_k(\varepsilon)$$

$$= -\frac{1}{2\pi i} \int\limits_{\mathcal{T}} [Q^*(w) + Q(w)] \frac{\beta(w)}{w^2}\, dw.$$

Thus

$$L(f_{22}^*) = -\operatorname{Re}\left\{ \int\limits_{|z|=r} P(z)\, \eta'[f(z), \varepsilon]\, h(z) \frac{dz}{2\pi i z}\right\} + \operatorname{Re} h_0 \operatorname{Re}\left\{\frac{1}{2\pi i} \int\limits_{\mathcal{T}} Q^*(w) \frac{\beta(w)}{w^2}\, dw\right\}.$$

Recalling (18.6), we find from the above that

$$L(f_2^*) = -\operatorname{Re}\left\{ \int\limits_{|z|=r} P(z)\, \eta'[f(z), \varepsilon]\, h(z) \frac{dz}{2\pi i z}\right\} + O(\varepsilon).$$

and upon using the differential equation (5.13) and definition (17.19) we conclude that

$$L(f_2^*) = \operatorname{Re}\left\{\frac{1}{2\pi i} \int\limits_{\mathscr{C}_r} \frac{Q(w)}{w^2} \eta'(w, \varepsilon)\, \alpha(w)\, dw\right\} + O(\varepsilon). \tag{18.7}$$

From (17.18) it follows that

$$L(\eta_1) = -\operatorname{Re}\left\{\frac{1}{2\pi i} \int\limits_{\mathcal{T}} \left[\omega_1(w) - \frac{1}{2} w'(\sigma) \chi\chi'\right] Q(w) \frac{dw}{w^2}\right\}.$$

But $(Q(w)/w^2) w'(\sigma) \chi\chi' dw$ is real-valued on $\mathcal{T}$, and so

$$L(\eta_1) = -\operatorname{Re}\left\{\frac{1}{2\pi i} \int\limits_{\mathcal{T}} \omega_1(w) \frac{Q(w)}{w^2}\, dw\right\}.$$

In accordance with the results of §17 we have

$$\omega_1(w) = \beta(w)\, g'(w, \varepsilon; \infty) + \beta(w) \frac{1}{2\pi i} \int\limits_{\mathcal{T}} \frac{w\beta'(t)}{t(t-w)}\, dt,$$

where

$$g(w, \varepsilon; \infty) = \int\limits_{|\zeta|=r} \frac{w^2}{w - f(\zeta)} \left(\frac{Df(\zeta)}{f(\zeta)}\right)^2 h_0(\zeta) \frac{d\zeta}{2\pi i \zeta}.$$

But $h_0(z) = h(z) + \operatorname{Re} h_0$ and $Df(z) h(z) = \alpha(w)$, $z = f^{-1}(w)$. Then

$$g(w, \varepsilon; \infty) = \operatorname{Re} h_0 w + \frac{1}{2\pi i} \int\limits_{\mathscr{C}_r} \frac{w^2}{(w-t)\, t^2} \alpha(t, \varepsilon)\, dt.$$

In the domain $\mathcal{G}(\mathcal{L}_r, \mathcal{T})$ we consider

$$\tilde{g}(w) = \frac{1}{2\pi i} \int\limits_{\mathscr{C}_r} \frac{w^2}{(w-t)\, t^2} \alpha(t)\, dt. \tag{18.8}$$

and we set

$$\widetilde{\omega}_1(w) = \beta(w)\widetilde{g}'(w) + \beta(w)\frac{1}{2\pi i}\int_{\mathcal{T}} \frac{w\beta'(t)}{t(t-w)}\,dt.$$

Then, recalling (18.2), we find that

$$L(\eta_1) = -\operatorname{Re}\left\{\frac{1}{2\pi i}\int_{\mathcal{T}} \widetilde{\omega}_1(w)\frac{Q(w)}{w^2}\,dw\right\} + \operatorname{Re} h_0 L(\eta).$$

Hence

$$L(\eta_1) = -\operatorname{Re}\left\{\frac{1}{2\pi i}\int_{\mathcal{T}} \omega_1(w)\frac{Q(w)}{w^2}\,dw\right\}.$$

From this latter formula and from (18.7) we finally deduce that

$$\begin{aligned} L(f_*) = L(f) + \varepsilon^2 \Big\{L(\tilde{f}_2) + \operatorname{Re}\Big[\frac{1}{2\pi i}\int_{\mathscr{C}_r} \eta'(w,\varepsilon)\,\alpha(w)\frac{Q(w)}{w^2}\,dw \\ - \frac{1}{2\pi i}\int_{\mathcal{T}} \widetilde{\omega}_1(w)\frac{Q(w)}{w^2}\,dw\Big]\Big\} + O(\varepsilon^3). \end{aligned} \tag{18.9}$$

§19. We shall begin this section by studying a variation of a fairly general type, and then we shall be concerned with a simplification of formula (4.18), which defines the second variation of a linear functional with respect to admissible functions of the form (2.2). The measure $\nu(dz)$ and the constant c_1 will be arbitrary. We first rewrite formula (4.18):

$$\begin{aligned} \delta^2 L(f) = \operatorname{Re}\Big\{\int_{\mathscr{C}}\int_{\mathscr{C}} \mathfrak{A}(z,\zeta)\,\nu(dz)\,\nu(d\zeta) - \int_{\mathscr{C}}\int_{\mathscr{C}} \mathfrak{B}(z,\bar{\zeta})\,\nu(dz)\,\overline{\nu(d\zeta)} \\ + 2ic_1\int_{\mathscr{C}} \mathfrak{C}(z)\,\nu(dz) - c_1^2\sum_1^{r} k^2\bar{\lambda}_k a_k\Big\}. \end{aligned}$$

According to Lemma 5.1 the functions $\mathfrak{A}(z,\zeta)$ and $\mathfrak{B}(z,\zeta)$ are analytic in the bicylinder $\{|z|<1\}$, and $\mathfrak{A}(z,0) = \mathfrak{B}(z,0) = 0$. From Lemma 5.2, the function $\mathfrak{C}(z)$ is regular in $|z|<1$, and $\mathfrak{C}(0) = 0$. Thus $\delta^2 L(t)$ depends only on the moments

$$h_k = \int_{\mathscr{C}} z^k \nu(dz), \quad k = 1, 2, \ldots,$$

of the measure $\nu(dz)$, whose support is contained in $|z| \leq \rho$. It is clear that

$$|h_k| < \rho^k v(\nu, \mathscr{C}), \qquad k = 1, 2, \cdots. \tag{19.1}$$

We set

$$h(z) = \sum_1^{\infty}(h_k z^{-k} - \bar{h}_k z^k) - ic_1. \tag{19.2}$$

It follows from (19.1) that $h(z)$ is regular in the annulus $r \leq |z| \leq r^{-1}$, where $r > \rho$. If in place of the measure $\nu(dz)$ we substitute the measure $h(z)dz/2\pi iz$

with support $\{z:|z|=r\}$, the quantity $\delta^2 L(f)$ remains unchanged. By doing this we get a new formula for the second variation. Set

$$I=\int_{|z|=r} h(z)\frac{dz}{2\pi iz}\left[\int_{|\zeta|=r}\mathfrak{A}(z,\zeta)h(\zeta)\frac{d\zeta}{2\pi i\zeta}-\int_{|\zeta|=r}\mathfrak{B}(z,\bar\zeta)\overline{h(\zeta)\frac{d\zeta}{2\pi i\zeta}}\right].$$

and

$$\mathcal{Y}(z)=\int_{|\zeta|=r}\mathfrak{A}(z,\zeta)h(\zeta)\frac{d\zeta}{2\pi i\zeta}-\int_{|\zeta|=r}\mathfrak{B}(z,\bar\zeta)\overline{h(\zeta)\frac{d\zeta}{2\pi i\zeta}}$$

where we assume that $|z|=r$. It is clear that

$$\mathcal{Y}(z)=\int_{|\zeta|=r_1}\mathfrak{A}(z,\zeta)h(\zeta)\frac{d\zeta}{2\pi i\zeta}-\int_{|\zeta|=r_1}\mathfrak{B}(z,\bar\zeta)\overline{h(\zeta)\frac{d\zeta}{2\pi i\zeta}},\tag{19.3}$$

where r_1 is any constant satisfying $r<r_1<1$. We also put

$$A(z,\zeta)=P(z)\left[\left(\frac{Df(\zeta)}{f(z)-f(\zeta)}\right)^2-\left(\frac{Df(\zeta)}{f(\zeta)}\right)^2\right]+P(\zeta)\left[\left(\frac{Df(z)}{f(\zeta)-f(z)}\right)^2-\left(\frac{Df(z)}{f(z)}\right)^2\right]$$

and

$$\mathfrak{A}(z,\zeta)=A(z,\zeta)+B(z,\zeta).$$

Since $|z|=r$, the function $B(z,\zeta)$ has no singularities on $|\zeta|=r_1$. We consider next

$$\mathcal{Y}_0(z)=\int_{|\zeta|=r_1}B(z,\zeta)h(\zeta)\frac{d\zeta}{2\pi i\zeta}-\int_{|\zeta|=r_1}\mathfrak{B}(z,\bar\zeta)\overline{h(\zeta)\frac{d\zeta}{2\pi i\zeta}}.$$

Set

$$B(z,\zeta)=B_0(z,\zeta)+\left(\frac{Df(\zeta)}{f(\zeta)}\right)^2[Q^*(f(\zeta),z)+Q^*(f(\zeta))].$$

From (4.19) we have that

$$\begin{aligned}B_0(z,\zeta)={}&-\frac{z\zeta}{(z-\zeta)^2}[P(z)+P(\zeta)]+\frac{zP^*(\zeta)-\zeta P(z)}{z-\zeta}\\&+\left(\frac{Df(z)}{f(z)}\right)^2[Q^*(f(z),\zeta)+Q^*(f(z))]-R(z)-R(\zeta)+\sum_1^n k^2\bar\lambda_k a_k.\end{aligned}\tag{19.4}$$

Put

$$\mathfrak{B}(z,\bar\zeta)=B_0^*(z,\bar\zeta)-\overline{\left(\frac{Df(\zeta)}{f(\zeta)}\right)^2Q^*\left(f(\zeta),\frac{1}{\bar z}\right)}.$$

Then from (4.20) it follows that

$$\begin{aligned}B_0^*(z,\bar\zeta)={}&\frac{z\bar\zeta}{(1-z\bar\zeta)^2}[P(z)+P(\zeta)]-\frac{z\bar\zeta}{1-z\bar\zeta}(P^*(z)+\overline{P^*(\zeta)})\\&-\left(\frac{Df(z)}{f(z)}\right)^2Q^*\left(f(z),\frac{1}{\bar\zeta}\right)+R\left(\frac{1}{\bar z}\right)+\overline{R\left(\frac{1}{\bar\zeta}\right)}.\end{aligned}\tag{19.5}$$

Taking into account that

$$P(\zeta) = \overline{P(\zeta)}, \qquad P^*(\zeta) = -\overline{P^*(\zeta)}$$

for $|\zeta| = 1$, we see from (19.4) and (19.5) that with $|\zeta| = 1$

$$B_0(z, \zeta) + B_0^*(z, \bar{\zeta}) = \left(\frac{Df(z)}{f(z)}\right)^2 Q^*(f(z)) + P^*(z) - R(z) + \overline{R\left(\frac{1}{\bar{z}}\right)} + \sum_1^n k^2\bar{\lambda}_k a_k,$$

which with the aid of (4.21) can be written as

$$B_0(z, \zeta) + B_0^*(z, \bar{\zeta}) = \mathfrak{C}(z). \tag{19.6}$$

Therefore

$$\mathcal{J}_0^*(z) = \int_{|\zeta|=r_1} B_0(z, \zeta)\, h(\zeta) \frac{d\zeta}{2\pi i\zeta} - \int_{|\zeta|=r_1} B_0^*(z, \bar{\zeta})\, \overline{h(\zeta) \frac{d\zeta}{2\pi i\zeta}}$$
$$= \int_{|\zeta|=1} [B_0(z, \zeta) + B_0^*(z, \bar{\zeta})]\, h(\zeta) \frac{d\zeta}{2\pi i\zeta},$$

since for $|\zeta| = 1$ we have $h(\zeta) = -\overline{h(\zeta)}$. It follows on the basis of (19.6) that

$$\mathcal{J}_0^*(z) = -ic_1\mathfrak{C}(z),$$

and so

$$\mathcal{J}_0(z) = -ic_1\mathfrak{C}(z) + \omega(z) + \int_{|\zeta|=r_1} \left(\frac{Df(\zeta)}{f(\zeta)}\right)^2 Q^*(f(\zeta))\, h(\zeta) \frac{d\zeta}{2\pi i\zeta},$$

where

$$\omega(z) = \int_{|\zeta|=r_1} \left(\frac{Df(\zeta)}{f(\zeta)}\right)^2 Q^*(f(\zeta), z)\, h(\zeta) \frac{d\zeta}{2\pi i\zeta} + \overline{\int_{|\zeta|=r_1} \left(\frac{Df(\zeta)}{f(\zeta)}\right)^2 Q^*\left(f(\zeta), \frac{1}{\bar{z}}\right) h(\zeta) \frac{d\zeta}{2\pi i\zeta}}.$$

The quasi-polynomial $\omega(z)$ is real valued on $|z| = 1$. Thus we have

$$I_0 = \int_{|z|=r} \mathcal{J}_0(z)\, h(z) \frac{dz}{2\pi iz} = -ic_1 \int_{|z|=r} \mathfrak{C}(z)\, h(z) \frac{dz}{2\pi iz}$$
$$- ic_1 \int_{|z|=r_1} \left(\frac{Df(z)}{f(z)}\right)^2 Q^*(f(z))\, h(z) \frac{dz}{2\pi iz} + \int_{|z|=r} \omega(z)\, h(z) \frac{dz}{2\pi iz}.$$

We note that

$$\int_{|z|=r} \omega(z)\, h(z) \frac{dz}{2\pi iz} = \overline{\int_{|z|=r} \omega(z)\, h(z) \frac{dz}{2\pi iz}} = iA,$$

where A is a real constant. On the basis of (4.21) it follows that

$$\int_{|z|=r} \mathfrak{C}(z)\, h(z) \frac{dz}{2\pi iz}$$
$$= \int_{|z|=r} \left[\left(\frac{Df(z)}{f(z)}\right)^2 Q^*(f(z)) + P^*(z) - R(z) + \overline{R\left(\frac{1}{\bar{z}}\right)} + \sum_1^n k^2\lambda_k a_k\right] h(z) \frac{dz}{2\pi iz}.$$

But

$$\int\limits_{|z|=r} \left[P^*(z) - R(z) + \overline{R\left(\frac{1}{\bar z}\right)}\right] h(z) \frac{dz}{2\pi i z}$$
$$= \int\limits_{|z|=1} [P^*(z) - R(z) + \overline{R(z)}]\, h(z) \frac{dz}{2\pi i z}.$$

Since the function on the square brackets above is purely imaginary, the integral itself is equal to some real constant, say B. Therefore

$$\int\limits_{|z|=r} \mathfrak{C}(z)\, h(z) \frac{dz}{2\pi i z} = \int\limits_{|z|=r} \left(\frac{Df(z)}{f(z)}\right)^2 Q^*(f(z))\, h(z) \frac{dz}{2\pi i z} - ic_1 \sum_1^n k^2 \bar{\lambda}_k a_k + B,$$

and so

$$I_0 = -2ic_1 \int\limits_{|z|=r} \mathfrak{C}(z)\, h(z) \frac{dz}{2\pi i z} + c_1^2 \sum_1^n k^2 \bar{\lambda}_k a_k + i(A + Bc_1). \tag{19.7}$$

But

$$\mathcal{Y}(z) = \int\limits_{|\zeta|=r_1} A(z, \zeta)\, h(\zeta) \frac{d\zeta}{2\pi i \zeta} + \mathcal{Y}_0(z),$$

and so

$$I = \int\limits_{|z|=r} h(z) \frac{dz}{2\pi i z} \int\limits_{|\zeta|=r_1} A(z, \zeta)\, h(\zeta) \frac{d\zeta}{2\pi i \zeta} + I_0.$$

Recalling (19.7), we see from this that

$$\delta^2 L(f) = \operatorname{Re}\left\{ \int\limits_{|z|=r} h(z) \frac{dz}{2\pi i z} \int\limits_{|\zeta|=r_1} A(z, \zeta)\, h(\zeta) \frac{d\zeta}{2\pi i \zeta} \right\}. \tag{19.8}$$

This formula can be simplified considerably. We set

$$\mathcal{I}_1(z) = \int\limits_{|\zeta|=r_1} A(z, \zeta)\, h(\zeta) \frac{d\zeta}{2\pi i \zeta}.$$

and we introduce, as in §17 (see (17.19)) the function

$$\alpha(w) = Df(z) h(z), \quad z = f^{-1}(w). \tag{19.9}$$

The function $\alpha(w)$ is regular in the domain $\mathcal{G}(\mathcal{L}_r, \mathcal{T})$, and we have determined some of its properties. The function

$$\left[\frac{Q(w)}{w^2}\right]^{1/2} \alpha(w) = \left[\left(\frac{Df(z)}{f(z)}\right)^2 Q(f(z))\right]^{1/2} h(z) = iP^{1/2}(z)\, h(z)$$

is bounded in $\mathcal{G}(\mathcal{L}_r, \mathcal{T})$. If w_j is a terminal vertex of the tree $\mathcal{T}$, then, since $P(z_j) = 0$, where $z_j = f^{-1}(w_j)$, we have

$$\left[\frac{Q(w_j)}{w_j^2}\right]^{1/2} \alpha(w_j) = 0.$$

From (17.24) it follows that for $w \in \mathcal{T}$

$$\operatorname{Im}\left\{\left[\frac{Q(w)}{w^2}\right]^{1/2} \alpha_{\pm}(w)\right\} = 0. \tag{19.10}$$

By (5.13) the function $A(z, \zeta)$ can be written in the form

$$A(z, \zeta) = -\left(\frac{Df(z)}{f(z)}\right)^2 Q(f(z)) \left[\left(\frac{Df(\zeta)}{f(z) - f(\zeta)}\right)^2 - \left(\frac{Df(\zeta)}{f(\zeta)}\right)^2\right]$$
$$- \left(\frac{Df(\zeta)}{f(\zeta)}\right)^2 Q(f(\zeta)) \left[\left(\frac{Df(z)}{f(\zeta) - f(z)}\right)^2 - \left(\frac{Df(z)}{f(z)}\right)^2\right].$$

Thus

$$\mathcal{J}_1(z) = -\frac{1}{2\pi i} \int_{\mathscr{C}_{r_1}} \left\{\left(\frac{Df(z)}{f(z)}\right)^2 Q(f(z)) \left[(f(z) - w)^{-2} - w^{-2})\right]\right.$$
$$\left. - \frac{Q(w)}{w^2} \left[\left(\frac{Df(z)}{f(z) - w}\right)^2 - \left(\frac{Df(z)}{f(z)}\right)^2\right]\right\} \alpha(w)\, dw.$$

Suppose that w lies inside the curve $\mathscr{C}_r$. In accordance with (17.20) we introduce the function

$$\xi(w) = \frac{1}{2\pi i} \int_{\mathscr{C}_r} \frac{w^2 \alpha(t)}{(w - t)\, t^2}\, dt. \tag{19.11}$$

As was shown in §17, $\xi(w)$ is regular off $\mathcal{T}$; and if the path of integration $\mathscr{C}_r$ is replaced by the arc $\mathscr{C}_{r_1}$, with $r_1 > r$, then we get an analytic continuation of $\xi(w)$ into the domain bounded by $\mathscr{C}_{r_1}$. We note that if $w \in \mathcal{G}(\mathscr{C}_r, \mathcal{T})$, then

$$\xi(w) + \alpha(w) = \frac{1}{2\pi i} \int_{\mathscr{C}_r} \frac{w^2 \alpha(t)}{(w - t)\, t^2}\, dt. \tag{19.12}$$

From this it follows that the function $[Q(w)/w^2]^{1/2}\xi(w)$ is bounded in the domain $\mathcal{G}(\mathscr{C}_r, \mathcal{T})$. Set

$$\xi_0(w) = \frac{1}{2\pi i} \int_{\mathscr{C}_r} \frac{Q(t)\, \alpha(t)}{(w - t)\, t^2}\, dt.$$

Clearly $\xi_0(w)$ is regular in $\mathcal{G}(\mathcal{T})$, and if $w \in \mathcal{G}(\mathscr{C}_r, \mathcal{T})$, then

$$\xi_0(w) + \frac{Q(w)}{w^2}\alpha(w) = \frac{1}{2\pi i} \int_{\mathscr{C}_r} \frac{Q(t)\, \alpha(t)}{(w - t)\, t^2}\, dt.$$

Since

$$\xi_0(w) - \frac{Q(w)}{w^2}\xi(w) = \frac{1}{2\pi i} \int_{\mathscr{C}_2} [Q(t) - Q(w)] \frac{\alpha(t)\, dt}{(w - t)\, t^2},$$

we have that

$$\xi_0(w) - \frac{Q(w)}{w^2}\xi(w) = -\sum_1^n \gamma_k w^{-k},$$

where $\sum_1^n \gamma_k w^{-k}$ is that part of the Laurent series of $(Q(w)/w^2)\xi(w)$ which has negative powers of w. Now we shall determine the quantities γ_k. If in a neighborhood of $w=0$ we set $\xi(w)=\sum_1^\infty \xi_k w^{k+1}$, then

$$\xi_k = -\frac{1}{2\pi i}\int_{\mathscr{C}_r} \alpha(t)\, t^{-k-2}\,dt. \tag{19.13}$$

It is clear that

$$\gamma_k = \sum_{l=k}^{n} A_l \xi_{l+1-k}, \qquad k=1,\dots,n. \tag{19.14}$$

Using the above functions we can write $\mathscr{I}_1(z)$ in the form

$$\mathscr{I}_1(z) = \frac{Df(z)}{f^2(z)}\, Q(f(z))\, D\xi\,[f(z)] + Df(z)\, D\xi_0[f(z)] + \left(\frac{Df(z)}{f(z)}\right)^2 \frac{1}{2\pi i}\int_{\mathscr{C}_{r_1}} \frac{Q(w)}{w^2}\,\alpha(w)\,dw.$$

But

$$D\xi_0[f(z)] = D\left\{\frac{Q(f(z))}{f^2(z)}\,\xi\,[f(z)]\right\} + Df(z)\sum_1^n k\gamma_k\,[f(z)]^{-k-1},$$

and

$$\frac{1}{2\pi i}\int_{\mathscr{C}_r} \frac{Q(w)}{w^2}\,\alpha(w)\,dw = -\gamma_1, \tag{19.15}$$

as follows from (19.13) and (19.14). Therefore

$$\mathscr{I}_1(z) = \frac{Df(z)}{f^2(z)}\, Q(f(z))\, D\,[\xi(f(z))] + D\left[\frac{Q(f(z))}{f^2(z)}\,\xi(f(z))\right] + (Df(z))^2\sum_2^n k\gamma_k\,[f(z)]^{-k-1}.$$

After making some obvious transformations, we arrive at

$$\mathscr{I}_1(z) = 2Df(z)\sqrt{\frac{Q(f(z))}{f^2(z)}}\, D\left[\sqrt{\frac{Q(f(z))}{f^2(z)}}\,\xi(f(z))\right] + (Df(z))^2\sum_2^n k\gamma_k\,[f(z)]^{-k-1}.$$

In turn, we see that

$$\int_{|z|=r} \mathscr{I}_1(z)\, h(z)\,\frac{dr}{2\pi i z} = \frac{1}{\pi i}\int_{\mathscr{C}_r}\sqrt{\frac{Q(w)}{w^2}}\,\alpha(w)\, d\left(\sqrt{\frac{Q(w)}{w^2}}\,\xi(w)\right) + \frac{1}{2\pi i}\int_{\mathscr{C}_r}\sum_2^n k\gamma_k w^{-k-1}\alpha(w)\,dw,$$

and, recalling (19.13), we finally conclude that

$$\int_{|z|=r} \mathcal{I}_1(z)\, h(z) \frac{dz}{2\pi i z} = \frac{1}{\pi i} \int_{\mathscr{C}_r} \sqrt{\frac{Q(w)}{w^2}}\, \alpha(w)\, d\left(\sqrt{\frac{Q(w)}{w^2}}\, \xi(w)\right)$$
$$- \sum_{1}^{n-1} (k+1)\, \xi_k \gamma_{k+1}.$$

For notational ease we set

$$\xi^*(w) = \sqrt{\frac{Q(w)}{w^2}}\, \xi(w), \qquad \xi^*_{\pm}(w) = \sqrt{\frac{Q(w)}{w^2}}\, \xi_{\pm}(w).$$

Our next task is to find a new expression for the integral

$$j = \frac{1}{\pi i} \int_{\mathscr{C}_r} \sqrt{\frac{Q(w)}{w^2}}\, \alpha(w)\, d\xi^*(w).$$

Let R be a large positive number, $K_R = \{w_k : |w| \leqq R\}$, and $\mathscr{T}_R = \mathscr{T} \cap K_R$. Let T_R be any arc of the circumference ∂K_R whose boundary points belong to $\mathscr{T}$. On T_R we have

$$\xi^*(w) = -\sqrt{\frac{Q(w)}{w^2}}\, \alpha(w) + \sqrt{\frac{Q(w)}{w^2}}\, \frac{1}{2\pi i} \int_{\mathscr{C}_r} \frac{w^2 \alpha(t)}{(w-t)\, t^2}\, dt,$$

since $\mathscr{C}_r \subset K_R$ for sufficiently large R. Thus on T_R we have

$$\frac{d}{dw}\, \xi^*(w) = -\frac{d}{dw}\left(\sqrt{\frac{Q(w)}{w^2}}\, \alpha(w)\right) + O(R^{-3/2}),$$

and so

$$\frac{1}{\pi i} \int_{T_R} \sqrt{\frac{Q(w)}{w^2}}\, \alpha(w)\, d\xi^*(w) = -\frac{1}{2\pi i} \int_{T_R} d\left(\frac{Q(w)}{w^2}\, \alpha^2(w)\right) + O(R^{-1/2}).$$

If w_1 and w_2 are boundary points of T_R, the last integral equals

$$-\frac{1}{2\pi i}\left[\frac{Q(w_2)}{w_2^2}\, \alpha^2(w_2) - \frac{Q(w_1)}{w_1^2}\, \alpha^2(w_1)\right].$$

This is a real number, as is shown by (19.10). Thus, changing the contour $\mathscr{C}_r$ suitably and letting $R \to \infty$, we find that

$$\operatorname{Re} j = \operatorname{Re}\left\{\frac{1}{\pi i} \int_{\mathscr{T}} \left[\sqrt{\frac{Q(w)}{w^2}}\, \alpha_+(w)\, d\xi^*_+(w) - \sqrt{\frac{Q(w)}{w^2}}\, \alpha_-(w)\, d\xi^*_-(w)\right]\right\}.$$

It follows upon using (19.10) that

$$\operatorname{Re} j = -\operatorname{Re}\left\{\frac{1}{\pi i} \int_{\mathscr{T}} \left[\sqrt{\frac{Q(w)}{w^2}}\, \alpha_+(w)\, \overline{d\xi^*_+(w)} - \sqrt{\frac{Q(w)}{w^2}}\, \alpha_-(w)\, \overline{d\xi^*_-(w)}\right]\right\},$$

from which we conclude that

$$\operatorname{Re} j = \operatorname{Re}\left\{\frac{1}{\pi i}\int_{\mathscr{T}} [\xi_+^*(w)\,\overline{d\xi_+^*(w)} - \xi_-^*(w)\,\overline{d\xi_-^*(w)}]\right\}$$
$$- \operatorname{Re}\left\{\frac{1}{\pi i}\int_{\mathscr{T}} \left[\sqrt{\frac{Q(w)}{w^2}}(\xi_+(w)+\alpha_+(w))\,\overline{d\xi_+^*(w)} - \sqrt{\frac{Q(w)}{w^2}}(\xi_-(w)+\alpha_-(w))\,\overline{d\xi_-^*(w)}\right]\right\}.$$

We denote the first integral in this formula by j_1. From (19.12) we have

$$\operatorname{Re} j = \operatorname{Re} j_1 - \operatorname{Re}\left\{\frac{1}{\pi i}\int_{\mathscr{T}} \sqrt{\frac{Q(w)}{w^2}}\,\xi_+(w)+\alpha_+(w))\,\overline{d\,[\xi_+^*(w)-\xi_-^*(w)]}\right\}.$$

From (19.12) and (19.10) we get

$$\overline{\xi_+^*(w)-\xi_-^*(w)} = -\sqrt{\frac{Q(w)}{w^2}}\,(\alpha_+(w)-\alpha_-(w)),$$

and so

$$\operatorname{Re} j = \operatorname{Re} j_1 + \operatorname{Re}\left\{\frac{1}{\pi i}\int_{\mathscr{T}} \sqrt{\frac{Q(w)}{w^2}}(\xi_+(w)+\alpha_+(w))\,d\left[\sqrt{\frac{Q(w)}{w^2}}(\alpha_+(w)-\alpha_-(w))\right]\right\};$$

and this relation, in combination with (19.12), implies that

$$\operatorname{Re} j = \operatorname{Re} j_1 + \operatorname{Re}\left\{\frac{1}{\pi i}\int_{\mathscr{T}} \left[\sqrt{\frac{Q(w)}{w^2}}(\xi_+(w)+\alpha_+(w))\,d\left[\sqrt{\frac{Q(w)}{w^2}}\,\alpha_+(w)\right] - \sqrt{\frac{Q(w)}{w^2}}(\xi_-(w)+\alpha_-(w))\,d\left[\sqrt{\frac{Q(w)}{w^2}}\,\alpha_-(w)\right]\right]\right\}.$$

We now proceed to integrate the last integral by parts. The result of this operation is a sum of terms of the form

$$\pm\lim_{w\to w_j}\left\{\frac{1}{\pi i}\sqrt{\frac{Q(w)}{w^2}}\,[\xi_\pm(w)+\alpha_\pm(w)]\sqrt{\frac{Q(w)}{w^2}}\,\alpha_\pm(w)\right\}, \tag{19.16}$$

where w_j is an arbitrary vertex of $\mathscr{T}$. If w_j is not terminal, then $Q(w_j)=0$, and it follows from (19.12) that

$$\lim_{w\to w_j}\sqrt{\frac{Q(w)}{w^2}}\,[\xi_\pm(w)+\alpha_\pm(w)] = 0.$$

As shown previously, the function $\sqrt{Q(w)/w^2}\,\alpha(w)$ is bounded in $\mathscr{G}(\mathscr{L}_r, \mathscr{T})$. Thus for nonterminal vertices w_j the expression (19.16) is zero. If w_j is a terminal vertex, we previously proved that

$$\sqrt{\frac{Q(w_j)}{w_j^2}}\,\alpha_\pm(w_j) = 0$$

and so (19.16) vanishes in this case also. Therefore

$$\operatorname{Re} j = \operatorname{Re} j_1 - \operatorname{Re}\left\{\frac{1}{\pi i}\int_{\mathscr{T}}\left[\sqrt{\frac{Q(w)}{w^2}}\alpha_+(w)d\left[\sqrt{\frac{Q(w)}{w^2}}(\xi_+(w)+\alpha_+(w))\right]\right.\right.$$
$$\left.\left. - \sqrt{\frac{Q(w)}{w^2}}\alpha_-(w)d\left[\sqrt{\frac{Q(w)}{w^2}}(\xi_-(w)+\alpha_-(w))\right]\right]\right\}.$$

In view of (19.10) this relation implies that

$$\operatorname{Re} j = \operatorname{Re} j_1 - \operatorname{Re}\left\{\frac{1}{\pi i}\int_{\mathscr{T}}\left[\sqrt{\frac{Q(w)}{w^2}}\alpha_+(w)\,d\xi_+^*(w)\right.\right.$$
$$\left.\left. - \sqrt{\frac{Q(w)}{w^2}}\alpha_-(w)\,d\xi_-^*(w)\right]\right\},$$

or, in other words, $2\operatorname{Re} j = \operatorname{Re} j_1$. Therefore

$$\operatorname{Re} j = \operatorname{Re}\left\{\frac{1}{2\pi i}\int_{\mathscr{T}}[\xi_+^*(w)\overline{d\xi_+^*(w)} - \xi_-^*(w)\overline{d\xi_-^*(w)}]\right\}.$$

On the basis of (19.8) we have

$$\delta^2 L(f) = \operatorname{Re}\left\{\frac{1}{2\pi i}\int_{\mathscr{T}}[\xi_+^*(w)\overline{d\xi_+^*(w)} - \xi_-^*(w)\overline{d\xi_-^*(w)}]\right.$$
$$\left. - \sum_1^{n-1}(k+1)\xi_k\gamma_{k+1}\right\}. \tag{19.17}$$

Using (19.14) we have that

$$\sum_1^{n-1}(k+1)\xi_k\gamma_{k+1} = \sum_1^{n-1}(k+1)\xi_k\sum_{l=k+1}^{n}A_l\xi_{l-k},$$

and upon using the easily verified identity

$$\sum_{k=1}^{n-1}(k+1)\xi_k\sum_{l=k+1}^{n}A_l\xi_{l-k} = \frac{1}{2}\sum_{k=1}^{n-1}\xi_k\sum_{l=1}^{n-k}(k+l+2)A_{k+l}\xi_l$$

we see that

$$\delta^2 L(f) = \operatorname{Re}\left\{\frac{1}{2\pi i}\int_{\mathscr{T}}[\xi_+^*(w)\overline{d\xi_+^*(w)} - \xi_-^*(w)\overline{d\xi_-^*(w)}]\right.$$
$$\left. - \frac{1}{2}\sum_{k=1}^{n-1}\xi_k\sum_{l=1}^{n-k}(k+l+2)A_{k+l}\xi_l\right\}. \tag{19.18}$$

This is our desired expression for the second variation of a linear functional. We can give an interpretation to the function $\xi(w)$. If we recall the results of §17, it is easy to verify that $\xi[f(z)] = \delta f(z)$. In fact this formula can be verified by direct computation, since the situation of §17 is very similar to the present situation. To see this, note that in §17 the constant c_1 was chosen so that $\operatorname{Im}\varphi_1(0) = 0$; in the present section the constant c_1 is arbitrary, but the function $h(z)$ is defined as in (19.2). According to (2.11),

$$\delta f(z; \nu; c_1) = \int_{\mathscr{C}} \left[\frac{f^2(z)}{f(z) - f(t)}\left(\frac{Df(t)}{f(t)}\right)^2 + \frac{Df(z)\,t}{t - z} - f(z)\right] \overline{\nu(dt)}$$
$$+ Df(z) \int_{\mathscr{C}} \frac{\bar{t}z}{1 - \bar{t}z} \overline{\nu(dt)} + ic_1 (Df(z) - f(z)).$$

Since the kernel

$$\frac{f^2(z)}{f(z) - f(t)}\left(\frac{Df(t)}{f(t)}\right)^2 + \frac{Df(z)\,t}{t - z} - f(z)$$

is a regular function of t in the disc $|t| < 1$, and since this function vanishes at $t = 0$, we have

$$\delta f(z) = \int_{|t|=r} \left[\frac{f^2(z)}{f(z) - f(t)}\left(\frac{Df(t)}{f(t)}\right)^2 + \frac{Df(z)\,t}{t - z} - f(z)\right] h(t) \frac{dt}{2\pi i t}$$
$$+ Df(z) \int_{|t|=r} \frac{\bar{t}z}{1 - \bar{t}z} h(t) \frac{dt}{2\pi i t} + ic_1 (Df(z) - f(z)).$$

If $|z| < r$, then

$$\int_{|t|=r} \left[\frac{Df(z)\,t}{t - z} - f(z)\right] h(t) \frac{dt}{2\pi i t} + Df(z) \int_{|t|=r} \overline{\frac{\bar{t}z}{1 - \bar{t}z} h(t) \frac{dt}{2\pi i t}}$$
$$= \int_{|t|=1} \left[\frac{Df(z)}{t - z} - f(z)\right] h(t) \frac{dt}{2\pi i t} + Df(z) \int_{|t|=1} \overline{\frac{\bar{t}z}{1 - \bar{t}z} h(t) \frac{dt}{2\pi i t}}$$
$$= [Df(z) - f(z)] \int_{|t|=1} h(t) \frac{dt}{2\pi i t} = -ic_1 [Df(z) - f(z)].$$

It follows that for $|z| < r$

$$\delta f(z) = \int_{|t|=r} \frac{f^2(z)}{f(z) - f(t)}\left(\frac{Df(t)}{f(t)}\right)^2 h(t) \frac{dt}{2\pi i t} = \frac{1}{2\pi i} \int_{\mathscr{C}_r} \frac{f^2(z)\,\alpha(w)}{(f(z) - w)\,w^2} dw,$$

and so $\delta f(z) = \xi[f(z)]$, as required.

We can also establish a connection between the quantities ξ_k defined by (19.13) and the first coefficient variation of $f(z)$. From the last equality above we have

$$\delta a_k = \frac{1}{2\pi i} \int_{\mathscr{C}_r} \alpha(w)\, q_k(w) \frac{dw}{w^2}.$$

Applying the formulas for $q_k(w)$ established in §10, we find that

$$\delta a_k = \sum_{l=1}^{k} a_k^{(l+1)} \xi_l, \qquad k = 1, 2, \ldots. \tag{19.19}$$

From this, just as in §10, we conclude that

$$\xi_k = \sum_{l=1}^{k} b_k^{(l+1)}\,\delta a_l, \qquad k = 1, 2, \ldots, \tag{19.20}$$

where $b_k^{(l)}$ is defined by the condition that $z^l = [f^{-1}(w)]^l = \sum_{k=l-1}^{\infty} b_k^{(l)} w^{k+1}$. In view of the above remarks, formula (19.18) is quite plausible from an intuitive point of view, since the first part of this formula is a quadratic functional of the first variation of $f(z)$. In the theory of variational problems on a linear space this conclusion would be quite trivial. However, for nonlinear spaces such as the class S this fact is not at all obvious.

Theorem 11. *For any admissible function of the class $\mathscr{D}_f^0$, the second variation of a linear functional is a quadratic functional, of the form* (19.18), *of the first variation of the extremal function $f(z)$.*

Remark 19.1. For nonlinear functionals of the form $\mathscr{F}(f) = F(x)$, formula (19.18) is changed in only a trivial way.

§20. We shall now prove that the formula for the second variation given in §19 for admissible functions of the class $\mathscr{D}_f^0$ continues to hold for the more general class $\mathscr{D}_f$. From (18.9) it follows that our problem is really to find another expression for

$$\mathscr{I}(\varepsilon) = L(\tilde{f}_2) + \operatorname{Re}\left\{\frac{1}{2\pi i}\int_{\mathscr{C}_r} \eta'(w,\varepsilon)\,\alpha(w)\,Q(w)\,\frac{dw}{w^2} - \frac{1}{2\pi i}\int_{\mathscr{T}} \tilde{\omega}_1(w)\,Q(w)\,\frac{dw}{w^2}\right\}.$$

Recalling the remarks made in §18, we see that $L(2\tilde{f}_2)$ is calculated according to formula (4.18). Since $\xi[f(z),\epsilon]$ is the first variation of $f(z)$, as defined by the measure $\nu(dz)$ and the constant c_1, we have from (19.18) that

$$\begin{aligned} L(2\tilde{f}_2) = \operatorname{Re}\Bigg\{\frac{1}{2\pi i}\int_{\mathscr{T}} [\xi_+^*(w,\varepsilon)\,\overline{d\xi_+^*(w,\varepsilon)} - \xi_-^*(w,\varepsilon)\,\overline{d\xi_-^*(w,\varepsilon)}] \\ - \frac{1}{2}\sum_{k=1}^{n-1}\xi_k(\varepsilon)\sum_{l=1}^{n-k}(k+l+2)\,A_{k+l}\xi_l(\varepsilon)\Bigg\}. \end{aligned} \tag{20.1}$$

Let

$$I_1 = -\frac{1}{2\pi i}\int_{\mathscr{T}} \frac{Q(w)}{w^2}\,\beta(w)\,dw\,\frac{1}{2\pi i}\int_{\mathscr{T}} \frac{w\beta'(t)}{t(t-w)}\,dt \tag{20.2}$$

and

$$I_2 = \frac{1}{2\pi i}\int_{\mathscr{C}_r} \eta'(w,\varepsilon)\,\frac{Q(w)}{w^2}\,\alpha(w)\,dw - \frac{1}{2\pi i}\int_{\mathscr{T}} \tilde{g}'(w)\,\frac{Q(w)}{w^2}\,\beta(w)\,dw. \tag{20.3}$$

Then, recalling (18.9), we find that

$$\mathscr{I}(\epsilon) = L(\tilde{f}_2) + \operatorname{Re}(I_1 + I_2). \tag{20.4}$$

We shall evaluate I_1 and I_2 each by two different methods. Beginning with I_1, we note that the formula

$$\eta'(w, \varepsilon) = \frac{1}{2\pi i} \int_{\mathcal{T}} \frac{w\beta'(t)}{t(t-w)} dt$$

follows easily from (17.17). Applying the Sohockiĭ-Plemelj formulas we obtain

$$\frac{1}{2}(\eta'_+(w, \varepsilon) + \eta'_-(w, \varepsilon)) = \frac{1}{2\pi i} \int_{\mathcal{T}} \frac{w\beta'(t)}{t(t-w)} dt$$

for $w \in \mathcal{T}$. Using this latter formula, we get that

$$I_1 = -\frac{1}{4\pi i} \int_{\mathcal{T}} \frac{Q(w)}{w^2} \beta(w)\, d\,[\eta_+(w, \varepsilon) + \eta_-(w, \varepsilon)]. \tag{20.5}$$

We now evaluate I_1 by a different method. Interchanging the order of integration in (20.2), we have

$$I_1 = -\frac{1}{2\pi i} \int_{\mathcal{T}} \beta'(t) \frac{dt}{2\pi i} \int_{\mathcal{T}} \frac{Q(w)}{w^2} \beta'(w) \left(\frac{1}{t-w} - \frac{1}{t}\right) dw$$

and we now integrate by parts. Noting that $\beta(w)$ vanishes at the vertices of the tree $\mathcal{T}$, we find that

$$I_1 = -\frac{1}{2\pi i} \int_{\mathcal{T}} \beta(t)\, d\left[\frac{1}{2\pi i} \int_{\mathcal{T}} \frac{Q(w)\beta(w)}{(w-t)w^2} dw\right] + \frac{1}{2\pi i} \int_{\mathcal{T}} \frac{\beta(t)}{t^2} dt \frac{1}{2\pi i} \int_{\mathcal{T}} \frac{Q(w)}{w^2} \beta(w)\, dw.$$

Consider the function

$$\Psi(w) = \frac{1}{2\pi i} \int_{\mathcal{T}} \frac{Q(t)\beta(t)}{(t-w)t^2} dt - \frac{Q(w)}{w^2} \eta(w, \varepsilon).$$

From the formulas for the jump discontinuities it follows that on $\mathcal{T}$ we have $\Psi_+(w) = \Psi_-(w)$, and $\Psi(w)$ can have a singular point only at $w = 0$, or at the vertices of the tree. But $\beta(w)$ is continuously differentiable and vanishes at the vertices, so $\Psi(w)$ is bounded in a neighborhood of each vertex, which implies that $\Psi(w)$ is regular everywhere, with the exception of the point $w = 0$. Set

$$Q(w) w^{-2} \eta(w, \varepsilon) = \sum_1^n \mu_k w^{-k} + \cdots,$$

where the dots represent the sum of all the terms involving nonnegative powers of w. Then

$$\Psi(w) = -\sum_1^n \mu_k w^{-k}. \tag{20.6}$$

We shall express the quantities μ_k in terms of the coefficients of the power series for $\eta(w, \epsilon)$. Put $\eta(w, \epsilon) = \sum_1^\infty \eta_k(\epsilon) w^{k+1}$. Then

(20.7) $$\mu_k = \sum_{l=k}^{n} A_l \eta_{l+1-k}(\varepsilon), \qquad k = 1, \ldots, n.$$

From (20.6) we have for $w \in \mathscr{T}$ that

$$\frac{1}{2\pi i} \int_{\mathscr{T}} \frac{Q(t)\beta(t)\,dt}{t^2(t-w)} = \frac{1}{2} Q(w) w^{-2} [\eta_+(w,\varepsilon) + \eta_-(w,\varepsilon)] - \sum_1^n \mu_k w^{-k}.$$

We note that

(20.8) $$\mu_1 = \frac{1}{2\pi i} \int_{\mathscr{T}} \frac{Q(w)}{w^2} \beta(w)\,dw.$$

Using these two latter equations, we find that

$$I_1 = -\frac{1}{4\pi i} \int_{\mathscr{T}} \beta(w)\, d\left[\frac{Q(w)}{w^2} (\eta_+(w,\varepsilon) + \eta_-(w,\varepsilon))\right] - \sum_{k=2}^{n} k\mu_k \eta_{k-1}(\varepsilon).$$

Combining this result with (20.5) and performing simple rearrangements, we find that

$$I_1 = -\frac{1}{4\pi i} \int_{\mathscr{T}} \beta(w) \sqrt{\frac{Q(w)}{w^2}}\, d[\eta_+^*(w,\varepsilon) + \eta_-^*(w,\varepsilon)] - \frac{1}{2} \sum_1^{n-1} (k+1)\mu_{k+1}\eta_k(\varepsilon),$$

where

$$\eta^*(w,\varepsilon) = \sqrt{\frac{Q(w)}{w^2}}\,\eta(w,\varepsilon), \qquad \eta_\pm^*(w,\varepsilon) = \sqrt{\frac{Q(w)}{w^2}}\,\eta_\pm(w,\varepsilon).$$

Set

$$j = \operatorname{Re}\left\{\frac{1}{4\pi i} \int_{\mathscr{T}} \beta(w) \sqrt{\frac{Q(w)}{w^2}}\, d[\eta_+^*(w,\varepsilon) + \eta_-^*(w,\varepsilon)]\right\}.$$

Since $\operatorname{Im}\{\beta(w)\sqrt{Q(w)/w^2}\} = 0$ and $\eta_+(w,\epsilon) - \eta_-(w,\epsilon) = \beta(w)$ on $\mathscr{T}$, we have that

$$j = -\operatorname{Re}\left\{\frac{1}{4\pi i} \int_{\mathscr{T}} [\eta_+^*(w,\varepsilon) - \eta_-^*(w,\varepsilon)]\, d\,\overline{[\eta_+^*(w,\varepsilon) + \eta_-^*(w,\varepsilon)]}\right\},$$

and so

$$j = -\operatorname{Re}\left\{\frac{1}{4\pi i} \int_{\mathscr{T}} [\eta_+^*(w,\varepsilon)\,\overline{d\eta_+^*(w,\varepsilon)} - \eta_-^*(w,\varepsilon)\,\overline{d\eta_-^*(w,\varepsilon)}] + \frac{1}{4\pi i} \int_{\mathscr{T}} [\eta_+^*(w,\varepsilon)\,\overline{d\eta_-^*(w,\varepsilon)} - \eta_-^*(w,\varepsilon)\,\overline{d\eta_+^*(w,\varepsilon)}]\right\}.$$

It is not difficult to see that the second integral in the last formula is purely imaginary. In fact, after integration by parts we get that

$$\frac{1}{4\pi i} \int_{\mathscr{T}} [\eta_+^*(w,\varepsilon)\,\overline{d\eta_-^*(w,\varepsilon)} - \eta_-^*(w,\varepsilon)\,\overline{d\eta_+^*(w,\varepsilon)}] =$$

$$= \frac{1}{4\pi i} \int_{\mathcal{T}} [\eta_+^*(w, \varepsilon) \overline{d\eta_-^*(w, \varepsilon)} + \overline{\eta_+^*(w, \varepsilon)}\, d\eta_-^*(w, \varepsilon)] + \frac{1}{4\pi i} A,$$

where A is a sum of terms not involving any integrals. At the vertices w_j we have $\beta(w_j) = 0$, and so $\eta_+^*(w_j, \epsilon) = \eta_-^*(w_j, \epsilon)$, which in turn implies that A is a sum of terms of the form $\pm|\eta^*(w_j, \epsilon)|^2$, and so in particular A is real. Combining all the above remarks, we conclude that

$$\operatorname{Re} I_1 = \frac{1}{2} \operatorname{Re}\Big\{\frac{1}{2\pi i} \int_{\mathcal{T}} [\eta_+^*(w, \varepsilon) \overline{d\eta_+^*(w, \varepsilon)} - \eta_-^*(w, \varepsilon) \overline{d\eta_-^*(w, \varepsilon)}] - \sum_{1}^{n-1} (k+1) \mu_{k+1} \eta_k(\varepsilon)\Big\}. \tag{20.9}$$

We now evaluate I_2 by two methods. From (18.8) we have for $w \in \mathcal{G}(\mathcal{C}_r, \mathcal{T})$ that

$$\tilde{g}'(w) = \frac{1}{2\pi i} \int_{\mathcal{T}} [t^{-2} - (t - w)^{-2}]\, \alpha(t)\, dt,$$

and so

$$\frac{1}{2\pi i} \int_{\mathcal{T}} \beta(w)\, \tilde{g}'(w) \frac{Q(w)}{w^2}\, dw = -\frac{1}{2\pi i} \int_{\mathcal{C}_r} \alpha(t)\, d\left[\frac{1}{2\pi i} \int_{\mathcal{T}} \frac{Q(w)\, \beta(w)}{w^2 (w - t)}\, dw\right] + \frac{1}{2\pi i} \int_{\mathcal{C}_r} \alpha(t) \frac{dt}{t^2} \frac{1}{2\pi i} \int_{\mathcal{T}} \frac{Q(w)}{w^2} \beta(w)\, dw.$$

It now follows from (20.6) and (20.7) that

$$\frac{1}{2\pi i} \int_{\mathcal{T}} \beta(w)\, \tilde{g}'(w) \frac{Q(w)}{w^2}\, dw = -\frac{1}{2\pi i} \int_{\mathcal{C}_r} \alpha(w)\, d\left[\frac{Q(w)}{w^2} \eta(w, \varepsilon)\right] - \frac{1}{2\pi i} \int_{\mathcal{C}_r} \alpha(w) \sum_{2}^{n} k\mu_k w^{-k-1}\, dw.$$

As before, set $\xi(w, \epsilon) = \sum_1^\infty \xi_k(\epsilon) w^{k+1}$. From (17.19) we get

$$\xi_k(\varepsilon) = -\frac{1}{2\pi i} \int_{\mathcal{C}_r} \alpha(w)\, w^{-k-2} dw.$$

Thus

$$\frac{1}{2\pi i} \int_{\mathcal{T}} \beta(w)\, \tilde{g}'(w) \frac{Q(w)}{w^2}\, dw = -\frac{1}{2\pi i} \int_{\mathcal{C}_{r'}} \alpha(w)\, d\left[\frac{Q(w)}{w^2} \eta(w, \varepsilon)\right] + \sum_{1}^{n-1} (k+1)\, \xi_k(\varepsilon)\, \mu_{k+1},$$

and hence

$$I_2 = \frac{1}{\pi i}\int_{\mathscr{C}_r} \alpha(w)\sqrt{\frac{Q(w)}{w^2}}\,d\eta^*(w,\varepsilon) - \sum_1^{n-1}(k+1)\,\xi_k(\varepsilon)\,\mu_{k+1}.$$

The integral over the level curve $\mathscr{L}_r$ can be transformed to an integral over the tree $\mathscr{T}$. To prove this, recall that in §19 we showed that the function $\alpha(w)\sqrt{Q(w)/w^2}$ was bounded in $\mathscr{G}(\mathscr{L}_r, \mathscr{T})$. In the course of the proof of Lemma 17.2 it was shown that the function $\eta'(w,\epsilon)$ is continuous in $\mathscr{G}(\mathscr{T})$ up to the boundary, and on $\mathscr{T}$ it satisfies (17.23). Thus, recalling (17.22), we find that

$$\frac{d}{dw}\left\{\sqrt{\frac{Q(w)}{w^2}}\,\eta(w,\varepsilon)\right\} = O(w^{-3/2}) \tag{20.10}$$

in $\mathscr{G}(\mathscr{L}_r, \mathscr{T})$. Therefore

$$I_2 = \frac{1}{\pi i}\int_{\mathscr{T}}\left[\sqrt{\frac{Q(w)}{w^2}}\,\alpha_+(w)\,d\eta_+^*(w,\varepsilon) - \sqrt{\frac{Q(w)}{w^2}}\,\alpha_-(w,\varepsilon)\,d\eta_-^*(w,\varepsilon)\right]$$
$$- \sum_1^{n-1}(k+1)\,\xi_k(\varepsilon)\,\mu_{k+1}.$$

Using (17.24), we deduce from the above that

$$\operatorname{Re} I_2 = -\operatorname{Re}\left\{\frac{1}{\pi i}\int_{\mathscr{T}}\left[\sqrt{\frac{Q(w)}{w^2}}\,\alpha_+(w)\,\overline{d\eta_+^*(w,\varepsilon)} - \sqrt{\frac{Q(w)}{w^2}}\,\alpha_-(w)\,\overline{d\eta_-^*(w,\varepsilon)}\right]\right.$$
$$\left. + \sum_1^{n=1}(k+1)\,\xi_k(\varepsilon)\,\mu_{k+1}\right\}. \tag{20.11}$$

Now we evaluate I_2 by another method. Noting that

$$\eta'(w,\varepsilon) = \frac{1}{2\pi i}\int_{\mathscr{T}}[(t-w)^{-2} - t^{-2}]\,\beta(t)\,dt,$$

we have

$$\frac{1}{2\pi i}\int_{\mathscr{C}_r}\eta'(w,\varepsilon)\,\alpha(w)\frac{Q(w)}{w^2}\,dw = \frac{1}{2\pi i}\int_{\mathscr{T}}\beta(t)\,d\left[\frac{1}{2\pi i}\int_{\mathscr{C}_r}\frac{Q(w)\,\alpha(w)}{w^2(w-t)}\,dw\right]$$
$$-\frac{1}{2\pi i}\int_{\mathscr{T}}\beta(t)\,t^{-2}\,dt\,\frac{1}{2\pi i}\int_{\mathscr{T}}\alpha(w)\,Q(w)\,\frac{dw}{w^2}.$$

In §19 we proved that

$$\frac{1}{2\pi i}\int_{\mathscr{T}}\frac{Q(w)\,\alpha(w)}{w^2(w-t)}\,dw = \frac{Q(t)}{t^2}\,\frac{1}{2\pi i}\int_{\mathscr{C}_r}\frac{t^2\alpha(w)\,dw}{w^2(w-t)} + \sum_1^n \gamma_k t^{-k}.$$

If $t\in\mathscr{G}(\mathscr{L}_r, \mathscr{T})$, then from (18.8) we have

$$\frac{1}{2\pi i}\int_{\mathscr{C}_r}\frac{t^2\alpha(w)}{w^2(w-t)}\,dw = -\tilde{g}(t).$$

Using these latter two relations and formula (19.15), we find that

$$\frac{1}{2\pi i}\int_{\mathscr{C}_r} \eta'(w,\varepsilon)\,\alpha(w)\frac{Q(w)}{w^2}\,dw = -\frac{1}{2\pi i}\int_{\mathscr{T}} \beta(t)\,d\left[\frac{Q(t)}{t^2}\tilde{g}(t)\right]$$
$$-\frac{1}{2\pi i}\int_{\mathscr{T}} \beta(t)\sum_2^n k\gamma_k t^{-k-1}\,dt.$$

Thus we have

$$I_2 = -\frac{1}{\pi i}\int_{\mathscr{T}} \sqrt{\frac{Q(w)}{w^2}}\,\beta(w)\,d\left[\sqrt{\frac{Q(w)}{w^2}}\,\tilde{g}(w)\right] - \sum_1^{n-1}(k+1)\gamma_{k+1}\eta_k(\varepsilon).$$

We now integrate by parts. Since $\beta(w)$ vanishes at the vertices of the tree, we find that

$$I_2 = \frac{1}{\pi i}\int_{\mathscr{T}} \sqrt{\frac{Q(w)}{w^2}}\,\tilde{g}(w)\,d\left[\sqrt{\frac{Q(w^2}{w^2}}\,\beta(w)\right] - \sum_1^{n-1}(k+1)\gamma_{k+1}\eta_k(\varepsilon)$$

and since on $\mathscr{T}$ the function $\sqrt{Q(w)/w^2}\,\beta(w)$ is real-valued we have that

$$\sqrt{\frac{Q(w)}{w^2}}\,\beta(w) = \overline{\eta_+^*(w,\varepsilon)} - \overline{\eta_-^*(w,\varepsilon)}.$$

Therefore

$$I_2 = \frac{1}{\pi i}\int_{\mathscr{T}} \sqrt{\frac{Q(w)}{w^2}}\,\tilde{g}(w)\,d\,[\overline{\eta_+^*(w,\varepsilon)} - \overline{\eta_-^*(w,\varepsilon)}] - \sum_1^{n-1}(k+1)\gamma_{k+1}\eta_k(\varepsilon).$$

Taking the real part and combining the result with (20.11), we find that

$$\operatorname{Re} I_2 = \operatorname{Re}\left\{\frac{1}{2\pi i}\int_{\mathscr{T}}\left[\sqrt{\frac{Q(w)}{w^2}}\,(\tilde{g}(w) - \alpha_+(w))\,d\overline{\eta_+^*(w,\varepsilon)}\right.\right.$$
$$\left.-\sqrt{\frac{Q(w)}{w^2}}\,(\tilde{g}(w) - \alpha_-(w))\,d\overline{\eta_-^*(w,\varepsilon)}\right] - \frac{1}{2}\sum_1^{n-2}(k+1)$$
$$\left.\times(\xi_k(\varepsilon)\mu_{k+1} + \eta_k(\varepsilon)\gamma_{k+1})\right\}.$$

Also, using (17.21) we see that

$$\operatorname{Re} I_2 = \operatorname{Re}\left\{\frac{1}{2\pi i}\int_{\mathscr{T}} [\xi_+^*(w,\varepsilon)\,d\overline{\eta_+^*(w,\varepsilon)} - \xi_-^*(w,\varepsilon)\,d\overline{\eta_-^*(w,\varepsilon)}]\right.$$
$$\left.-\frac{1}{2}\sum_1^{n-1}(k+1)(\xi_k(\varepsilon)\mu_{k+1} + \eta_k(\varepsilon)\gamma_{k+1})\right\}.$$

After integrating by parts we get that

$$\frac{1}{2\pi i}\int_{\mathscr{T}} [\xi_+^*(w,\varepsilon)\,d\overline{\eta_+^*(w,\varepsilon)} - \xi_-^*(w,\varepsilon)\,d\overline{\eta_-^*(w,\varepsilon)}]$$
$$= -\frac{1}{2\pi i}\int_{\mathscr{T}} [\overline{\eta_+^*(w,\varepsilon)}\,d\xi_+^*(w,\varepsilon) - \overline{\eta_-^*(w,\varepsilon)}\,d\xi_-^*(w,\varepsilon)] + \frac{1}{2\pi i}A, \tag{20.12}$$

where A is a sum of terms of the form $\xi_{\pm}^*(w_j,\epsilon)\overline{\eta_{\pm}^*(w_j,\epsilon)}$, and where w_j are the vertices of $\mathscr{T}$. If w_j is not a terminal vertex, then $\eta_{\pm}(w_j)=0$. Above we showed that $[\alpha(w)\sqrt{Q(w)/w^2}]_{\pm}$ is bounded. From (17.21) it follows that $\xi_{\pm}^*(w,\epsilon)$ is bounded. Thus in the expression for A only the terms involving terminal vertices can survive. But at the terminal vertices $\alpha(w_j)=0$, and so $\xi_+(w_j,\epsilon)=\xi_-(w_j,\epsilon)$ in view of (17.21). A bit earlier we proved that $\eta_+(w_j,\epsilon)=\eta_-(w_j,\epsilon)$. Since in A the terms $\xi_+^*(w_j,\epsilon)\eta_+^*(w_j,\epsilon)$ and $\xi_-^*(w_j,\epsilon)\overline{\eta_-^*(w_j,\epsilon)}$ appear with opposite signs, it follows that in fact $A=0$. Thus from (20.12) we have

$$\begin{aligned}\operatorname{Re} I_2 = \operatorname{Re}\Big\{\frac{1}{4\pi i}\int_{\mathscr{T}} [\xi_+^*(w,\varepsilon)\,\overline{d\eta_+^*(w,\varepsilon)} + \eta_+^*(w,\varepsilon)\,\overline{d\xi_+^*(w,\varepsilon)}\\ - \xi_-^*(w,\varepsilon)\,\overline{d\eta_-^*(w,\varepsilon)} - \eta_-^*(w,\varepsilon)\,\overline{d\xi_-^*(w,\varepsilon)}]\\ - \frac{1}{2}\sum_1^{n-1}(k+1)(\xi_k(\varepsilon)\mu_{k+1}+\eta_k(\xi)\gamma_{k+1})\Big\}.\end{aligned} \tag{20.13}$$

Using (19.14) and (20.7), we find that

$$\begin{aligned}&\sum_1^{n-1}(k+1)(\xi_k(\varepsilon)\mu_{k+1}+\eta_k(\varepsilon)\gamma_{k+1})\\ &= \sum_{k=1}^{n-1}(k+1)\sum_{l=k+1}^{n} A_l(\xi_k(\varepsilon)\eta_{l-k}(\varepsilon)+\eta_k(\varepsilon)\xi_{l-k}(\varepsilon)).\end{aligned} \tag{20.14}$$

We set

$$\zeta(w,\epsilon)=\xi(w,\epsilon)+\eta(w,\epsilon),$$

$$\zeta(w,\varepsilon)=\sum_1^\infty \zeta_k(\varepsilon)w^{k+1},$$

$$\zeta^*(w,\varepsilon)=\sqrt{\frac{Q(w)}{w}}\,\zeta(w,\varepsilon).$$

Using (20.1), (20.7), (20.9), (20.13) and (20.14), we conclude that

$$\begin{aligned}\mathscr{I}(\varepsilon)=\frac{1}{2}\operatorname{Re}\Big\{\frac{1}{2\pi i}\int_{\mathscr{T}}[\zeta_+^*(w,\varepsilon)\,\overline{d\zeta_+^*(w,\varepsilon)} - \zeta_-^*(w,\varepsilon)\,\overline{d\zeta_-^*(w,\varepsilon)}]\\ - \frac{1}{2}\sum_{k=1}^{n-1}\zeta_k(\varepsilon)\sum_{l=1}^{n-k}(k+l+2)A_{k+l}\zeta_l(\varepsilon)\Big\}.\end{aligned}$$

Thus, in view of (18.9),

$$L(f_*)=L(f)+\epsilon^2\mathscr{I}(\epsilon)+O(\epsilon^3).$$

We now define the second variation of the functional L with respect to the class $\mathscr{D}_f$ of admissible functions in the following manner:

$$\delta^2 L(f)=2\lim_{\varepsilon_k\to 0}\varepsilon_k^{-2}[L(f_*)-L(f)],$$

where the limit is taken over the subsequence $\{\epsilon_k\}$ whose existence was established in Lemma 17.2. We shall show that this definition makes sense, i.e. that the limit exists, and we shall find a formula for the second variation. We set

$$\zeta(w) = \xi(w) + \eta(w),$$

where $\xi(w)$ and $\eta(w)$ are defined by Lemma 17.1. Then

$$\zeta[f(z)] = \delta f(z).$$

In accordance with the lemma,

$$\lim_{\varepsilon_k \to 0} \zeta(w, \varepsilon_k) = \zeta(w).$$

In view of the lemma, the functions $\eta_\pm(w, \epsilon_k)$ and $d\eta_\pm(w, \epsilon_k)/dw$ converge uniformly on $\mathscr{T}$ to the functions $\eta_\pm(w)$ and $d\eta_\pm(w)/dw$, respectively. From (17.21) it follows that on the set $\mathscr{T}\setminus\mathscr{K}_\delta$ ($\mathscr{K}_\delta$ as defined in Lemma 17.2) the functions $\xi^*_\pm(w, \epsilon_k)$ and $d\xi^*_\pm(w, \epsilon_k)/dw$ converge uniformly to $\xi^*_\pm(w)$ and $d\xi^*_\pm(w)/dw$, respectively. Thus on $\mathscr{T}\setminus\mathscr{K}_\delta$ we have that $\xi^*_\pm(w, \epsilon_k)$ and $d\xi^*_\pm(w, \epsilon_k)/dw$ converge uniformly to $\xi^*_\pm(w)$ and $d\xi^*_\pm(w)/dw$, respectively. From (17.19) it follows that $\xi^*_\pm(w, \epsilon_k)$ is uniformly bounded on $\mathscr{T}$. In §17 we showed that

$$\sqrt{\frac{Q(w)}{w^2}}\, \alpha_\pm(w) = i\,[\sqrt{P(z)}\, h(z)]_\pm,$$

and so

$$\frac{d}{dw}\left[\sqrt{\frac{Q(w)}{w^2}}\, \alpha_\pm(w, \varepsilon_k)\right] = i\left[\frac{1}{f'(z)}\, \frac{d}{dz}\,(\sqrt{P(z)}\, h(z, \varepsilon))\right]_\pm. \tag{20.15}$$

From this we have in a neighborhood of any vertex w_j, except the base of the tree, that

$$\frac{d}{dw}\left[\sqrt{\frac{Q(w)}{w^2}}\, \alpha_\pm(w, \varepsilon_k)\right] = O(|w - w_j|^{1/2}), \tag{20.16}$$

and moreover the constant is independent of ϵ_k. In a neighborhood of the base it follows from (20.15) that

$$\frac{d}{dw}\left[\sqrt{\frac{Q(w)}{w^2}}\, \alpha_\pm(w, \varepsilon_k)\right] = O(w^{-3/2}) \tag{20.17}$$

uniformly with respect to ϵ_k. From (17.21), (20.16) and (17.23) we conclude that

$$\frac{d}{dw}\,\zeta^*_\pm(w, \varepsilon_k) = O(|w - w_j|^{-1/2})$$

uniformly with respect to ϵ_k. From (17.21), (20.17) and (17.23) we have in a neighborhood of the base that

$$\frac{d}{dw}\,\zeta^*_\pm(w, \varepsilon) = O(w^{-3/2})$$

uniformly with respect to ϵ_k. Combining these results, we deduce that

$$\lim_{\varepsilon_k \to 0} \frac{1}{2\pi i} \int_{\mathscr{T}} [\zeta_+^*(w, \varepsilon_k) \overline{d\zeta_+^*(w, \varepsilon_k)} - \zeta_-^*(w, \varepsilon_k) \overline{d\zeta_-^*(w, \varepsilon_k)}] = \frac{1}{2\pi i} \int_{\mathscr{T}} [\zeta_+^*(w) \overline{d\zeta_+^*(w)} - \zeta_-^*(w) \overline{d\zeta_-^*(w)}]. \tag{20.18}$$

Since $\lim_{\epsilon_k \to 0} \zeta_l(\epsilon_k) = \zeta_l$, $l = 1, \cdots, n$, where ζ_l are the coefficients of the power series for $\zeta(w)$, we see upon recalling (20.18) that

$$\delta^2 L(f) = \operatorname{Re}\left\{\frac{1}{2\pi i} \int_{\mathscr{T}} [\zeta_+^*(w) \overline{d\zeta_+^*(w)} - \zeta_-^*(w) \overline{d\zeta_-^*(w)}] - \frac{1}{2} \sum_{k=1}^{n-1} \zeta_k \sum_{l=1}^{n-k} (k + l + 2) A_{k+l} \zeta_l\right\}. \tag{20.19}$$

This formula for the second variation of a function L with respect to the class $\mathscr{D}_f$ of admissible functions agrees with formula (19.18), which held only for the class $\mathscr{D}_f^0$.

Theorem 12. *The second variation of a linear functional with respect to the class $\mathscr{D}_f$ of admissible functions is a quadratic functional, of the type* (20.19), *of the first variation of the extremal function $f(z)$.*

Remark 20.1. The connection between the first coefficient variation of $f(z)$ and the coefficients of the power series for the function $\zeta(w)$ is given by formula (19.19) with ξ_l replaced by ζ_l. An immediate consequence of Lemmas 17.1, 17.2 and Theorem 12 is the following:

Lemma 20.1. *Suppose $x^0 \in \mathscr{R}$ and $d_{n-1}(x^0) \neq 0$. If there is given any sequence of points in the set $\mathfrak{D}^{x^0}$ which converges to x^0, then there exists a subsequence $\{x^k\}$ with $\lim_{k\to\infty} x^k = x_0$ satisfying the following conditions.*

1) *The limit*

$$\lim_{x^k \to x^0} \frac{f(z, x^k) - f(z, x^0)}{|x^k - x^0|} = \delta f(z, x^0)$$

exists in the disc $|z| < 1$.

2) *If $f(z, x^k) = \sum_{l=0}^{\infty} a_l^k z^{l+1}$, $a_0^k = 1$, $k = 0, 1, \cdots,$ then*

$$\lim_{x^k \to x^0} \frac{1}{|x^k - x^0|^2} \operatorname{Re} \sum_{l=1}^{n} \lambda_l(x^0)(a_l^k - a_l^0)$$
$$= \frac{1}{2} \operatorname{Re}\left\{\frac{1}{2\pi i} \int_{\mathscr{T}} [\zeta_+^*(w) \overline{d\zeta_+^*(w)} - \zeta_-^*(w) \overline{d\zeta_-^*(w)}] - \frac{1}{2} \sum_{k=1}^{n-1} \zeta_k \sum_{l=1}^{n-k} (k + l + 2) A_{k+l}^0 \zeta_l\right\},$$

where $\zeta(w) = \delta f(z, x^0)$ and $z = f^{-1}(w, x^0)$.

CHAPTER 4

AN INVESTIGATION OF THE SECOND VARIATION OF LINEAR FUNCTIONALS

In the preceding chapter we established the important formula (20.19) for the second variation of a linear functional. This chapter will be devoted to an investigation of extremal problems involving the quadratic functional (20.19). The theory presented in this chapter will be in some ways analogous to the classical theory of Jacobi. In the preceding chapter we also introduced the concept of the first variation of an extremal function with respect to the class $\mathscr{D}_f$ of admissible functions. The set of first variations gives a very complete characterization of the extremal functions, and the study of this set will lead us to answers for many questions concerned with the smoothness of the boundary of the coefficient region. However, the study of these topics cannot be pursued until after we have introduced an appropriate topology on the set of variations.

§21. Let $f(z)$ be an extremal function, $f(z) \in S_n$, and let $\mathscr{T}$ be its boundary tree. We consider the set of all first variations of $f(z)$ defined by admissible functions of the class $\mathscr{D}_f$. For the first variation we have two equivalent representations in formula (2.11) and in the formula $\delta f(z) = \xi[f(z)]$ where $\xi(w)$ is defined by (19.11). It will be more convenient for our purposes to consider the first variation as a function of the variable w. The function $\xi(w)$ is regular in the domain $\mathscr{G}(\mathscr{T})$, which henceforth shall be denoted simply by $\mathscr{G}$.

DEFINITION 21.1. The set $\mathscr{A}_0$ is the set of all functions $\xi(w)$ representable in the form

$$\xi(w) = \frac{1}{2\pi i} \int_{\mathscr{C}_r} \frac{w^2 \alpha(t)}{(w-t)\, t^2}\, dt, \tag{21.1}$$

where $\mathscr{C}_r$ is a level curve of the function $|f^{-1}(w)|$ and $r = r(\xi) < 1$. The function $\alpha(w)$ is regular in $\mathscr{G}(\mathscr{C}_r, \mathscr{T})$ and has the form

$$\alpha(w) = Df(z)\, h(z), \quad z = f^{-1}(w), \tag{21.2}$$

where $h(z)$ is regular in the annulus $r \leqq |z| \leqq r^{-1}$ and satisfies

$$\operatorname{Re} h(z) = 0 \tag{21.3}$$

on the circumference $|z| = 1$.

The set of first variations, considered as a function of w, is contained in $\mathscr{A}_0$. Conversely, each function $\xi(w) \in \mathscr{A}_0$ defines a first variation with respect to an admissible function of the class $\mathscr{D}_f^0$. To prove this, consider the measure $\nu(dz) = h(z)dz/2\pi iz$ with support $\mathscr{C} = \{z : |z| = r\}$. This measure defines an admissible function of the form (2.2). Using this admissible function and setting $c_1 = i \int_{\mathscr{C}} \nu(dz)$, we define a first variation $\delta f(z)$ by formula (2.11). Then

$$\xi[f(z)] = \delta f(z).$$

Throughout this chapter we shall assume $d_{n-1/2}(f) > 0$. We note the following properties of the set $\mathscr{A}_0$:

1. If $\xi(w) \in \mathscr{A}_0$, then on $\mathscr{T}$ we have

$$\operatorname{Im}\left\{\sqrt{\frac{Q(w)}{w^2}}\,[\xi_+(w) - \xi_-(w)]\right\} = 0. \tag{21.4}$$

This was proved in §17.

2. $\mathscr{A}_0$ is a linear manifold over the reals.

3. Let $\xi(w) \in \mathscr{A}_0$, $\xi(w) = \sum_1^\infty \xi_k w^{k+1}$. Then

$$\operatorname{Re} \sum_1^n A_k \xi_k = 0. \tag{21.5}$$

To prove this, we have $\xi[f(z)] = \delta f(z) = \sum_1^\infty \delta a_k z^{k+1}$. Since $\operatorname{Re}\sum_1^n \overline{\lambda}_k \delta a_k = 0$, (21.5) follows from (19.20) and (10.3).

4. Let $\xi_1, \cdots, \xi_n$ be arbitrary complex numbers satisfying (21.5). Then there exists $\xi(w) \in \mathscr{A}_0$ such that $\xi(w) = \xi_1 w^2 + \cdots + \xi_n w^{n+1} + \cdots$.

In order to prove this we consider the orthonormal basis $\lambda, t^1, \cdots, t^{2n-1}$ in R^{2n}. Using Lemma 3.4 we construct for each vector t^j a measure $\nu_j(dz) = \chi_j(z)\,dz/2\pi iz$ which has as its support $\mathscr{C} = \{z : |z| = r_0\}$, where r_0 is an absolute constant. By the lemma we have

$$\frac{1}{2\pi} \int\limits_{|z|=r_0} |\chi_j(z)|^2 \left|\frac{dz}{z}\right| < d_{n-1/2}^{-2}(f).$$

The first variation constructed with the measure $\nu_j(dz)$ (with $c_1 = 0$) defines a function $\xi_j(w) \in \mathscr{A}_0$. Set $\alpha_j(w) = Df(z)h_j(z)$, $z = f^{-1}(w)$. Then, if $\alpha_{jk} = \int_{\mathscr{C}} z^k \nu_j(dz)$, $k = 1, 2, \cdots$, we have $h_j(z) = \sum_1^\infty (\alpha_{jk} z^{-k} - \overline{\alpha}_{jk} z^k)$. If $r_1 = \frac{1}{2}(1 + r_0)$, then, in view of the above inequality, for z in the annulus $r_1 \leqq |z| \leqq r_1^{-1}$ we have $|h_j(z)| < M_0 d_{n-1/2}^{-1}(f)$, where M_0 is an absolute constant. Thus from (19.12) it follows that for $|f^{-1}(w)| \geqq r_1$

$$|\xi_j(w)| < (M_0 |Df(z)| + M_1 |f(z)|)\, d_{n-1/2}^{-1}(f) < M_2 (1 - |z|)^{-3} d_{n-1/2}^{-1}(f).$$

Thus in $\mathscr{G}$ we have that

$$|\xi_j(w)| < M\,[1 - |f^{-1}(w)|]^{-3} d_{n-1/2}^{-1}(f), \tag{21.6}$$

where M is an absolute constant.

Using the quantities $\xi_1, \cdots, \xi_n$, we determine the quantities $t_k^0 = \delta a_k$, $k = 1, \cdots, n$, via formula (19.19). Set $t^0 = (t_1^0, \cdots, t_n^0)$. Since $(\lambda, t^0) = 0$, we have that $t^0 = \sum_1^{2n-1} \beta_j t^j$. It is clear the β_j are linear functions of $\operatorname{Re}\xi_k$ and $\operatorname{Im}\xi_k$, $k = 1, \cdots, n$. The desired function can now be expressed as $\xi(w) = \sum_1^{2n-1} \beta_j \xi_j(w)$. It follows easily from the estimates proved in §10 for $a_k^{(l)}$ that

$$|\beta_j| < \alpha_0^n \left(\sum_1^n |\xi_k|^2\right)^{1/2}, \qquad 1 \leq j \leq 2n-1,$$

where α_0 is an absolute constant. We shall use the notation $\mathfrak{z} = (\xi_1, \cdots, \xi_n)$ for the vector consisting of the first n coefficients of the function $\xi(w) \in \mathscr{A}_0$. By (21.6) we also have inside any level curve $\mathscr{C}_r$ that

$$|\xi(w)| < \alpha^n (1-r)^{-3} d_{n-1/2}^{-1}(f) |\mathfrak{z}|, \tag{21.7}$$

where α is an absolute constant.

5. Let $\mathscr{G}_r = \mathscr{G} \setminus \{w : |w| < r\}$. If $r < 1/4$, the relation

$$\frac{1}{2\pi i} \int_{\mathscr{T}} (\xi_+^* \overline{d\xi_+^*} - \xi_-^* \overline{d\xi_-^*}) = \frac{1}{2\pi i} \int_{|w|=r} \xi^* \overline{d\xi^*} - \frac{1}{\pi} \int_{\mathscr{G}_r} \left| \frac{d\xi^*}{dw} \right|^2 d\sigma \tag{21.8}$$

holds for the function $\xi(w) \in \mathscr{A}_0$. (In the integral on the right, $d\sigma$ is the area element in the w-plane.)

In order to prove this, let $\mathscr{K}_\delta$ be the system of discs defined in §16. In the domain $\mathscr{G}_r \setminus \mathscr{K}_\delta = \mathscr{G}_{r,\delta}$ we can apply Green's formula to the integral

$$\frac{1}{\pi} \int_{\mathscr{G}_{r,\delta}} \left| \frac{d\xi^*}{dw} \right|^2 d\sigma.$$

The existence of a zero of the function $Q(w)w^{-2}$ is not a difficult problem to overcome. We connect each zero w_j by a path l_j with $\partial \mathscr{G}_{r,\delta}$, and we take the paths to be nonintersecting. We construct discs $K_{j,\epsilon}$ centered at w_j with radius ϵ. After having deleted these discs from $\mathscr{G}_{r,\delta}$ and after having divided the resulting domain by the paths l_j, we obtain a domain $\mathscr{G}_{r,\delta,\epsilon}$ in which the function $\xi^*(w)$ is regular. Green's formula then gives that

$$\frac{1}{\pi} \int_{\mathscr{G}_{r,\delta,\varepsilon}} \left| \frac{d\xi^*}{dw} \right|^2 d\sigma = -\frac{1}{2\pi i} \int_{\partial \mathscr{G}_{r,\delta,\varepsilon}} \xi^* \overline{d\xi^*}.$$

We now let $\epsilon \to 0$, and since the integrals over opposite sides of the slits cancel each other, we find that

$$\frac{1}{\pi} \int_{\mathscr{G}_{r,\delta}} \left| \frac{d\xi^*}{dw} \right|^2 d\sigma = -\frac{1}{2\pi i} \int_{\partial \mathscr{G}_{r,\delta}} \xi^* \overline{d\xi^*}.$$

In $\mathscr{G}(\mathscr{C}_r, \mathscr{T})$ the function $\xi^*(w)$ is bounded, and in a neighborhood of the vertices w_j its derivatives satisfy the inequalities given at the end of §20. Thus in the above formula we may take the limit as $\delta \to 0$, and we conclude that (21.8) holds, as required.

Define an integer n^* by

$$n^* = \begin{cases} \frac{1}{2} n, & n \text{ even}, \\ \frac{1}{2}(n+1), & n \text{ odd}. \end{cases}$$

If $\xi_1 = \cdots = \xi_{n^*-1}$, then in (21.8) we can set $r = 0$, and so

$$\frac{1}{2\pi i} \int_{\mathcal{T}} [\xi_+^* \overline{d\xi_+^*} - \xi_-^* \overline{d\xi_-^*}] = -\frac{1}{\pi} \int_{\mathcal{G}} \left| \frac{d\xi^*}{dw} \right|^2 d\sigma.$$

In this case

$$\delta^2 L(f) = -\frac{1}{\pi} \int_{\mathcal{G}} \left| \frac{d\xi^*}{dw} \right|^2 d\sigma + \frac{1}{2}\left(n - 2\left[\frac{n}{2}\right] - 1\right)(n+2)^2 \operatorname{Re} A_n \xi^2_{\left[\frac{n}{2}\right]}.$$

Relation (21.8) suggests a natural metric suitable for $\mathcal{A}_0$. It is obvious that this metric would be closely related to the second variation. In order to overcome the difficulties involved with the presence, on the right-hand side of (21.8), of the integral over the circle $|w| = r$, we must prove some propositions of a somewhat technical nature. These propositions will be considered in the following sections.

§22. Let the function $\zeta(w)$ be regular in the disc $|w| < r_0$ and have there the expansion

$$\zeta(w) = \sum_{n^*}^{\infty} \zeta_k w^{k+1}. \tag{22.1}$$

By $R(w)$ we denote a rational function of the form

$$R(w) = w^{-n-2} \sum_{0}^{\nu} \alpha_k w^k, \qquad \alpha_0 \neq 0.$$

Set $\zeta^*(w) = \sqrt{R(w)}\,\zeta(w)$, and set $\theta = 1$ if n is even, $\theta = 0$ if n is odd. Also set

$$\mathcal{Y}^2 = \frac{1}{\pi} \int_{|w|<r_0} \left| \frac{d\zeta^*}{dw} \right|^2 d\sigma + \frac{\theta}{2} |\alpha_0 \zeta_{n^*}^2|. \tag{22.2}$$

Proposition 22.1. *If $\mathcal{Y} < \infty$, then the function $\zeta(w)$ has a limit almost everywhere on the circle $|w| = r_0$. For any $p > 0$,*

$$\left(\frac{1}{2\pi} \int_{|w|=r_0} |\zeta^*(w)|^p \left| \frac{dw}{w} \right| \right)^{1/p} < B_p \mathcal{Y}.$$

The constant B_p depends only on p.

Proof. Let $r < r_0$ be such that in the annulus $r \leq |w| < r_0$ the function $R(w)$ has no zeros. Denote by w_j, $j = 1, \cdots, l$, the zeros of $R(w)$ lying in the disc $|w| < r$. Consider the functions

$$q_0(w;r) = \left(\frac{w}{r}\right)^{l/2}\prod_1^l (w^{-1} - w_j^{-1}), \qquad q_1(w,r) = \left(\frac{r}{w}\right)^{l/2}\prod_1^l\left(\frac{w}{r^2} - \frac{1}{\overline{w}_j}\right).$$

Set $Q_0(w;r) = q_0(w;r)q_1(w;r)$ and $R_1(w;r) = R(w)\,[Q_0(w;r)]^{-1}$. The identity

$$Q_0(w;r) = |q_0(w;r)|^2 \tag{22.3}$$

clearly holds on the circle $|w| = r$. Set $h(w;r) = R_1^{1/2}(w;r)\zeta(w)$. Then $\zeta^*(w) = Q_0^{1/2}(w;r)h(w;r)$. Using the identity (22.3) we easily verify that

$$\zeta^*\overline{d\zeta^*} = \frac{1}{2}|h^2|\,dQ_0 + Q_0 h\overline{dh}$$

on the circle $|w| = r$. Differentiating (22.3) and substituting into the last equation, we find after rearranging terms that

$$\zeta^*\overline{d\zeta^*} = \frac{1}{2}\,q_0 h\overline{d\,(q_0 h)} + \frac{1}{2}\,\overline{q}_0 h d\,(q_0\overline{h}).$$

But if $|w| = r$, then $q_0\overline{(w;r)} = q_1(w;r)$, and so

$$\zeta^*\overline{d\zeta^*} = \frac{1}{2}\,q_0 h\overline{d\,(q_0 h)} + \frac{1}{2}\,q_1 h\overline{d\,(q_1 h)}. \tag{22.4}$$

It is not difficult to verify that for small $|w|$ we have the estimates

$$q_0(w;r)\,h(w;r) = (-1)^{l/2}\left(\alpha_0\prod_1^l\overline{w}_j\right)^{1/2} r^{-l/2}\zeta_{n^*}w^{-\frac{n}{2}+n^*} + O\left(w^{-\frac{n}{2}+n^*+1}\right)$$

and

$$q_1(w;r)\,h(w;r) = (-1)^{l/2}\left(\alpha_0\prod_0^l\overline{w}_j^{-1}\right)^{1/2} r^{l/2}\zeta_{n^*}w^{-\frac{n}{2}+n^*} + O\left(w^{-\frac{n}{2}+n^*+1}\right). \tag{22.5}$$

If $n^* = \frac{1}{2}n$, then the functions $q_j(w;r)h(w;r)$, $j = 0, 1$, are regular in $|w| < r_0$. If $n^* = \frac{1}{2}(n+1)$, then the functions $w^{-1/2}q_j(w;r)h(w;r)$, $j = 0, 1$, are regular in $|w| < r_0$. Set $\delta = 0$ when $n^* = \frac{1}{2}n$ and $\delta = \frac{1}{2}$ when $n^* = \frac{1}{2}(n+1)$. It is clear that in $|w| < r_0$ we have

$$q_j(w;r)\,h(w;r) = \sum_0^\infty a_{jk}(r)\,w^{k+\delta}, \qquad j = 0, 1.$$

From (22.4) it follows that

$$\frac{1}{2\pi i}\int_{|w|=r}\zeta^*\overline{d\zeta^*} = -\frac{1}{2}\sum_0^\infty (k+\delta)\left(|a_{0k}(r)|^2 + |a_{1k}(r)|^2\right) r^{2k+2\delta},$$

and so upon using Green's formula we find that

$$\frac{1}{\pi}\int_{|w|<r}\left|\frac{d\zeta^*}{dw}\right|^2 d\sigma = \frac{1}{2}\sum_0^\infty (k+\delta)\left(|a_{0k}(r)|^2 + |a_{1k}(r)|^2\right) r^{2k+2\delta},$$

which in turn implies, due to the arbitrariness of r, that

$$\frac{\theta}{2}\,|\,\alpha_0\zeta_{n^*}^2\,| + \frac{1}{2}\sum_0^\infty (k+\delta)\,(|\,a_{0k}(r_0)\,|^2 + |\,a_{1k}(r_0)\,|^2)\,r_0^{2k+2\delta} \leqslant \mathcal{Y}^2. \tag{22.6}$$

The dependence on r_0 will not be explicitly included in the notation used below. It follows from the first of relations (22.5) that

$$|\,a_{00}\,|^2 = \left|\,\alpha_0\zeta_{n^*}^2 \prod_1^l \frac{w_j}{r_0}\right| \leqslant |\,\alpha_0\zeta_{n^*}^2\,|.$$

Thus

$$(\delta+\theta)\,|\,a_{00}\,|^2\,r_0^{2\delta} + \sum_1^\infty (k+\delta)\,|\,a_{0k}\,|^2\,n_0^{2k+2\delta} \leqslant 2\mathcal{Y}^2,$$

and, since $\delta+\theta \geqq \frac{1}{2}$, we have

$$\sum_0^\infty (k+1)\,|\,a_{0k}\,|^2\,r_0^{2k+2\delta} < 4\mathcal{Y}^2.$$

From this inequality it follows that the function

$$u(\varphi) = \sum_0^\infty a_{0k} r_0^{k+\delta} e^{i(k+1)\varphi}$$

has a fractional derivative, of order $\frac{1}{2}$, which is square integrable. Thus, in view of known results (see [9], §9.83),

$$\left(\frac{1}{2\pi}\int_0^{2\pi} |\,u(\varphi)\,|^p\,d\varphi\right)^{1/p} < B_p\mathcal{Y}, \tag{22.7}$$

where the constant B_p depends only on p. Thus the function

$$r_0^{-1+\delta}\sum_0^\infty a_{0k} w^{k+1} = \left(\frac{w}{r_0}\right)^{1-\delta} q_0(w;r_0)\,h(w;r_0)$$

belongs to the Hardy class H^p in the disc $|w| < r_0$, and so it has the limit $u(\varphi)$ almost everywhere on the circle $|w| = r_0$. Thus $\zeta(w)$ has a limit almost everywhere on $|w| = r_0$, and $|\zeta^*(w)| = |u(\varphi)|$. This relation and the inequality (22.7) imply the inequality stated in the proposition. Q.E.D.

We now assume that $\nu \leqq 2n$ in the definition of $R(w)$.

Proposition 22.2. *If the hypotheses of Proposition* 22.1 *are satisfied and if $r_0 \leqq 1$, then for any $r < r_0$*

$$\max_{|w|\leqslant r} |\,\zeta(w)\,| < C_r\mathcal{Y}. \tag{22.8}$$

Proof. Let $r^* = \frac{1}{2}(r + r_0)$. In the annulus $r^* \leqq |w| \leqq r_0$ there exists an annulus $r_1 - \Delta \leqq |w| \leqq r_1 + \Delta$, $\Delta = (r_0 - r)/8n$, containing no zeros of $R(w)$. From a theorem of Hardy it follows that

$$\frac{1}{2\pi}\int_{|w|=r_1}|\zeta^*(w)|^2\left|\frac{dw}{w}\right|=\frac{r_1^{-n-2}}{2\pi}\int_{|w|=r_1}|w^{n+2}R(w)\,\zeta^2(w)|\left|\frac{dw}{w}\right|$$
$$<\left(\frac{r_0}{r_1}\right)^{n+2}\frac{1}{2\pi}\int_{|w|=r_0}|\zeta^*(w)|^2\left|\frac{dw}{w}\right|.$$

From this we obtain that

$$\frac{1}{2\pi}\int_{|w|=r_1}|\zeta(w)|^2\left|\frac{dw}{w}\right|<\left(\frac{r_0}{r_1}\right)^{n+2}\{\min_{|w|=r_1}|R(w)|\}^{-1}B_2^2\mathcal{Y}^2.$$

If $m_1=\min|R(w)|$ and $M_1=\max_{|w|=r_1}|R(w)|$, then it is easily seen that $m_1M_1^{-1}\geqq((r_0-r)/12nr_0)^{2n}$. From the obvious inequality

$$M_1^2\geqslant\frac{1}{2\pi}\int_{|w|=r_0}|R(w)|^2\left|\frac{dw}{w}\right|$$

it follows that $M_1\geqq r_1^{n-2}\mu_0$, where $\mu_0=(\sum_0^n|\alpha_k|^2)^{1/2}$. Thus

$$m_1\geqslant\left(\frac{r_0-r)}{12nr_0}\right)^{2n}r^{n-2}\mu_0.$$

For $|w|\leqq r$ we see from Cauchy's formula that

$$\zeta(w)=\frac{1}{2\pi i}\int_{|w|=r^*}\frac{\zeta(t)}{t-w}dt.$$

Applying Schwarz' inequality, we conclude that

$$|\zeta(w)|<\left(\frac{r_0}{r_0-r}\right)^{1/2}\left(\frac{1}{2\pi}\int_{|w|=r^*}|\zeta(w)|^2\left|\frac{dw}{w}\right|\right)^{1/2}.$$

Using the above inequalities and assuming without loss of generality that $r\geqq\frac{1}{2}r_0$, we finally deduce that

$$|\zeta(w)|<B(n)\,r_0^{\frac{n+3}{2}}(r_0-r)^{-n-\frac{1}{2}}\mu_0^{-\frac{1}{2}}\mathcal{Y},$$

etc.

Let $w_1,\cdots,w_l$, $|w_1|\leqq\cdots\leqq|w_l|$, be those zeros of $R(w)$ lying in the disc $|w|<r_0$. In the annulus $|w_l|<|w|<r_0$ the function $\zeta^*(w)$ is analytic. We slit the annulus by the radius $\arg w=0$ and we fix a branch of this function, which we shall continue to denote by $\zeta^*(w)$.

Proposition 22.3. *Let $\zeta(w)=\sum_1^\infty\zeta_kw^{k+1}$ be regular in the disc $|w|<r_0$. Assume that the function $\zeta(w)-\sum_1^{n^*-1}\zeta_kw^{k+1}=\zeta_0(w)$ satisfies the conditions of Proposition 22.1. Then in the $L^p(0,2\pi)$ metric, $p<\infty$, $\lim_{r\to r_0}\zeta^*(re^{i\varphi})=\zeta^*(r_0e^{i\varphi})$.*

Proof. It is sufficient to prove the existence of the limit for the function $\zeta_0^*(w)$. Set $\zeta_1(w)=\zeta_0^*(w)\psi(w)$, where

$$\psi(w) = \left(\prod_1^l (r_0^2 - \bar{w}_j w)^{\epsilon}(w - w_j)^{-1}\right)^{1/2}\left(\frac{r_0}{w}\right)^{\delta}.$$

In the disc $|w| < r_0$ the function $\zeta_1(w)$ is regular and, according to Proposition 22.1, possesses almost everywhere on $|w| = r_0$ a limit which satisfies $|\zeta_1(w)| = |\zeta_0^*(w)|$. Proposition 22.1 implies, via classical results, that $\zeta_1(w) \in H^p$. Therefore $\lim_{r\to r_0} \zeta_1(re^{i\varphi}) = \zeta_1^*(r_0 e^{i\varphi})$, the limit taken in the L^p metric. It follows trivially from this that $\lim_{r\to r_0} \zeta_0^*(re^{i\varphi}) = \zeta_0^*(r_0 e^{i\varphi})$, the limit again taken in the $L^p(0, 2\pi)$ metric for any $p < \infty$.

We now proceed to construct a scalar product which we shall subsequently apply to $\mathscr{A}_0$. We shall exclude from consideration the Koebe functions $f(z) = z(1 - e^{i\theta}z)^{-2}$. Thus the tree $\mathscr{T}$ must be in the complement of the disc $|w| \leq 1/4$. In (21.8) we set $r = 1/4$. Let $\zeta(w) = \sum_1^\infty \zeta_k w^{k+1}$ be a function regular in $\mathscr{G}$. We shall estimate the integral

$$\frac{1}{2\pi i}\int_{|w|=1/4} \zeta^* \overline{d\zeta^*},$$

although we must first transform it. Set $p(w) = \sum_1^{n^*-1} \zeta_k w^{k+1}$ and $\zeta_0(w) = \zeta(w) - p(w)$. As in the previous section, it is easily seen that

$$\frac{1}{2\pi i}\int_{|w|=1/4} \zeta_0^* \overline{d\zeta_0^*} = -\frac{1}{\pi}\int_{|w|<1/4} \left|\frac{d\zeta_0^*}{dw}\right|^2 d\sigma.$$

Thus

$$\operatorname{Re}\left\{\frac{1}{2\pi i}\int_{|w|=1/4} \zeta^* \overline{d\zeta^*}\right\} = -\frac{1}{\pi}\int_{|w|<1/4} \left|\frac{d\zeta^*}{dw}\right|^2 d\sigma + \operatorname{Re}\left\{\frac{1}{\pi i}\int_{|w|=1/4} \zeta_0^* \overline{dp^*} + \frac{1}{2\pi i}\int_{|w|=1/4} p^* \overline{dp^*}\right\}.$$

Set

$$J^2 = \frac{1}{\pi}\int_{|w|<1/4} \left|\frac{d\zeta_0^*}{dw}\right|^2 d\sigma + \frac{\theta}{2}|A_n \zeta_{n^*}^2|. \tag{22.9}$$

Put

$$\mu = \left(\sum_1^n |A_k|^2\right)^{1/2}. \tag{22.10}$$

and introduce the notation $\sigma_l = (\sum_1^l |\zeta_l|^2)^{1/2}$, $\sigma = \sigma_{n^*-1}$.

Proposition 22.4. *With the above notation,*

$$\left|\frac{1}{\pi i}\int_{|w|=1/4} \zeta_0^* \overline{dp^*}\right| < \alpha^n \mu^{1/2} \sigma J,$$

where α is an absolute constant.

Proof. By Hölder's inequality

$$\left|\frac{1}{\pi i}\int_{|w|=1/4}\zeta_0^*\overline{dp^*}\right|<\frac{1}{2}\left\{\frac{1}{2\pi}\int_{|w|=1/4}|\zeta_0^*(w)|^4\left|\frac{dw}{w}\right|\right\}^{1/4}\times\left\{\frac{1}{2\pi}\int_{|w|=1/4}\left|\frac{dp^*}{dw}\right|^{4/3}\left|\frac{dw}{w}\right|\right\}^{3/4}.$$

Thus, by Proposition 22.1,

$$\left|\frac{1}{2\pi}\int_{|w|=1/4}\zeta_0^*\overline{dp^*}\right|<\frac{1}{2}B_4JJ_1, \tag{22.11}$$

where

$$J_1=\left\{\frac{1}{2\pi}\int_{|w|=1/4}\left|\frac{dp^*}{dw}\right|^{4/3}\left|\frac{dw}{w}\right|\right\}^{3/4}.$$

We now estimate J_1. Using Schwarz' inequality, it is easy to prove that for $|w|=1/4$ we have

$$\left|\sum_1^{n^*-1}\zeta_k w^{k+1}\right|<\frac{1}{4}\sigma,\qquad \left|\sum_1^{n^*-1}(k+1)\zeta_k w^k\right|<\sigma.$$

From this it follows that

$$\left|\frac{dp^*}{dw}\right|<\left(\frac{1}{4}\left|\frac{d}{dw}\sqrt{\frac{Q(w)}{w^2}}\right|+\left|\sqrt{\frac{Q(w)}{w^2}}\right|\right)\sigma.$$

Also we have the identity

$$\frac{d}{dw}\sqrt{\frac{Q(w)}{w^2}}=\frac{1}{2}w^{-\frac{n+2}{2}}\sum_1^{n-1}\left(\frac{w^nQ(w)}{w-w_k}\right)^{1/2}(w-w_k)^{-1/2}-\frac{n+2}{2}w^{-1}\left(\frac{Q(w)}{w^2}\right)^{1/2},$$

where w_k are the zeros of $w^nQ(w)$. It is easy to show, by considering the two cases $|w_k|\geqq 1/2$ and $|w_k|<1/2$, that for $|w|=1/4$

$$\left|\frac{w^nQ(w)}{w-w_k}\right|<8\mu.$$

Therefore

$$\left|\frac{dp^*}{dw}\right|<\left[2^{n-1}\sqrt{8\mu}\sum_1^{n-1}|w-w_k|^{-1/2}+\frac{n+4}{2}\left|\frac{Q(w)}{w^2}\right|^{1/2}\right]\sigma.$$

By Minkowski's inequality,

$$J_1<\left[2^{n-1}\sqrt{8\mu}\sum_1^{n-1}\left(\frac{1}{2\pi}\int_{|w|=1/4}|w-w_k|^{-2/3}\left|\frac{dw}{w}\right|\right)^{3/4}+\frac{n+4}{2}\left(\frac{1}{2\pi}\int_{|w|=1/4}\left|\frac{Q(w)}{w^2}\right|^{2/3}\left|\frac{dw}{w}\right|\right)^{3/4}\right]\sigma.$$

It is not difficuit to show that for any w_k

$$\left(\frac{1}{2\pi}\int_{|w|=1/4}|w-w_k|^{-2/3}\left|\frac{dw}{w}\right|\right)^{3/4}<4\sqrt{2}.$$

Applying Schwarz' inequality, we find that $|w^nQ(w)|<\mu\sqrt{16/15}$ for $|w|$

$=1/4$, and so

$$J_1 < (3n+2)2^{n+2}\sqrt{\mu}\sigma. \tag{22.12}$$

The desired result follows from (22.11) and (22.12). Q.E.D.

PROPOSITION 22.5. *With the above notation,*

$$\left|\frac{1}{2\pi i}\int_{|w|=1/4} p^*\overline{dp^*}\right| < (3n+2)\,2^{2n+3}\,\mu\sigma^2.$$

PROOF. From Hölder's inequality we get

$$\left|\frac{1}{2\pi i}\int_{|w|=1/4} p^*(w)\,\overline{dp^*(w)}\right| < \frac{1}{4}\left(\frac{1}{2\pi}\int_{|w|=1/4} |p^*(w)|^4\left|\frac{dw}{w}\right|\right)^{1/4} J_1.$$

But $|p^*(w)| < (1/4)\sigma|Q(w)/w^2|^{1/2} < 2^n\sqrt{16/15}\,\sigma\sqrt{\mu} < 2^{n+1}\sigma\sqrt{\mu}$, if $|w| = 1/4$. This last inequality and (22.12) together imply the conclusion of the proposition.

Set $n_0 = [n/2]$.

PROPOSITION 22.6. *There exists an absolute constant α_0 such that*

$$-\operatorname{Re}\left\{\frac{1}{2\pi i}\int_{|w|=1/4}\zeta^*\overline{d\zeta^*}\right\} + \alpha_0^n\sum_1^{n_0}|\zeta_k|^2 > \frac{1}{2}\left[\frac{1}{\pi}\int_{|w|<1/4}\left|\frac{d\zeta_0^*}{dw}\right|^2 d\sigma + \sum_1^{n_0}|\zeta_k|^2\right].$$

PROOF. Assuming $\theta = 0$ for odd n, we have

$$\begin{aligned} I &= -\operatorname{Re}\left\{\frac{1}{2\pi i}\int_{|w|=1/4}\zeta^*\overline{d\zeta^*}\right\} + \alpha_0^n\sum_1^{n_0}|\zeta_k|^2 \geqslant J^2 + \left(\alpha_0^n - \frac{1}{2}\mu\right)\sum_1^{n_0}|\zeta_k|^2 \\ &\quad - \operatorname{Re}\left\{\frac{1}{\pi i}\int_{|w|=1/4}\zeta^*\overline{dp^*} + \frac{1}{2\pi i}\int_{|w|=1/4}p^*\overline{dp^*}\right\}. \end{aligned}$$

Applying Propositions 22.4 and 22.5, we find that

$$I > J^2 + \left(\alpha_0^n - \frac{1}{2}\mu\right)\sigma_{n_0}^2 - \alpha^n\sqrt{\mu}\,\sigma_{n_0}J - (3n+2)\,2^{2n+3}\mu\sigma_{n_0}^2.$$

But $\mu \leqq \alpha_1^n$ by Proposition 10.1, where α_1 is an absolute constant. Therefore

$$I > J^2 + (\alpha_0^n - \alpha_2^n)\,\sigma_{n_0}^2 - \alpha_3^n\sigma_{n_0}J \geqslant \frac{1}{2}[J^2 + \sigma_{n_0}^2], \tag{22.13}$$

if $\alpha_3^{2n} \leqq 2(\alpha_0^n - \alpha_2^n - \frac{1}{2})$. Taking $\alpha_0 = \alpha_2 + \alpha_3^2 + 1$, we reach the desired conclusion.

REMARK 22.1. In formula (21.8) we set $r = 1/4$. This choice of r can in fact be replaced by any absolute constant, say 1/8, which is less than 1/4. Such a choice of r does not in any way affect the above results, or any later result based on the above results. If we choose $r < 1/4$, then there is no need to exclude the Koebe function from consideration, as we did above.

§23. In this section we shall introduce three real Hilbert spaces, one of which will be the space of first variations of the function $f(z)$. In this chapter we shall also consider the space of those functions $\xi(w)$ regular in $\mathscr{G}$ and satisfying $\xi(0)=\xi'(0)=0$. Considering this space as a real linear space, we shall study the bilinear form

$$\text{(23.1)}\qquad \operatorname{Re}\left\{\frac{1}{\pi}\int\limits_{|w|>1/4}\frac{d\xi^*}{dw}\overline{\frac{d\eta^*}{dw}}\,d\sigma-\frac{1}{2\pi i}\int\limits_{|w|=1/4}\xi^*\overline{d\eta^*}+\vartheta\sum_1^n\xi_k\overline{\eta_k}\right\},$$

where $\vartheta=\alpha_0^n$, and where the double integral is defined by

$$\frac{1}{\pi}\int\limits_{|w|>1/4}\frac{d\xi^*}{dw}\overline{\frac{d\eta^*}{dw}}\,d\sigma=\lim\int\limits_{\mathscr{G}_r}\frac{d\xi^*}{dw}\overline{\frac{d\eta^*}{dw}}\,d\sigma,$$

where $\mathscr{G}_r=\mathscr{G}(\mathscr{L}_r)\backslash\{w:|w|\leq 1/4\}$.

This form is symmetric and, as shown by Proposition 22.6, positive. Hence the form (23.1) satisfies all the conditions for a scalar product, and so we shall denote it by (ξ,η). In this way we have defined a pre-Hilbert space $\mathscr{L}$. As usual we set $\|\xi\|=\sqrt{(\xi,\xi)}$. We shall now show that the form (23.1) is nonsingular. Suppose in fact that $(\xi,\xi)=0$. Then, according to Proposition 22.6, $\xi_0^*(w)=C$, and $\xi_1=\cdots=\xi_n=0$. Thus $\xi^*(w)=C$ and $\xi(w)=C(w^2/Q(w))^{1/2}$. For small w we deduce that $\xi(w)=CA_n^{-1/2}w^{n^*+1}(1+\cdots)$, and so $C=0$, which implies $\xi(w)\equiv 0$, as desired.

From the above it follows that $\|\xi\|$ is a norm. If $\xi\in\mathscr{A}_0$, this norm is connected in a very simple way with the second variation. To see this, we compare (19.18) and (21.8) with $r=1/4$, and we find that

$$\text{(23.2)}\qquad \delta^2L(f,\xi)=-\|\xi\|^2+\vartheta\sum_1^n|\xi_k|^2-\frac{1}{2}\operatorname{Re}\sum_1^n\xi_k\sum_{l=1}^{n-k}(k+l+2)\,A_{k+l}\xi_l.$$

The functions $\xi(w)\in\mathscr{L}$ possess several properties, some of which will be investigated below. First we prove the simple

Proposition 23.1. *Let $\xi_1,\cdots,\xi_{n^*-1}$ be arbitrary complex constants. Then there exists $\tilde{\xi}(w)\in\mathscr{A}_0$ satisfying the following conditions.*

1. $$\xi(w)=\xi_1w^2+\cdots+\xi_{n^*-1}w^{n^*}+\cdots.$$
2. $$\text{(23.3)}\qquad \|\tilde{\xi}\|<\tilde{C}\sigma.$$
3. *In the domain* $\mathscr{G}(\mathscr{L}_r)$,

$$\text{(23.4)}\qquad |\tilde{\xi}(w)|<\tilde{C}_r\sigma,$$

where $\sigma=(\sum_1^{n^*-1}|\xi_k|^2)^{1/2}$, and where the constants $\tilde{C}$ and $\tilde{C}_r$ are independent of ξ_k.

Proof. We write $\xi_j=-\gamma\bar{A}_j(\sum_{n^*}^n|A_k|^2)^{-1}, n^*\leq j\leq n$, where $\gamma=\operatorname{Re}\sum_1^{n^*-1}A_k\xi_k$. Then $\xi_1,\cdots,\xi_n$ satisfy (21.5), and so there exists a function $\tilde{\xi}(w)=\sum_1^{2n-1}\beta_j\xi_j(w)$ which satisfies condition 1 above. From (23.2) it follows that $\|\xi_j\|<C$. Thus $\sum_1^n|\xi_k|^2<\sigma^2(\sum_1^n|A_k|^2)(\sum_{n^*}^n|A_k|^2)^{-1}$, and then upon recalling the estimates for the quantities $|\beta_j|$, as derived in §21, we deduce (23.3). Also, (23.4) follows immediately from (21.7). Q.E.D.

Lemma 23.1. *Let $\xi \in \mathscr{E}$. Then the following assertions are true.*

1. *The limits $\xi_+(w)$ and $\xi_-(w)$ exist almost everywhere on $\mathscr{T}$.*
2. *For any $p < \infty$ and $\rho > 3/2$,*

$$\left\{\frac{1}{2\pi}\int_{\mathscr{T}} |\xi^*_\pm(w)|^p \frac{|dw|}{|w|^\rho}\right\}^{1/p} < B_p(\mathscr{T},\rho)\|\xi\|. \tag{23.5}$$

3. *In $\mathscr{G}(\mathscr{L}_r)$*

$$|\xi(w)| < C(\mathscr{T},r)\|\xi\|. \tag{23.6}$$

The constant $B_p(\mathscr{T},\rho)$ depends only upon p, ρ and the extremal function $f(z)$. The constant $C(\mathscr{T},r)$ depends only on r and $f(z)$.

Proof. Let $\tilde{\xi}(w)$ be the function constructed in the preceding proposition. If $\eta(w) = \xi(w) - \tilde{\xi}(w)$, then $\eta(w) = \eta_{n^*}w^{n^*+1} + \cdots$. According to (23.3) and the triangle inequality, $\|\eta\| \leqq \|\xi\| + \tilde{C}\sigma < (1 + \vartheta^{-1}\tilde{C})\|\xi\|$. Note that

$$\|\eta\|^2 = \frac{1}{\pi}\int_{\mathscr{G}}\left|\frac{d\eta^*}{dw}\right|^2 d\sigma + \vartheta\sum_{n^*}^{n}|\eta_k|^2.$$

Making the change of variable $w = f(z)$, we have that

$$\eta^*(f(z)) = i\sqrt{\frac{P(z)}{z^2}}\,\frac{\eta(f(z))}{f'(z)},$$

or, setting $\zeta(z) = i\eta(f(z))(f'(z))^{-1}$, we have that

$$\eta^*(f(z)) = \sqrt{\frac{P(z)}{z^2}}\,\zeta(z).$$

We claim that the function $\zeta(z)$ satisfies the hypothesis of Proposition 22.1, with $R(z) = z^{-2}P(z)$, $r_0 = 1$. In order to prove this, we expand $\zeta(z)$ in a series of the form $\zeta(z) = i\eta_{n^*}z^{n^*+1} + \cdots$ and note that

$$J^2 = \frac{1}{\pi}\int_{|z|<1}\left|\frac{d\zeta^*(z)}{dz}\right|^2 d\sigma_z + \frac{\theta}{2}|B_n\eta_{n^*}^2| = \frac{1}{\pi}\int_{\mathscr{G}}\left|\frac{d\eta^*}{dw}\right|^2 d\sigma + \frac{\theta}{2}|B_n\eta_{n^*}^2|,$$

where B_n is the coefficient of z^{-n} in the quasipolynomial $P(z)$. But $B_n = -A_n = \overline{\lambda}_n$, and so $|B_n| \leqq 1 < \vartheta$. From Proposition 22.6 we conclude that $J^2 \leqq \|\eta\|^2 < \infty$, and so the conditions of Proposition 22.1 are indeed satisfied. Thus the function $\zeta(z)$ has a limit almost everywhere on the circumference $|z| = 1$. The circumference $|z| = 1$ is divided into the $2m$ arms given in (16.1). Let $I = \bigcup_j I_j$ and $I' = \bigcup_j I'_j$. Sets of measure zero lying in I and I' correspond to sets of measure zero on $\mathscr{T}$. Thus the function $\eta(w)$ has left and right "limits" almost everywhere on $\mathscr{T}$. Since $\xi(w) = \eta(w) + \tilde{\xi}(w)$, and $\tilde{\xi} \in \mathscr{A}_0$, this must also be true of $\xi(w)$. Making the change of variable $z = f^{-1}(w)$ in the inequality

$$\frac{1}{2\pi}\int_{|z|=1}|\zeta^*(z)|^p\left|\frac{dz}{z}\right| < B_p^p J^p,$$

we find that

$$\frac{1}{2\pi}\int_{\mathscr{T}}\left[|\eta_+^*(w)|^p\left|\frac{dz_+}{dw}\right| + |\eta_-^*(w)|^p\left|\frac{dz_-}{dw}\right|\right]|dw| < B_p^p J^p.$$

No matter which vertex w_j of $\mathscr{T}$ we consider, there exists a constant $\gamma > 0$ such that

$$|dz_\pm/dw| \geqq C_0|w - w_j|^\gamma.$$

If $w_j = \infty$, then the right-hand side above must be changed to $|w|^{-\gamma}$. Without loss of generality, we assume that $\gamma > 1$. Upon applying the Schwarz and Hölder inequalities, we find that

$$\frac{1}{2\pi}\int_{\mathscr{T}}|\eta_+^*(w)|^p\frac{|dw|}{|w|^\rho} = \frac{1}{2\pi}\int_{\mathscr{T}}|\eta_+^*|^p\left|\frac{dz_+}{dw}\right|^{\frac{1}{4\gamma}}\left|\frac{dz_+}{dw}\right|^{-\frac{1}{4\gamma}}\frac{|dw|}{|w|^\rho}$$

$$\leqslant\left(\frac{1}{2\pi}\int_{\mathscr{T}}\left|\frac{dz_+}{dw}\right|^{-\frac{1}{2\gamma}}|w|^{-\rho}|dw|\right)^{1/2}\left(\frac{1}{2\pi}\int_{\mathscr{T}}|\eta_+^*|^{2p}\left|\frac{dz_+}{dw}\right|^{\frac{1}{2\gamma}}\frac{|dw|}{|w|^\rho}\right)^{1/2}$$

$$\leqslant\left(\frac{1}{2\pi}\int_{\mathscr{T}}\left|\frac{dz_+}{dw}\right|^{-\frac{1}{2\gamma}}\frac{|dw|}{|w|^\rho}\right)^{1/2}\left(\frac{1}{2\pi}\int_{\mathscr{T}}|w|\right)^{-\rho\frac{2\gamma}{2\gamma-1}}|dw|^{\frac{2\gamma-1}{4\gamma}}$$

$$\times\left(\frac{1}{2\pi}\int_{\mathscr{T}}|\eta_+^*|^{4p\gamma}\left|\frac{dz_+}{dw}\right||dw|\right)^{\frac{1}{4\gamma}}.$$

and so we conclude that

$$\left\{\frac{1}{2\pi}\int_{\mathscr{T}}|\eta_+^*(w)|^p\frac{|dw|}{|w|^\rho}\right\}^{1/p} < C_{\gamma,\rho}B_{4p\gamma}J. \tag{23.7}$$

The same inequality also holds for the function $\eta_-^*(w)$. We can establish an analogous inequality for the function $\tilde{\xi}(w)$. First, recalling Proposition 23.1, we have that $\tilde{\xi}(w) = \sum_1^{2n-1}\beta_j\xi_j(w)$. Noting that on $\mathscr{T}$ we have, in view of (19.12),

$$\left|\sqrt{\frac{Q(w)}{w^2}}(\alpha_j(w))_\pm\right| = |(\sqrt{P(z)}\,\dot{h}_j(\dot{z}))_\pm| < M_1 d_{n-1/2}^{-1}(f),$$

we find that $|(\xi_j^*(w)_\pm)| < M_2$. Recalling the inequalities we established in §21 for $|\beta_j|$, we find that $|\tilde{\xi}_\pm^*(w)| < C\sigma < C\vartheta^{-1}\|\xi\|$, and so

$$\left(\frac{1}{2\pi}\int_{\mathscr{T}}|\tilde{\xi}_\pm^*(w)|^p\frac{|dw|}{|w|^\rho}\right)^{1/p} < C_1\|\xi\|. \tag{23.8}$$

Now (23.5) follows from (23.7) and (23.8) upon using Minkowski's inequality. For the proof of (23.6) we use Proposition 22.2. For $|z| \leqq r$ we have $|\zeta(z)| <$

C_rJ, and so $|\eta(w)| < C_r|f'(z)|J < C_r(1-r)^{-3}J$ in $\mathscr{G}(\mathscr{L}_r)$. Upon combining this inequality and (23.4), we arrive at (23.6), as desired. Q.E.D.

By $\mathfrak{H}$ we shall denote the subset of $\mathscr{L}$ consisting of those elements for which (21.4) is satisfied almost everywhere on $\mathscr{T}$.

Lemma 23.2. *$\mathfrak{H}$ is a real Hilbert space.*

Proof. Clearly all we need show is that $\mathfrak{H}$ is complete. Let $\{\xi_k(w)\}$ be a Cauchy sequence. By Lemma 23.1 we have in $\mathscr{G}(\mathscr{L}_r)$, $r<1$, that

$$|\xi_k(w) - \xi_l(w)| < C_r\|\xi_k - \xi_l\|.$$

Thus $\{\xi_k(w)\}$ converges in $\mathscr{G}$ to some regular function $\xi_0(w)$. By Proposition 22.6

$$\frac{1}{2\pi}\int_{|w|>1/4}\left|\frac{d\zeta^*}{dw}\right|^2 d\sigma < \|\zeta\|^2.$$

Thus, if $\mathscr{G}_r = \mathscr{G}(\mathscr{L}_r)\setminus\{w:|w|<1/4\}$ and $k,l \geqq N(\epsilon)$, then

$$\frac{1}{2\pi}\int_{\mathscr{G}_r}\left|\frac{d(\xi_k-\xi_l)^*}{dw}\right|^2 d\sigma \leqslant \|\xi_k - \xi_l\|^2 < \varepsilon,$$

and so as $l \to \infty$ we find that

$$\frac{1}{2\pi}\int_{\mathscr{G}_r}\left|\frac{d(\xi_k-\xi_0)^*}{dw}\right|^2 d\sigma < \varepsilon.$$

In view of the arbitrariness of r it now follows that

$$\frac{1}{2\pi}\int_{|w|>1/4}\left|\frac{d(\xi_k-\xi_0)^*}{dw}\right|^2 d\sigma < \varepsilon.$$

Since the coefficients of the power series for $\xi_k(w)$ converge to the corresponding coefficients of $\xi_0(w)$, and $\xi_k(w)$ and $\xi_k'(w)$ tend respectively to $\xi_0(w)$ and $\xi_0'(w)$ in the disc $|w|<1/4$, from the last inequality above we conclude that $\|\xi_k - \xi_0\| < 2\epsilon$ if $k \geqq N_1(\epsilon)$. Thus $\lim_{k\to\infty}\xi_k = \xi_0$, since ϵ is arbitrary. Applying (23.5) to the function $\xi_k - \xi_0$ with $p=2$, we find that

$$\frac{1}{2\pi}\int_{\mathscr{T}}|(\xi_k-\xi_0)_\pm^*|^2\frac{|dw|}{|w|^\rho} < B_2^2(\mathscr{T},\rho)\|\xi_k-\xi_0\|^2.$$

We can now find, by well-known theorems, a subsequence $\{k_l\}$ such that the functions $(\xi_{k_l}-\xi_0)_\pm^*$ tend to zero almost everywhere on $\mathscr{T}$. Since ξ_{k_l} satisfies condition (21.4) almost everywhere on $\mathscr{T}$, it follows that ξ_0 possesses this same property, and so $\xi_0 \in \mathfrak{H}$. Q.E.D.

In passing we remark that $\mathscr{A}_0 \subset \mathfrak{H}$.

Definition 23.1. The closure of $\mathscr{A}_0$ will be called the *Hilbert space of first variations* of the extremal function $f(z)$, and it will be denoted by $\mathscr{A}$. It is clear that

(23.9) $$\mathscr{A} \subseteq \mathfrak{H}.$$

The spaces $\mathscr{A}$ and $\mathfrak{H}$ are of course defined by the tree $\mathscr{T}$, or, in other words, by the points $x \in \partial \mathscr{V}_n$. With regard to the smoothness of the boundary of the coefficient region, it is important to know when $\mathscr{A} = \mathfrak{H}$. Whether or not these spaces coincide depends on properties of $\mathscr{T}$.

To see this, let w_j be vertices (of a tree $\mathscr{T}$) which are zeros of multiplicity κ_j of $Q(w)$. Then in $\mathfrak{H}$ we have elements of the form

(23.10) $$w^2(w-w_j)^{-l}, \qquad iw^2(w-w_j)^{-l}, \qquad 1 \leqq l \leqq [\kappa_j/2].$$

Denote by $\mathfrak{R}_j$ the linear span of the functions (23.10). If we consider $w_j = \infty$ to be the base of the tree, and if κ_0 is the multiplicity of $w_j = \infty$ as a zero of $Q(w)$, then $\mathfrak{H}$ contains the functions

(23.11) $$w^{l+1}, \qquad iw^{l+1}, \qquad 1 \leqq l \leqq [\kappa_0/2].$$

Denote the linear span of the functions (23.11) by $\mathfrak{R}_0$. Set

(23.12) $$\mathfrak{R} = \mathfrak{R}_0 \oplus \mathfrak{R}_1 \oplus \cdots.$$

The space $\mathscr{A}$ might not contain the linear manifold $\mathfrak{R}$, and the question of when $\mathfrak{R} \subset \mathscr{A}$ is connected with some very subtle properties of the tree $\mathscr{T}$. Set $\mathscr{R}_j = \mathfrak{R}_j \cap \mathscr{A}$, $j = 0, 1, \cdots$, and $\mathscr{R} = \mathfrak{R} \cap \mathscr{A}$.

From the function-theoretic point of view the space $\mathscr{A}$ is rather complex in structure. In order to elucidate this structure, it is useful to introduce still another Hilbert space. Denote by H_0 the real linear manifold of functions of the form

(23.13) $$\xi(w) = \frac{1}{2\pi i} \int_{\mathscr{T}} \frac{w^2 \psi(t)}{(w-t)\, t^2}\, dt,$$

where $\psi(w)$ satisfies the following conditions.

1. On each branch of the tree $\mathscr{T}$, $\psi(w)$ is twice continuously differentiable and vanishes in some neighborhood of the terminal vertices of the branch.

2.

(23.14) $$\operatorname{Im}\left\{\sqrt{\frac{Q(w)}{w^2}}\, \psi(w)\right\} = 0.$$

We shall show that $H_0 \subset \mathfrak{H}$.

Proposition 23.2. *Any function $\xi \in H_0$ satisfies*

(23.15) $$\|\xi\|^2 < A \int_{\mathscr{T}} \left[\left|\frac{\psi(w)}{w^2}\right|^2 + \left|\frac{\psi'(w)}{w}\right|^2\right] |dw|,$$

where A is independent of ξ. Also,

$$\xi'(w) = \frac{1}{2\pi i} \int_{\mathscr{T}} \frac{w \psi'(t)}{(w-t)\, t}\, dt.$$

Clearly $\xi(w)$ and $\xi'(w)$ are regular in $\mathscr{G}$ and continuous up to the boundary. For large $|w|$ we have $\xi(w) = O(|w|)$ and $\xi'(w) = O(1)$. As in the proof of (21.8), we conclude that

$$\frac{1}{\pi}\int_{|w|\geqslant 1/4}\left|\frac{d\xi^*}{dw}\right|^2 d\sigma = -\frac{1}{2\pi i}\int_{\mathscr{T}}[\xi_+^*\overline{d\xi_+^*} - \xi_-^*\overline{d\xi_-^*}] + \frac{1}{2\pi i}\int_{|w|=1/4}\xi^*\overline{d\xi^*}.$$

Thus

$$\|\xi\|^2 = -\operatorname{Re}\left\{\frac{1}{2\pi i}\int_{\mathscr{T}}[\xi_+^*\overline{d\xi_+^*} - \xi_-^*\overline{d\xi_-^*}]\right\} + \vartheta\sum_1^n|\xi_k|^2.$$

From (20.2) and (20.9) it follows that

$$j = \operatorname{Re}\left\{\frac{1}{2\pi i}\int_{\mathscr{T}}[\xi_+^*\overline{d\xi_+^*} - \xi_-^*\overline{d\xi_-^*}]\right\}$$

$$= \operatorname{Re}\left\{\frac{1}{2\pi i}\int_{\mathscr{T}}\frac{Q(w)}{w^2}\psi(w)\,dw\,\frac{1}{\pi i}\int_{\mathscr{T}}\frac{w\psi'(t)}{t(w-t)}\,dt + \sum_1^{n-1}(k+1)\mu_{k+1}\xi_k\right\}.$$

If $w\in\mathscr{T}$, then $|Q(w)| < C|w|^{-1}$. Applying Schwarz' inequality to the integral and the sum, and using (13.20) and (20.7), we find that

$$|j| < A_0\left\{\left(\int_{\mathscr{T}}\left|\frac{\psi(w)}{w^2}\right|^2|dw|\right)^{1/2}\left(\int_{\mathscr{T}}\left|\frac{\psi'(w)}{w}\right|^2|dw|\right)^{1/2} + \sum_1^{n-1}|\xi_k|^2\right\}.$$

Since

$$\xi_k = -\frac{1}{2\pi i}\int\psi(w)\,w^{-k-2}\,dw, \qquad k\geqslant 1,$$

we get

$$|\xi_k|^2 < C\int_{\mathscr{T}}\left|\frac{\psi(w)}{w^2}\right|^2|dw|, \qquad k = 1, \ldots, n.$$

Inequality (23.15) follows from the latter two inequalities. Q.E.D.

Since by (23.14) the function $\xi(w)$ satisfies (21.4), we have that $H_0\subset\mathfrak{H}$. By H we denote the closure of H_0. Clearly $H\subseteq\mathfrak{H}$. We now give a less obvious property of H.

Proposition 23.3. $H\subseteq\mathscr{A}$.

Proof. Divide the circle $|z| = 1$ into $2m$ arcs by (16.1). Let $\xi(w)\in H_0$. On $|z| = 1$ we define a function $h(z)$ as follows: $h(z) = \frac{1}{2}\psi(f(z))/Df(z)$ if $z\in\bigcup_j I_j$, and $h(z) = -\frac{1}{2}\psi(f(z))/Df(z)$ if $z\in\bigcup_j I_j'$. Since $\psi(f(z))$ vanishes in some neighborhood of the end points of each of the arcs in (16.1), $h(z)$ is a twice continuously differentiable function. Let $\epsilon > 0$ be given. From (23.14) it follows that $h(z)$ assumes only purely imaginary values. Construct a purely imaginary trigonometric polynomial $S_\epsilon(z)$ such that

$$|h(z) - S_\epsilon(z)| < \epsilon, \quad |h'(z) - S_\epsilon'(z)| < \epsilon.$$

Denote by z_j the end points of the arcs (16.1). Let $R_\epsilon(z)$ be the interpolation polynomial of minimal degree defined by the conditions $R_\epsilon(z_j) = -S_\epsilon(z_j)$, $j = 1, 2, \cdots$. Since $|S_\epsilon(z_j)| < \epsilon$, we have $|R_\epsilon(z_j)| < C_1\epsilon$ and $|R'_\epsilon(z_j)| < C_2\epsilon$, where the constants C_1 and C_2 depend only on the original division of the circumference by the arcs (16.1). With $T_\epsilon(z) = S_\epsilon(z) + R_\epsilon(z)$, we have $|h(z) - T_\epsilon(z)| < (C_1 + 1)\epsilon$ and $|h'(z) - T'_\epsilon(z)| < (C_2 + 1)\epsilon$. Since $h(z_j) = T_\epsilon(z_j) = 0$, it follows from the latter estimates that

$$|h(z) - T_\varepsilon(z)| < C\varepsilon \prod_j |z - z_j|, \tag{23.16}$$

where C is independent of ϵ. Let $\alpha_\epsilon(w) = Df(z)\,T_\epsilon(z)$, $z = f^{-1}(w)$. Then $\alpha_\epsilon(w)$ is regular in any domain $\mathscr{G}(\mathscr{L}_r, \mathscr{T})$, $r < 1$. For $w \in \mathscr{G}(\mathscr{L}_r)$ set

$$\xi_\varepsilon(w) = \frac{1}{2\pi i} \int_{\mathscr{C}_r} \frac{w^2 \alpha_\varepsilon(t)}{(w - t)\, t^2}\, dt.$$

By definition $\xi_\epsilon(w) \in \mathscr{A}_0$. We shall show that in $\mathscr{G}(\mathscr{L}_r, \mathscr{T})$ we have $\alpha_\epsilon(w) = O(w)$. Indeed, if z_j is an algebraic singularity of $f(z)$, $f(z_j) \neq \infty$, then $Df(z) = O((z - z_j)^{\alpha_j - 1})$, where $\alpha_j > 0$. Hence $Df(z)\,T_\epsilon(z) = O((z - z_j)\,Df(z)) = O((z - z_j)^{\alpha_j})$. If z_j is an algebraic pole, then $f(z) = a_j(z - z_j)^{-\alpha_j}(1 + \cdots)$ and $Df(z) = O((z - z_j)^{-\alpha_j - 1})$. Hence $Df(z)\,T_\epsilon(z) = O((z - z_j)\,Df(z)) = O((z - z_j)^{-\alpha_j}) = O(w)$.

Thus we have that

$$\xi_\varepsilon(w) = \frac{1}{2\pi i} \int_{\mathscr{T}} \frac{w^2}{(w - t)\, t^2} \left[(\alpha_\varepsilon(t))_+ - (\alpha_\varepsilon(t))_-\right] dt.$$

In order to estimate $\|\xi - \xi_\epsilon\|$, we consider the function $\chi(w) = \psi(w) - [(\alpha_\epsilon(w))_+ - (\alpha_\epsilon(w))_-]$. The identity

$$\chi(w) = Df(z_+)\,[h(z_+) - T_\varepsilon(z_+)] + Df(z_-)[T_\varepsilon(z_-) - h(z_-)] \tag{23.17}$$

is easily verified. In view of the above and (23.16) we get that $\|\chi(w)\| < C_3|w|\epsilon$, where C_3 is independent of ϵ. Differentiating (23.17), we find that

$$\chi'(w) = z_+[h'(z_+) - T'_\varepsilon(z_+)] + z_-[T'_\varepsilon(z_-) - h'(z_-)] + \frac{D^2 f(z_+)}{Df(z_+)}[h(z_+) - T_\varepsilon(z_+)] + \frac{D^2 f(z_-)}{Df(z_-)}[T_\varepsilon(z_-) - h(z_-)].$$

Since $D^2 f(z)/Df(z) = O((z - z_j)^{-1})$ in a neighborhood of z_j, we see that $|\chi'(w)| < C_4\epsilon$, where C_4 is independent of ϵ. Finally we note that at any vertex of the tree other than at the base we have $\chi(w_j) = 0$. If we can show that (23.15) holds for $\xi(w) - \xi_\epsilon(w)$, it will follow that $\|\xi - \xi_\epsilon\| < C_5\epsilon$, where C_5 is independent of ϵ. Thus $\xi \in \mathscr{A}$, and so $H \subseteq \mathscr{A}$, as desired. Q.E.D.

All that now remains to complete the proof is to show that inequality (23.15) holds for $\xi(w) - \xi_\epsilon(w)$. This follows from

REMARK 23.1. Let the function $\xi(w)$ be represented by the integral (23.13), where $\psi(w)$ satisfies the following conditions.

1. On each branch not emanating from the base, $\psi(w)$ is continuous and vanishes at the vertices of the branch; $\psi(w)$ has a bounded derivative on

each branch of the tree, and the derivative is continuously differentiable on the open branch. On a branch emanating from the base, $\psi(w)$ is continuous and satisfies $|\psi(w)| < C|w|$, and $\psi(w)$ vanishes at any vertex different from the base.

2. $\psi(w)$ satisfies condition (23.14).

Then inequality (23.15) holds, and $\xi \in H$.

Let $\mathscr{K}_\delta$ be the system of discs constructed in §16. It is easily seen that there exists a real-valued twice continuously differentiable function $\theta_\delta(w)$ with the following properties: a) $\theta_\delta(w) = 1$ if $w \in \mathscr{T} \setminus \mathscr{K}_{2\delta}$: b) $\theta_\delta(w) = 0$ if $w \in \mathscr{T} \cap \mathscr{K}_\delta$; and c) $|\theta_\delta'(w)| < C\delta^{-1}$ in each disc $K_{j2\delta}$ centered at the vertex w_j (w_j is not the base of the tree), and $|\theta_\delta'(w)| < C|w|^{-1}$, $|w| \geqq 1/2\delta$, where the constant C does not depend on δ. Set $\psi_\delta(w) = \theta_\delta(w)\psi(w)$ and define, via (23.13), the function $\xi_\delta(w) \in H_0$. If $\delta_1 < \delta_2$, then

$$\|\xi_{\delta_1} - \xi_{\delta_2}\|^2 < A \int\limits_{\mathscr{T} \cap \mathscr{K}_{2\delta_2}} \left\{ \left| \frac{\psi(w)}{w^2} (\theta_{\delta_1}(w) - \theta_{\delta_2}(w)) \right|^2 \right.$$

$$\left. + \left| \frac{\psi'(w)}{w} (\theta_{\delta_1}(w) - \theta_{\delta_2}(w)) + \frac{\psi(w)}{w} (\theta_{\delta_1}'(w) - \theta_{\delta_2}'(w)) \right|^2 \right\} |dw|,$$

as follows from (23.15). Since $\psi(w) = O(w - w_j)$ in a neighborhood of each terminal vertex w_j, it follows that $\|\xi_{\delta_1} - \xi_{\delta_2}\| = o(1)$. Hence the sequence $\xi_\delta(w)$ converges to some function $\xi_0(w) \in H$. By (23.6), $\xi_\delta(w)$ converges uniformly in any domain $\mathscr{G}(\mathscr{L}_r)$ to $\xi_0(w)$. Since

$$\xi_\delta(w) = \frac{1}{2\pi i} \int\limits_{\mathscr{T}} \frac{w^2 \psi(t)\, \theta_\delta(t)}{t^2 (w - t)}\, dt,$$

we find upon taking the limit as $\delta \to 0$ that $\xi_0(w)$ satisfies (23.13). Thus $\xi_0(w) \equiv \xi(w)$. Applying (23.15) to the function $\psi_\delta(w)$ and taking the limit as $\delta \to 0$, we also conclude that this inequality holds for $\psi(w)$. Hence we have established the inclusion relations

$$H \subseteq \mathscr{A} \subseteq \mathfrak{H}. \tag{23.18}$$

Lemma 23.3. *The first variations generated by the admissible functions of class $\mathscr{D}_f$ are elements of $\mathscr{A}$.*

Proof. In §17 we proved that $\delta f(z) = \xi[f(z)] + \eta[f(z)]$, where $\xi \in \mathscr{A}_0$, and (by Remark 23.1) $\eta \in H$. Thus $\xi + \eta \in \mathscr{A}$. Q.E.D.

This lemma justifies the name given to $\mathscr{A}$. We set $R_j = \mathscr{R}_j \cap H$, $j = 0, 1, \cdots$, and $R = \bigcup_j R_j$. It is easy to describe the finite-dimensional subspace R, and this will be done a bit later. To conclude this section, we prove a lemma which sharpens Lemma 23.1 on one point.

Lemma 23.4. *If $\xi \in \mathscr{A}$, then*

$$|\xi(w)| < C_{n,r}\|\xi\|$$

in $\mathscr{G}(\mathscr{L}_r)$, where

$$C_{n,r} = B_0(n)(1 - r)^{-n - \frac{7}{2}} (1 + d_{n-1/2}^{-1}(f)). \tag{23.19}$$

PROOF. Set $\xi(w) = \sum_1^\infty \xi_k w^{k+1}$ and $\mathfrak{z} = (\xi_1, \cdots, \xi_n)$. Relation (21.5) holds for any element of $\mathscr{A}_0$. Thus it holds for $\xi(w)$. Let $\tilde{\xi}(w)$ be the function constructed in the definition of §21.4, and set $\eta = \xi - \tilde{\xi}$. Then $\eta(w) = \eta_{n+1} w^{n+2} + \cdots$, and $\|\eta\| \leq \|\xi\| + \|\tilde{\xi}\|$. We now proceed as in the proof of Lemma 23.1. We set $\zeta(z) = i\eta[f(z)]/f'(z)$ and $R(z) = z^{-2}P(z)$. It is clear that

$$\|\eta\|^2 = \frac{1}{\pi}\int_{\mathscr{G}} \left|\frac{d\eta^*}{dw}\right|^2 d\sigma = \frac{1}{\pi}\int_{|z|<1} \left|\frac{d\zeta^*}{dz}\right|^2 d\sigma_z = J^2.$$

By Proposition 22.2

$$|\zeta(z)| \leq B(n)(1-|z|)^{-n-1/2}\mu^{-1/2}\|\eta\|.$$

In our case we have

$$\mu^2 = \frac{1}{2\pi}\int_{|z|=1} |P(z)|^2 |dz|.$$

Thus

$$\mu^2 > \frac{r_0^{2n}}{2\pi}\int_{|z|=r_0} |P(z)|^2 \left|\frac{dz}{z}\right| = \frac{r_0^{2n}}{2\pi}\int_{|z|=r_0} \left|\left(\frac{Df(z)}{f(z)}\right)^2 Q(f(z))\right|^2 \left|\frac{dz}{z}\right|$$

$$> \min_{|z|=r_0}\left|\frac{Df(z)}{f^2(z)}\right| \frac{r_0^{2n}}{2\pi}\int_{\mathscr{C}_{r_0}} |Q(w)|^2 |dw|.$$

From this, with the aid of the distortion theorems and Proposition 10.1, it follows that $\mu \geq \alpha_1^n$, where α_1 is an absolute constant. Therefore

$$|\eta(f(z))| < B_1(n)(1-|z|)^{-n-\frac{7}{2}}\|\eta\|,$$

where $B_1(n)$ depends only on n. Since $|\mathfrak{z}| \leq \|\xi\|$, it follows from the last inequality and (21.7) that for $w \in \mathscr{G}(\mathscr{C}_r)$,

$$|\xi(w)| < (1-r)^{-n-\frac{7}{2}}(B_1(n)\|\eta\| + \alpha^n \alpha_{n-1/2}^{-1}(f)\|\xi\|).$$

In order to obtain our final result, we must estimate $\|\tilde{\xi}\|$. In accordance with §21.4 we have $\tilde{\xi}(w) = \sum_1^{2n-1}\beta_j \xi_j(w)$. Thus $\|\tilde{\xi}\| < \sum_1^{2n-1}|\beta_j|\,\|\xi_j(w)\| < \alpha_0^n|\mathfrak{z}|\sum_1^{2n-1}\|\xi_j\|$. We estimate $\|\xi_j\|$ by setting $\xi_j(w) = \sum_1^\infty \xi_{jk} w^{k+1}$, where

$$\xi_{jk} = -\frac{1}{2\pi i}\int_{\mathscr{C}_{r_1}} \alpha_j(w)\, w^{-k-2}\, dw.$$

From the distortion theorem and the estimate for $h_j(z)$ given in §21.4 we find that

$$|\xi_{jk}| < M_0 d_{n-1/2}^{-1}\frac{1}{2\pi}\int_{|z|=r_1} |Df(z)|^2 |f(z)|^{-k-2}\left|\frac{dz}{z}\right| < \alpha_2^n d_{n-1/2}^{-1}(f).$$

Relation (23.2), combined with the above inequalities and Proposition 10.1, implies that

$$\|\xi_j\|^2 < |\delta^2 L(f, \xi_j)| + \alpha_3^n d_{n-1/2}^{-2}(f).$$

But $\delta^2 L(f;\xi_j) = \delta^2 L(f;\nu_j)$ (see §21.4). Applying (4.18) (with the constant $c_1 = 0$) and Proposition 5.11, we get $\|\xi_j\| < B_2(n) d_{n-1/2}^{-1}(f)$, where $B_2(n)$ is a constant depending only on n. Thus we have proved that $\|\tilde{\xi}\| < B_3(n) d_{n-1/2}^{-1}(f)\|\xi\|$. It now follows that

$$|\xi(w)| < B_0(n)(1-r)^{-n-\frac{7}{2}}(1 + d_{n-1/2}^{-1}(f))\|\xi\|,$$

as required. Q.E.D.

§24. We shall now address ourselves to the question of investigating the structure of the spaces $\mathfrak{H} \ominus \mathscr{A}$ and $\mathfrak{H} \ominus H$. For convenience we shall use the following terminology.

DEFINITION 24.1. The multiplicity of the vertex w_j of $\mathscr{T}$ as a zero of $Q(w)$ will be called the *index* of the vertex w_j. The index will be denoted by κ_j. It is clear that for any vertex w_j we have $\nu_j \leqq \kappa_j + 2$, where ν_j is the order of the vertex w_j with respect to the tree $\mathscr{T}$. If $\eta_j = 0$, then w_j is a terminal vertex and $\nu_j = 1$. On the circle $|z| = 1$ there correspond to w_j those points $z_{j1}, \cdots, z_{j\nu}$ which are the end points of the arcs (16.1). The function $P(z)$ can vanish on $|z| = 1$ only at the points z_{jk}. Let $2\mu_{jk}$ be the multiplicities of these zeros. The disc $K_{j\delta} \subset \mathscr{K}_\delta$ (see §16 for the definition of $\mathscr{K}_\delta$) is divided into ν_j curvilinear sectors σ_{jk}, provided δ is sufficiently small. As before, denote by τ_{jk} the image of σ_{jk} under the mapping $z = f^{-1}(w)$. From (5.13) it follows that

$$f(z) = w_j + \alpha_{jk}(z - z_{jk})^{r_{jk}} + \ldots, \tag{24.1}$$

where

$$r_{jk} = \frac{2\mu_{jk} + 2}{\varkappa_j + 2}. \tag{24.2}$$

If $w_j = \infty$, that is, if the vertex under consideration is the base of the tree, then (24.1) must be replaced by

$$f(z) = \alpha_{jk}(z - z_{jk})^{-r_{jk}} + \ldots, \tag{24.3}$$

where

$$r_{jk} = \frac{2\mu_{jk} + 2}{\varkappa_j}. \tag{24.4}$$

For convenience we shall always denote the base of the tree by the subscript $j = 0$. In each sector σ_{jk} we have the expansion

$$z - z_{jk} = \beta_{0k}(w - w_j)^{\frac{1}{r_{jk}}} + \ldots, \tag{24.5}$$

and for the base of the tree we have

$$z - z_{0k} = \beta_{0k} w^{-\frac{1}{r_{0k}}} + \ldots. \tag{24.6}$$

We shall show that any element $\xi \in \mathfrak{H}$ which is orthogonal to H must satisfy a differential equation. First, however, we will reformulate the expression defining the scalar product for two elements $\xi \in \mathfrak{H}$ and $\eta \in \mathscr{A}_0$.

PROPOSITION 24.1. *Let $\xi \in \mathfrak{H}$ and $\eta \in \mathscr{A}_0$. The scalar product of these two elements is given by*

$$(24.7)\qquad (\xi, \eta) = \operatorname{Re}\left\{-\frac{1}{2\pi i}\int_{\mathscr{T}}[\xi_+^*\overline{d\eta_+^*} - \xi_-^*\overline{d\eta_-^*}] + \vartheta\sum_1^n \xi_k\overline{\eta}_k\right\}.$$

PROOF. Set $\xi_0(z) = i\xi[f(z)][f'(z)]^{-1}$, $\eta_0(z) = i\eta[f(z)][f'(z)]^{-1}$, so that $\xi^*[f(z)] = \xi_0^*(z)$, where $\xi_0^*(z) = \sqrt{P(z)/z^2}\xi_0(z)$; analogous relations hold also for $\eta(w)$. Set

$$\mathscr{G}_r = \mathscr{G}(\mathscr{C}_r)\setminus\{w : |w| \leqslant 1/4\}, \qquad G_r = f^{-1}\{\mathscr{G}_r\}, \text{ and } \mathscr{C} = \left\{z : |f(z)| = \frac{1}{4}\right\}.$$

Choose r so that $1 - r$ is sufficiently small. Then

$$\frac{1}{\pi}\int_{\mathscr{G}_r}\frac{d\xi^*}{dw}\overline{\frac{d\eta^*}{dw}}\,d\sigma = \frac{1}{\pi}\int_{G_r}\frac{d\xi_0^*}{dz}\overline{\frac{d\eta_0^*}{dz}}\,d\sigma_z$$

$$= -\frac{1}{2\pi i}\int_{|z|=r}\xi_0^*(z)\overline{d\eta_0^*(z)} + \frac{1}{2\pi i}\int_{\mathscr{C}}\xi_0^*(z)\overline{d\eta_0^*(z)}$$

by Green's formula. From (19.12), which holds for any element of $\mathscr{A}_0$, it follows that for $r_\eta < |z| < 1$

$$(24.8)\qquad \eta_0^*(z) = -i\sqrt{P(z)}\,h(z) + \sqrt{Q(f(z))}\,\frac{1}{2\pi i}\int_{\mathscr{C}_{r_\eta}}\frac{f(z)\,\alpha(t)}{(f(z)-t)\,t^2}\,dt.$$

Since we may assume without loss of generality that $P(z)$ does not vanish in the annulus $r_\eta < |z| < 1$ and that the zeros of $P(z)$ on $|z| = 1$ are of even multiplicity, it follows that for $r_\eta < |z| < 1$ we have $(d/dz)(\sqrt{P(z)}\,h(z)) = O(1)$.

The derivative of the second term in (24.8) can be easily estimated if $z \in \tau_{jk}$. Let $j \neq 0$ and $\kappa_j \geqq 1$; using (24.1) and (24.2), it is not difficult to show that this derivative is $O((z - z_{jk})^{\lambda_{jk}})$, where $\lambda_{jk} = (\mu_{jk}\kappa_j - 2)/(\kappa_j + 2)$. If $\kappa_j = 0$, then the vertex is terminal, and the derivative is $O(z - z_j)$. If however $j = 0$, it follows from (24.3) that the derivative is $O((z - z_{0k})^{\mu_{0k}})$. Thus we have shown that $d\eta_0^*(z)/dz = O((z - z_{jk})^{-\beta})$, $z \in \tau_{jk}$, where $\beta < 1$. Hence for any $q > 1$ and for $r_\eta < r < 1$ we have $d\eta_0^*(re^{i\theta})/dz \in L^q(0, 2\pi)$; the L^q norm of $d\eta_0^*(re^{i\theta})/dz$ is bounded by a constant independent of r. By Proposition 22.3, $\xi^*(re^{i\theta})$ converges in $L^p(0, 2\pi)$, with $p = q/(q-1)$, to $\xi^*(e^{i\theta})$. We shall require the identity

$$j(r) = \frac{1}{2\pi i}\int_{|z|=r}\xi_0^*(z)\overline{d\eta_0^*(z)} = \frac{1}{2\pi i}\int_0^{2\pi}\xi_0^*(e^{i\theta})\overline{\frac{d\eta_0^*(re^{i\theta})}{d\theta}}\,d\theta$$

$$+\frac{1}{2\pi i}\int_0^{2\pi}[\xi_0^*(re^{i\theta}) - \xi_0^*(e^{i\theta})]\overline{\frac{d\eta_0^*(re^{i\theta})}{d\theta}}\,d\theta.$$

Applying Hölder's inequality with conjugate indices p and q to the second integral, we see from what has been proved above that this integral is $o(1)$.

The function $d\eta_0^*(z)/dz$ is regular in $\{z: r_\eta < |z| \leq 1\} \setminus \bigcup_{j,k} \tau_{jk\delta}$ for any $\delta > 0$. Thus from the inequalities satisfied by this function in any domain $\tau_{jk\delta}$ and from the fact that $\xi^*(e^{i\theta}) \in L^p$ it follows that

$$\lim_{r \to 1} j(r) = \frac{1}{2\pi i} \int_0^{2\pi} \xi_0^*(e^{i\theta}) \overline{d\eta_0^*(e^{i\theta})}.$$

Dividing this integral into the sum of two integrals over the sets $I = \bigcup_j I_j$ and $I' = \bigcup_j I'_j$, we find that (24.7) holds, as required.

Proposition 24.2. *If ξ and η satisfy the conditions of the previous proposition, then the integral*

$$\frac{1}{2\pi i} \int_{\mathscr{T}} [\xi_+^* \overline{d\eta_+^*} - \xi_-^* \overline{d\eta_-^*}] = \frac{1}{2\pi i} \int_{|z|=1} \xi_0^*(z) \overline{d\eta_0^*(z)}$$

exists, and for any $r < 1$ it equals $(1/2\pi i) \int_{\mathscr{C}_r} \xi^* \overline{d\eta^*}$.

Remark 24.1. Formula (24.7) holds if $\eta \in H_0$. The proof of Proposition 24.1 used the fact that $\eta \in \mathscr{A}_0$ only in the study of $d\eta_0^*/dz$. For this, in turn, formula (24.8) was the essential requirement, and in particular it was essential in that it showed that the properties of $d\eta_0^*/dz$ were completely defined by the second term of (24.8). But for the function

$$\sqrt{Q(f(z))} \frac{1}{2\pi i} \int_{\mathscr{T}} \frac{f(z) \psi(t)\, dt}{(f(z) - t)\, t^2}$$

the above comments continue to be valid, since $\psi(t)$ vanishes in a neighborhood of the vertices of $\mathscr{T}$.

Remark 24.2. Proposition 24.2 holds if $\xi \in \mathfrak{H}$ and $\eta \in H_0$.

Suppose that H is a proper subspace of $\mathfrak{H}$, and let ξ be a vector orthogonal to H. Then the function $\xi(w)$ satisfies a differential equation, as we shall now show. In the process of our argument, we shall encounter the quantities

$$\xi_+^*(w) = \sqrt{\frac{Q(w)}{w^2}}\, \xi_+(w), \qquad \xi_-^*(w) = \sqrt{\frac{Q(w)}{w^2}}\, \xi_-(w),$$

where in each case we take the same branch of the square root.

If $\eta \in H_0$, then $\eta(w)$ can be represented via formula (23.13). We shall also be making use of (24.7). Applying (23.14) and the formulas for the jump discontinuities, we find that

$$\xi_+^* \overline{d\eta_+^*} - \xi_-^* \overline{d\eta_-^*} = -\frac{1}{2} (\xi_+^* + \xi_-^*)\, d\left(\sqrt{\frac{Q(w)}{w^2}}\, \psi(w)\right) + (\xi_+^* - \xi_-^*)\, \overline{d\left(\sqrt{\frac{Q(w)}{w^2}}\, J(w)\right)},$$

where

$$J(w) = \frac{1}{2\pi i} \int_{\mathscr{T}} \frac{w^2 \psi(t)\, dt}{(w - t)\, t^2}, \qquad w \in \mathscr{T}.$$

Thus

$$(\xi, \eta) = \operatorname{Re}\left\{\frac{1}{2\pi i}\int_{\mathscr{T}}\left[\frac{1}{2}(\xi_+^* + \xi_-^*)\, d\left(\sqrt{\frac{Q(w)}{w^2}}\,\psi(w)\right)\right.\right.$$
$$\left.\left. + (\xi_+^* - \xi_-^*)\, d\left(\sqrt{\frac{Q(w)}{w^2}}\,J(w)\right)\right] + \vartheta\sum_1^n \xi_k\bar{\eta}_k\right\},$$

and so

$$(\xi, \eta) = \operatorname{Re}\left\{\frac{1}{2\pi i}\int_{\mathscr{T}}\left[(\xi_+^* + \xi_-^*)\, d\left(\sqrt{\frac{Q(w)}{w^2}}\,\psi(w)\right) + \xi_+^* d\eta_+^* - \xi_-^* d\eta_-^*\right] + \vartheta\sum_1^n \bar{\xi}_k\eta_k\right\}.$$

Using Remark 24.2, we get

$$J = \frac{1}{2\pi i}\int_{\mathscr{T}}[\xi_+^* d\eta_+^* - \xi_-^* d\eta_-^*] + \vartheta\sum_1^n \bar{\xi}_k\eta_k = \frac{1}{2\pi i}\int_{\mathscr{C}_r}\xi^* d\eta^* + \vartheta\sum_1^n \bar{\xi}_k\eta_k$$
$$= -\frac{1}{2\pi i}\int_{\mathscr{C}_r}\left[\eta^* d\xi^* - \vartheta\sum_1^n \bar{\xi}_k w^{-k-2}\eta(w)\, dw\right].$$

Defining the quantities α_k by the expansion

$$\sqrt{\frac{Q(w)}{w^2}}\,\frac{d\xi^*}{dw} = \sum_0^{n-1}\alpha_k w^{-k-2} + \alpha_{-1}w^{-1} + \dots, \tag{24.9}$$

we see that

$$\alpha_k = \frac{1}{2}\sum_{l=1}^{n-k}(l-k)\, A_{k+l}\xi_l, \qquad k = 0, 1, \dots, n-1. \tag{24.10}$$

We next consider the following polynomial in $1/w$:

$$\tilde{Q}(w) = \frac{1}{2}\left(\vartheta\sum_1^n \bar{\xi}_k w^{-k} - \sum_1^{n-1}\alpha_k w^{-k}\right). \tag{24.11}$$

Clearly

$$J = \frac{1}{\pi i}\int_{\mathscr{C}_r}\tilde{Q}(w)\,\eta(w)\,\frac{dw}{w^2}.$$

Letting $r \to 1$ in this formula and using our above results, we find that

$$(\xi, \eta) = \operatorname{Re}\left\{\frac{1}{2\pi i}\int_{\mathscr{T}}\left[(\xi_+^* + \xi_-^*)\, d\left(\sqrt{\frac{Q(w)}{w^2}}\,\psi(w)\right) - 2\,\frac{\tilde{Q}(w)}{w^2}\,\psi(w)\, dw\right]\right\}.$$

Since $(\xi, \eta) = 0$ for any element $\eta \in H_0$, it follows from this latter relation and (23.14) that on each branch $\gamma \subset \mathscr{T}$ we have

$$\operatorname{Im}\left\{\xi_+^*(w) + \xi_-^*(w) + 2\int_{w_\gamma}^{w}\frac{\tilde{Q}(w)}{w^2}\left(\frac{w^2}{Q(w)}\right)^{1/2} dw\right\} = C_\gamma, \tag{24.12}$$

where w_γ is some point of the branch other than the vertex. Hence the function $\operatorname{Im}\{\xi_+^*(w) + \xi_-^*(w)\}$, as a function of arc length on the branch, is infinitely differentiable on each open branch. The functions $\operatorname{Im}\xi_+^*(w)$ and $\operatorname{Im}\xi_-^*(w)$ possess this same property since $\xi_+^* - \xi_-^*$ is real. Set $\xi_0(z) = \xi[f(z)][Df(z)]^{-1}$, so that $\xi^*(f(z)) = i\sqrt{P(z)}\,\xi_0(z)$, and in particular we have that $\xi^*_\pm(w) = i\sqrt{P(z_\pm)}\,\xi_0(z_\pm)$, where $w = f(z_\pm)$. Therefore $\operatorname{Re}\xi_0(z)$, for $|z| = 1$, is an infinitely differentiable function everywhere except at the end points of the intervals (16.1). Thus the conjugate function $\operatorname{Im}\xi_0(z)$ will also be differentiable everywhere on $|z| = 1$, with the exception of the above-mentioned points. In particular, the function $\xi^*_\pm(w)$ is differentiable on any open branch. Substituting $\xi^*_+(w)$ for $\xi^*_-(w)$ in (24.12) and differentiating with respect to arc length, we find that

$$\operatorname{Im}\left\{\left(\frac{dw}{ds}\right)^2\left[\sqrt{\frac{Q(w)}{w^2}}\,\frac{d\xi^*_-(w)}{dw} + \frac{\widetilde{Q}(w)}{w^2}\right]\right\} = 0.$$

This equation continues to hold if we replace $\xi^*_+(w)$ by $\xi^*_-(w)$. Putting $\Phi(w) = \sqrt{Q(w)/w^2}\,d\xi^*(w)/dw + \widetilde{Q}(w)/w^2$, we see from the above remarks that

$$\operatorname{Im}\{(Df(z))^2\Phi[f(z)]\} = 0, \tag{24.13}$$

if $|z| = 1$ and z is not an end point of the intervals in (16.1). In a neighborhood of $w = 0$ we have, in accord with (24.9),

$$\Phi(w) = \frac{1}{2}w^{-2}\left[\vartheta\sum_1^n \bar{\xi}_k w^{-k} + \sum_1^{n-1}\alpha_k w^{-k} + 2\alpha_0\right] + \cdots.$$

We define the quantities β_k, $k = -n, \cdots, n$, $\beta_k = \bar{\beta}_{-k}$; so that

$$\left(\frac{Df(z)}{f(z)}\right)^2\left\{\frac{\vartheta}{2}\sum_1^n \bar{\xi}_k[f(z)]^{-k} + \frac{1}{2}\sum_1^{n-1}\alpha_k[f(z)]^{-k} + \alpha_0\right\} + \sum_{-n}^{n}\beta_k z^{-k} = i\alpha + O(z), \tag{24.14}$$

where α is real. Set

$$\widetilde{\mathscr{P}}(z) = \sum_{-n}^{n}\beta_k z^{-k}. \tag{24.15}$$

Since $\widetilde{\mathscr{P}}(z)$ is real for $|z| = 1$, it follows that (24.13) can be written in the form $\operatorname{Im}\{(Df(z))^2\Phi[f(z)] + \widetilde{\mathscr{P}}(z)\} = 0$. Denoting by $F(z)$ the expression in curly brackets, we see via (24.14) that $F(z)$ is regular in $|z| < 1$, and on the open arcs (16.1) we have $\operatorname{Im}F(z) = 0$. Thus $F(z)$ can be analytically continued across each of these arcs; of course $\xi[f(z)]$ also possesses this property. We set

$$F_1(z) = \frac{1}{i}\int_0^z [F(z) - F(0)]\frac{dz}{z}.$$

On the open intervals (16.1) the function $F_1(z)$ is regular, and $\operatorname{Im}F_1(z)$ is a linear function of $\theta = \arg z$ on each arc I_j or I'_j. We shall show that $F_1(e^{i\theta}) \in L(0, 2\pi)$. To prove this we write

$$F(z) = -\sqrt{P(z)}\,z\frac{d}{dz}\left(\sqrt{P(z)}\,\xi_0(z)\right) + \left(\frac{Df(z)}{f(z)}\right)^2\widetilde{Q}(f(z)) + \widetilde{\mathscr{P}}(z).$$

The sum of the last two terms is regular on the open arcs (16.1). In a neighborhood of each point z_j we can expand the last two terms in a series of fractional powers of $z-z_j$, and the first term of the expansion has the form $a_j(z-z_j)^{-\alpha_j}$, where $\alpha_j<2$. (Formulas (24.1)–(24.4) justify the above statements.) Let $p(z)$ be the sum of the first n terms of the power series for $\xi_0(z)$. Set $\zeta(z)=\xi_0(z)-p(z)$. Since $\zeta(z)$ satisfies the conditions of Proposition 22.1, we have $P^{1/2}(e^{i\theta})\,\zeta(e^{i\theta})\in L^p(0,2\pi)$ for any $p<\infty$. We also note that

$$\int_0^z P^{1/2}(z)\frac{d}{dz}(P^{1/2}(z)\zeta(z))\,dz = P(z)\zeta(z)-\frac{1}{2}\int_0^z P'(z)\zeta(z)\,dz.$$

Since $P(z)$ has only zeros of even multiplicity on $|z|=1$, we have that $P'(e^{i\theta})\,\zeta(e^{i\theta})\in L^p(0,2\pi)$. Combining all these remarks, we conclude that $F_1(e^{i\theta})\in L^p(0,2\pi)$, and so by a classical theorem we have

$$F_1(z)=\frac{1}{2\pi}\int_0^{2\pi}\operatorname{Im}F_1(e^{i\theta})\frac{e^{i\theta}+z}{e^{i\theta}-z}\,d\theta.$$

Therefore

$$F(z)=F(0)-\frac{1}{\pi}\int_0^{2\pi}\operatorname{Im}F_1(e^{i\theta})\frac{e^{i\theta}z}{(e^{i\theta}-z)^2}\,d\theta.$$

Integrating by parts and recalling that $F(0)=i\alpha$ and $(d/d\theta)\operatorname{Im}F_1(e^{i\theta})=-\alpha$ for $e^{i\theta}\neq z_j$, we find that

$$F(z)=-i\sum_j\omega_j\frac{z+z_j}{z-z_j},$$

where the ω_j are real. Set

$$\widetilde{R}(z)=\widetilde{\mathscr{P}}(z)+i\sum_j\omega_j\frac{z+z_j}{z-z_j}.$$

We finally conclude that $\xi(w)$ satisfies the differential equation

$$\sqrt{\frac{Q(w)}{w^2}}\frac{d\xi^*(w)}{dw}+\frac{1}{w^2}\widetilde{Q}(w)+\frac{\widetilde{R}(z)}{(Df(z))^2}=0,\qquad z=f^{-1}(w). \tag{24.16}$$

In the course of the above proof we established that $\xi[f(z)]$ can be analytically continued across each of the open arcs (16.1). Hence $\xi(w)$ is analytically continuable across each branch of $\mathscr{T}$ from each side.

Let $\mathscr{K}_\delta$ be the system of discs defined in §16, and let $K_{j\delta}\in\mathscr{K}_\delta$. The disc $K_{j\delta}$ is divided by that branch of $\mathscr{T}$ emanating from w_j into ν_j curvilinear sectors $\sigma_{jk\delta}$. At the risk of abusing the terminology, we shall call the set $K_{j\delta}\setminus\mathscr{T}$ a *neighborhood* at the point w_j. Set

$$S(w)=w^{-2}\widetilde{Q}(w)+\frac{\widetilde{R}(z)}{(Df(z))^2},\qquad z=f^{-1}(w). \tag{24.17}$$

PROPOSITION 24.3. *In a neighborhood of any vertex w_j, $j\neq 0$, $S(w)=O((w-w_j)^{\alpha_j})$, where $\alpha_j>\kappa_j/2-1$. In a neighborhood of w_0, $S(w)=O(w^{-\alpha_0})$, where $\alpha_0>\kappa_0/2+2$.*

Proof. Since $\xi \in \mathfrak{H}$, it follows from (24.16) that

$$\int\limits_{|w|>1/4} |S(w)|^2 \left|\frac{w^2}{Q(w)}\right| d\sigma < \infty. \tag{24.18}$$

From (24.5) and (24.6) we find that in each sector $\sigma_{jk\delta}$, for δ sufficiently small, we have $S(w) = A_{jk}(w - w_j)^{\gamma_{jk}} + \cdots$. Thus inequality (24.18) implies $\gamma_{jk} > \kappa_j/2 - 1$ and $\gamma_{0k} > \kappa_0/2 + 2$. Q.E.D.

We shall now derive a somewhat more manageable formula for the scalar product of the function $\xi(w)$ defined by (24.16) and a function $\eta(w) \in \mathfrak{H}$ which satisfies: a) $\eta^*(w)$ is bounded for $w \in \mathscr{K}_\delta$; b) in $\mathscr{G} \setminus \mathscr{K}_\delta$, $\eta(w)$ is continuous up to the boundary for any $\delta > 0$ sufficiently small. Set $\mathscr{G}_\delta = \mathscr{G} \setminus (\mathscr{K}_\delta \cup \{w : |w| \leq 1/4\})$, let l_δ be that component of the boundary of $\mathscr{G}_\delta$ different from $|w| = 1/4$, and put $C_\delta = \partial \mathscr{K}_\delta \setminus \mathscr{T}$. Also put $l_\delta^0 = l_\delta \setminus C_\delta$.

Proposition 24.4. *If $\xi(w) \in \mathfrak{H}$ is a solution of equation* (24.16) *and if $\eta(w) \in \mathfrak{H}$ satisfies conditions* a) *and* b), *then*

$$(\xi, \eta) = \lim_{\delta \to 0} \operatorname{Re} \left\{ \frac{1}{2\pi i} \int\limits_{C_\delta} \eta(w) \left[w^{-2} \widetilde{Q}(w) - \frac{\widetilde{R}(z)}{(Df(z))^2} \right] dw \right\}. \tag{24.19}$$

Proof. Applying Green's formula as in §21, we get

$$\frac{1}{\pi} \int\limits_{\mathscr{G}_\delta} \frac{d\eta^*}{dw} \overline{\frac{d\xi^*}{dw}} d\sigma = -\frac{1}{2\pi i} \int\limits_{l_\delta} \eta^* \overline{d\xi^*} + \frac{1}{2\pi i} \int\limits_{|w|=1/4} \eta^* \overline{d\xi^*}.$$

Consider first

$$j_\delta = \frac{1}{2\pi i} \int\limits_{l_\delta^0} \eta^* \overline{d\xi^*}.$$

This integral can be written as

$$j_\delta = \frac{1}{2\pi i} \int\limits_{\mathscr{T} \setminus \mathscr{K}_\delta} [\eta_+^* \overline{d\xi_+^*} - \overline{\eta_+^* \xi_-^*}] = -\frac{1}{2\pi i} \int\limits_{\mathscr{T} \setminus \mathscr{K}_\delta} \left[\eta_+^*(w) \overline{\left(\frac{S(w)}{\sqrt{\frac{Q(w)}{w^2}}} \right)_+ dw} - \eta_-^*(w) \overline{\left(\frac{S(w)}{\sqrt{\frac{Q(w)}{w^2}}} \right)_- dw} \right],$$

whence

$$j_\delta = -\frac{1}{2\pi i} \int\limits_{\mathscr{T} \setminus \mathscr{K}_\delta} \left\{ \sqrt{\frac{Q(w)}{w^2}} (\eta_+(w) - \eta_-(w)) \overline{\frac{\widetilde{Q}(w)}{w^2} \sqrt{\frac{w^2}{Q(w)}} dw} \right.$$
$$\left. + \eta_+^*(w) \overline{\frac{\widetilde{R}(z_+)}{(Df(z_+))^2} \sqrt{\frac{w^2}{Q(w)}} dw} - \eta_-^*(w) \overline{\frac{\widetilde{R}(z_-)}{(Df(z_-))^2} \sqrt{\frac{w^2}{Q(w)}} dw} \right\}.$$

Since the expressions

$$\frac{R(z_\pm)}{(Df(z_\pm))^2} \sqrt{\frac{w^2}{Q(w)}} dw \quad \text{and} \quad \sqrt{\frac{Q(w)}{w^2}} (\eta_+(w) - \eta_-(w))$$

are real on $\mathcal{T}$, we have

$$\operatorname{Re} j_\delta = \operatorname{Re}\left\{\frac{1}{2\pi i}\int\limits_{\mathcal{T}\setminus\mathcal{K}_\delta}\left[(\eta_+(w)-\eta_-(w))\frac{Q(w)}{w^2}dw-\eta_+(w)\frac{R(z_+)}{(Df(z_+))^2}dw\right.\right.$$
$$\left.\left.+\eta_-(w)\frac{R(z_-)}{(Df(z_-))^2}dw\right]\right\}=\operatorname{Re}\left\{\frac{1}{2\pi i}\int\limits_{l_\delta^0}\eta(w)\left[\frac{\widetilde{Q}(w)}{w^2}-\frac{R(z)}{(Df(z))^2}\right]dw\right\},$$

and so

$$\frac{1}{\pi}\int\limits_{\mathcal{G}_\delta}\frac{d\eta^*}{dw}\overline{\frac{d\xi^*}{dw}}d\sigma=\operatorname{Re}\left\{-\frac{1}{2\pi i}\int\limits_{l_\delta}\eta(w)\left[\frac{\widetilde{Q}(w)}{w^2}-\frac{\widetilde{R}(z)}{(Df(z))^2}\right]dw\right.$$
$$-\frac{1}{2\pi i}\int\limits_{C_\delta}\eta^*(w)\,S(w)\sqrt{\frac{w^2}{Q(w)}}dw+\frac{1}{2\pi i}\int\limits_{C_\delta}\eta(w)\left[\frac{\widetilde{Q}(w)}{w^2}-\frac{\widetilde{R}(z)}{(Df(z))^2}\right]dw$$
$$\left.+\frac{1}{2\pi i}\int\limits_{|w|=1/4}\eta^*\overline{d\xi^*}\right\}.$$

Note that

$$\frac{1}{2\pi i}\int\limits_{l_\delta}\eta(w)\left[\frac{\widetilde{Q}(w)}{w^2}-\frac{\widetilde{R}(z)}{(Df(z))^2}\right]dw=\frac{1}{2\pi i}\int\limits_{\mathcal{C}_r}\eta(w)\left[\frac{\widetilde{Q}(w)}{w^2}-\frac{\widetilde{R}(z)}{(Df(z))^2}\right]dw,$$

where $r<1$ is arbitrary. Since $\eta^*(w)=o(1)$ in $\mathcal{K}_\delta$, it follows from Proposition 24.3 that

$$\frac{1}{2\pi i}\int\limits_{C_\delta}\eta^*(w)\,S(w)\sqrt{\frac{w^2}{Q(w)}}dw=o(1).$$

Thus

$$(\xi,\eta)=\operatorname{Re}\left\{-\frac{1}{2\pi i}\int\limits_{\mathcal{C}_r}\eta(w)\left[\frac{\widetilde{Q}(w)}{w^2}-\frac{\widetilde{R}(z)}{(Df(z))^2}\right]dw+\vartheta\sum_1^n\bar{\xi}_k\eta_k\right\}$$
$$+\lim_{\delta\to 0}\operatorname{Re}\left\{\frac{1}{2\pi i}\int\limits_{C_\delta}\eta(w)\left[\frac{\widetilde{Q}(w)}{w^2}-\frac{\widetilde{R}(z)}{(Df(z))^2}\right]dw\right\}.$$

In this formula consider the first term, which we denote by J. It is clear that

$$J=\operatorname{Re}\left\{\frac{1}{2\pi i}\int\limits_{\mathcal{C}_r}\eta(w)\left[-\frac{\widetilde{Q}(w)}{w^2}+\frac{\widetilde{R}(z)}{(Df(z))^2}+\vartheta\sum\bar{\xi}_k w^{-k-2}\right]dw\right\}.$$

We now change to the variable z, with $w=f(z)$. Denoting by $t(z)$ the expression in the square brackets, we have

$$t(z)=\frac{1}{(Df(z))^2}\left\{\left(\frac{Df(z)}{f(z)}\right)^2\left[\widetilde{Q}(f(z))+\vartheta\sum_1^n\bar{\xi}_k\,[f(z)]^{-k}\right]+\widetilde{R}(z)\right\}.$$

Using (24.11) and the definition of $\widetilde{R}(z)$, we find that

$$t(z)=\frac{1}{(Df(z))^2}\left\{\frac{1}{2}\left(\frac{Df(z)}{f(z)}\right)^2\left[\sum_1^{n-1}\alpha_k\,[f(z)]^{-k}+\vartheta\sum_1^n\bar{\xi}_k\,[f(z)]^{-k}\right.\right.$$
$$\left.+\mathcal{P}(z)+i\sum_j\omega_j\frac{z+z_j}{z-z_j}\right\}$$

Applying (24.14) and (24.15), we see that

$$t(z)=\frac{1}{(Df(z))^2}\Big[i\alpha-\alpha_0-i\sum_j\omega_j+O(z)\Big].$$

Therefore $J=0$, and this implies that formula (24.19) is valid. Q.E.D.

We can deduce some interesting corollaries from (24.19). First we note that if $\eta\in\mathscr{A}_0$, then the conditions of Proposition 24.4 are satisfied.

With the aid of Proposition 24.3, formula (24.19) can be written as

$$(\xi,\eta)=-\lim_{\delta\to 0}\operatorname{Re}\Big\{\frac{1}{\pi i}\int_{C_\delta}\eta(w)\frac{\widetilde{R}(z)}{(Df(z))^2}\,dw\Big\}.$$

The function $\eta(w)$ is defined by (21.1) and (21.2). Using (19.12) and introducing the notation

$$J(w)=\frac{1}{2\pi i}\int_{\mathscr{C}_r}\frac{w^2\alpha(t)}{(w-t)\,t^2}\,dt,\qquad w\in\mathscr{G}(\mathscr{C}_r,\mathscr{T}),$$

we see that

$$(\xi,\eta)=\lim_{\delta\to 0}\operatorname{Re}\Big\{\frac{1}{\pi i}\int_{C_\delta}h(z)\frac{\widetilde{R}(z)}{Df(z)}\,dw-\frac{1}{\pi i}\int_{C_\delta}J(w)\frac{\widetilde{R}(z)}{(Df(z))^2}\,dw\Big\}.$$

Upon applying Proposition 24.3 to the second integral, we find that

$$(\xi,\eta)=\lim_{\delta\to 0}\operatorname{Re}\Big\{\frac{1}{\pi i}\int_{C_\delta}h(z)\frac{\widetilde{R}(z)}{Df(z)}\,dw+\frac{1}{\pi i}\int_{C_\delta}J(w)\frac{\widetilde{Q}(w)}{w^2}\,dw\Big\},$$

and so

$$(\xi,\eta)=\lim_{\delta\to 0}\operatorname{Re}\Big\{\frac{1}{\pi i}\int_{C_\delta}h(z)\frac{\widetilde{R}(z)}{Df(z)}\,dw\Big\}.$$

In order to evaluate this limit, we set $w=f(z)$. Under this transformation each curvilinear sector $\sigma_{jk}=\sigma_{jk\delta}$ corresponds to a quasineighborhood $\tau_{jk}=\tau_{jk\delta}$ of the point z_{jk}, and C_δ corresponds to $\bigcup_{j,k}\partial_0\tau_{jk\delta}$, where $\partial_0\tau_{jk\delta}=\partial\tau_{jk\delta}\cap\{z:|z|<1\}$. Thus we have

$$(\xi,\eta)=\lim_{\delta\to 0}\operatorname{Re}\Big\{\sum_{j,k}\frac{1}{\pi i}\int_{\partial_0\tau_{jk\delta}}h(z)\widetilde{R}(z)\frac{dz}{z}\Big\}.$$

Using the definition of $\widetilde{R}(z)$ and reindexing the quasineighborhoods $\tau_{jk\delta}$ over a single index set, we find that

$$(\xi,\eta)=\lim_{\delta\to 0}\operatorname{Re}\Big\{\sum_j\frac{1}{\pi i}\int_{\partial_0\tau_{j\delta}}h(z)\Big[\mathscr{P}(z)+i\sum_l\omega_l\frac{z+z_l}{z-z_l}\Big]\frac{dz}{z}\Big\},$$

and so

$$(\xi,\eta)=\lim_{\delta\to 0}\operatorname{Re}\Big\{\sum_j\frac{\omega_j}{\pi}\int_{\partial_0\tau_{j\delta}}h(z)\frac{z+z_j}{z-z_j}\frac{dz}{z}\Big\}.$$

From (24.5) it follows that $dz/(z-z_j)=(i+O(1))\,d\theta$ on $\partial_0\tau_{j\delta}$, where θ varies from $\theta_1 = o(1)$ to $\theta_2 = -\pi + o(1)$. Thus $(\xi, \eta) = -2\,\mathrm{Re}\{\sum_j i\omega_j h(z_j)\}$. Since $ih(z_j)$ is real,

$$(\xi,\eta) = -2i\sum_j \omega_j h(z_j). \tag{24.20}$$

Lemma 24.1. *Suppose that $\mathscr{A}$ is a proper subspace of $\mathfrak{H}$. Any function $\xi(w) \in \mathfrak{H}$ orthogonal to $\mathscr{A}$ satisfies in $\mathscr{G}$ the differential equation*

$$\sqrt{\frac{Q(w)}{w^2}}\frac{d\xi^*(w)}{dw} + w^{-2}\widetilde{Q}(w) + \frac{\widetilde{P}(z)}{(Df(z))^2} = 0, \qquad z = f^{-1}(w), \tag{24.21}$$

where $\widetilde{Q}(w)$ is defined by (24.11) *and* (24.10), *and $\widetilde{\mathscr{P}}(z)$ by* (24.15) *and* (24.14).

Proof. Since $H_0 \subset \mathscr{A}$, we see that $\xi(w)$ satisfies (24.16). If $\eta \in \mathscr{A}_0$, then in accord with (24.20) we have $\sum_j \omega_j h(z_j) = 0$, and since $h(z_j)$ are arbitrary purely imaginary quantities, it follows that $\omega_j = 0$. Q.E.D.

§25. We now reexamine the question of the structure of the spaces $\mathscr{A}$ and $\mathfrak{H}$. We shall first give some consequences of (24.19). As we noted above, this formula can be written as

$$(\xi,\eta) = \lim_{\delta\to 0}\mathrm{Re}\left\{\frac{1}{\pi i}\int_{C_\delta}\eta(w)\widetilde{Q}(w)\frac{dw}{w^2}\right\}. \tag{25.1}$$

Let $\eta(w) \in \mathfrak{R}_j$, $\eta(w) = Aw^2(w-w_j)^{-l}$, where A is arbitrary. From (25.1) we find that

$$(\xi,\eta) = -\lim_{\delta\to 0}\mathrm{Re}\left\{\frac{1}{\pi i}\int_{|w-w_j|=\delta}\frac{A\widetilde{Q}(w)}{(w-w_j)^l}dw\right\},$$

and so

$$(\xi,\eta) = -\frac{2}{(l-1)!}\mathrm{Re}\{A\widetilde{Q}^{(l-1)}(w_j)\}. \tag{25.2}$$

If $j=0$, then $\eta(w) = Aw^{l+1}$. Having set $\widetilde{Q}(w) = \sum_1^n \widetilde{A}_k w^{-k}$, we arrive at the formula

$$(\xi,\eta) = 2\,\mathrm{Re}\{A\widetilde{A}_l\}, \tag{25.3}$$

analogous to the above.

Lemma 25.1. *In order that the space $\mathscr{A}$ of first variations coincide with $\mathfrak{H}$, it is necessary and sufficient that $\mathfrak{R} \in \mathscr{A}$.*

Proof. The necessity is obvious, and so we shall prove only the sufficiency. Let $\xi(w) \in \mathfrak{H}$ be orthogonal to $\mathscr{A}$. Then $\xi(w)$ satisfies (24.21). By hypothesis $\xi(w)$ is orthogonal to every function $\eta(w) \in \mathfrak{R}$. Thus it follows from (25.2) and (25.3) that each vertex w_j, $j \neq 0$, is a zero of $\widetilde{Q}(w)$ of multiplicity at least $[\kappa_j/2]$, and the base w_0 of the tree is a zero of $\widetilde{Q}(w)$ of multiplicity at least $[\kappa_0/2]+1$. Still using the notation $S(w)$ for the function $w^{-2}\widetilde{Q}(w) + \widetilde{\mathscr{P}}(z)/(Df(z))^2$, $z = f^{-1}(w)$, we find from (24.21) that in any sector σ_{jk}

$$\xi^*(w) = C_{jk} - \int_{w_j}^{w} S(w) \left(\frac{w^2}{Q(w)} \right)^{1/2} dw$$

(the integral converges as a result of the estimates of Proposition 24.3). Thus if $w \in \sigma_{jk}$, then $\xi^*(w) = O(1)$, and so Proposition 24.4 can be applied to the function $\xi(w)$. According to (25.1) we have

$$(\xi, \xi) = \lim_{\delta \to 0} \operatorname{Re} \left\{ \frac{1}{\pi i} \int_{C_\delta} \xi^*(w) \frac{\widetilde{Q}(w)}{w^2} \left(\frac{w^2}{Q(w)} \right)^{1/2} dw \right\}.$$

Recalling the previous remarks concerning the multiplicities of the zeros w_j of $\widetilde{Q}(w)$, we see that in fact $(\xi, \xi) = 0$, so $\xi(w) \equiv 0$. Hence $\mathscr{A} = \mathfrak{H}$. Q.E.D.

The subspace $\mathfrak{R}$ can be expressed as the direct sum of the subspace $\mathscr{R}$ and its complement $\mathfrak{R}^*$:

$$\mathfrak{R} = \mathscr{R} \oplus \mathfrak{R}^*.$$

Lemma 25.2. *The space $\mathfrak{H}$ is the direct sum of $\mathscr{A}$ and $\mathfrak{R}^*$.*

Proof. Since $\mathfrak{R} \in \mathscr{A} \oplus \mathfrak{R}^*$, we find upon applying the proof of the previous lemma to $\mathscr{A} \oplus \mathfrak{R}^*$ that $\mathfrak{H} = \mathscr{A} \oplus \mathfrak{R}^*$. Q.E.D.

The question of when $\mathfrak{R}_j \subset \mathscr{A}$ is closely connected with the structure of $\mathscr{T}$ in a neighborhood of the vertices w_j. For ease of notation we shall introduce a classification of the vertices of the tree $\mathscr{T}$. Let w_j be a vertex of $\mathscr{T}$, and let κ_j be its index. With w_j there are also associated the ν_j even integers $2\mu_{jk}$, $k = 1, \cdots, \nu_j$, which were introduced at the beginning of the previous section as the multiplicities of z_{jk} as zeros of $P(z)$. The quantities $\mu_{jk} + 1$, $k = 1, \cdots, \nu_j$, have a simple geometric interpretation. Let $\sigma_{jk\delta}$ be the curvilinear sectors into which $K_{j\delta}$ is divided by the branches of the tree emanating from w_j, and let $\tau_{jk\delta}$ be the image of $\sigma_{jk\delta}$ under the mapping $z = f^{-1}(w)$. Then $z_{jk} \in \partial \tau_{jk\delta}$. The branch bounding $\sigma_{jk\delta}$ forms at the point w_j an angle φ_{jk} which is a multiple of $2\pi / (\kappa_j + 2)$. It is easy to see that in fact $\varphi_{jk} = 2\pi(\mu_{jk} + 1) / (\kappa_j + 2)$.

Definition 25.1. A vertex w_j with index $\kappa_j \geqq 2$ will be called *singular* in the following cases: a) it is terminal, b) it is not terminal, but all the quantities $\mu_{jk} > 0$, and the greatest common divisor of $2\mu_{j1} + 2, \cdots, 2\mu_{j\nu_j} + 2$, $\kappa_j + 2$ is greater than 1. If a nonterminal vertex is the base of the tree, then instead of the above we require that the greatest common divisor of $2\mu_{01} + 2, \cdots, 2\mu_{0\nu_0} + 2$, κ_0 be greater than 1.

If w_j does not meet the above conditions, it will be called *nonsingular*. The degree and index of a singular nonterminal vertex satisfy the inequality $\nu_j \leqq (\kappa_j + 2)/2$ if it is not the base of the tree, and $\nu_0 \leqq \kappa_0/2$ if it is the base. The term "singular vertex" is justified by the fact that the set of points $x \in \partial \mathscr{V}_n$ for which $\mathscr{T}_x$ contains at least one singular vertex has dimension less than $2n - 1$, a fact which we shall prove later on.

Proposition 25.1. *If the vertex w_j is nonsingular, then $\mathfrak{R}_j \subset \mathscr{A}$.*

Proof. Let $\eta \in \mathfrak{R}_j$ be arbitrary. Represent η in the form $\eta = \zeta + \xi$, where $\zeta \in \mathscr{A}$ and ξ is orthogonal to $\mathscr{A}$. Then $\xi(w)$ satisfies (24.21). We shall show that w_j is a root of $\widetilde{Q}(w)$ of multiplicity at least $[\kappa_j/2]$. Since for $\kappa_j = 1$ this statement is trivially true, we assume $\kappa_j \geqq 2$. Using (5.13), we write $S(w)$ as

$$S(w) = w^{-2}\left[\widetilde{Q}(w) - Q(w)\frac{\widetilde{\mathscr{P}}(z)}{P(z)}\right].$$

In each sector σ_{jk} relations (24.5) and (24.6) hold, and so if z_{jk} is a zero of $\widetilde{\mathscr{P}}(z)$ of multiplicity s_{jk}, then for $w \in \sigma_{jk}$

$$S(w) = w^{-2}\,[\widetilde{Q}(w) - Q(w)(A_{jk}(w - w_j)^{-\frac{2\mu_{jk}-s_{jk}}{r_{jk}}} + \dots)].$$

But w_j is a zero of $Q(w)$ of multiplicity κ_j, and so this relation implies that

$$S(w) = w^{-2}\,[\widetilde{Q}(w) + B_{jk}(w - w_j)^{\varkappa_j - \frac{2\mu_{jk}-s_{jk}}{r_{jk}}} + \dots].$$

If κ_j is even, then, since w_j is a nonsingular vertex, we have for at least one value of k, say k_0, that $\mu_{jk_0} = 0$. Thus in σ_{jk_0} we have

$$S(w) = w^{-2}\,[\widetilde{Q}(w) + B_{jk_0}(w - w_j)^{\varkappa_j + \frac{s_{jk_0}}{\varkappa_j + 2}} + \dots],$$

and since the inequality of Proposition 24.3 must also be satisfied, we see that the multiplicity of w_j as a zero of $\widetilde{Q}(w)$ must be at least $\kappa_j/2$. If κ_j is odd, then at least one of the numbers

$$\frac{2\mu_{jk} - s_{jk}}{r_{jk}} = \frac{\varkappa_j + 2}{2\mu_{jk} + 2}(2\mu_{jk} - s_{jk})$$

must be a noninteger. Then, according to Proposition 24.3, w_j is a zero of $\widetilde{Q}(w)$ of multiplicity at least $\kappa_j/2 - 1$, that is to say, not less than $\kappa_j/2 - 1/2 = [\kappa_j/2]$. If w_j is the base of the tree, then in place of (24.5) we must use (24.6), and the rest of the proof remains unchanged.

Once we have established this fact, we have that $(\xi, \xi) = (\xi, \eta - \zeta) = (\xi, \eta)$, and from (25.2) or (25.3) it follows that $(\xi, \xi) = 0$, so that $\xi \equiv 0$, and in particular $\eta = \zeta \in \mathscr{A}$. Q.E.D.

The following lemma is a trivial consequence of Lemma 25.1 and the previous proposition.

Lemma 25.3. *If all vertices of the tree $\mathscr{T}$ are nonsingular, then $\mathscr{A} = \mathfrak{H}$.*

The converse of this lemma is false, since our classification of the vertices is quite coarse. If the vertex w_j is not terminal and of odd index, the containment of $\mathfrak{R}_j$ in $\mathscr{A}$ depends on more subtle properties of $\mathscr{T}$ than we have at hand in our classification of vertices. However, for our purposes it is sufficient to show that if the tree has at least one nonterminal singular vertex or a singular vertex with even index, then $\mathscr{A}$ is a proper subspace of $\mathfrak{H}$. At the same time we shall show that the set of points $x \in \partial\mathscr{V}_n$ for which $\mathscr{A}_x \subset \mathfrak{H}_x$ is not empty.

We showed above that $H \subseteq \mathscr{A}$. We now determine the complement of H in $\mathscr{A}$. Let w_j be a nonterminal vertex of even index. If w_j is nonsingular, we expand the subspace $\mathscr{R}_j$ into a direct sum $\mathscr{R}_j = R_j \oplus \mathscr{R}_j{}^*$, where $\mathscr{R}_j{}^*$ is the complement of R_j. Form the direct sum

$$\sum_j \oplus \mathscr{R}_j^*. \tag{25.4}$$

where the sum is over all nonterminal nonsingular vertices with even index. Now let w_j be a nonterminal singular vertex with even index $\kappa_j > 4$. Denote by $\mathfrak{R}_j^1$ the linear span of functions of the form (23.10), under the condition that $1 \leqq l \leqq \kappa_j/2 - 1$. If w_j is the base of the tree, however, we take the linear span of the functions (23.11) such that $1 \leqq l \leqq \kappa_j/2 - 1$. As before we set $\mathscr{R}_j^1 = \mathfrak{R}_j^1 \cap \mathscr{A}$ and $R_j^1 = \mathfrak{R}_j^1 \cap H$. Let $\mathscr{R}_j^1 = R_j^1 \oplus \mathscr{R}_j{}^*$. Form the direct sum

$$\sum_j{}' \oplus \mathscr{R}_j^*, \tag{25.5}$$

where the sum is over all nonterminal singular vertices with even index. The direct sum of the subspaces (25.4) and (25.5) will be denoted by $\mathscr{R}^*$.

Lemma 25.4. $\mathscr{A} = H \oplus \mathscr{R}^*$.

Proof. Let w_j be a nonterminal singular vertex with even index $\kappa_j > 4$. Set $\mathfrak{R}_j^1 = \mathscr{R}_j^1 \oplus \widetilde{\mathfrak{R}}_j$ and form the direct sum $\widetilde{\mathfrak{R}} = \sum_j' \oplus \widetilde{\mathfrak{R}}_j$. Consider the spaces $H \oplus \mathscr{R}^* \oplus \widetilde{\mathfrak{R}}$ and $\mathscr{A} \oplus \widetilde{\mathfrak{R}}$. Clearly $H \oplus \mathscr{R}^* \oplus \widetilde{\mathfrak{R}} \subseteq \mathscr{A} \oplus \widetilde{\mathfrak{R}}$. Suppose that we have strict inclusion, and pick $\xi \in \mathscr{A} \oplus \widetilde{\mathfrak{R}}$ orthogonal to $H \oplus \mathscr{R}^* \oplus \widetilde{\mathfrak{R}}$. Clearly $\xi(w)$ satisfies the differential equation (24.16). We have from the above that

$$\widetilde{R}(z) = \widetilde{\mathscr{P}}(z) + i \sum_l \omega_l \frac{z + z_l}{z - z_l}.$$

We shall show that all the quantities w_l are zero. Let w_l be a terminal vertex and z_l its image. In accord with (24.5), $z - z_l = \beta_l (w - w_l)^{1/2} + \cdots$, and by (24.1) and (24.5) $(Df(z))^2 = \gamma_l (w - w_l) + \cdots$. Thus if $|w - w_l|$ is small, then $|\widetilde{R}(z) [Df(z)]^{-2}| \geq C_l |w_l| \, |w - w_l|^{-3/2}$. From this inequality and from (24.3) it follows that $w_l = 0$. If the base of the tree is a terminal vertex, analogous reasoning shows that $w_l = 0$. We thus conclude that

$$\widetilde{R}(z) = \widetilde{\mathscr{P}}(z) + i \sum_j \sum_{k=1}^{\nu_j} \omega_{jk} \frac{z + z_{jk}}{z - z_{jk}},$$

where the sum with respect to j is over all the nonterminal vertices. Let w_j be a nonterminal vertex. Using (5.13), we find that

$$\widetilde{R}(z) \, [Df(z)]^{-2} = - w^{-2} Q(w) \widetilde{R}(z) \, [P(z)]^{-1}.$$

If $w \in \sigma_{jk}$, then from (24.5) and (24.2) it follows that $\widetilde{R}(z) \, [Df(z)]^{-2} = \omega_{jk} \gamma_{jk} (w - w_j)^{s_{jk}} + \cdots$, where $s_{jk} = \kappa_j - (\kappa_j + 2)(2\mu_{jk} + 1)/(2\mu_{jk} + 2)$. Since $s_{jk} \leqq \frac{1}{2}\kappa_j - 1$, the fact that the inequality of Proposition 24.3 must hold forces us to conclude that s_{jk} is a whole number. If κ_j is odd, then s_{jk} is not an integer, and so, for a vertex w_j with odd index, $\omega_{jk} = 0$, $k = 1, \cdots, \nu_j$. All that remains to be considered are the vertices with even index. Since $s_{jk} \geqq 0$ implies that $\kappa_j \geqq 2(2\mu_{jk} + 1)$, we have $\kappa_j > 4$ if w_j is a singular vertex. If w_j is a nonsingular vertex with even index, then $\mathfrak{R}_j \subset H \oplus \mathscr{R}^*$, since $\mathfrak{R}_j = \mathscr{R}_j$ by Proposition 25.1. Thus $\xi(w)$ is orthogonal to $\mathfrak{R}_j$, and from (25.2) (or (25.3), if $w_j = w_0$ is the base) it follows that w_j is a zero of $\widetilde{Q}(w)$ of multiplicity at least $(\kappa_j/2)(\kappa_0/2 + 1)$. Since $s_{jk} \leqq \frac{1}{2}\kappa_j - 1$ $(s_{0k} \leqq \kappa_0/2)$, Proposition 24.3 implies $\omega_{jk} = 0$, $k = 1, \cdots, \nu_j$.

Let w_j be a singular vertex with $\kappa_j > 4$. Since $\mathfrak{R}_j^1 < H \oplus \mathscr{R}^* \oplus \widetilde{\mathfrak{R}}$, it follows that $\xi(w)$ is orthogonal to $\mathfrak{R}_j^1$; and, in view of (25.2) or (25.3), $Q(w)$ has a zero at w_j of multiplicity at least $\frac{1}{2}\kappa_j - 1$. On the other hand, $s_{jk} < \frac{1}{2}\kappa_j - 1$ for $\mu_{jk} \neq 0$, and so from Proposition 24.3 we have $\omega_{jk} = 0$, $k = 1, \cdots, \nu_j$. We can now conclude the proof of the lemma quite easily. We have established formula (24.20) for the scalar product of a solution of equation (24.16) with any function $\eta \in \mathscr{A}_0$. From this we see that $\xi(w)$ is orthogonal to $\mathscr{A}$, and since by construction $\xi(w)$ is orthogonal to $\widetilde{\mathfrak{R}}$ and also an element of $\mathscr{A} \oplus \widetilde{\mathfrak{R}}$. it follows that $\xi(w) \equiv 0$. Hence $H \oplus \mathscr{R}^* \oplus \widetilde{\mathfrak{R}} = \mathscr{A} \oplus \widetilde{\mathfrak{R}}$. Therefore $\mathscr{A} = H \oplus \mathscr{R}^*$. Q.E.D.

COROLLARY 25.1. *If w_j is a terminal vertex or a nonterminal vertex with odd index, or a nonterminal vertex which is singular and of even index $\kappa_j \leqq 4$, then $\mathscr{R}_j = R_j$.*

COROLLARY 25.2. *If w_j is a nonterminal singular vertex with even index $\kappa_j > 4$, then $\mathscr{R}_j = R_j \oplus \mathscr{R}_j^*$.*

It is quite easy to investigate the structure of the subspace R_j. Let w_j be a vertex of index $\kappa_j \geqq 2$. Choose δ sufficiently small, and suppose that the disc $K_{j\delta}$ is such that $K_{j\delta} \subset \mathscr{K}_\delta$. Let $w = w_k(s)$ be the equation of the branch $\gamma_k \subset \mathscr{T}$ emanating from w_j; assume that $s \geqq 0$ and $w_k(0) = w_j$. In accord with Remark 16.2 we can choose the parameter s so that $w_k(s)$ is an analytic function of s at the point $s = 0$. Let the arc $\gamma_k \cap K_{j\delta}$ correspond to values of s such that $0 \leqq s \leqq s_\delta$. Without loss of generality we assume that $w_k(s)$ is regular for $|s| \leqq s_\delta$. Then if $|w - w_j| \geqq C_\delta$, the function $w^2[w - w_k(s)]^{-1}(w_k'(s)/w_k(s))^2$ will also be regular there. We set

$$i\,\frac{w^2}{w - w_k(s)}\left(\frac{w_k'(s)}{w_k(s)}\right)^2 = \sum_{r=1}^{\infty} s^{r-1} F_{jkr}(w).$$

It is easily seen that $F_{jkr}(w)$ is a rational function with a pole of order r at w_j, and possibly a first order pole at $w = \infty$. We also see that

$$F_{jkr}(w) = w^2\left\{i\,\frac{[w_k'(0)]^{r+1}}{w_j^2}(w - w_j)^{-r} + \cdots\right\}, \tag{25.6}$$

where by the dots we mean the sum of lower powers of $(w - w_j)^{-1}$. If the vertex w_j is the base (i.e. $j=0$, in our notation), then we may assume that $w_k(\infty)=\infty$, and by (16.2) the function $[w_k(1/\sigma)]^{-1}$ is regular in a neighborhood of $\sigma=0$. Hence we have

$$i\frac{w^2}{w-w_k(s)}\left(\frac{w_k'(s)}{w_k(s)}\right)^2=\sum_{r=1}^{\infty}s^{-r-2}F_{0kr}(w),$$

and it is easily seen that

$$F_{0kr}(w)=w\left\{-i\frac{w^r}{[w_k'(\infty)]^r}+\cdots\right\}, \tag{25.7}$$

where again the dots indicate the sum of lower powers.

Denote by $\widetilde{R}_j$ the linear span (over the reals) of the functions $F_{jkr}(w)$, $k=1,\cdots,\nu_j$; $r=1,\cdots,[\kappa_j/2]$.

Proposition 25.2. $R_j\subset\widetilde{R}_j$.

Proof. Let $\xi(w)$ be any element of R_j. Then there exists a sequence $\{\xi_l(w)\}$, $\xi_l(w)\in H_0$, $l=1,2,\cdots$, converging to $\xi(w)$. By (23.13)

$$\xi_l(w)=\frac{1}{2\pi i}\int_{\mathscr{T}}\frac{w^2\psi_l(t)}{(w-t)t^2}dt$$

for $w\in\mathscr{G}$. Applying inequality (23.5) to the function $\xi_l(w)-\xi(w)$ and using Minkowski's inequality and the formula for the jumps, we find that

$$\left\{\frac{1}{2\pi}\int_{\mathscr{T}}\left|\sqrt{\frac{Q(w)}{w^2}}\;\psi_l(w)\right|^p\frac{|dw|}{|w|^\rho}\right\}^{1/p}<2A_p(\mathscr{T},\rho)\,\|\xi_l-\xi\|. \tag{25.8}$$

Set $\mathscr{T}_\delta=\mathscr{T}\cap\mathscr{K}_\delta$ and assume that $w\in\mathscr{G}(\mathscr{C}_r)$, where $r<1$ is some fixed quantity. With the aid of the last inequality we find that

$$\xi_l(w)=\frac{1}{2\pi i}\int_{\mathscr{T}_\delta}\frac{w^2\psi_l(t)}{(w-t)t^2}dt+o(1). \tag{25.9}$$

If w_m is a vertex of $\mathscr{T}$ with index $\kappa_m<1$, then, using (25.8), we easily see that

$$\frac{1}{2\pi i}\int_{\mathscr{T}\cap K_{m\delta}}\frac{w^2\psi(t)}{(w-t)t^2}dt=o(1).$$

Suppose the vertex w_m has index $\kappa_m\geqq 2$, and that $\gamma_1,\cdots,\gamma_{\nu_m}$ are the branches emanating from w_m. For $w\in\gamma_k$ set $\psi_l(w)=\varphi_{lk}(s)w_k'(s)$. Then by (23.14) the function φ_{lk} is real-valued. Taking δ sufficiently small, we get that

$$\frac{1}{2\pi i}\int_{\mathscr{T}\cap K_{m\delta}}\frac{w^2\psi_l(t)}{(t-w)t^2}dt'=-\sum_{k=1}^{\nu_m}\frac{\theta_k}{2\pi}\int_0^{s_k}\left[\sum_{r=1}^{\left[\frac{\kappa_m}{2}\right]}s^{r-1}F_{mkr}(w)+O\left(s^{\left[\frac{\kappa_m}{2}\right]}\right)\right]\varphi_{lk}(s)ds,$$

where θ_k equals $+1$ or -1, depending on whether the orientation of the branch agrees with the orientation induced by the parameter s. It is easy to verify that

$$\int_0^{s_k} s^{\left[\frac{\varkappa_m}{2}\right]} |\varphi_{lk}(s)|\, ds < C \int_{\gamma_k \cap K_{m\delta}} \left| \sqrt{\frac{\theta(w)}{w^2}}\, \psi_l(w) \right| \frac{|dw|}{|w - w_m|^{q_m}},$$

where $q_m = \kappa_m/2 - [\kappa_m/2]$. Applying Hölder's inequality with conjugate indices p and q, $p > 2$, we see as a consequence of (25.8) that

$$\int_0^{s_k} s^{\left[\frac{\varkappa_m}{2}\right]} |\varphi_{lk}(s)|\, ds = o(1).$$

The obvious changes must be made above if w_m is the base. Setting

$$a_{lmkr} = -\frac{\theta_k}{2\pi} \int_0^{s_k} \varphi_{lk}(s)\, s^{r-1} ds,$$

we get from (25.9) that

$$\xi_l(w) = \sum_m \sum_{k=1}^{\nu_m} \sum_{r=1}^{\left[\frac{\varkappa_m}{2}\right]} a_{lmkr} F_{mkr}(w) + o(1),$$

where the sum is over all vertices for which $\kappa_m \geqq 2$. Applying (23.6) to the function $\xi(w) - \xi_l(w)$, we find that $\xi(w) - \xi_l(w) = O(1)$ if $w \in \mathcal{G}(\mathcal{C}_r)$. Therefore, if $w \in \mathcal{G}(\mathcal{C}_r)$,

$$\xi(w) - \sum_m \sum_{k=1}^{\nu_m} \sum_{r=1}^{\left[\frac{\varkappa_m}{2}\right]} a_{lmkr} F_{mkr}(w) = o(1).$$

Since $\xi \in R_j$, this latter relation can be written as

$$\sum_m T_{lm}(w) = o(1), \tag{25.10}$$

where

$$T_{lm}(w) = w^2 \sum_{k=1}^{\left[\frac{\varkappa_m}{2}\right]} b_{lmk} (w - w_m)^{-k},$$

if w_m is not the base. For the base we must replace $(w - w_m)^{-1}$ by w, and the summation goes from 0 to $[\kappa_0/2] - 1$. From (25.10) it is easy to see that $T_{lm}(w) = o(1)$. To prove this, let $p_r(w) = \prod_{m \neq r} (w^{-1} - w_m^{-1})^{[\kappa_m/2]}$. The function $w^{-3} p_r(w) \sum_m T_{lm}(w) = \Phi_{lr}(w)$ has a unique pole w_r of order $[\kappa_r/2]$. By the residue theorem

$$b_{lrk} = -\frac{w_r^{2k-1}}{p_r(w_r)} \frac{1}{2\pi i} \int_{\mathcal{C}_r} \left(\frac{1}{w_r} - \frac{1}{w} \right)^{k-1} \Phi_{lr}(w)\, dw,\ k = 1, \ldots, \left[\frac{\varkappa_r}{2}\right].$$

Thus $b_{lkk} = o(1)$, $k = 1, \cdots, [\kappa_r/2]$, and so $T_{lm}(w) = o(1)$ for $w \in \mathcal{G}(\mathcal{C}_r)$.

Therefore

$$\xi(w) = \sum_{k=1}^{\nu_j} \sum_{r=1}^{\left[\frac{\varkappa_j}{2}\right]} a_{ljkr} F_{jkr}(w) + o(1).$$

It follows that $\xi \in \widetilde{\mathscr{R}}_j$, as required. Q.E.D.

LEMMA 25.5. *If the tree $\mathscr{T}$ has at least one singular terminal vertex or a singular nonterminal vertex with even index, then $\mathscr{A}$ is a proper subspace of $\mathfrak{H}$.*

PROOF. We first consider the case of a terminal vertex w_j. By Corollary 25.1 and Proposition 25.2, $\mathscr{R}_j \subseteq \widetilde{R}_j$. We shall estimate the dimension of $\widetilde{R}_j$. In (25.6) we take $k = 1$, and, assuming without loss of generality that $w_1'(0) = e^{i\varphi}$, we find that

$$F_{jr}(w) = w^2\{ iw_j^{-2} e^{i(r+1)\varphi}(w - w_j)^{-r} + \cdots \}.$$

Clearly no linear combination of the functions $F_{jr}(w)$, $1 \leqq r \leqq [\kappa_j/2]$, can give the function $(w/w_j)^2 e^{i(r+1)\varphi}(w - w_j)^{-[\kappa_j/2]}$. Hence $\dim \widetilde{R}_j < 2[\kappa_j/2]$, and so $\dim \mathscr{R}_j < 2[\kappa_j/2] = \dim \mathfrak{R}_j$.

Now suppose that w_j is a nonterminal singular vertex with even index. We shall suppose that the sector σ_{jk} is bounded by the branches γ_k and γ_{k+1}, and also that $\gamma_{\nu_j+1} = \gamma_1$. We have already remarked that at w_j the branches γ_k and γ_{k+1} form the angle $2\pi(\mu_{jk}+1)/(\kappa_j+2)$. Thus if $w_k'(0) = e^{i\varphi_k}|w_k'(0)|$, we have $\varphi_{k+1} - \varphi_k = 2\pi(\mu_{jk}+1)/(\kappa_j+2)$. Without loss of generality we assume $|w_k'(0)| = 1$. Then a linear combination of the functions (25.6) with $r = \kappa_j/2$ has the form

$$i\left(\frac{w}{w_j}\right)^2 (w - w_j)^{-\frac{\varkappa_j}{2}} \sum_{1}^{\nu_j} \alpha_k e^{i\frac{\varkappa_j+2}{2}\varphi_k} + \ldots = i\left(\frac{w}{w_j}\right)^2 (w - w_j)^{-\frac{\varkappa_j}{2}} e^{i\varphi_1\frac{\varkappa_j+2}{2}} \{\alpha_1$$

$$+ \alpha_2 e^{\pi i(\mu_{j1}+1)} + \ldots + \alpha_{\nu_j} e^{\pi i(\mu_{j\nu_j-1}+1)}\} + \ldots,$$

where the α_k are real. Since the expression in the curly brackets is real, no linear combination of the functions (25.6) can lead to the function

$$e^{i\varphi_1\frac{\varkappa_j+2}{2}}\left(\frac{w}{w_j}\right)^2 (w - w_j)^{-\frac{\varkappa_j}{2}}. \tag{25.11}$$

Thus the subspace R_j does not contain the function (25.11), and by Corollaries 25.1 and 25.2 this function is not contained in $\mathscr{R}_j$. Hence $\dim \mathscr{R}_j < \kappa_j = \dim \mathfrak{R}_j$. This proves the lemma.

At this point we conclude our investigation of the spaces H, $\mathscr{A}$ and $\mathfrak{H}$, and we turn our attention to the examination of extremal problems for the quadratic functional (20.19).

§26. Let $\mathfrak{z}_1, \cdots, \mathfrak{z}_n$ be any complex numbers such that $\operatorname{Re}\sum_1^n A_k \mathfrak{z}_k = 0$. For convenience we write $\mathfrak{z} = (\mathfrak{z}_1, \cdots, \mathfrak{z}_n)$, and as usual we shall consider $\mathfrak{z}$ as a real $2n$-dimensional vector. Denote by $\mathfrak{Z}$ the subset of $\mathscr{A}$ consisting of those functions $\xi(w) = \sum_1^\infty \xi_k w^{k+1}$ for which $\xi_k = \mathfrak{z}_k$, $k = 1, \cdots, n$. By §21.4, $\mathfrak{Z}$ is not empty. It is clear that $\mathfrak{Z}$ is a closed set in the plane. Set $\mathfrak{Z}_0 = \mathfrak{Z} \cap \mathscr{A}_0$. We shall require the following simple result.

Proposition 26.1. *$\mathfrak{Z}$ is the closure of $\mathfrak{Z}_0$.*

Proof. Let $\xi(w) \in \mathfrak{Z}$, and choose a subsequence $\{\zeta_k(w)\}$, $\zeta_k(w) \in \mathscr{A}_0$, $k = 1, 2, \cdots$, which converges to $\xi(w)$. Let $\zeta_k(w) = \sum_1^\infty \zeta_l^k w^{l+1}$. With $\eta^k = (\mathfrak{z}_1 - \zeta_1^k, \cdots, \mathfrak{z}_n - \zeta_n^k)$, it is clear that $|\eta^k| \to 0$ as $k \to \infty$. Via §21.4 we construct a function

$$\eta_k(w) = (\mathfrak{z}_1 - \zeta_1^k) w^2 + \cdots + (\mathfrak{z}_n - \zeta_n^k) w^{n+1} + \cdots.$$

Since $\eta_k(w) = \sum_1^{2n-1} \beta_{kj} \xi_j(w)$, it follows from the inequalities satisfied by β_{kj}, $j = 1, \cdots, 2n-1$, that $\|\eta_k\| < C_0 |\eta^k| \sum_1^{2n-1} \|\xi_k\|$. Using inequality (23.22), we find that $\|\eta_k\| < C|\eta^k|$. We now need note only that $\zeta_k(w) + \eta_k(w) \in \mathfrak{Z}_0$, $k = 1, 2, \cdots$, and $\lim_{k\to\infty} (\zeta_k + \eta_k) = \xi$. This proves the proposition.

Consider the following extremal problem.

Problem 1. Find the infimum for $\|\xi\|$ over elements of $\mathfrak{Z}$.

Since $\mathfrak{Z}$ is a set in the plane, we know that there exists $\zeta(w) \in \mathfrak{Z}$ such that

$$\inf_{\xi \in \mathfrak{Z}} \|\xi\| = \|\zeta\|, \tag{26.1}$$

and the function $\zeta(w)$ is unique, as follows from the properties of the norm on $\mathscr{A}$. Indeed, if there exists another function $\zeta_1(w)$ satisfying (26.1), then, since $\frac{1}{2}(\zeta + \zeta_1) \in \mathfrak{Z}$, we have that $\|\frac{1}{2}(\zeta + \zeta_1)\| < \frac{1}{2}\|\zeta\| + \frac{1}{2}\|\zeta_1\| = \inf_{\xi\in\mathfrak{Z}} \|\xi\|$, which is a contradiction. From Proposition 26.1 it follows that

$$\inf_{\xi \in \mathfrak{Z}} \|\xi\| = \inf_{\xi \in \mathfrak{Z}_0} \|\xi\|. \tag{26.2}$$

The extremal function $\zeta(w)$ satisfies a differential equation, as we shall now show. This differential equation is similar to (24.21), and we could use the results of §24 to derive it. However, this method presents some nontrivial difficulties, and we prefer to use another way.

Relation (26.2) implies that

$$\|\zeta\| = \inf_{\xi \in \mathfrak{Z}_0} \|\xi\|.$$

Thus from (23.2) it is clear that Problem 1 is equivalent to the following problem.

Problem 2. Find the supremum for the second variation in the class of admissible functions $\mathscr{D}_f^0$, subject to the condition that the first coefficient variations $\delta a_k(f) = t_k$, $k = 1, \cdots, n$, $t = (t_1, \cdots, t_n)$, are fixed.

It is clear that the connection between $t_1, \cdots, t_n$ and $\mathfrak{z}_1, \cdots, \mathfrak{z}_n$ is given by (19.19). Using the remarks in §21, we shall see that in the investigation of Problem 2 it is possible to vary both the measure $\nu(dz)$ defining the admissible function (2.2), and the constant c_1. For this it is convenient to use formula (4.18), which defines the second variation of a linear functional. We shall denote the upper bound for the second variation by $D[f,t]$. Suppose that the sequence $\{\xi_k(w)\}\subset \mathfrak{Z}_0$ converges to $\zeta(w)$. The existence of such a subsequence is guaranteed by Proposition 26.1. Each function $\xi_k(w)$, as we saw in §21, defines a measure $\nu_k(dz)$ with support $\mathscr{C}_k=\{z:|z|=r_k\}$ and a constant c_{1k} such that, after having defined $\delta f(z;\nu_k,c_{1k})$ by formula (2.11), we have $\delta f(z;\nu_k,c_1)=\xi_k[f(z)]$. Let $\tilde\nu(dz)$ be a measure with support $\mathscr{C}$ contained in $|z|<\inf_k r_k$, and suppose that this measure is subject to the conditions $\delta a_l(f;\tilde\nu)=0$, $l=1,\cdots,n$. In this way we are actually putting $\tilde c_1=0$. Taking $\epsilon>0$, we form the measure $\tilde\nu_k(dz)=\nu_k(dz)+\epsilon\tilde\nu(dz)$, and we also set $\tilde c_{1k}=c_{1k}$. Then

$$\delta^2L(f;\tilde\nu_k,\tilde c_{1k})\leqq D[f,t].$$

Applying formula (4.18), we find that

$$\delta^2 L(f;\nu_k,c_{1k})+\varepsilon^2\delta^2L(f,\tilde\nu)+2\varepsilon\mathcal{A}_k\leqslant D[f,t], \tag{26.3}$$

where

$$\mathcal{A}_k=\operatorname{Re}\Big\{\int_{\mathscr{C}}\tilde\nu(dz)\Big[\int_{\mathscr{C}_k}\mathfrak{A}(z,t)\,\nu_k(dt)-\int_{\mathscr{C}_k}\mathfrak{B}(z,\bar t)\,\overline{\nu_k(dt)}+ic_{ik}\mathfrak{C}(z)\Big]\Big\}.$$

Since for $|z|=r_k$ and $\nu_k(dz)=h_k(z)\,dz/2\pi iz$, we see, after having set

$$J_k(z)=\int_{\mathscr{C}_k}\mathfrak{A}(z,t)\,\nu_k(dt)-\int_{\mathscr{C}_k}\mathfrak{B}(z,\bar t)\,\overline{\nu_k(dt)},\qquad |z|<r_k,$$

that we have encountered, and in fact evaluated, a similar expression in §19 (see (19.3)). Set $J_k(z)=J_{0k}(z)+J_{1k}(z)$. Using the results of §19, we have

$$J_{0k}(z)=-ic_{1k}\mathfrak{C}(z)+\omega_k(z)+\int_{|t|=r_k}\left(\frac{Df(t)}{f(t)}\right)^2Q^*[f(t)]\,h_k(t)\,\frac{dt}{2\pi it},$$

where

$$\omega_k(z)=\int_{|t|=r_k}\left(\frac{Df(t)}{f(t)}\right)^2Q^*[f(t),z]\,h_k(t)\,\frac{dt}{2\pi it}$$
$$+\overline{\int_{|t|=r_k}\left(\frac{Df(t)}{f(t)}\right)^2Q^*\left[f(t),\frac{1}{\bar z}\right]h_k(t)\,\frac{dt}{2\pi it}}.$$

Changing variables via the formula $t=f^{-1}(w)$, we find

$$\omega_k(z)=\frac{1}{2\pi i}\int_{\mathscr{C}_{r_k}}Q^*(w,z)\frac{\alpha_k(w)}{w^2}\,dw+\overline{\frac{1}{2\pi i}\int_{\mathscr{C}_{r_k}}Q^*\left(w,\frac{1}{\bar z}\right)\frac{\alpha_k(w)}{w^2}\,dw}.$$

From this and (4.12) we have

$$\omega_k(z)=\sum_{p=2}^{n}\bar\lambda_p\sum_{l=1}^{p-1}(p-l+1)\,t_{p-l}z^{-l}+\overline{\sum_{p=2}^{n}\bar\lambda_p\sum_{l=1}^{p-1}(p-l-1)\,t_{p-l}\bar z^{l}}. \tag{26.4}$$

In this formula the right-hand side is independent of k, and so we shall henceforth drop the index k from $\omega(z)$. We can easily verify by methods similar to the above that

$$\int\limits_{|t|=r_k} \left(\frac{Df(t)}{f(t)}\right)^2 Q^*[f(t)]\, h_k(t)\, \frac{dt}{2\pi i t} = \sum_1^n l\bar{\lambda}_l t_l = A_0. \tag{26.5}$$

With regard to $J_{1k}(z)$, we have, in accord with the results established in §19, that

$$\begin{aligned} J_{1k}(z) = 2Df_{\mathfrak{s}}(z) \sqrt{\frac{Q(f(z))}{f^2(z)}}\, D_{\mathfrak{s}}\left[\sqrt{\frac{Q(f(z))}{f^2(z)}}\, \xi_k(f(z))\right] \\ + (Df(z))^2 \sum_2^n l\gamma_l [f(z)]^{-l-1}, \end{aligned} \tag{26.6}$$

where γ_l is defined by (19.14). From (23.6) it follows that the sequence $\{\xi_k(w)\}$ converges in the domain $\mathscr{G}$ to a function $\zeta(w)$. Thus from (26.4)–(26.6) it follows that $\mathscr{A} = \lim_{k\to\infty} \mathscr{A}_k$ exists. Since $\delta^2 L[f; \nu_x, c_{1k}] \to D[f,t]$ as $k \to \infty$, (26.3) implies

$$\epsilon^2 \delta^2 L(f, \tilde{\nu}) + 2\epsilon \mathscr{A} \leq 0,$$

and, since ϵ is arbitrary, we have $\mathscr{A} = 0$. We shall write out this equation completely. Using (26.4)–(26.6), we get

$$\begin{aligned} \operatorname{Re}\left\{\int\limits_{\mathscr{C}} \left[2Df(z) \sqrt{\frac{Q(f(z))}{f^2(z)}}\, D\left(\sqrt{\frac{Q(f(z))}{f^2(z)}}\, \zeta(f(z))\right)\right.\right. \\ \left.\left. + D(f(z))^2 \sum_2^n l\gamma_l [f(z)]^{-l-1} + \omega(z) + A_0\right] \tilde{\nu}(dz)\right\} = 0. \end{aligned} \tag{26.7}$$

In this equation it is easy to change the measure $\tilde{\nu}(dz)$ to any measure $\nu(dz)$ having the same support. Incidentally, note that we may assume that the support $\mathscr{C}$ lies in an arbitrary disc $|z| \leq \rho < 1$, since each of the quantities r_k, $k = 1, 2, \cdots$, can be taken as near as desired to 1, and thus $\inf_k r_k$ can be made as close to 1 as desired. We set

$$v_l = \int\limits_{\mathscr{C}} [\mathfrak{A}_l(z)\, \nu(dz) + \mathfrak{B}_l(\bar{z})\, \overline{\nu(dz)}], \qquad l = 1, \ldots, n. \tag{26.8}$$

By Lemma 3.4 there exists a measure $\nu_1(dz) = \Phi(z)\, dz/2\pi i z$, $|z| = r_0$, for which $\delta a_l(f; \nu_1) = v_l$, $l = 1, \cdots, n$ (the constant $c_1 = 0$). We note that by construction $\Phi(z)$ is linear in $\operatorname{Re} v_l$ and $\operatorname{Im} v_l$, $l = 1, \cdots, n$. Denote by $F(z)$ the expression in the square brackets under the integral sign in (26.7). Applying (26.7) to the measure $\tilde{\nu}(dz) = \nu(dz) - \nu_1(dz)$, we get that

$$\operatorname{Re}\left\{\int\limits_{\mathscr{C}} F(z)\, \nu(dz)\right\} - \operatorname{Re} \int\limits_{\mathscr{C}} F(z)\, \nu_1(dz) = 0.$$

By the above remarks the terms being subtracted above are linear in $\operatorname{Re} v_l$ and $\operatorname{Im} v_l$, $l = 1, \cdots, n$, and so we can write the above in the form $\operatorname{Re} \sum_1^n \bar{\mu}_k v_k$, where $\mu_1, \cdots, \mu_n$ are constants defined by the function $\zeta(w)$. Using (26.8), we get that

$$\operatorname{Re}\left\{\int_{\mathscr{C}}\left[F(z)-\sum_1^n(\bar{\mu}_l\,\mathfrak{A}_l(z)+\mu_l\overline{\mathfrak{B}_l(\bar{z})})\right]\nu(dz)\right\}=0.$$

From this equation, in view of the arbitrariness of $\nu(dz)$, we get the desired differential equation

$$\sqrt{\frac{Q(w)}{w^2}}\,\frac{d\zeta^*(w)}{dw}+w^{-2}\widetilde{Q}(w)+\frac{\widetilde{\mathscr{P}}(z)}{(Df(z))^2}=0,\qquad z=f^{-1}(w), \tag{26.9}$$

where

$$\widetilde{Q}(w)=\frac{1}{2}\sum_1^{n-1}(k+1)\gamma_{k+1}w^{-k}-\frac{1}{2}\sum_1^n\bar{\mu}_kq_k(w) \tag{26.10}$$

and

$$\begin{aligned}\widetilde{\mathscr{P}}(z)=\frac{1}{2}(\omega(z)+A_0)-\frac{1}{2}\sum_{l=1}^n\Bigg[\bar{\mu}_l\sum_{k=1}^l(l-k+1)a_{l-k}z^{-l}\\ +\mu_l\sum_{k=1}^l(l-k+1)\bar{a}_{l-k}z^l+l\bar{\mu}_la_l\Bigg].\end{aligned} \tag{26.11}$$

We shall show that for $|z|=1$

$$\operatorname{Im}\widetilde{\mathscr{P}}(z)=0. \tag{26.12}$$

To prove this, we have from (26.9) and the fact that $\zeta\in\mathscr{A}$ that

$$\int_{|w|>1/4}|S(w)|^2\left|\frac{w^2}{Q(w)}\right|d\sigma<\infty, \tag{26.13}$$

where, as before, $S(w)=w^{-2}\widetilde{Q}(w)+\widetilde{\mathscr{P}}(z)/(Df(z))^2$. The tree $\mathscr{T}$ has at least one terminal vertex w_j. If z_j is its image, then in a neighborhood of z_j we have $Df(z)=\alpha(z-z_j)+\cdots$, $\widetilde{\mathscr{P}}(z)/(Df(z))^2=\beta\widetilde{\mathscr{P}}(z_j)/(z-z_j)^2+\cdots$. Hence (26.13) can only be satisfied if $\widetilde{\mathscr{P}}(z_j)=0$. From this it follows that $\operatorname{Im}\{A_0-\sum_1^n l\bar{\mu}_la_l\}=0$, and this in turn implies (26.12). It is not difficult to show that in fact equation (26.9) is a variant of the Euler-Schiffer equation (5.13), although we shall not pursue this point.

We now derive a formula for the scalar product of $\zeta(w)$ with an arbitrary function $\eta(w)\in\mathscr{A}$. For this purpose we express the function $\widetilde{Q}(w)$ in a slightly different manner. For the function $\zeta(w)$ we define the quantities α_k in accord with (24.9). Then α_k, $k=0,\cdots,n-1$, can be computed via (24.10), with $\xi_1,\cdots,\xi_n$ replacing $\zeta_1,\cdots,\zeta_n$. We get that

$$\widetilde{Q}(w)=\frac{1}{2}\left[\sum_1^n(\vartheta\bar{\zeta}_k-\bar{\mathfrak{r}}_k)w^{-k}-\sum_1^{n-1}\alpha_kw^{-k}\right]. \tag{26.14}$$

The quantities $\mathfrak{r}_1,\cdots,\mathfrak{r}_n$ are defined by comparing the right-hand sides of (26.10) and (26.14). In (26.9) we substitute $w=f(z)$ and use (24.9). We deduce the identity

$$\frac{1}{f^2(z)}\left[-\widetilde{Q}(f(z))+\sum_1^n(\vartheta\bar{\zeta}_k-\bar{\mathfrak{r}}_k)[f(z)]^{-k}+\alpha_0\right]+\frac{\widetilde{\mathscr{P}}(z)}{(Df(z))^2}+\frac{\mathfrak{P}(z)}{z}=0, \tag{26.15}$$

where $\mathfrak{P}(z)$ is a series with no negative powers of z. Considering that $\zeta(w)\in\mathfrak{H}$ satisfies (26.9), and that $\eta(w)$ satisfies the conditions of Proposition 24.4, we shall deduce a formula for the scalar product (ζ,η) which generalizes (24.19). It is clear that all the arguments used in the proof of this latter formula continue to hold, with the exception of the evaluation of the quantity J. In the present case we have

$$J=\operatorname{Re}\left\{\frac{1}{2\pi i}\int_{\mathscr{C}_r}\left[-w^{-2}\widetilde{Q}(w)+\frac{\widetilde{\mathscr{P}}(z)}{(Df(z))^2}+\vartheta\sum_1^n\bar{\zeta}_k w^{-k-2}\right]\eta(w)\,dw\right\}.$$

Using the identity

$$J=\operatorname{Re}\left\{\frac{1}{2\pi i}\int_{\mathscr{C}_r}\left[-w^{-2}\widetilde{Q}(w)+\frac{\widetilde{\mathscr{P}}(z)}{(Df(z))^2}+\sum_1^n(\vartheta\bar{\zeta}_k-\bar{\mathfrak{r}}_k)w^{-k-2}\right]\eta(w)\,dw\right\}$$
$$+\operatorname{Re}\left\{\frac{1}{2\pi i}\int_{\mathscr{C}_r}\sum_1^n\bar{\mathfrak{r}}_k w^{-k-2}\eta(w)\,dw\right\}$$

and making the change of variable $w=f(z)$ in the first integral, we get from (26.15) that

$$J=\operatorname{Re}\left\{\frac{1}{2\pi i}\int_{\mathscr{C}_r}\sum_1^n\bar{\mathfrak{r}}_k w^{-k-2}\eta(w)\,dw\right\}=\operatorname{Re}\sum_1^n\bar{\mathfrak{r}}_k\eta_k.$$

Therefore we have

$$(\zeta,\eta)=\operatorname{Re}\sum_1^n\bar{\mathfrak{r}}_k\eta_k+\lim_{\delta\to 0}\left\{\frac{1}{2\pi i}\int_{C_\delta}\left[w^{-2}\widetilde{Q}(w)-\frac{\widetilde{\mathscr{P}}(z)}{(Df(z))^2}\right]\eta(w)\,dw\right\}. \tag{26.16}$$

Now assume that $\eta\in\mathscr{A}_0$. Then the second term of the above formula has already been evaluated, and it equals the right-hand side of (24.20). Since $w_j=0$, we have

$$(\zeta,\eta)=\operatorname{Re}\sum_1^n\bar{\mathfrak{r}}_k\eta_k. \tag{26.17}$$

This formula has been proved under the assumption that $\eta\in\mathscr{A}_0$, and consequently it is also true in the more general case when $\eta\in\mathscr{A}$. The results given above are summarized in the following lemma.

Lemma 26.1. *The extremal functions for either Problem* 1 *or Problem* 2 *satisfy equation* (26.9), *with* $\widetilde{Q}(w)$ *defined by* (26.14), *and with* $\widetilde{\mathscr{P}}(z)$ *defined by* (26.15) *and the additional condition* $\operatorname{Im}\widetilde{\mathscr{P}}(z)=0$ *for* $|z|=1$.

It is very simple now to prove

Lemma 26.2. *Let the function* $\zeta(w) = \mathfrak{z}_1 w^2 + \mathfrak{z}_2 w^3 + \cdots \in \mathscr{A}$ *satisfy* (26.9), *with* $\widetilde{Q}(w)$ *and* $\widetilde{\mathscr{P}}(z)$ *defined as in the previous lemma. Define the set* $\mathfrak{Z}$ *by the vector* $\mathfrak{z} = (\mathfrak{z}_1, \cdots, \mathfrak{z}_n)$. *Then* $\|\zeta\| = \inf_{\xi \in \mathfrak{Z}} \|\xi\|$.

Proof. Any function $\xi \in \mathfrak{Z}$ can be represented in the form $\xi(w) = \zeta(w) + \eta(w)$, where $\eta(w) = \eta_{n+1} w^{n+2} + \cdots \in \mathscr{A}$. Thus $(\zeta, \eta) = 0$ by (26.17). Hence $\|\xi\|^2 = \|\zeta\|^2 + \|\eta\|^2 \geqq \|\zeta\|^2$. Q.E.D.

To conclude this section we note the following immediate corollary of Lemma 23.3 and relation (26.2).

Lemma 26.3. *The supremum for the second variation remains unchanged if in Problem* 2 *it is computed over the class* $\mathscr{D}_f$ *instead of the class* $\mathscr{D}_f^0$ *of admissible functions.*

§27. The elements of the space $\mathscr{A}$ are essentially the first variations of the function $f(z)$. The extremal functions for Problem 1 will be called extremal first variations, or simply extremal variations. The extremal variation determined by the vector $\mathfrak{z}$ will be denoted by $\zeta(w, \mathfrak{z})$. The extremal variations form a finite-dimensional subspace $\mathscr{W} \subset \mathscr{A}$. It is easy to verify that each element of the subspace $\mathscr{R}$ satisfies equation (26.9). Thus, by Lemma 26.2, $\mathscr{R} \subset \mathscr{W}$. Denote by Z the linear subspace of R^{2n} defined by the condition $\operatorname{Re} \sum_1^n A_k \mathfrak{z}_k = 0$ if $\mathfrak{z} = (\mathfrak{z}_1, \cdots, \mathfrak{z}_n) \in Z$. Each element $\mathfrak{z} \in Z$ uniquely defines an extremal variation $\zeta(w, \mathfrak{z}) \in \mathscr{W}$. For arbitrary real numbers α and β it is easily seen that if $\mathfrak{z}_1, \mathfrak{z}_2 \in Z$, then $\alpha\zeta(w, \mathfrak{z}_1) + \beta\zeta(w, \mathfrak{z}_2)$ satisfies equation (29.9), and so by Lemma 26.2 we have $\alpha\zeta(w, \mathfrak{z}_1) + \beta\zeta(w, \mathfrak{z}_2) \in \mathscr{W}$. Since the solution of Problem 1 is unique, we must have that

$$\zeta(w, \alpha\mathfrak{z}_1 + \beta\mathfrak{z}_2) = \alpha\zeta(w, \mathfrak{z}_1) + \beta\zeta(w, \mathfrak{z}_2). \tag{27.1}$$

We now make an important remark. The vector $\mathfrak{r} = (\mathfrak{r}_1, \cdots, \mathfrak{r}_n)$, which in the end determines the functions $\widetilde{Q}(w)$ and $\widetilde{\mathscr{P}}(z)$, is not uniquely defined, since to equation (26.9) it is possible to add equation (5.13), after having multiplied it by $\alpha[Df(z)]^{-2}$ (α is any real constant), and the form of equation (26.9) is not changed as a result of this operation. Via this process the vector $\mathfrak{r}$ becomes $\mathfrak{r} - \alpha\overline{A}$, where $\overline{A} = (A_1, \cdots, A_n)$. In order to remove this ambiguity with regard to the choice of $\mathfrak{r}$, we shall require that

$$\operatorname{Re} \sum_1^n A_k \mathfrak{r}_k = 0;$$

our previous remarks insure that this condition can in fact be met. In this way we have defined on Z a single-valued function with values in Z. From (27.1) it follows that $\mathfrak{r}(\mathfrak{z})$ is a linear function; that is,

$$\mathfrak{r}(\alpha\mathfrak{z}_1 + \beta\mathfrak{z}_2) = \alpha\mathfrak{r}(\mathfrak{z}_1) + \beta\mathfrak{r}(\mathfrak{z}_2).$$

We extend this function to all of R^{2n} by defining $\mathfrak{r}(\alpha\overline{A} + \mathfrak{z}) = \alpha\overline{A} + \mathfrak{r}(\mathfrak{z})$ for $\mathfrak{z} \in Z$. It is clear that with this definition Z becomes an invariant subspace of the linear operator $\mathfrak{r}(\mathfrak{z})$. Let $\mathfrak{r}$ be the matrix of the linear operator $\mathfrak{r}(\mathfrak{z})$, i.e. $\mathfrak{r}(\mathfrak{z}) = \mathfrak{r}\mathfrak{z}$. In order to avoid confusion in notation, we denote the scalar product of $x, y \in R^{2n}$ by $[x, y]$. Then (26.17) can be written in the form

$$(\zeta, \eta) = [\mathfrak{r}\mathfrak{z}, \mathfrak{Y}], \tag{27.2}$$

where $\mathfrak{Y} = (\eta_1, \cdots, \eta_n)$. We shall now show that the matrix $\mathfrak{r}$ is symmetric. Let $\eta \in \mathscr{W}$. Then $(\zeta, \eta) = (\eta, \zeta) = [\mathfrak{r}\mathfrak{Y}, \mathfrak{z}] = [\mathfrak{z}, \mathfrak{r}\mathfrak{Y}]$, and so $[\mathfrak{r}\mathfrak{z}, \mathfrak{Y}] = [\mathfrak{z}, \mathfrak{r}\mathfrak{Y}]$ for any vectors $\mathfrak{z}, \mathfrak{Y} \in Z$. Since the vector $\overline{A}$ is orthogonal to Z, it is easy to verify that the latter relation holds for arbitrary vectors $x, y \in R^{2n}$. From (27.2) we find the following elegant formula for the norm of the extremal variation:

$$\|\zeta\|^2 = [\mathfrak{r}\mathfrak{z}, \mathfrak{z}]. \tag{27.3}$$

From this it easily follows that the matrix $\mathfrak{r}$ is positive definite.

We now examine $D[f, t]$. Recalling that the vectors t and $\mathfrak{z}$ are connected via (19.19) (see §21 for the definition of t), and that the second variation and the norm in $\mathscr{A}$ are connected via (23.2), it is easy to prove that $D[f, t]$ is a quadratic form in $\operatorname{Re} t_k$ and $\operatorname{Im} t_k$, $k = 1, \cdots, n$. We write this form as $[D(f)t, t]$, where $D(f)$ is a symmetric matrix. This matrix has a simple geometric interpretation, as we shall see in the next chapter. Indeed, $D[f, t] = [D(f)t, t]$. We stress again that the vector t is not arbitrary, but must satisfy the condition

$$[\lambda, t] = \operatorname{Re} \sum_1^n \overline{\lambda}_k t_k = 0.$$

Set

$$\Lambda = \max_{[\lambda, t] = 0, |t| = 1} [D(f)t, t]. \tag{27.4}$$

By a theorem of Weierstrass, $\Lambda = [D(f)t^0, t^0]$, $|t^0| = 1$ and $[\lambda, t^0] = 0$. Thus if we denote by $\mathfrak{z}^0$ the vector gotten from t^0 by the transformation (19.20), and if we denote by $\mathfrak{Z}^0$, as above, the set of all $\xi(w) \in \mathscr{A}_0$ whose power series expansions have the form $\xi(w) = \mathfrak{z}_1^0 w^2 + \cdots + \mathfrak{z}_n^0 w^{n+1} + \cdots$, then $\sup_{\xi \in \mathfrak{Z}^0} \delta^2 L(f; \xi) = \Lambda$. Thus

$$\Lambda \leqslant \sup_{\xi \in \mathscr{H}} \frac{\delta^2 L(f; \xi)}{|t|^2}.$$

Since the opposite inequality is trivially true, we have

$$\Lambda = \sup_{\xi \in \mathscr{H}} \frac{\delta^2 L(f; \xi)}{|t|^2} = \sup_{\xi \in \mathscr{H}_0} \frac{\delta^2 L(f; \xi)}{|t|^2}. \tag{27.5}$$

Thus

$$\Lambda = \sup_{\nu, c_1} \frac{\delta^2 L(f; \nu, c_1)}{\sum_1^n |\delta a_k(\nu, c_1)|^2}. \tag{27.6}$$

It is easy to show that the function $\zeta(w)$ which yields the upper bound (27.5) satisfies the equation (26.9), where

$$\widetilde{Q}(w) = \frac{1}{2}\sum_{1}^{n-1}(k+1)\gamma_{k+1}w^{-k} - \frac{\Lambda}{2}\sum_{1}^{n}\bar{t}_k^0 q_k(w), \tag{27.7}$$

with $t_1^0, \cdots, t_n^0$ the first coefficient variations of $f(z)$ as defined by the first variation $\zeta[f(z)]$.

§28. The extremal Problem 1 can also be formulated in the space $\mathfrak{H}$. By $\mathfrak{Z}^*$ we denote the set of elements $\xi \in \mathfrak{H}$ whose power series expansions have the form

$$\xi(w) = \mathfrak{z}_1 w^2 + \cdots + \mathfrak{z}_n w^{n+1} + \cdots.$$

Problem 3. Determine the infimum of $\|\xi\|$ over the set $\mathfrak{Z}^*$.

Our remarks concerning the existence and uniqueness of a function yielding the greatest lower bound continue to be true in the present situation. An element $\zeta \in \mathfrak{H}$ which attains the greatest lower bound for Problem 3 will be called an extremal element. The set of extremal elements will be denoted by $\mathfrak{W}$. A function ζ belongs to $\mathfrak{W}$ if and only if for any function $\eta \in \mathfrak{H}$ with power series of the form $\eta(w) = \eta_{n+1}w^{n+2} + \cdots$ we have $(\zeta, \eta) = 0$. It is easily verified that any extremal element satisfies an equation of the form (26.9). To prove this, let $\zeta \in \mathfrak{W}$, $\zeta = \xi + \eta$, where $\xi \in \mathscr{A}$ and η is orthogonal to $\mathscr{A}$. Then if $\eta^0(w) = \eta_{n+1}^0 w^{n+2} + \cdots, \eta^0 \in \mathscr{A}$, we have $(\zeta, \eta^0) = 0$. Hence $(\xi, \eta^0) + (\eta, \eta^0) = (\xi, \eta^0) = 0$. Thus $\xi \in \mathscr{W}$, and so $\xi(w)$ satisfies equation (26.9). Since $\eta(w)$ satisfies (24.21), $\zeta = \xi + \eta$ satisfies an equation of the form of (26.9). Conversely, any function $\zeta \in \mathscr{A}$ satisfying (26.9) is an extremal variation; for the space $\mathfrak{H}$ the analogous statement is false, unless of course $\mathscr{A}$ and $\mathfrak{H}$ coincide.

Proposition 28.1. *If $\zeta(w) \in \mathfrak{W}$, then $\zeta(w)$ is a solution of equation* (26.9), *where the function $\widetilde{Q}(w)$ satisfies the following conditions: any vertex w_j, $w_j \neq \infty$, of the tree $\mathscr{T}$ is a zero of $\widetilde{Q}(w)$ of multiplicity at least $[\kappa_j/2]$, and the base of the tree is a zero of multiplicity at least $[\kappa_0/2]+1$. Conversely, if $\zeta(w) \in \mathfrak{H}$ is a solution of* (26.9) *with $\widetilde{Q}(w)$ satisfying the above conditions, then $\zeta(w) \in \mathfrak{W}$.*

Proof. For $\zeta \in \mathfrak{W}$ we showed earlier that $\zeta(w)$ satisfies (26.9). Let $\eta(w) = Aw^2(w - w_j)^{-l}$ be an element of $\mathfrak{R}_j$. According to §21.4 we determine $\xi(w) \in \mathscr{A}_0$ so that $\eta(w) + \xi(w) = \eta_{n+1}^0 w^{n+2} + \cdots$. This is obviously possible, since the first n coefficients $\eta_1, \cdots, \eta_n$ of $\eta(w)$ satisfy condition (21.5). Note that $(\zeta, \eta + \xi) = 0$. Applying (26.16) and transforming the second term of the result with the aid of (24.3), we find that

$$(\zeta, \eta) = \operatorname{Re}\sum_{1}^{n}\bar{r}_k\eta_k + \lim_{\delta\to 0}\left\{\frac{1}{\pi i}\int_{C_\delta}\bar{Q}(w)\,\frac{\eta(w)'}{w^2}\,dw\right\}. \tag{28.1}$$

Thus

(28.2) $$(\zeta, \eta) = \operatorname{Re} \sum_{1}^{n} \bar{r}_k \eta_k - \frac{2}{(l-1)!} \operatorname{Re} \{A \widetilde{Q}^{(l-1)}(w_j)\}.$$

Computing (ζ, ξ) by (26.17) and comparing the result with the previous equality, we find that

$$\operatorname{Re}\{A \widetilde{Q}^{(l-1)}(w_j)\} = 0.$$

An analogous result holds when we are considering the base of the tree. Since A is arbitrary, the first statement of the proposition is proved.

Now suppose that $\zeta(w)$ satisfies equation (26.9) with $\widetilde{Q}(w)$ as in the hypothesis above. If $\eta \in \mathfrak{H}$, then by Lemma 25.2 we have $\eta = \xi + \eta^0$, where $\xi \in \mathscr{A}$ and $\eta^0 \in \mathfrak{R}^*$. By (26.17), $(\zeta, \xi) = \operatorname{Re} \sum_1^n \bar{r}_k \xi_k$. Applying (28.1) to (ζ, η^0) and noting that the second term in this formula vanishes (due to the properties of $\widetilde{Q}(w)$), we find that $(\zeta, \eta^0) = \operatorname{Re} \sum_1^n \bar{\mathfrak{r}}_k \eta_k^0$. Therefore

$$(\zeta, \eta) = \operatorname{Re} \sum_{1}^{n} \bar{\mathfrak{r}}_k \eta_k.$$

If $\eta(w) = \eta_{n+1} w^{n+2} + \cdots$, then $(\zeta, \eta) = 0$, and so $\zeta \in \mathfrak{W}$, as required. Q.E.D.

REMARK 28.1. In the course of the proof we showed that formula (26.17) holds when $\zeta \in \mathfrak{W}$ and η is an arbitrary element of $\mathfrak{H}$. With the aid of Proposition 28.1 we can extend the results of §27 to the set of extremal elements $\mathfrak{W}$. Since it follows from Lemma 25.2 that any function $\xi(w) = \sum_1^\infty \xi_k w^{k+1} \in \mathfrak{H}$ satisfies

$$\operatorname{Re} \sum_{1}^{n} A_k \xi_k = 0,$$

the linear manifold $\mathfrak{W}$ is isomorphic to Z. Denote the matrix of the linear transformation $\mathfrak{r}(\mathfrak{z})$ by $\mathscr{L}$. Instead of (27.3), we get that the norm of an extremal element is given by the formula

(28.3) $$\| \zeta \|^2 = [\mathscr{L} \mathfrak{z}, \mathfrak{z}].$$

With the aid of (23.2) we extend the quadratic functional $\delta^2 L(f; \xi)$ to the whole space $\mathfrak{H}$. The least upper bound of this functional on $\mathfrak{Z}^*$ will be denoted by $\mathfrak{D}[f, t]$, where the vector t is connected with the vector $\mathfrak{z}$ by the mapping (19.19). As before, we have

$$\mathfrak{D}[f, t] = [\mathfrak{D}(f) t, t],$$

where $\mathfrak{D}(f)$ is a symmetric matrix. In conclusion we make a remark which might shed some light on the role of the space $\mathfrak{R}$. It is easy to see that

(28.4) $$\mathfrak{R} \subset \mathfrak{W}.$$

Indeed, each element $\xi \in \mathfrak{R}$ satisfies a differential equation of the form (26.9) in which the function $\widetilde{Q}(w)$ satisfies the conditions of Proposition 28.1, and so $\xi \in \mathfrak{W}$. From (28.4) we then easily see that the following statement is true.

Proposition 28.2. *If* $\mathscr{W} = \mathfrak{W}$, *then* $\mathscr{A} = \mathfrak{H}$.

Proof. It follows from the hypothesis and (28.4) that $\mathfrak{R} \subset \mathscr{W} \subset \mathscr{A}$, and so by Lemma 25.1 $\mathscr{A} = \mathfrak{H}$. Q.E.D.

§29. Our previous investigations have been concerned with a fixed extremal function $f(z) \in S_n$ for which $d_{n-1/2}(f) > 0$; we introduced above the important functionals $D[f,t]$ and $\mathfrak{D}[f,t]$. Since S_n and $\partial\mathscr{V}_n$ are homeomorphic, we see as above that at each point x of the set $\mathscr{R}$ ($\mathscr{R} = \{x : x \in \partial\mathscr{V}_n, d_{n-1/2}(x) > 0\}$) the quantities $D[f(z,x),t]$ and $\mathfrak{D}[f(z,x),t]$ depend also on a vector of the tangent subspace to $\partial\mathscr{V}_n$ at x. Denote by $\mathfrak{Y}$ the set of tangent vectors to $\mathscr{R}$; that is, the set of pairs $\mathfrak{y} = (x,t)$, where $x \in \mathscr{R}$ and t is a vector of the tangent subspace to $\partial\mathscr{V}_n$ at x. The set $\mathfrak{Y}$ will be considered as a subset of R^{4n-1}. Hoping that the notation will not be confusing, we also set

$$D[f(z, x), t] = D[x, t] = D(\mathfrak{y}), \qquad \mathfrak{D}[f(z, x), t] = \mathfrak{D}[x, t] = \mathfrak{D}(\mathfrak{y}).$$

We shall study the functions $D(\mathfrak{y})$ and $\mathfrak{D}(\mathfrak{y})$ as functions defined on the points of the set $\mathfrak{Y}$.

Lemma 29.1. *The function* $D(\mathfrak{y})$ *is lower semicontinuous on* $\mathfrak{Y}$.

Proof. Let $m(\mathfrak{y})$ be the lower limit of $D(\mathfrak{y})$ at the point $\mathfrak{y}$. The relationship $m(\mathfrak{y}) = D(\mathfrak{y})$ on $\mathfrak{Y}$ will follow if we can show that for any sequence $\{\mathfrak{y}^k\}$ converging to $\mathfrak{y}$ we have

$$D(\mathfrak{y}) \leqslant \varliminf_{k\to\infty} D(\mathfrak{y}^k).$$

To prove this inequality we begin by taking an arbitrary point $\mathfrak{y} = (x,t)$ in $\mathfrak{Y}$ and a sequence $\{\mathfrak{y}^k\}$, $\mathfrak{y}^k = (x^k, t^k)$, $k = 1, 2, \cdots$, converging to $\mathfrak{y}$. Then $\lim_{k\to\infty} x^k = x$ and $\lim_{k\to\infty} t^k = t$. Since Problems 1 and 2 are equivalent, we can find for any $\epsilon > 0$ a measure $\nu(dz)$ with support $\mathscr{C}$ and a constant c_1 such that

$$\delta^2 L[f(z,x); \nu, c_1] \geqq D[x,t] - \epsilon. \tag{29.1}$$

Using this measure and constant we determine the first variation of the function $f(z,x^k)$, $k = 1, 2, \cdots$. Let $\delta a_l[f(z,x_k); \nu, c_1]$, $l = 1, \cdots, n$, be the first coefficient variations of $f(z,x^k)$. Since $\lim_{k\to\infty} f(z,x^k) = f(z,x)$, in $|z| < 1$, it follows from (2.13) that

$$\lim_{k\to\infty} \delta a_l[f(z, x^k); \nu, c_1] = \delta a_l[f(z, x); \nu, c_1] = t_l, \qquad l = 1, \ldots, n.$$

Since $\lim_{k\to\infty} t^k = t$, as $k \to \infty$ we have

$$\Delta_k^2 = \sum_{l=1}^{n} | t_l^k - \delta a_l[f(z, x^k); \nu, c_1] |^2 \to 0. \tag{29.2}$$

According to Lemma 3.4 we define a measure $\mu_k(dz) = \eta_k(z)dz/2\pi iz$ with support $\mathscr{C}^*=\{z:|z|=r^*\}$ such that

$$\delta a_l\,[f(z,x^k);\ \mu_k]=t_l^k-\delta a_l\,[f(z,x^k);\ \nu,c_1]\qquad l=1,\ldots,n.$$

From (3.24) we find that

$$\frac{1}{2\pi}\int_{|z|=r^*}|\eta_k(z)|^2\left|\frac{dz}{z}\right|\leqslant\Delta_k^2d_{n-1/2}^{-2}(x^k).$$

Since $d_{n-1/2}(x)$ is continuous on $\mathscr{R}$, we have $d_{n-1/2}^{-1}(x^k)\leqq B$ for $k=1,2,\cdots$. Thus

$$\frac{1}{2\pi}\int_{|z|=r^*}|\eta_k(z)|^2\left|\frac{dz}{z}\right|\leqslant B^2\Delta_k^2.\tag{29.3}$$

The measure $\nu_k(dz)=\nu(dz)+\mu_k(dz)$ and the constant c_1 determine a first and second variation of $f(z,x^k)$. By construction $\delta a_l[f(z,x^k);\nu_k,c_1]=t_l^k$, $l=1,\cdots,n$, and so

$$\delta^2L[f(z,x^k);\nu_k,c_1]\leqq D[x^k,t^k].\tag{29.4}$$

From (4.18) it follows in a simple way that

$$\delta^2L\,[f(z,\ x^k);\ \nu_k,c_1]=\delta^2L[f(z,\ x^k);\nu,\ c_1]+\delta^2L[f(z,\ x^k);\ \mu_k]+2\mathscr{A}_k,$$

where

$$\mathscr{A}_k=\operatorname{Re}\Bigl\{\int_{\mathscr{C}}\nu(dz)\Bigl[\int_{\mathscr{C}^*}\mathfrak{A}(z,\zeta;\ x^k)\,\mu_k(d\zeta)-\int_{\mathscr{C}^*}\mathfrak{B}(z;\bar\zeta;\ x^k)\,\overline{\mu_k(d\zeta)}\Bigr]$$
$$+\,ic_1\int_{\mathscr{C}^*}\mathfrak{C}(z,\ x^k)\,\mu_k(dz)\Bigr\}.$$

From Proposition 5.1 it now follows that the sequence $\{\mathfrak{A}(z,\zeta;x^k)\}$ converges to $\mathfrak{A}(z,\zeta;x)$ if $|z|\leqq\rho$ and $|\zeta|\leqq\rho$, where $\rho<1$ is arbitrary. This follows since, for $|z|\leqq\rho$, $f(z,x)$ is continuous on $\partial\mathscr{V}_n$ and $\lambda(x)$ is continuous on $\mathscr{R}$. Similarly the sequence $\{\mathfrak{A}(z,\zeta;x^k)\}$ is uniformly bounded if $|z|\leqq\rho$ and $|\zeta|\leqq\rho$. Analogous statements hold for the sequences $\{\mathfrak{B}(z,\bar\zeta;x^k)\}$ and $\{\mathfrak{C}(z,x^k)\}$. Therefore, in view of (29.2) and (29.3),

$$\lim_{k\to\infty}\delta^2L\,[f(z,\ x^k);\ \nu_kc_1]=\delta^2L\,[f(z,\ x);\ \nu,\ c_1].$$

Now, according to (29.1) and (29.4),

$$D\,[x,\ t]-\varepsilon\leqslant\varliminf_{k\to\infty}D\,[x^k,\ t^k].$$

Allowing $\epsilon\to0$, we deduce the required inequality. Q.E.D.

LEMMA 29.2. *The function* $\mathfrak{D}[\mathfrak{y}]$ *is lower semicontinuous on* $\mathfrak{Y}$.

PROOF. Recalling the definition of $\mathfrak{D}[f,t]$, we see that it is sufficient to prove that the function $\|\zeta(w,\mathfrak{z})\|$ is lower semicontinuous, where $\zeta(w,\mathfrak{z})$ is an extremal function for Problem 3. In order to clearly indicate the dependence on x, we shall use a subscript x on the norm symbol, and this will also remind us that the norm is to be taken in the space $\mathfrak{H}_x$. To begin, let $\mathfrak{y}$ be any point of $\mathfrak{Y}$, and suppose the sequence $\{\mathfrak{y}^k\}$ converges to $\mathfrak{y}$. Determine the vector $\mathfrak{z}^k$ by the vector t^k with the aid of (19.20), where the coefficients are determined from the function $f(z, x^k)$ as in §19. Then $\mathfrak{z}^k = \mathfrak{A}^k t^k$, where $\mathfrak{A}^k$ is the matrix of the transformation (19.20). Since $\lim_{k\to\infty} f(z, x^k) = f(z, x)$ for $|z| < 1$, we have $\lim_{k\to\infty} \mathfrak{A}^k = \mathfrak{A}$, where the matrix $\mathfrak{A}$ corresponds to the function $f(z, x)$ in the manner described above. Also, $\lim_{k\to\infty} \mathfrak{z}^k = \mathfrak{z}$, where $\mathfrak{z} = \mathfrak{A}t$. The vector $\mathfrak{z}^k$ determines a set $\mathfrak{Z}_{x^k} \subset \mathfrak{H}_{x^k}$; let $\zeta(w, \mathfrak{z}^k)$ be an extremal element of $\mathfrak{Z}_{x^k}$. With this notation, we must prove that

$$\| \zeta(w, \mathfrak{z}) \|_x \leqslant \underline{\lim}_{k\to\infty} \| \zeta(w, \mathfrak{z}^k) \|_{x^k}.$$

Without loss of generality we assume the existence of the finite limit $\lim_{k\to\infty} \| \zeta(w, \mathfrak{z}^k) \|_{x^k} < \infty$. (If the limit does not exist, we merely choose a subsequence on which the limit exists and is finite.) Let $r < 1$ be arbitrary, and set $r_0 = 2r - 1$ and $r_1 = \frac{1}{2}(r+1)$. As usual, set $\mathscr{C}_{r,x} = \{w : |f^{-1}(w, x)| = r\}$. For $k \geqq k_0(r)$ the level curve $\mathscr{C}_{r,x^k}$ will lie in the complement of $\mathscr{G}(\mathscr{C}_{r,x})$. Therefore by Lemma 23.4 and the inequality $d_{n-1/2}^{-1}(x^k) \leqq B$ we see that the sequence $\{\zeta(w, \mathfrak{z}^k)\}_{k_0}^{\infty}$, $w \in \mathscr{G}(\mathscr{C}_{r,x})$, is uniformly bounded. Hence we can extract a subsequence converging uniformly on $\mathscr{G}(\mathscr{C}_{r_0,x})$. Using this remark, we find by a well-known method a subsequence $\{\zeta(w, \mathfrak{z}^{k_i})\}$ converging in $\mathscr{G}_x$ to a regular function $\xi(w)$. For notational ease, we assume that the initial sequence $\{\zeta(w, \mathfrak{z}^k)\}$ converges in $\mathscr{G}_x$ to $\zeta(w)$. Since $\zeta(w, x^k) = \mathfrak{z}_1^k w^2 + \cdots + \mathfrak{z}_n^k w^{n+1} + \cdots$, we have $\xi(w) = \mathfrak{z}_1 w^2 + \cdots + \mathfrak{z}_n w^{n+1} + \cdots$. Let $\mathscr{G}_r = \mathscr{G}(\mathscr{C}_{r,x}) \cap \{w : |w| \geqq 1/4\}$. Since $Q(w, x^k) \to Q(w, x)$ as $k \to \infty$, provided $w \neq 0$, we have for $k \geqq k_0$ that

$$\frac{1}{\pi} \int_{\mathscr{G}_r} \left| \frac{d\xi^*(w)}{dw} \right|^2 d\sigma \leqslant \frac{1}{\pi} \int_{\mathscr{G}_r} \left| \frac{d\zeta^*(w, \mathfrak{z}^k)}{dw} \right|^2 d\sigma + \varepsilon,$$

where $\epsilon > 0$ is as small as desired. From this we obtain

$$\frac{1}{\pi} \int_{\mathscr{G}_r} \left| \frac{d\xi^*(w)}{dw} \right|^2 d\sigma < \frac{1}{\pi} \int_{|w|>1/4} \left| \frac{d\zeta^*(w, \mathfrak{z}^k)}{dw} \right|^2 d\sigma + \varepsilon.$$

and so

$$\frac{1}{\pi} \int_{\mathscr{G}_r} \left| \frac{d\xi^*(w)}{dw} \right|^2 d\sigma \leqslant \underline{\lim}_{k\to\infty} \frac{1}{\pi} \int_{|w|>1/4} \left| \frac{d\zeta^*(w, \mathfrak{z}^k)}{dw} \right|^2 d\sigma.$$

Thus, since $r < 1$ was arbitrary,

$$\frac{1}{\pi} \int_{|w|>1/4} \left| \frac{d\xi^*(w)}{dw} \right|^2 d\sigma \leqslant \underline{\lim}_{k\to\infty} \frac{1}{\pi} \int_{|w|>1/4} \left| \frac{d\zeta^*(w, \mathfrak{z}^k)}{dw} \right|^2 d\sigma.$$

From this inequality and the fact that $\lim_{k\to\infty} \mathfrak{z}^k = \mathfrak{z}$ it follows that $\| \xi(w) \|_x \leqq \underline{\lim}_{k\to\infty} \| \zeta(w, \mathfrak{z}^k) \|_{x^k}$, and so

$$\|\xi(w)\|_x \leqslant \lim_{k\to\infty} \|\zeta(w;\ \mathfrak{z}^k)\|_{x^k},$$

as required.

We shall show that $\xi(w)$ satisfies condition (21.4), or in other words, that $\xi\in\mathfrak{H}_x$. The functions $\zeta(w,\mathfrak{z}^k)$ satisfy equation (26.9), which we write as

$$\sqrt{\frac{Q(w,x^k)}{w^2}}\frac{d}{dw}\left[\sqrt{\frac{Q(w,x^k)}{w^2}}\zeta(w,\mathfrak{z}^k)\right]+\frac{1}{w^2}\widetilde{Q}(w,\mathfrak{z}^k)+\frac{\widetilde{\mathscr{P}}(z,\mathfrak{z}^k)}{(Df(z,x^k))^2}=0. \tag{29.5}$$

We shall show first that the sequence $\{\mathfrak{r}(\mathfrak{z}^k)\}$ is bounded. Supposing that this is not the case, we can assume without loss of generality that $|\mathfrak{r}(\mathfrak{z}^k)|\to\infty$, and that the sequence $\{\mathfrak{r}(\mathfrak{z}^k)/|\mathfrak{r}(\mathfrak{z}^k)|\}$ converges to a vector $\mathfrak{r}^0$. We divide (29.5) by $|\mathfrak{r}(\mathfrak{z}^k)|$ and pass to the limit as $k\to\infty$, with $|z|\leqq r<1$. As a result we get an equation

$$w^{-2}Q_0(w)+\frac{\mathscr{P}_0(z)}{(Df(z,x))^2}=0,\qquad w=f(z,x),$$

satisfied by $f(z,x)$, where $Q_0(w)=\sum_1^n \mathfrak{r}_l^0 w^{-l}$. Since for $|z|=1$ we have $\operatorname{Im}\widetilde{\mathscr{P}}(z,\mathfrak{z}^k)=0$, we see that $\mathscr{P}_0(z)$ is real-valued on $|z|=1$. Since $\operatorname{Re}\sum_1^n A_l(x^k)\mathfrak{r}_l(\mathfrak{z}^k)=0$, we have that $\operatorname{Re}\sum_1^n A_l(x)\mathfrak{r}_l^0=0$. On the other hand, $d_{n-1/2}(x)>0$ implies that $\mathfrak{r}_l^0=C\overline{A}_l(x)$, $l=1,\cdots,n$, where C is real, and $C\neq 0$. Thus $\operatorname{Re}\sum_1^n A_l(x)\mathfrak{r}_l^0=C\sum_1^n|A_l(x)|^2\neq 0$, which is a contradiction. Hence the sequence $\{\mathfrak{r}(\mathfrak{z}^k)\}$ is bounded, and we can in fact assume that it converges (take a subsequence, if necessary). Then, taking the limit in (29.5), for $|z|\leqq r<1$ we get

$$\sqrt{\frac{Q(w,x)}{w^2}}\frac{d}{dw}\left(\sqrt{\frac{Q(w,x)}{w^2}}\xi(w)\right)+\frac{1}{w^2}\widetilde{Q}(w)+\frac{\widetilde{\mathscr{P}}(z)}{(Df(z,x))^2}=0. \tag{29.6}$$

Now divide $|z|=1$ into the arcs (16.1), as defined by $f(z,x)$. Corresponding to the tree $\mathscr{T}_x$, we construct the system $\mathscr{K}_\rho$ of circles $K_{j\rho}$ as in §16; let G_ρ be the domain defined in §16. By Proposition 16.7 the sequence $\{f(z,x^k)\}$ converges uniformly in G_ρ to $f(z,x)$ for any $\rho>0$. It is easy to see that in G_ρ the sequence $\{Df(z,x^k)\}$ converges uniformly to $Df(z,x)$. To prove this, note that G_ρ is mapped by $f(z,x)$ to $\mathscr{G}_x\setminus\mathscr{K}_\rho$, and so $Q(f(z,x),x)\neq 0$ for $z\in G_\rho$. Hence for $k\geqq k(\rho)$ we have $Q(f(z,x^k),x^k)\neq 0$ for $z\in\partial G_\rho$, and since $Df(z,x^k)=i\,[P(z,x^k)f(z,x^k)/Q(f(z,x^k),x^k)]^{1/2}$, the sequence $\{Df(z,x^k)\}$ converges uniformly on ∂G_ρ to $Df(z,x)$. It now follows that it converges uniformly in G_ρ. We now change to the variable z in equations (29.5) and (29.6) by setting $w=f(z,x^k)$ and $w=f(z,x)$, respectively. Then, if $\zeta_0(z,\mathfrak{z}^k)=\zeta[f(z,x^k),\mathfrak{z}^k]\times[Df(z,x^k)]^{-1}$ and $\xi_0(z)=\xi[f(z,x)][Df(z,x)]^{-1}$, we find that

$$\begin{aligned}&\sqrt{P(z,x^k)}\,z\frac{d}{dz}\left[\sqrt{P(z,x^k)}\,\zeta_0(z,\mathfrak{z}^k)\right]\\&\qquad-\left(\frac{Df(z,x^k)}{f(z,x^k)}\right)^2\widetilde{Q}\,[f(z,x^k),\mathfrak{z}^k]-\widetilde{\mathscr{P}}(z,\mathfrak{z}^k)=0,\end{aligned}\tag{29.7}$$

and

$$(29.8)\quad \sqrt{P(z,x)}\, z\frac{d}{dz}\,[\sqrt{P(z,x)}\,\xi_0(z)] - \left(\frac{Df(z,x)}{f(z,x)}\right)^2 \widetilde{Q}\,[f(z,x)] - \widetilde{\mathfrak{P}}(z) = 0.$$

Since $\lim_{k\to\infty} \zeta_0(z,\mathfrak{z}^k) = \xi_0(z)$ uniformly in any disc $|z| \leqq r < 1$, and since $|P(z,x)| \leqq C_0$ for $z \in \partial G_\rho$, it is an easy matter, using equations (29.7) and (29.8), to show that the sequence $\{\zeta_0(z,\mathfrak{z}^k)\}$ converges uniformly in G_ρ to $\xi_0(z)$. In addition, (29.7) implies that $\zeta_0'(z,\mathfrak{z}^k) = O(1)$ for $z \in \partial G_\rho$ and $k \geqq k(\rho)$. Let $\eta(z,\mathfrak{z}^k) = \zeta[f(z,x^k),\mathfrak{z}^k]$ and $\eta(z) = \xi[f(z,x)]$. From the above it follows that the sequence $\{\eta(z,\mathfrak{z}^k)\}$ converges uniformly in G_ρ to $\eta(z)$. Recalling that in G_ρ we have $D^2f(z,x^k) = O(1)$ for $k \geqq k(\rho)$, we find that $\eta'(z,\mathfrak{z}^k) = O(1)$ for $k \geqq k(\rho)$ and $z \in \partial G_\rho$. Relation (21.4) is satisfied everywhere on the tree $\mathscr{T}_{x^k}$ except possibly at its vertex. If $w \in \mathscr{T}_{x^k}$ and $w = f(u,\mathfrak{z}^k) = f(v,\mathfrak{z}^k)$ where $|u| = |v| = 1$, then (21.4) is equivalent to

$$(29.9)\quad \operatorname{Re}\left\{\frac{\sqrt{P(u,x^k)}}{Df(u,x^k)}\,[\eta(u,\mathfrak{z}^k) - \eta(v,\mathfrak{z}^k)]\right\} = 0.$$

Set $I_{j\rho} = I_j \cap G_\rho$ and $I'_{j\rho} = I'_j \cap G_\rho$, $j = 1,\cdots,m$. Let γ_j be any branch of the tree $\mathscr{T}_x$; in view of the definition of the arcs I_j and I'_j, these arcs will be the images of γ_j under the mapping $z = f^{-1}(w,x)$. If z_0 is an interior point of $I_{j\rho}$, then $f(z_0,x) = w_0 \in \gamma_j$. This point corresponds to a point $\theta(z_0) \in I'_{j\rho}$ for which $f(\theta(z_0),x) = w_0$. In our previous notation, $z_0 = z_+$ and $\theta(z_0) = z_-$. Due to our choice of z_0, we can find a disc $K_0 = \{w : |w - w_0| \leqq h\}$ of small radius h which does not intersect the discs of the set $\mathscr{K}_\rho$ and which contains no zeros of $Q(w,x)$. For $k \geqq k_0(\epsilon)$ we have

$$|f(z_0,x^k) - f(z_0,x)| \leqq \epsilon, \qquad |f(\theta(z_0),x^k) - f(\theta(z_0),x)| \leqq \epsilon,$$

and so $f(z_0,x^k)$, $f(\theta(z_0),x^k) \in K_0$. The points $w_0^k = f(z_0,x^k)$ and $w_1^k = f(\theta(z_0),x^k)$ belong to the tree $\mathscr{T}_{x^k}$. We shall show that in fact they lie on the same branch. Suppose the opposite. The disc K_0 intersects some of the trajectories of the differential $Q(w,x^k)(dw/w)^2$, since they are the branches Γ_l^k of the graph of this differential. Each of these branches intersects K_0 in some arc $\Gamma_l^k \cap K_0$, which will be as close as desired to the arc $\gamma_j \cap K_0$ if k is sufficiently large. From the point w_0 we construct an orthogonal trajectory of the differential $Q(w,x^k)(dw/w)^2$. Then it will intersect all the arcs $\Gamma_l^k \cap K_0$, since $Q(w,x^k) \neq 0$ if $w \in K_0$. From this it follows that the arcs Γ_l^k will be bounded by arcs of the image domain, and moreover no two of these arcs can lie in one and the same component of the boundary of some part of the image domain. If the points w_0^k and w_1^k lie on different arcs, then the tree $\mathscr{T}_{x^k}$ would be disconnected, which is a contradiction. Thus we have shown that they both lie on some branch γ_{jk} of $\mathscr{T}_{x^k}$. We now show that on the arc $I'_{j\rho}$ there exists a point $\theta_k(z_0)$ such that $f(\theta_k(z_0),x^k) = w_0^k$. To prove this, note that the distance in the metric $ds = |Q^{1/2}(w,x^k)dw/w|$ between w_0^k and w_1^k is at most $C|w_1^k - w_0^k|$, where C is a constant independent of k, and so this distance is at most $2C\epsilon$. On the other hand, the distance from w_1^k to some point $w \in \gamma_{jk}$ can be determined via the integral

$$\int_{\theta(z_0)}^{z} |P^{1/2}(z, x^k)| \frac{dz}{iz}, \tag{29.10}$$

where the path is along the unit circle and where the upper limit is the image of w. Since on $I'_{j\rho}$ we have the estimate $P(z, x^k) \geqq C_0$, on $I'_{j\rho}$ there exist points z_1 and z_2 for which the integral (29.10) is greater than $2C\epsilon$ and less than $-2C\epsilon$ (since $\theta(z_0)$ is an interior point of $I'_{j\rho}$ and ϵ can be taken arbitrarily small). Thus there exists a point $\theta_k(z_0) \in I'_{j\rho}$ such that $f(\theta_k(z_0), x^k) = f(z_0, x^k)$. Next we prove that the sequence $\{\theta_k(z_0)\}$ converges. From this sequence we can extract a convergent subsequence $\{\theta_{k_l}(z_0)\}$. Since on $I'_{j\rho}$ we have $Df(z, x^k) = O(1)$ for $k \geqq k(\rho)$, it follows that $\lim_{l\to\infty} f(\theta_{k_l}(z_0), x^{k_l}) = f(\theta^*, x)$, where $\theta^* = \lim_{l\to\infty} \theta_{k_l}(z_0)$. But, on the other hand,

$$\lim_{l\to\infty} f(\theta_{k_l}(z_0), x^{k_l}) = \lim_{l\to\infty} f(z_0, x^{k_l}) = f(z_0, x).$$

Thus $\theta^* = \theta(z_0)$, and so $\lim_{k\to\infty} \theta_k(z_0) = \theta(z_0)$ exists. In (29.9) we set $u = z_0$ and $v = \theta_k(z_0)$. Taking the limit and recalling our earlier relation $\eta'(z, \mathfrak{z}^k) = O(1)$ for $k \geqq k(\rho)$, we obtain that

$$\operatorname{Re}\left\{\sqrt{\frac{P(z_0, x)}{Df(z_0, x)}}\,[\eta(z_0) - \eta[\theta(z_0)]]\right\} = 0.$$

Since ρ is arbitrary, this relation holds on the arc I_j, and so $\xi(w)$ satisfies condition (21.4), and therefore $\xi(w) \in \mathfrak{H}_x$. Thus $\|\zeta(w, \mathfrak{z})\|_x \leqq \|\xi(w)\|_x$, and so $\|\zeta(w, \mathfrak{z})\|_x \leqq \lim_{k\to\infty} \|\zeta(w, \mathfrak{z}^k)\|_{x^k}$. Q.E.D.

Lemma 29.3. *A point $\mathfrak{y}_0$ satisfying $D[\mathfrak{y}_0] = \mathfrak{D}[\mathfrak{y}_0]$ is a point of continuity of the functions $D[\mathfrak{y}]$ and $\mathfrak{D}[\mathfrak{y}]$.*

Proof. Let $m(\mathfrak{y})$ ($\mathfrak{m}(\mathfrak{y})$) and $M(\mathfrak{y})$ ($\mathfrak{M}(\mathfrak{y})$) be the upper and lower limits of $D[\mathfrak{y}]$ ($\mathfrak{D}[\mathfrak{y}]$) at $\mathfrak{y}$. Since $D[\mathfrak{y}] \leqq \mathfrak{D}[\mathfrak{y}]$, we have $m(\mathfrak{y}) \leqq \mathfrak{m}(\mathfrak{y})$ and $M(\mathfrak{y}) \leqq \mathfrak{M}(\mathfrak{y})$. Thus $D[\mathfrak{y}_0] = m(\mathfrak{y}_0) \leqq \mathfrak{m}(\mathfrak{y}_0) \leqq \mathfrak{M}(\mathfrak{y}_0) = \mathfrak{D}[\mathfrak{y}_0] = D[\mathfrak{y}_0]$, so $m(\mathfrak{y}_0) = M(\mathfrak{y}_0)$ and $\mathfrak{m}(\mathfrak{y}_0) = \mathfrak{M}(\mathfrak{y}_0)$. Q.E.D.

Without stopping to give a proof, we note that the following result also holds.

Lemma 29.4. *A point $\mathfrak{y}_0$ satisfying $D[\mathfrak{y}_0] < \mathfrak{D}[\mathfrak{y}_0]$ is a point of discontinuity of at least one of the functions $D[\mathfrak{y}]$ or $\mathfrak{D}[\mathfrak{y}]$.*

§30. To conclude this chapter we prove an inequality for the second variation which will, in the future, play an important role. In agreement with (23.2) and (23.1) we have for $\xi(w) \in \mathscr{A}$

$$\delta^2 L(f, \xi) = \operatorname{Re}\left\{-\frac{1}{\pi}\int_{|w|>1/4} \left|\frac{d\xi^*}{dw}\right|^2 d\sigma + \frac{1}{2\pi i}\int_{|w|=1/4} \xi^* \overline{d\xi^*} - \frac{1}{2} S\right\},$$

where

$$S = \sum_{k=1}^{n-1} \xi_k \sum_{l=1}^{n-k} (k+l+2) A_{k+l} \xi_l.$$

The following identity is easily verified:

(30.1) $$S = \sum_{k,\, l=1}^{\left[\frac{n}{2}\right]} (k+l+2) A_{k+l} \xi_k \xi_l + 2 \sum_{k<\frac{n}{2}} \xi_k \sum_{l>\frac{n}{2}}^{n-k} (k+l+2) A_{k+l} \xi_l.$$

Set

$$\sigma_l = \left\{ \sum_{k=1}^{l} |\xi_k|^2 \right\}^{1/2}, \quad \tau_l = \left\{ \sum_{k=l+1}^{n} |\xi_k|^2 \right\}^{1/2}, \quad \mu = \left\{ \sum_{1}^{n} |A_k|^2 \right\}^{1/2}.$$

Applying Schwarz' inequality to the right-hand side of (30.1) and using Proposition 10.1, we get that

(30.2) $$|S| < \alpha^n [\sigma_{n_0}^2 + \sigma_{n_0} t_{n_0}],$$

where $n_0 = [n/2]$ and α is an absolute constant; absolute constants will be denoted below by $\alpha_0, \alpha_1, \cdots$.

From (22.13) we have

(30.3) $$\operatorname{Re}\left\{ \frac{1}{2\pi i} \int_{|w|=1/4} \xi^* \overline{d\xi^*} \right\} < -\frac{1}{2} J^2 + \alpha_0^n \sigma_{n_0}^2.$$

Let r_0 be a quantity such that $1/8 < r_0 < 1/4$. By a theorem of Hardy,

$$\frac{1}{2\pi} \int_{|w|=r_0} | w^n Q(w) \xi_0^2(w) | \left| \frac{dw}{w} \right| < \frac{1}{2\pi} \int_{|w|=1/4} | w^n Q(w) \xi_0^2(w) | \left| \frac{dw}{w} \right|$$
$$= \frac{4^{-n-2}}{2\pi} \int_{|w|=1/4} | \xi_0^*(w) |^2 \left| \frac{dw}{w} \right|.$$

Thus from Proposition 22.1 we get

$$\frac{1}{2\pi} \int_{|w|=r_0} | w^n Q(w) \xi_0^2(w) | \left| \frac{dw}{w} \right| < 4^{-n-2} B_2^2 J^2.$$

Choose r_0 as in Proposition 9.6. Then

$$\frac{1}{2\pi} \int_{|w|=r_0} | w^n Q(w) \xi_0^2(w) | \left| \frac{dw}{w} \right| \geqslant \alpha_1^{-n} M \frac{1}{2\pi} \int_{|w|=r_0} | \xi_0(w) |^2 \left| \frac{dw}{w} \right|,$$

where $M = \max_{|w|=r_0} |w^n Q(w)|$. We saw earlier that $M \geqq \alpha_2^n \mu$. Thus, by Proposition 10.1, $M \geqq \alpha_3^n$. From the above inequalities it follows that

$$\frac{1}{2\pi} \int_{|w|=r_0} | \xi_0(w) |^2 \left| \frac{dw}{w} \right| < \alpha_4^n J^2,$$

and so

$$\sum_{n^*}^{n} |\xi_k|^2 r_0^{2k+2} < \alpha_4^n J^2.$$

Since $n^* \leqq [n/2]+1$ and $r_0 \geqq 1/8$, we have $t_{n_0} < \alpha_5^n J$. From the latter inequality and from (30.2) and (30.3) it is easy to see that

$$\delta^2 L(f, \xi) < \alpha_6^n \sigma_{n_0}^2 - \alpha_7^n \sigma_n^2. \tag{30.4}$$

The function $\xi[f(z)]$ is the first variation of the extremal function $f(z)$. In our previous notation we have $\xi[f(z)] = \sum_1^\infty \delta a_k z^{k+1}$. Also, the inequality

$$\alpha_8^{-n} \sum_1^r |\delta a_k|^2 < \sum_1^r |\xi_k|^2 < \alpha_8^n \sum_1^r |\delta a_k|^2, \qquad r = 1, \ldots, n, \tag{30.5}$$

is easily verified. Indeed, the quantities $\xi_1, \cdots, \xi_n$ and $\delta a_1, \cdots, \delta a_n$ are connected by the mappings (19.19) and (19.20), which we reproduce here for the sake of the reader:

$$\delta a_k = \sum_{l=1}^{k} a_k^{(l+1)} \xi_l, \qquad \xi_k = \sum_{l=1}^{k} b_k^{(l+1)} \delta a_l.$$

In the course of proving Proposition 10.1 we showed that

$$|a_k^{(l+1)}| < \alpha_3^{k+l+2} \quad \text{and} \quad |b_k^{(l+1)}| < \alpha_3^{k+l+2}.$$

The inequalities (30.5) follow easily from these estimates.

Lemma 30.1. *If $\xi \in \mathscr{A}$ and the vector λ is normalized, then the second variation satisfies*

$$\delta^2 L(f, \xi) \leqslant \alpha^n \sum_1^{[\frac{n}{2}]} |\delta a_k|^2 - \beta^n \sum_{[\frac{n}{2}]+1}^{n} |\delta a_k|^2, \tag{30.6}$$

where α and β are absolute constants.

This inequality is a trivial consequence of (30.4) and (30.5).

Corollary 30.1. *Inequality* (30.6) *holds for the functions $D[f,t]$ and $\mathfrak{D}[f,t]$.*

We need only prove that (30.6) holds for $\mathfrak{D}[f,t]$. We determine the second variation for $\xi \in \mathfrak{H}$ via formula (23.2). However, inequality (30.4) is an immediate consequence of (23.2), and since in the course of the proof we never used the fact that $\xi \in \mathscr{A}$, it follows that (30.6) holds for $\mathfrak{D}[f,t]$.

It is clear that the second variation cannot be bounded from below, even if we fix the first variations δa_k, $k = 1, \cdots, n$. The following lemma will shed some light on this question.

Lemma 30.2. *Suppose that $d_{n-1/2}(f) \neq 0$ for the extremal function $f(z)$. For any quantities t_k, $k = 1, \cdots, n$, satisfying $\operatorname{Re} \sum_1^n \bar{\lambda}_k t_k = 0$ there exists a first* *variation of* $f(z)$ *such that* $\delta f(z)=\xi[f(z)]=\sum_1^n t_k z^{k+1}+\cdots$ *and* $\delta^2 L(f,\xi)> -\alpha_*^n d_{n-1/2}^{-2}(f)\sum_1^n|t_k|^2$, *where* α_* *is an absolute constant.*

Proof. We define via Lemma 3.4 a measure $\nu(dz)$ with support $\mathscr{L}=\{z:|z|=r_*\}$ and such that the first variation $\delta f(z;\nu)$ (with the constant $C_1=0$!) satisfies $\delta f(z;\nu)=\sum_1^n t_k z^{k+1}+\cdots$. Set $\xi(w)=\delta f(z;\nu)$ and $z=f^{-1}(w)$. Since $\nu(dz)=\eta(z)\,dz/2\pi iz$, it follows from (3.22) and (4.18) that

$$\delta^2 L(f,\xi)>-d_{n-1/2}^{-2}(f)\,[\max_{|z|=|t|=r_*}|\mathfrak{A}(z,t)|+\max_{|z|=|t|=r_*}|\mathfrak{B}(z,\bar{t})|]\sum_1^n|t_k|^2.$$

Applying Proposition 5.1 to the functions $f(z)$ and $g(z)=z(1-z)^{-2}$, $\lambda(y)=(1,0,\cdots,0)$ and recalling that r_* is an absolute constant, we can show quite easily that the term in the square brackets on the right-hand side of the last inequality is at most α_*^n. Q.E.D.

Corollary 30.1 and this last lemma imply

Lemma 30.3. *If* $d_{n-1/2}(f)\neq 0$, *then*

$$-\alpha_*^n d_{n-1/2}^{-2}(f)\sum_1^n|t_k|^2\leqslant D[f,t]\leqslant\mathfrak{D}[f,t]\leqslant\alpha^n\sum_1^{[\frac{n}{2}]}|t_k|^2-\beta^n\sum_{[\frac{n}{2}]+1}^n|t_k|^2.$$

(30.7)

CHAPTER 5

ON THE SMOOTHNESS OF THE BOUNDARY OF THE COEFFICIENT REGION

In §12 we introduced for any point $x^0 \in \mathscr{R}$ the local coordinates (y, τ) and showed that on the disc $\mathfrak{U} = \{\mathfrak{y} : \mathfrak{y} = (y, 0),\ |y| < R, R \leq R_0\}$ there exists a continuously differentiable function $\psi(y)$ whose graph, the set $\mathfrak{W}$, is a neighborhood of the point x^0 in $\partial\mathscr{V}_n$. In other words, in local coordinates the equation of the boundary of $\mathscr{V}_n$ in a neighborhood of x^0 has the form $\tau = \psi(y)$. In the theory of variational problems for functionals of the form $J[f] = F(x)$, an important question is whether the second differential of such a function $\psi(y)$ exists. If at the point y the second differential of $\psi(y)$ exists, it is almost obvious geometrically that it is very closely connected with the functions $D[x,t]$ and $\mathfrak{D}[x,t]$ introduced in the previous chapter. We shall explore this connection on the basis of the theory developed below.

§31. Our notation will be the same as the notation of §12. We shall revert to the usual notation of round brackets for the scalar product, as opposed to the notation used in the last section. Consider the operator $\operatorname{grad}\psi(y) = (\partial\psi/\partial y_1, \cdots, \partial\psi/\partial y_{2n-1})$, and set $|\operatorname{grad}\psi| = [\sum_1^{2n-1}(\partial\psi/\partial y_k)^2]^{1/2}$. With $v = (v_1, \cdots, v_{2n-1}) \in R^{2n-1}$ arbitrary, set

$$(\operatorname{grad} \psi(y),\ v) = \sum_1^{2n-1} \frac{\partial \psi(y)}{\partial y_k} v_k.$$

We now introduce upper and lower second order derivatives of the function $\psi(y)$ with respect to the vector v; the upper derivative is defined to be

$$\overline{D}_v^2 = \overline{\lim_{\varepsilon \to +0}} \frac{2}{\varepsilon^2} [\psi(y + \varepsilon v) - \psi(y) - \varepsilon(\operatorname{grad}\psi(y), v)],$$

and the lower derivative is defined to be

$$\underline{D}_v^2 = \underline{\lim}_{\varepsilon \to +0} \frac{2}{\varepsilon^2} [\psi(y + \varepsilon v) - \psi(y) - \varepsilon(\operatorname{grad}\psi(y), v)].$$

This is the so-called second derivative of de la Vallée-Poussin. If $\underline{D}_v^2\psi(y) = \overline{D}_v^2\psi(y)$, then $\psi(y)$ has at y a second de la Vallée-Poussin derivative in the direction v, and this common value is denoted by $D_v^2\psi(y)$. We note that, as was shown by Denjoy, for a function of one variable the second de la Vallée-Poussin derivative always exists on a set of positive measure, regardless of whether or not the ordinary second derivative exists anywhere. Let $(y, 0)$ be any point of the disc $\mathfrak{U}$, and let $x \in \partial\mathscr{V}_n$ be its image:

$$x = x^0 + U\mathfrak{y}, \qquad \mathfrak{y} = (y, \psi(y)). \tag{31.1}$$

The normal vector at $(y,0)$ to the graph of $\psi(y)$ will be defined by the formula $\mu(y) = (|\operatorname{grad}\psi(y)|^2+1)^{-1/2}(-\partial\psi/\partial y_1, \cdots, -\partial\psi/\partial y_{2n-1}, 1)$, or, in a more compact form, by

$$\mu(y) = (|\operatorname{grad}\psi(y)|^2+1)^{-1/2}(-\operatorname{grad}\psi(y), 1). \tag{31.2}$$

If t is any tangent vector to $\partial\mathscr{V}_n$ at x, then in the local coordinate system this vector has the form (v,σ), where $v=(v_1,\cdots,v_{2n-1})$ and $\sigma = (\operatorname{grad}\psi(y), v)$. The connection between these vectors is given by (12.3):

$$(v,\sigma) = U't. \tag{31.3}$$

By $\mathfrak{Y}^0$ we denote the set of all vectors tangent to the space $\mathfrak{W}$. In the local coordinate system $\mathfrak{Y}^0$ is the set of pairs (y,v), where $y=(y_1,\cdots,y_{2n-1})$, $|y|<R$, and $v\in R^{2n-1}$ is arbitrary. Since the functions $D[x,t]$ and $\mathfrak{D}[x,t]$ are quadratic forms in the coordinates of the vector t, we see upon considering the restriction of these functions to $\mathfrak{D}^0$ that $D[x,t] = (\widetilde{D}(y)v, v)$, where $\widetilde{D}(y)$ is a symmetric matrix, where v is the vector connected with t by relation (31.3), and where y is connected with x by (31.1). In the same way, $\mathfrak{D}[x,t] = (\widetilde{\mathfrak{D}}(y)v,v)$. We set

$$(D(y)v,v) = (|\operatorname{grad}\psi(y)|^2+1)^{1/2}(\widetilde{D}(y)v,v),$$

and similarly $(\mathfrak{D}(y)v,v) = (|\operatorname{grad}\psi(y)|^2+1)^{1/2}(\widetilde{\mathfrak{D}}(y)v,v)$.

PROPOSITION 31.1. *For any point* $(y,v)\in\mathfrak{Y}^0$,

$$(D(y)v,v) \leqq D_v^2\psi(y). \tag{31.4}$$

PROOF. Corresponding to the vectors y and v we determine the vector t via (31.3), and via (31.1) we determine the image x of the point $(y,0)\in\mathfrak{U}$, so $x\in\partial\mathscr{V}_n$. Consider any variation of the function $f(z,x)$ as defined by an admissible function (2.2). Let the measure $\nu(dz)$ and the constant c_1 be such that the first coefficient variations satisfy

$$\delta a_l[f(z,x);\nu,c_1] = t_l, \qquad l=1,\cdots,n.$$

Denote the variation of $f(z,x)$ by $f_*(z,x;\epsilon)$, and set $f_*(z,x;\epsilon) = \sum_0^\infty a_l^*(x,\epsilon)z^{l+1}$, $a_0^*(x,\epsilon)=1$. For $|\epsilon|\leqq\epsilon_0$ we can expand $a_l(x,\epsilon)$ in a convergent power series in ϵ. If $f(z,x) = \sum_0^\infty a_l(x)z^{l+1}$, $a_0(x)=1$, then

$$a_l^*(x,\varepsilon) = a_l(x) + \varepsilon t_l + \frac{\varepsilon^2}{2}\delta^2 a_l(x;\nu,c_1) + \ldots, \quad l=1,\ldots,n. \tag{31.5}$$

We set $x^*(\epsilon) = (a_1^*(x,\epsilon),\cdots,a_n^*(x,\epsilon))$ and we determine y^* and τ^* from the relation $(y^*,\tau^*) = U(x^*(\epsilon)-x^0)$. If ϵ is sufficiently small, then $|y^*| < R$. Since $x^*(\epsilon)\in\mathscr{V}_n$, we have $\tau^*-\psi(y^*)\leqq 0$ and so

$$\tau^* - \tau - (\operatorname{grad}\psi(y),\ y^*-y) - [\psi(y^*) - \psi(y) - (\operatorname{grad}\psi(y),\ y^*-y)] \leqslant 0.$$

But

$$(|\operatorname{grad}\psi(y)|^2+1)^{-1/2}\,[\tau^*-\tau-(\operatorname{grad}\psi(y),\ y^*-y)]=(\mu(y),\ \mathfrak{y}^*-\mathfrak{y}),$$

where $\mathfrak{y}^*=(y^*,\tau^*)$ and $\mathfrak{y}=(y,\tau)$. Since $(\mu(y),\mathfrak{y}^*-\mathfrak{y})=(U'\lambda(x),\mathfrak{y}^*-\mathfrak{y})$ $=(\lambda(x),x^*(\epsilon)-x^0)$, we have

$$[\,|\operatorname{grad}\psi(y)|^2+1]^{-1/2}\,[\tau^*-\tau-(\operatorname{grad}\psi(y),\ y^*-y)]$$
$$=\operatorname{Re}\sum_1^n\bar{\lambda}_k(x)\,(\overset{*}{a}_k(x,\varepsilon)-a_k(x))=\frac{\varepsilon^2}{2}\delta^2L\,[f(z,x);\ \nu_1,\ c_1]+O(\varepsilon^3).$$

On the other hand, it follows from (31.5) that

$$y^*=y+\varepsilon v+\frac{\varepsilon^2}{2}\tilde{y}(\varepsilon).$$

Therefore, in view of the continuous differentiability of $\psi(y)$,

$$\psi(y^*)-\psi(y)-(\operatorname{grad}\psi(y),\ y^*-y)=\psi(y+\varepsilon v)-\psi(y)$$
$$-\varepsilon\,(\operatorname{grad}\psi(y),\ v)+o(\varepsilon^2).$$

Hence

$$\frac{\varepsilon^2}{2}(|\operatorname{grad}\psi(y)|^2+1)^{1/2}\delta^2L\,[f(z,x);\ \nu,\ C_1]-[\psi(y+\varepsilon v)-\psi(y)$$
$$-\varepsilon\,(\operatorname{grad}\psi(y),\ v)]+o(\varepsilon^2)\leqslant 0.$$

From this we deduce the inequality

$$(|\operatorname{grad}\psi(y)|^2+1)^{1/2}\,\delta^2L\,[f(z,x);\ \nu,\ c_1)]\leqslant\underline{D}_v^2\,\psi(y),$$

and (31.4) now follows from this. Q.E.D.

Now it is not too difficult to understand the significance of the function $(D(y)v,v)$. In fact, the equality sign in (31.4) holds on a set of full measure in $|y|\leqq R$. However, before presenting a proof of this fact, we must first establish some auxiliary results.

Consider the set $\mathscr{R}\cap\{x:x\in\partial\mathscr{V}_n,d_{n-1}(x)=0\}$. If x belongs to this set, then $\lambda_n(x)=0$. We prove this fact by contradiction. If $\lambda_n(x)\neq 0$, then for $f(z,x)$ there exist two non-collinear associated vectors $\lambda(x)$ and $\lambda^1(x)$. It then follows that $d_{n-1/2}(x)=0$, which is a contradiction, and so $\lambda_n(x)$ must be zero. Now, if $x\in\partial\mathscr{V}_n$ and $\lambda_n(x)=0$, then $f(z,x)\in S_{n-1}$. Set $T_{n-1}=$ $\{x:x\in\partial\mathscr{V}_n,f(z,x)\in S_{n-1}\}$, so that $\mathscr{R}\cap T_{n-1}=\{x:x\in\partial\mathscr{V}_n,d_{n-1}(x)=0\}$. Denote by $\mathfrak{U}_{n-1}$ the projection of the set $\mathfrak{W}\cap T_{n-1}$ onto the disc $\mathfrak{U}$.

Proposition 31.2. *The set* $\mathfrak{U}_{n-1}$ *has Lebesgue measure zero.*

Proof. We begin with the following remarks. For any point $x^0\in\mathscr{R}$ we can find a neighborhood $V=\{x:|x-x^0|\leqq\rho\}$ such that, given $\epsilon>0$ sufficiently small, there exists an ϵ-covering of V containing at most $C_\rho\epsilon^{-2n+1}$ sets, where the constant C_ρ is independent of ϵ. In order to prove this, determine ρ such that $V\subseteq\mathfrak{W}$. Let V_0 be the projection of V, $V_0\subseteq\mathfrak{U}$. Set $C_0=$ $\max_{|y|\leqq R}|\operatorname{grad}\psi(y)|$, $\epsilon_0=\epsilon(C_0^2+1)^{1/2}$. Let $\alpha_1^0,\cdots,\alpha_N^0$ be an ϵ_0-covering of V_0. It is clear that we can construct this covering so that $N\leqq C_\rho\epsilon^{-2n+1}$. Denote the image of α_k^0 by α_k. Clearly $\bigcup_k \alpha_k = V$. We shall show that the diameter of α_k is at most ϵ. To prove this fact, let $x^1, x^2 \in \alpha_k$, and let $(y^1, 0)$, $(y^2, 0)$ be their projections on $\mathfrak{U}$; we set $\mathfrak{y}^j = (y^j, \psi(y^j)), j = 1, 2$. Since $|x^1 - x^2| = |\mathfrak{y}^1 - \mathfrak{y}^2| = [|y^1 - y^2|^2 + [\psi(y^1) - \psi(y^2)]^2]^{1/2}$, we have $|x^1 - x^2| \leq \epsilon_0 \sqrt{C_0^2 + 1} = \epsilon$.

We now pass to the construction of an ϵ-covering for the set $\mathfrak{U}_{n-1}$, and we must also estimate the cardinality of the covering. We project $\mathscr{V}_n$ orthogonally onto the coordinate plane $P^{2n-2} = \{x : x_{2n-1} = x_{2n} = 0\}$. This projection can be considered as the realization of $\mathscr{V}_{n-1}$, and we identify it with $\mathscr{V}_{n-1}$. The projection of the set T_{n-1} is $\partial \mathscr{V}_{n-1}$, and moreover the projection establishes a one-to-one correspondence between these two sets. We shall show that in fact this correspondence is a homeomorphism. Denoting the projection of $x \in \mathscr{V}_n$ by $\hat{x}$, we note that

$$x_{2n-1} + ix_{2n} = a_n = \frac{1}{2\pi i} \int_{|z|=1/2} f(z, \hat{x}) \frac{dz}{z^{n+2}}.$$

Thus $x_{2n-1} + ix_{2n}$ is a continuous function on $\partial \mathscr{V}_{n-1}$, and so x is a continuous function on $\partial \mathscr{V}_{n-1}$, which in turn implies that T_{n-1} and $\partial \mathscr{V}_{n-1}$ are homeomorphic. We now consider the projection of the closed set $\mathfrak{W} \cap T_{n-1}$, and we denote this projection by $\hat{\mathfrak{W}}$. According to the previous remarks, for each point $\hat{x} \in \hat{\mathfrak{W}}$ there exists a neighborhood $\hat{V}$ admitting an ϵ-covering with cardinality $O(\epsilon^{-2n+3})$. Thus there exists an ϵ-covering $\hat{\alpha}_1, \cdots, \hat{\alpha}_M$ of the set $\hat{\mathfrak{W}}$ such that $M \leq C(\hat{\mathfrak{W}})\epsilon^{-2n+3}$. No matter what the $\hat{\alpha}_k$, $k = 1, \cdots, M$, happen to be, the variation of the function $x_{2n-1} + ix_{2n}$ on α_k does not exceed some quantity h, where $h \to 0$ when $\epsilon \to 0$. Therefore the image $\hat{\alpha}_k$ admits a second ϵ-covering with cardinality at most $C(h/\epsilon)^2$, where C is independent of both ϵ and k. Hence the set $\mathfrak{W} \cap T_{n-1}$ has a second ϵ-covering $\alpha_1, \cdots, \alpha_N$, where $N < C(\mathfrak{W})\epsilon^{-2n+1}h^2$.

The covering defined above determines, via the projection of the sets α_k onto the disc $\mathfrak{U}$, a covering of the sets $\mathfrak{U}_{n-1}$. Let α_k^0 be the projection of α_k, and let $\delta(\alpha_k^0)$ be the diameter of α_k^0. By construction $\delta(\alpha_k^0) \leq 2\epsilon$, and so we have

$$\sum_1^N [\delta(\alpha_k^0)]^{2n-1} < 2^{2n-1} C(\mathfrak{W}) h^2.$$

Denote by $m_p(X)$ the p-dimensional Hausdorff measure of a set X. Since $h \to 0$ when $\epsilon \to 0$, the latter inequality implies that $m_{2n-1}(\mathfrak{U}_{n-1}) = 0$. As is well known, it now follows from this fact that the Lebesgue measure of $\mathfrak{U}_{n-1}$ is zero. This completes the proof.

We shall next consider the set of trees $\{\mathscr{T}_x : x \in \mathscr{R}\}$. For each tree of this set, the index of each of its vertices has been unambiguously defined.

DEFINITION 31.1. The trees $\mathscr{T}_{x^1}$ and $\mathscr{T}_{x^2}$, where $x^1, x^2 \in \mathscr{R}$, are called *strongly isomorphic* if corresponding vertices have equal indices and if the bases of the trees correspond to each other.

It is clear that the concept of strong isomorphism defines an equivalence relation, since it is obviously reflexive, symmetric, and transitive. Thus the set of all trees corresponding to extremal functions $f(z,x)$, $x \in \mathscr{R}$, is partitioned into equivalence classes. Clearly the number of such classes is finite, and for each tree $\mathscr{T}_x$, $x \in \partial \mathscr{V}_n$, the number of branches does not exceed some constant depending only on n. Denote these equivalence classes by $\mathfrak{D}_1, \cdots, \mathfrak{D}_N$; then $\{\mathscr{T}_x : x \in \mathscr{R}\} = \bigcup_1^N \mathfrak{D}_j$. Each class $\mathfrak{D}_j$ determines in a unique way a set $O_j \in \mathscr{R}$, where $O_j = \{x : \mathscr{T}_x \in \mathfrak{D}_j\}$. The sets $O_1, \cdots, O_N$ are pairwise disjoint, and $\mathscr{R} = \bigcup_1^N O_j$. Let x^1 be any point of O_j, and let $V(\epsilon) = \{x : |x - x^1| < \epsilon\}$ be a neighborhood of x^1. For sufficiently small ϵ we have $V(\epsilon) \subset \mathscr{R}$.

Proposition 31.3. *If ϵ is sufficiently small, then $O_j \cap V(\epsilon) \subseteq \mathfrak{D}_{x^1}$.*

Proof. Given the tree $\mathscr{T}_{x^1}$, we construct the system of discs $\mathscr{K}_\rho$ defined in §16, where we choose ρ sufficiently small. Let $x \in O_j \cap V(\epsilon)$. In each disc $K_{j\rho}$ the function $Q(w,x)$ has only as many zeros as the multiplicity of the zero $w_j(x^1)$. By Proposition 16.7, all the vertices of the tree $\mathscr{T}_{x^1}$ lie inside the discs of the system $\mathscr{K}_\rho$. If in a particular disc $K_{j\rho}$ the function $Q(w,x)$ had at least two distinct zeros, then the sum of the indices of the vertices of the tree $\mathscr{T}_x$ would be less than the corresponding sum for $\mathscr{T}_{x^1}$, which is clearly impossible. Hence in each disc $K_{j\rho}$ there is one zero of $Q(w,x)$ of the same multiplicity as $w_j(x^1)$. Therefore the vertex $w_j(x^1)$ of the tree $\mathscr{T}_{x^1}$ corresponds to the vertex $w_j(x)$ (of the tree $\mathscr{T}_x$) which lies in the disc $K_{j\rho}$ and which has the same index. We shall show that the degree, or order, of the vertices $w_j(x)$ and $w_j(x^1)$ is the same. By the construction of the system $\mathscr{K}_\rho$, the circumference $\partial K_{j\rho}$ intersects each branch of $\mathscr{T}_{x^1}$ emanating from the vertex $w_j(x^1)$, exactly once. From Proposition 16.7 it thus follows that the degree of the vertex $w_j(x^1) \in \mathscr{T}_{x^1}$ is not greater than the degree of the vertex $w_j(x) \in \mathscr{T}_x$. Since we know that the sum of the degrees of the vertices of $\mathscr{T}_{x^1}$ is in fact the same as the sum for $\mathscr{T}_x$, it follows immediately that the degrees of $w_j(x^1)$ and $w_j(x)$ are also the same. Hence we have shown that $\mathscr{T}_x$ satisfies all the requirements of Definition 16.1, and so $x \in \mathfrak{D}_{x^1}$. In other words, $O_j \cap V(\epsilon) \subseteq \mathfrak{D}_{x^1}$. Q.E.D.

For the convenience of the reader we now state a theorem of A. N. Kolmogorov, I. Ja. Verčenko, and F. Roger. Suppose that we are given a set E in the space R^p. We say that the contingency of the set E at the point $\alpha \in E$ is the whole space if it contains all rays emanating from α. Analogously, we say that the contingency of the set E at α is a half-space if there exists a hyperplane Π passing through α such that the contingency consists of all those rays emanating from α and passing on the same "side" of Π. The hyperplane Π is called the extreme tangential hyperplane.

Theorem. *Let $\hat{E}$ be a subset of E with the property that at any point of $\hat{E}$ the contingency of E is not the whole space. Then $\hat{E}$ is the sum of either finitely or countably many sets of finite $(p-1)$-dimensional Hausdorff measure. At* *each point of $\hat{E}$, with the possible exception of a set of $(p-1)$-dimensional Hausdorff measure zero, either the set E has a tangent hyperplane, or the contingency of E is a half-space.*

This theorem was proved in the case $p=2$ by Kolmogorov and Verčenko [**10, 11**]. It was generalized to other values of p by Roger [**12, 13**].

To apply this theorem to our problem, we denote by $\mathfrak{Q}_j$ the projection of the set $O_j \cap \mathfrak{W}$ onto the disc $\mathfrak{U}$; by construction we have $\mathfrak{U} = \bigcup_j \mathfrak{Q}_j$. The set $\mathfrak{Q}_j$ is a subset of the $(2n-1)$-dimensional tangent space to $\partial \mathscr{V}_n$ at x^0. Let $\hat{\mathfrak{Q}}_j$ be the subset of $\mathfrak{Q}_j$ such that at each point of $\hat{\mathfrak{Q}}_j$ the contingency of $\mathfrak{Q}_j$ is not the whole space. From the above theorem it follows that $\hat{\mathfrak{Q}} = \bigcup_j \hat{\mathfrak{Q}}_j$ is the sum of at most countably many sets of finite $(2n-2)$-dimensional measure. Hence, as a subset of the disc $\mathfrak{U}$, $\hat{\mathfrak{Q}}$ has Lebesgue measure zero, and therefore, in accord with Proposition 31.2, the set $\mathscr{S} = \hat{\mathfrak{Q}} \cup \mathfrak{U}_{n-1}$ has Lebesgue measure zero.

Proposition 31.4. *For any vector v, we have on the set $\mathfrak{U} \setminus \mathscr{S}$ that*

$$D_v^2 \psi(y) = (D(y)v, v). \tag{31.6}$$

Proof. Let v be any normalized vector, and let $(y,0) \in \mathfrak{U} \setminus \mathscr{S}$. Then $(y,0) \in \mathfrak{Q}_j$, and by the construction of the set $\mathscr{S}$ the contingency of $\mathfrak{Q}_j$ at $(y,0)$ is the whole space. Therefore there exists a sequence $\{y^l\}$, $(y^l,0) \in \mathfrak{Q}_j$, $l=1,2,\cdots$, such that $(y^l-y)|y^l-y|^{-1} = v^l \to v$ as $l \to \infty$. Set $x^l = x^0 + U(y^l, \psi(y^l))$, $l=1,2,\cdots$, and $x = x^0 + U(y, \psi(y))$. Without loss of generality we assume that $\sup_l |y^l - y|$ is sufficiently small, and then by Proposition 31.3 we have $\{x^l\} \subset \mathfrak{D}_x$. Since $\sim \mathfrak{U}_{n-1} \subset \mathscr{S}$, we have $d_{n-1}(x) \neq 0$, and so we can apply Lemma 20.1 to the sequence $\{f(z,x^l)\}$. We set $f(z,x^l) = \sum_{k=0}^{\infty} a_k^l z^{k+1}$, $f(z,x) = \sum_0^\infty a_k z^{k+1}$. With the aid of the relation

$$\operatorname{Re} \sum_1^n \bar{\lambda}_k(x)(a_k^l - a_k) = [\,|\operatorname{grad}\psi(y)|^2 + 1]^{-1/2}\, [\psi(y^l) - \psi(y) - (\operatorname{grad}\psi(y), y^l - y)]$$

we get, via Lemma 20.1, that

$$\frac{1}{2}\delta^2 L[f(z,x), \zeta] = [\,|\operatorname{grad}\psi(y)|^2 + 1]^{-1/2} \lim_{l_m \to \infty} |x^{l_m} - x|^{-2} [\psi(y^{l_m}) - \psi(y) - (\operatorname{grad}\psi(y), y^{l_m} - y)].$$

We shall now reformulate the right-hand side of this relation. If $h_l = |y^l - y|$, then $y^l = y + h_l v + h_l(v^l - v)$. Thus

$$\psi(y^l) - \psi(y) - (\operatorname{grad}\psi(y), y^l - y) = \psi(y + h_l v) - \psi(y) - h_l(\operatorname{grad}\psi(y), v) + \Psi_l,$$

where

$$\Psi_l = \psi(y + h_l v + h_l(v^l - v)) - \psi(y + h_l v) - h_l(\operatorname{grad}\psi(y), v^l - v).$$

According to Lemma 11.1 the vector $\lambda(x)$ satisfies a Lipschitz condition in a sufficiently small neighborhood of any point x^* for which $d_{n-1}(x^*) \neq 0$. If $(y^*, 0)$ is the projection of x^* onto the disc $\mathfrak{U}$, then the vector $\mu(y)$ satisfies a Lipschitz condition in a sufficiently small neighborhood of y^*, and so grad $\psi(y)$ also satisfies a Lipschitz condition in such a neighborhood. Recalling that $d_{n-1}(x) \neq 0$, we get $\Psi_l = O(h_l^2 |v^l - v|)$. Since $|x^l - x|^2 = h_l^2(1 + \sigma^2 + o(1))$, where $\sigma = (\operatorname{grad} \psi(y), v)$, we have

$$\lim_{l_m \to \infty} |x^{l_m} - x|^{-2} [\psi(y^{l_m}) - \psi(y) - (\operatorname{grad} \psi(y), y^{l_m} - y)]$$

$$= (\sigma^2 + 1)^{-1} \lim_{l_m \to \infty} h_{l_m}^{-2} (\psi(y + l_m v) - \psi(y) - h_{l_m} (\operatorname{grad} \psi(y), v)]$$

$$\geqslant \frac{1}{2} (\sigma^2 + 1)^{-1} \underline{D}_v^2 \psi(y).$$

Thus

$$[|\operatorname{grad} \psi(y)|^2 + 1]^{1/2} (\sigma^2 + 1)\, \delta^2 L[f(z, x), \zeta] \geqslant \underline{D}_v^2 \psi(y).$$

Next, using the vector v, we determine the vector t via (31.3). If

$$\tilde{t}_k = \lim_{l_m \to \infty} (a_k^{l_m} - a_k) |x^{l_m} - x|^{-1}, \qquad k = 1, \cdots, n,$$

and $\tilde{t} = (\tilde{t}_1, \cdots, \tilde{t}_n)$, then $t = (\sigma^2 + 1)^{1/2} \tilde{t}$. Therefore, if we set $\xi(w) = (\sigma^2 + 1)^{1/2} \zeta(w)$ and determine the vector $\mathfrak{z}$ by using the vector t in (19.20), we find that $\xi(w) \in \mathfrak{Z}$, where $\mathfrak{Z} \subset \mathscr{A}_x$ is the plane set defined in §26. Thus we have

$$(\sigma^2 + 1)\, \delta^2 L[f(z, x), \zeta] = \delta^2 L[f(z, x), \xi] \leqslant D(x, t),$$

and so $(D(y)v, v) \geq \underline{D}_v^2 \psi(y)$. This inequality and (31.4) imply that (31.6) holds for any normalized vector v, and then we finally conclude that (31.6) holds for an arbitrary vector v, since $(D(y)v, v)$ is a quadratic form of the components of v.

From this proposition and Lemma 11.1 it is not difficult to deduce

Lemma 31.1. *For any vector v, the second de la Vallée-Poussin derivative of $\psi(y)$ in the direction of v exists almost everywhere on $\mathfrak{U}$, and*

$$D_v^2 \psi(y) = (D(y)v, v). \tag{31.7}$$

The exceptional set of measure zero is independent of the vector v.

Proof. Let x^1 be the image of the point $(y^1, 0)$. If $d_{n-1}(x^1) \neq 0$, then grad $\psi(y)$ satisfies a Lipschitz condition in a small neighborhood of y^1, as follows from Lemma 11.1. Therefore in $\mathfrak{U} \setminus \mathfrak{U}_{n-1}$ we have

$$\varlimsup_{y^* \to y} |\operatorname{grad} \psi(y^*) - \operatorname{grad} \psi(y)|\, |y^* - y|^{-1} < \infty.$$

Since $\mathfrak{U}_{n-1}$ has measure zero, the theorem of Rademacher and Stepanov implies that grad $\psi(y)$ is differentiable almost everywhere. At any point of differentiability of grad $\psi(y)$, the second de la Vallée-Poussin derivative exists in the direction of any vector v. The statement of the lemma follows from this fact and Proposition 31.4.

§32. As is easily seen, Lemma 11.1 played an essential role in the preceding arguments. Now, based on the previous results, we can give a sharpened form of Lemma 11.1 itself. We first require an auxiliary result. We set

$$\Delta_{n-1/2} = \Delta_{n-1/2}(x^0,\mathfrak{U}) = \min_{x\in\mathfrak{W}} d_{n-1/2}(x), \; C_* = \max_{|y|\leq R} |\operatorname{grad} \psi(y)|$$

and

$$\mathcal{M} = \alpha_0^n (1 + C_*^2)^{1/2}(1 + \Delta_{n-1/2}^{-2}), \tag{32.1}$$

where α_0 is an absolute constant. Let T be any convex body contained in the disc $\mathfrak{U}$.

Proposition 32.1. *Suppose that in T the vector-valued function* $\operatorname{grad}\psi(y)$ *satisfies a Lipschitz condition with some constant $C < \infty$. Then for any $y^1, y^2 \in T$*

$$|\operatorname{grad} \psi (y^1) - \operatorname{grad} \psi (y^2)| < \mathcal{M}|y^1 - y^2|. \tag{32.2}$$

Proof. Let y^1 and y^2 be arbitrary interior points of the set T, and let $\mathcal{I}$ be a straight line segment joining these points. We construct a convex body T_0 such that the segment $\mathcal{I}$ lies inside T_0 and such that the distance from the boundary of T_0 to the boundary of T is > 0. It is easily verified that

$$\Psi(y, v, \varepsilon) = \psi(y + \varepsilon v) - \psi(y) - \varepsilon(\operatorname{grad}\psi(y), v)$$
$$= \int_0^\varepsilon (\operatorname{grad}\psi(y + \chi v) - \operatorname{grad}\psi(y), v)\, d\chi.$$

For any $y \in T_0$ we have $|(2/\epsilon^2)\Psi(y, v, \epsilon)| < C|v|^2$ provided $\epsilon \leq \epsilon_0$, and so, according to Lemma 31.1, the sequence $\{(2/\epsilon^2)\Psi(y, v, \epsilon)\}$ is bounded and converges almost everywhere on T_0. We then conclude from well-known theorems of analysis for functions $\Phi(y)$ integrable on T_0 that for any such integrable function

$$\lim_{\varepsilon\to+0} \frac{2}{\varepsilon^2} \int_{T_0} \Psi(y, v, \varepsilon)\Phi(y)\, dy = \int_{T_0} (D(y)v, v)\Phi(y)\, dy. \tag{32.3}$$

Let $K(y)$ be a nonnegative continuously differentiable function with support contained in the sphere $|y| \leq 1$. Set

$$K_\delta(y) = K\left(\frac{y}{\delta}\right)\left(\int_{|y|\leq\delta} K\left(\frac{y}{\delta}\right)dy\right)^{-1}.$$

If $F(y)$ is an integrable function for $|y| \leq R$, we denote by $F_\delta(y)$ its integral with respect to the weight function $K_\delta(y)$:

$$F_\delta(y) = \int_{|\mathfrak{z}|\leq\delta} K_\delta(\mathfrak{z}) F(\mathfrak{z} + y)\, d\mathfrak{z}.$$

For sufficiently small δ and $|y| < R$ this latter integral is well defined. From (32.3) it follows that

$$\lim_{\varepsilon\to+0} \frac{2}{\varepsilon^2} \Psi_\delta(y, v, \varepsilon) = (D_\delta(y) v, v).$$

But $\Psi_\delta(y, v, \epsilon) = \psi_\delta(y+\epsilon v) - \psi_\delta(y) - \epsilon(\operatorname{grad}\psi_\delta(y), v)$, and since $\psi_\delta(y)$ is twice continuously differentiable, we have that

$$\lim_{\varepsilon\to 0} \frac{2}{\varepsilon^2} \Psi_\delta(y, v, \varepsilon) = \sum_{k,\, l=1}^{2n-1} \frac{\partial^2 \psi_\delta(y)}{\partial y_k \partial y_l} v_k v_l,$$

and so

$$\sum_{k,\, l=1}^{2n-1} \frac{\partial^2 \psi_\delta(y)}{\partial y_k \partial y_l} v_k v_l = (D_\delta(y), v, v).$$

Denote by $d_{kl}(y)$ the elements of the matrix $D(y)$, and by $d_{kl,\delta}(y)$ its average in the sense defined above. In view of the arbitrariness of the vector v we have

$$\frac{\partial^2 \psi_\delta(y)}{\partial y_k \partial y_l} = d_{kl,\delta}(y) \qquad k, l = 1, \ldots, 2n-1. \tag{32.4}$$

Set $v^1 = (y^2 - y^1)|y^2 - y^1|^{-1}$, and let v^2, with $|v^2| = 1$, be arbitrary. Then (32.4) implies that

$$(\operatorname{grad}\psi_\delta(y^2) - \operatorname{grad}\psi_\delta(y^1),\ v^2) = \int_0^\varepsilon (D_\delta(y_1 + \chi v^1) v^1,\ v^2)\, d\chi, \tag{32.5}$$

where $\epsilon = |y^2 - y^1|$. We shall estimate the integrand. If v is arbitrary and if t is connected with v via relation (31.3), then $|t| \leqq (1 + |\operatorname{grad}\psi(y)|^2)^{1/2}|v| \leqq (1 + C_*^2)^{1/2}|v|$. From this and Lemma 30.3 we have

$$(D(y) v, v) \leqslant \frac{1}{3} \mathscr{M} |v|^2,$$

where $\mathscr{M}$ is as defined in (32.1). It then follows from the identity

$$(D(y) v^1, v^2) = \frac{1}{2} [(D(y) v^3, v^3) - (D(y) v^1, v^1) - (D(y) v^2, v^2)],$$

where $v^3 = v^1 + v^2$, that

$$|(D(y) v^1, v^2)| \leqq \mathscr{M}.$$

Now (32.5) implies that

$$|(\operatorname{grad}\psi_\delta(y^2) - \operatorname{grad}\psi_\delta(y^1), v^2)| \leqq \mathscr{M}|y^2 - y^1|.$$

Hence, recalling that v^2 is an arbitrary normalized vector,

$$|\operatorname{grad}\psi_\delta(y^2) - \operatorname{grad}\psi_\delta(y^1)| \leqq \mathscr{M}|y^2 - y^1|.$$

Allowing δ to go to 0, we see from the continuity of $\operatorname{grad}\psi(y)$ that (32.2) holds. This inequality has been proved when y^1 and y^2 are interior points of T, but since $\operatorname{grad}\psi(y)$ is continuous on T, the inequality also holds in the general case.

THEOREM 13. *The vector-valued function* $\operatorname{grad}\psi(y)$ *satisfies a Lipschitz condition on* $\mathfrak{U}$ *with constant* $\mathscr{M}$.

PROOF. In the course of the proof we shall always denote the projections of the points $x, x^1, x^2, \cdots$ by $y, y^1, y^2, \cdots$. Before beginning the proof, we make the following simple observation. If for $x^1, x^2 \in \mathfrak{W}$ we have $|\lambda(x^1) - \lambda(x^2)| \leqq C_0|x^1 - x^2|$, then $|\operatorname{grad}\psi(y^1) - \operatorname{grad}\psi(y^2)| \leqq C_1|y^1 - y^2|$, where the constant C_1 depends only on C_0. This is proved by noting that

$$|x^1 - x^2| = [|y^1 - y^2|^2 + (\psi(y^1) - \psi(y^2))^2]^{1/2} < (1 + C_*^2)^{1/2}|y^1 - y^2|,$$

and so $|\mu(y^1) - \mu(y^2)| = |\lambda(x^1) - \lambda(x^2)| < C_0(1 + C_*^2)^{1/2}|y^1 - y^2|$. The desired inequality now follows easily from formula (31.2).

We divide the disc $\mathfrak{U}$ into subsets $\mathfrak{U}^1, \mathfrak{U}^2, \cdots$ as follows. A point y is to belong to $\mathfrak{U}^k$ if $d_n(x) = \cdots = d_{n-k+1}(x) = 0$ but $d_{n-k}(x) \neq 0$. We note that $d_1(x) \neq 0$ on $\mathfrak{W}$. In fact, if $d_1(x^1) = 0$, then it is easily seen that $f(z, x^1) = z(1 - e^{i\theta}z)^{-2}$. But for the function $f(z, x^1)$ (this is called the Koebe function) we have $d_{n-1/2}(x^1) = 0$, and so $x^1 \notin \mathscr{R}$. Thus we have shown that there exists a smallest k_0, where $1 \leqq k_0 \leqq n-1$, such that $d_{n-k_0}(x) \neq 0$ for $x \in \mathfrak{W}$. Note that if $d_{n-k}(x) \neq 0$ and $d_{n-k+1}(x) = 0$, then $y \in \mathfrak{U}^k$, since from the condition $d_{n-1/2}(x) \neq 0$ it follows that $d_{n-k+2}(x) = \cdots = d_n(x) = 0$. Some of the sets $\mathfrak{U}^j$, $1 \leqq j \leqq k$, could be empty. Let $\mathfrak{U}^{j_1}, \cdots, \mathfrak{U}^{j_r}$, with $j_1 < j_2 < \cdots < j_r$, be those which are not empty. Clearly $j_1 = 1$, since $\mathfrak{U}^1 = \mathfrak{U} \setminus \mathfrak{U}_{n-1}$; and $j_r = k_0$. The sets $\mathfrak{U}^{j_k}$ are pairwise disjoint, and their union is $\mathfrak{U}$.

For convenience we shall call a convex body T a body of rank 1 if the set of interior points of T does not intersect the sets $\mathfrak{U}^{j_1}, \cdots, \mathfrak{U}^{j_r}$. A convex body T will be called a body of rank k, where $k > 1$, if the set of interior points of T does not intersect $\mathfrak{U}^{j_{k+1}}, \cdots, \mathfrak{U}^{j_r}$, and if it is not a body of rank $k-1$. Clearly the disc $\mathfrak{U}$ is a body of rank r. We shall show that if inequality (32.2) holds for any body of rank $\leqq k-1$, then it holds also for any body of rank k. We first note that it is sufficient to consider the case when T does not intersect $\mathfrak{U}^{j_{k+1}}, \cdots, \mathfrak{U}^{j_r}$, since the continuity of the function $\operatorname{grad}\psi(y)$ allows us to deduce the general case immediately from the above special case. Thus, by these remarks, we may assume that $d_{n-j_k}(x) \geqq C^* > 0$ for any point $y \in T$. Let $y^1, y^2 \in T$, and let $\mathscr{I}$ be the segment joining these points. We first consider the case when the set $\mathscr{E} = \mathscr{I} \cap \mathfrak{U}^{j_k}$ is not empty. Define the points y^3 and y^4 by the condition

$$|y^3 - y^1| = \inf_{y \in \mathscr{E}} |y - y^1|, \quad |y^4 - y^2| = \inf_{y \in \mathscr{E}} |y - y^2|.$$

Clearly $y^3, y^4 \in \mathfrak{U}^{j_k}$, since $d_{n-j_k}(y^l) \neq 0$ and $d_{n-j_k+1}(y^l) = 0$, $l = 3, 4$. Therefore $f(z, x^3), f(z, x^4) \in S_{n-j_k+1}$. We shall apply Lemma 11.1 to these functions. Since $d_{n-j_k+1/2}(x) \geqq d_{n-1/2}(x)$, we have $|\lambda(x^3) - \lambda(x^4)| < C_0|x^3 - x^4|$, where the constant C_0 is independent of x^3 and x^4. On the basis of the remarks made above we have

$$|\operatorname{grad}\psi(y^3) - \operatorname{grad}\psi(y^4)| < C_1|y^3 - y^4|. \tag{32.6}$$

When $y^3 \neq y^1$, we consider the segment $\mathscr{I}_1$ joining these points. The set of interior points of $\mathscr{I}_1$ does not intersect the sets $\mathfrak{U}^{j_k}, \cdots, \mathfrak{U}^{j_r}$. Thus for any segment $\mathscr{I}_1^0$ contained strictly inside $\mathscr{I}_1$ we can find a body T_0 of rank $\leqq k-1$ such that $\mathscr{I}_1^0 \subset T_0$. If we denote by u and v the end points of the segment $\mathscr{I}_1^0$, then by the above $|\operatorname{grad}\psi(u) - \operatorname{grad}\psi(v)| < \mathscr{M}|u - v|$. Letting $u \to y^1$ and $v \to y^3$, we get

$$|\operatorname{grad}\psi(y^1) - \operatorname{grad}\psi(y^3)| \leqq \mathscr{M}|y^1 - y^3|.$$

Clearly an analogous inequality holds for the points y^2 and y^4. Thus from these inequalities and from (32.6) we find that

$$|\operatorname{grad}\psi(y^1) - \operatorname{grad}\psi(y^2)| < (C_1 + \mathscr{M})|y^1 - y^2|. \tag{32.7}$$

If we suppose now that $\mathscr{I} \cap \mathfrak{U}^{j_k} = \emptyset$, then by using the reasoning which we applied to the segment $\mathscr{I}_1$ above we conclude that inequality (32.7) holds in this case also. Thus for points of T the conditions of Proposition 32.1 are satisfied, and so inequality (32.2) is true on T. To conclude the proof of the theorem, we need only show that inequality (32.2) is true for any convex body of rank 1. To prove this fact, let T be any such body, and, as explained earlier, we assume without loss of generality that T does not intersect the sets $\mathfrak{U}^{j_2}, \cdots, \mathfrak{U}^{j_r}$. Since $d_{n-1}(x) \geqq C^* > 0$ for any $y \in T$, it follows from Lemma 11.1 and the remarks made at the beginning of the proof that $|\operatorname{grad}\psi(y^1) - \operatorname{grad}\psi(y^2)| \leqq C_1|y^1 - y^2|$ for any $y^1, y^2 \in T$. Hence, according to Proposition 32.1, we conclude that (32.2) holds on T. This completes the proof of the theorem.

Let $\mathscr{R}_\epsilon$ be that subset of $\mathscr{R}$ such that $d_{n-1/2}(x) \geqq \epsilon$ for $x \in \mathscr{R}_\epsilon$.

THEOREM 14. *At each point x of the set $\mathscr{R}$ there exists a tangent hyperplane to $\partial\mathscr{V}_n$ which is orthogonal to the vector $\lambda(x)$. The vector-valued function $\lambda(x)$ satisfies a Lipschitz condition on any set $\mathscr{R}_\epsilon$.*

PROOF. The first part of the theorem has already been proved in Lemma 12.1. Hence all we need prove is the statement concerning the Lipschitz condition.

At each point $x \in \mathscr{R}_\epsilon$ we can construct a parametric disc of maximal radius R_x. Since $\mathscr{R}_\epsilon$ is closed, $\inf_{x\in\mathscr{R}_\epsilon} R_x = R_\epsilon > 0$. Since $d_{n-1/2}(x)$ is continuous on $\partial\mathscr{V}_n$, we can determine a maximal r_ϵ such that $|x^1 - x^2| \leqq r_\epsilon$ implies $|d_{n-1/2}(x^1) - d_{n-1/2}(x^2)| \leqq \epsilon/2$.

If $\sup|\operatorname{grad}\psi(y)| > 1$ on the disc $\mathfrak{U}_x$, then, applying (32.2) to the points $y = 0$ and y^1 for which $|\operatorname{grad}\psi(y^1)| \geqq 1$, we find that $|y^1| \geqq 2^{-1/2}d_0^{-n}\epsilon^2(1+\epsilon^2)^{-1}$. From this it follows that at each point $x^0 \in \mathscr{R}_\epsilon$ we can construct a parametric disc $\mathfrak{U}_{x^0}$ such that:

1) $|\operatorname{grad}\psi(y)| \leqq 1$ on $\mathfrak{U}_{x^0}$;
2) $\mathfrak{W}_{x^0} \subset \mathscr{R}_{\epsilon_0}$;
3) $\{x : |x - x^0| \leqq \rho_\epsilon\} \in \mathfrak{W}_{x^0}$, where ρ_ϵ depends only on ϵ.

Let $x^0, x^1 \in \mathscr{R}_\epsilon$, $|x^0 - x^1| \leqq \rho_\epsilon$; construct in a neighborhood of x^0 a local coordinate system and apply Theorem 13. We find that $|\lambda(x^1) - \lambda(x^0)| = |\mu(y^1) - \mu(0)| \leqq \sqrt{2}|\operatorname{grad}\psi(y^1)| < 2\alpha_0^n(1 + 4\epsilon^{-2})|y^1|$, and so $|\lambda(x^1) - \lambda(x^0)| \leqq M_\epsilon|x^1 - x^0|$, where $M_\epsilon = 2\alpha_\epsilon^n(1 + 4\epsilon^{-2})$, since $|y^1| \leqq |x^1 - x^0|$. If instead of $|x^1 - x^0| \leqq \rho_\epsilon$ we have $|x^1 - x^0| > \rho_\epsilon$, then $|\lambda(x^1) - \lambda(x^0)| \leqq 2 < 2/\rho_\epsilon|x^1 - x^0|$, and thus in all cases we have $|\lambda(x^1) - \lambda(x^0)| \leqq \max(2/\rho_\epsilon, M_\epsilon)|x^1 - x^0|$. Q.E.D.

We introduced earlier the second de la Vallée-Poussin derivatives. We now introduce the second derivatives in the direction of the vector v; we have the lower derivative

$$\underline{d}_v^2\,\psi(y) = \varliminf_{\varepsilon\to+0} \varepsilon^{-1}\,[(\operatorname{grad}\psi(y + \varepsilon v), v) - (\operatorname{grad}\psi(y), v)]$$

and the upper derivative

$$\overline{d}_v^2\,\psi(y) = \varlimsup_{\varepsilon\to+0} \varepsilon^{-1}\,[(\operatorname{grad}\psi(y + \varepsilon v), v) - (\operatorname{grad}\psi(y), v)].$$

If $\underline{d}_v^2\psi(y) = \overline{d}_v^2\psi(y)$, then the function $\psi(y)$ has at y a second derivative in the direction v, and we denote this derivative by $d_v^2\psi(y)$. It is easily seen that

$$\underline{d}_v^2\,\psi(y) \leqslant \underline{D}_v^2\psi(y) \leqslant \overline{D}_v^2\,\psi(y) \leqslant \overline{d}_v^2\,\psi(y).$$

From Theorem 13 it follows that $\underline{d}_v^2\psi(y)$ and $\overline{d}_v^2\psi(y)$ are finite. From the results so far obtained it is not difficult to deduce an elegant inequality satisfied by these derivatives. According to Theorem 13, formula (32.5) holds for any y and y^1 with $|y| < R$ and $|y^1| < R$, provided δ is sufficiently small. In this formula we take $v^2 = v^1 = v$, where we consider the vector v to be fixed. Then

$$\int_0^\varepsilon (D_\delta(y + \chi v)\,v,\, v)\,d\chi = (\operatorname{grad}\psi_\delta(y^1), v) - (\operatorname{grad}\psi_\delta(y))$$

$$\leqslant \int_0^\varepsilon (\mathfrak{D}_\delta(y + \chi v)\,v,\, v)\,d\chi,$$

where $\epsilon = |y^1 - y|$. Taking the limit as $\delta \to 0$ and recalling (see Lemma 29.1) that the function $(D(y)v, v)$ is lower semicontinuous, we have

$$\int_0^\varepsilon (D(y + \chi v)\,v,\, v)\,d\chi \leqslant (\operatorname{grad}\psi(y^1), v) - (\operatorname{grad}\psi(y), v).$$

Analogously, with the aid of Lemma 29.2 we get that

$$(\operatorname{grad}\psi(y^1), v) - (\operatorname{grad}\psi(y), v) \leqslant \int_0^\varepsilon (\mathfrak{D}(y + \chi v)\,v,\, v)\,d\chi.$$

From these inequalities, via the same type of arguments, we get that

$$(D(y)v, v) \leqslant \underline{d}_v^2\psi(y) \leqslant \bar{d}_v^2\psi(y) \leqslant (\mathfrak{D}(y)v, v). \tag{32.8}$$

From this it is not difficult to deduce

LEMMA 32.1. *If* $(D(y)v, v) = (\mathfrak{D}(y)v, v)$, *then* $\psi(y)$ *has at the point* y *a second derivative in the direction* v.

In order that the equality

$$(D(y)v, v) = (\mathfrak{D}(y)v, v)$$

be satisfied at y for any vector v, it is necessary and sufficient that

$$\mathscr{A}_x = \mathfrak{H}_x,$$

as follows from Proposition 28.2.

The subset of those $x \in \mathscr{R}$ for which $\mathscr{A}_x$ is a proper subset of $\mathfrak{H}_x$ will be denoted by $\mathscr{A}_2$.

Due to insufficient space, we cannot give a complete account of the structure of $\mathscr{A}_2$, so we shall limit ourselves to the proof of the following lemma.

LEMMA 32.2. $\mathscr{A}_2$ *is a nowhere dense* F_δ *set.*

PROOF. Let $x^1 \in \mathscr{A}_2$. Let $w_1(x^1), \cdots, w_k(x^1)$ be a list of all the singular vertices of the tree $\mathscr{T}_{x^1}$ as well as of all those vertices whose degree is less than $\kappa + 2$, where κ is the index of the vertex. Consider the quadratic differential $Q(\rho w, x^1)(dw/w)^2$, where $\rho < 1$. The graph of this differential is a simple expansion of the graph of the differential $Q(w, x^1)(dw/w)^2$. Thus the tree $\mathscr{T}_{x^1}$ is mapped to some tree $\mathscr{T}_{x^1\rho}$ with exterior conformal radius $1/\rho$. We choose ρ very close to 1, and at each vertex $(1/\rho)w_j(x^1)$, $1 \leqq j \leqq k$, of the tree $\mathscr{T}_{x^1\rho}$ we carry out the following construction. We adjoin to the tree some new branches emanating from each of these vertices so that the degree of the vertex $(1/\rho)w_j(x^1)$ will be equal to $\kappa_j + 2$. This addition of branches can be carried out in such a way that the exterior conformal radius of the resulting tree is equal to 1. This can be done since the conformal radius is a monotone decreasing function as the domain under consideration is contracted. According to Theorem 5 the function $g(z)$, mapping $|z| < 1$ onto the complement of the tree constructed above and satisfying $g(0) = 0$ and $g'(0) > 0$, will be an extremal function. By x^ρ we denote the point of $\partial\mathscr{V}_n$ corresponding to this function; thus $g(z) = f(z, x^\rho)$. We also write the differential equation satisfied by $f(z, x^\rho)$ in the form

$$\left(\frac{Df(z, x^\rho)}{f(z, x^\rho)}\right)^2 Q[f(z, x^\rho), x^\rho] + P(z, x^\rho) = 0.$$

By construction, the only zeros on $|z| = 1$ of the quasipolynomial $P(z, x^\rho)$ are zeros of multiplicity two. We can find ϵ_ρ such that for any $x \in \mathscr{R}$, $|x - x^\rho| < \epsilon_\rho$, the corresponding quasipolynomial $P(z, x)$ has on $|z| = 1$ only second order zeros, and each zero is the image of a terminal vertex of the tree $\mathscr{T}_x$.

Indeed, we see from Proposition 16.7 that if $|x - x^\rho| < \epsilon_\rho$, the tree $\mathscr{T}_x$ has no fewer terminal vertices than does $\mathscr{T}_{x^\rho}$. Each terminal vertex of $\mathscr{T}_x$ is responsible for at least a double zero of $P(z, x)$ lying on $|z| = 1$; thus the number of distinct double zeros of $P(z, x)$ lying on $|z| = 1$ is not less than the corresponding number of zeros for the quasipolynomial $P(z, x^\rho)$. On the other hand, the number of such zeros of $P(z, x)$ cannot exceed the corresponding number for $P(z, x^\rho)$, and so it follows that on $|z| = 1$ this function has the same number of double zeros. Therefore we have shown that $\mathscr{T}_x$ cannot have singular vertices, and this in turn means that $\{x : |x - x^\rho| < \epsilon_\rho\}$ is contained in the complement of $\mathscr{A}_2$. Since $f(z, x^\rho)$ converges to $f(z, x^1)$ as $\rho \to 1$ (the domain $\mathscr{G}_{x^\rho}$ converges to $\mathscr{G}_{x^1}$) we have that $\mathscr{A}_2$ is nowhere dense.

We now prove the second part of the lemma. Denote by $\mathscr{A}_2^l$ that subset of $\mathscr{A}_2$ consisting of those $x \in \mathscr{R}_{1/l} \cap \mathscr{A}_2$ for which the tree $\mathscr{T}_x$ satisfies the following condition: any vertex w_j, other than the base, such that $|w_j| < l$, is at at least a distance $1/l$ from any other vertex and from any zero of $Q(w, x)$. Clearly

$$\mathscr{A}_2 = \bigcup_{l \geqslant 1} \mathscr{A}_2^l.$$

We shall show that for each l the set $\mathscr{A}_2^l$ is closed, from which it will follow that $\mathscr{A}_2$ is an F_δ set. Clearly the set $\mathscr{A}_2^l$ is nonempty for $l \geqq l_0$. Let the sequence $\{x^k\} \subset \mathscr{A}_2^l$ converge to x^0. It is clear that if $x^0 \in \mathscr{A}_2$, then $x^0 \in \mathscr{A}_2^l$ for some l. By the conditions given in the definition of the space $\mathfrak{H}_{x^k}$ there exists a normalized vector $\zeta_k(w)$ orthogonal to $\mathscr{A}_{x^k}$. According to Lemma 24.1 the function $\zeta_k(w)$ satisfies the differential equation

$$\sqrt{\frac{Q(w, x^k)}{w^2}}\,\frac{d\zeta_k^*(w)}{dw} + \frac{1}{w^2}\,\widetilde{Q}(w, x^k) + \frac{\widetilde{\mathscr{P}}(z, x^k)}{(Df(z, x^k))^2} = 0, \quad z = f^{-1}(w, x_k), \tag{32.9}$$

in $\mathscr{G}_{x^k}$. We are now in the same situation as in the proof of Lemma 29.2, when we were given a uniformly bounded sequence $\{\zeta_k(w)\}$, $\zeta_k(w) \in \mathscr{A}_{x^k}$, $k \geqq 1$, and we showed that we could extract a subsequence $\{\zeta_{k_m}(w)\}$ which converges in $\mathscr{G}_{x^0}$ to a function $\zeta_0(w)$ belonging to the space $\mathfrak{H}_{x^0}$. Without loss of generality we assume that the sequence $\{\zeta_k(w)\}$ actually converges to $\zeta_0(w)$. It is easy to show that $\widetilde{Q}(w, x^k)$ and $\widetilde{\mathscr{P}}(z, x^k)$ converge respectively to $\widetilde{Q}(w)$ and $\widetilde{\mathscr{P}}(z)$, provided $w \neq 0$ and $z \neq 0, \infty$. Thus

$$\sqrt{\frac{Q(w, x^0)}{w^2}} \cdot \frac{d\zeta^*(w)}{dw} + \frac{1}{w^2}\,\widetilde{Q}(w) + \frac{\widetilde{\mathscr{P}}(z)}{(Df(z, x^0))^2} = 0, \qquad z = f^{-1}(w, x^0)$$

and so, if $\widetilde{Q}(w) \not\equiv 0$, it follows from (24.20) that $(\zeta, \eta) = 0$ for any function $\eta \in \mathscr{A}_{x^0}$ (note that $w_j = 0$). From this we have that $\mathscr{A}_{x^0}$ is a proper subspace of $\mathfrak{H}_{x^0}$, and so $x^0 \in \mathscr{A}_2$. Thus all that remains to complete the proof is to show that $Q(w) \not\equiv 0$. From (24.9)–(24.11) we see that it is sufficient to show that $\inf_k |\mathfrak{z}^k| > 0$, where $\mathfrak{z}^k$ is the vector whose components are the first n coefficients of the power series for $\zeta_k(w)$. Assuming the opposite, we get that $\lim_{k \to \infty} \mathfrak{z}^k = 0$. As before, we set

$$S_k(w) = \frac{1}{w^2}\widetilde{Q}(w, x^k) + \frac{\widetilde{\mathscr{P}}(z, x^k)}{(Df(z, x^k))^2}, \qquad z = f^{-1}(w, x^k).$$

We now take δ sufficiently small, and for the tree $\mathscr{T}_{x^0}$ we construct the system of discs $\mathscr{K}_\delta$. In each disc $K_{j\delta}$ there exists a vertex $w_j(x^k) \in \mathscr{T}_{x^k}$, provided $k \geqq k_0$. That this fact is true for nonterminal vertices follows since in the disc $K_{j\delta}$ there is a zero of the function $Q(w, x^k)$ with the same multiplicity as the zero $w_j(x^0)$ of the function $Q(w, x^0)$. This zero will be a vertex of $\mathscr{T}_{x^k}$. If the disc $K_{j\delta}$ corresponds to a terminal vertex, then the above statement follows from Proposition 16.7. Hence we have that for $k \geqq k_0$ all the vertices of $\mathscr{T}_{x^k}$ lie in the system of discs $\mathscr{K}_\delta$. Define $r < 1$ so that the function $Q(w, x^0) \neq 0$ in the domain $\mathscr{G}(\mathscr{C}_r, \mathscr{T}_{x^0}) \backslash \mathscr{K}_\delta$, where $\mathscr{C}_r = \{w : |f^{-1}(w, x^0)| = r\}$. Since for $w \neq 0$ we have $\lim_{k\to\infty} Q(w, x^k) = Q(w, x^0)$, it follows for $k \geqq k_0$ that

$$|Q(w, x^k)| \geqq c_1 > 0, \qquad w \in \mathscr{G}(\mathscr{C}_r, \mathscr{T}_{x^0}) \backslash \mathscr{K}_\delta.$$

By Proposition 16.7 the sequence $\{Df(z, x^k)\}$ converges to $Df(z, x^0)$ on the image of $\mathscr{G}(\mathscr{C}_r, \mathscr{T}_{x^k}) \backslash \mathscr{K}_\delta$. Hence as $k \to \infty$

$$\int_{\mathscr{G}_{r,k}} |S_k(w)|^2 \left| \frac{w^2}{Q(w, x^k)} \right| d\sigma \to 0,$$

where $\mathscr{G}_{r,k} = \mathscr{G}(\mathscr{C}_r, \mathscr{T}_{x^k}) \backslash \mathscr{K}_\delta$. Therefore, if we recall that for $k \geqq k_0$

$$\|\zeta\|_{x^k} = \operatorname{Re}\left\{ \frac{1}{\pi} \int_{\mathscr{G}(\mathscr{C}_r, \mathscr{T}_{x^k})} \left| \frac{d\zeta^*}{dw} \right|^2 d\sigma - \frac{1}{2\pi i} \int_{\mathscr{C}_r} \zeta^* \overline{d\zeta^*} + \vartheta \sum_1^n |\zeta_k|^2 \right\},$$

we get

$$\|\zeta_k\|_{x^k} = \frac{1}{\pi} \int_{\mathscr{K}_\delta} |S_k(w)|^2 \left| \frac{w^2}{Q(w, x^k)} \right| d\sigma + o(1).$$

The disc $K_{j\delta}$ is divided into finitely many domains $\sigma_{j\nu}(x^k)$ by the branches of the tree $\mathscr{T}_{x^k}$; the number of these domains is independent of k, provided $k \geqq k_0$. In each such domain $\sigma_{j\nu}(x^k)$ an expansion of the form (24.5) or (24.6) holds. If $z_{j\nu}(x^k)$ is the image of the vertex $w_j(x^k)$ under the mapping $z = f^{-1}(w, x^k)$, then $z_{j\nu}(x^k)$ is a zero of $P(z, x^k)$, where the multiplicity of the zero is independent of k for $k \geqq k_0$. By assumption there exists a disc $|z - z_{j\nu}(x^k)| \leqq \Delta$ in which $P(z, x^k)$ vanishes only at $z_{j\nu}(x^k)$. Consider the equation

$$\left(\frac{Df(z, x^k)}{f(z, x^k)} \right)^2 Q[f(z, x^k), x^k] + P(z, x^k) = 0.$$

Since $P(z, x^k) \to P(z, x^0)$ and $Q(w, x^k) \to Q(w, x^0)$, and since the multiplicity of the zero $w_j(x^k)$ is independent of k for $k \geqq k_0$, there exists a disc $|w - w_j(x^0)| \leqq \Delta_0$ in which the function $f^{-1}(w, x^k)$ can be expanded in a series of the form (24.5), where the exponent of $w - w_j(x^k)$ is independent of k. Hence

$$z - z_{j\nu}(x^k) = \sum_{p \geqslant 0} \beta_{j\nu p}(x^k)(w - w_j(x^k))^{t_p}, \qquad t_0 = \frac{1}{r_{j\nu}}.$$

It is easy to prove that $|\beta_{j\nu p}(x^k)| < C\Delta_1^{-t_p}$, where C and Δ_1 are independent of k. Thus in $\sigma_{j\nu}(x^k)$ we have

$$S_k(w) = \sum_{p\geqslant 0} A_{j\nu p}(x^k)(w - w_j(x^k))^{\gamma_{j\nu p}},$$

where $\gamma_{j\nu 0} > \kappa_j/2 - 1$. For any fixed p, $A_{j\nu p}(x^k) \to 0$ as $k \to \infty$, and $|A_{j\nu p}(x^k)| < C\Delta_2^{-\gamma_{j\nu p}}$, as follows from the estimates for $\beta_{j\nu p}(x^k)$.

The same type of results hold for the base of the tree, the only difference being that $w - w_j(x^k)$ must be replaced by $1/w$. From this we conclude that $\|\zeta_k\|_{x^k} = o(1)$, which is an obvious contradiction. Thus we finally conclude that $\widetilde{Q}(w) \not\equiv 0$, and so $\zeta_0(w) \not\equiv 0$. Thus $\mathscr{A}_{x^0}$ is a proper subset of $\mathfrak{W}$. This completes the proof of the lemma, as explained previously.

Although this next fact was not proved above, we note that

$$\dim \mathscr{A}_2 \leqq 2n - 3.$$

Also, we note that the projection of the set $\mathfrak{W} \cap \mathscr{A}_2$ onto the disc $\mathfrak{U}$ has Lebesgue measure zero.

We shall prove these facts elsewhere. For convenience we summarize the results given above.

Lemma 32.3. *If y belongs to the complement of the projection of the set $\mathscr{A}_2 \cap \mathfrak{W}$ onto the disc $\mathfrak{U}$, then at y the function ψ has a second differential. If y is an interior point of the complement, then ψ is twice continuously differentiable in a small neighborhood of y.*

Proof. If y^1 lies in the complement of the projection of $\mathscr{A}_2 \cap \mathfrak{W}$, then the functions $d_{kl}(y)$, $k = 1, \cdots, 2n-1$, are continuous at y. From (32.5) it follows that

$$(\operatorname{grad}\psi(y^2) - \operatorname{grad}\psi(y^1), v^2) = \varepsilon(D(y^1)v^1, v^2) + 0(\varepsilon).$$

Thus the vector-valued function $\operatorname{grad}\psi(y)$ is differentiable at y^1, and

$$\frac{\partial^2\psi(y^1)}{\partial y_k^1\,\partial y_l^1} = \frac{\partial^2\psi(y^1)}{\partial y_l^1\,\partial y_k^1} = d_{kl}(y^1).$$

If y^1 is an interior point of the complement, then $\partial^2\psi(y)/\partial y_k\partial y_l$ is continuous in a small neighborhood of y^1, since $d_{kl}(y)$ is continuous there. Q.E.D.

Lemma 32.4. *Let v be any vector, and suppose that the vector t is determined by v via relation* (31.3). *Then*

$$\overline{d}_v^2\psi(y) \leqslant (1 + |\operatorname{grad}\psi(y)|^2)^{1/2}\left\{\alpha^n \sum_{k=1}^{\left[\frac{n}{2}\right]} |t_k|^2 - \beta^n \sum_{k=\left[\frac{n}{2}\right]+1}^{n} |t_k|^2\right\}. \tag{32.10}$$

Proof. This inequality is a trivial consequence of (30.7) and (32.8).

§33. We established above that there is a close connection between the structure of the tree $\mathscr{T}_x$ of the extremal function $f(z, x)$, $x \in \mathscr{R}$, and the smoothness of the boundary in a neighborhood of x. When the tree $\mathscr{T}_x$ has a particularly simple structure, we can sharpen the results derived above.

Suppose the equation for the extremal function $f(z,x)$, $x \in \mathscr{R}$, has the form

$$Q(w,x)\left(\frac{Dw}{w}\right)^2 + P(z,x) = 0,$$

where

$$Q(w,x) = \sum_1^n A_k(x) w^{-k}, \qquad P(z,x) = \sum_{-n}^n B_k(x) z^{-k}.$$

Denote by $\mathscr{R}_1^*$ the subset of $\mathscr{R}$ consisting of those points x for which the following conditions are satisfied:

1) $\mathscr{T}_x$ has one branch.
2) $A_1(x) \neq 0$.
3) A vertex of $\mathscr{T}_x$ different from the base is not a zero of $Q(w,x)$.

Schaeffer and Spencer investigated this set in [2], where in particular they showed that $\mathscr{R}_1^*$ is not empty. Clearly $\mathscr{R}_1^*$ is open. In fact, if $x^0 \in \mathscr{R}_1^*$, then $d_{n-1/2}(x^0) \neq 0$. Thus for sufficiently small $\epsilon > 0$ we have $d_{n-1/2}(x) \neq 0$ if $|x - x^0| < \epsilon$. Since $B_k(x)$ is a continuous function on $\mathscr{R}$, the quasipolynomial $P(z,x)$ has one double zero on $|z| = 1$, provided $|x - x^0| < \epsilon$. Therefore the tree $\mathscr{T}_x$ has only one branch, and the terminal vertex of $\mathscr{T}_x$ will not be a zero of $Q(w,x)$. Also, it is obvious that $A_1(x) \neq 0$, since only one branch emanates from the base.

In a neighborhood of the point $x^0 \in \mathscr{R}_1^*$ we introduce the local coordinates (y,τ) and we assume that the image of the disc $\mathfrak{U}$ is the set $\mathfrak{W} \subset \mathscr{R}_1^*$. Consider the restriction of the extremal function $f(z,x)$ to the set $\{z : |z| < 1\} \times \mathfrak{W}$; since $\mathfrak{U}$ and $\mathfrak{W}$ are homeomorphic, we have defined on $\{z : |z| < 1\} \times \mathfrak{U}$ a function $F(z,y) \equiv f(z,x)$, where $x = x_0 + V(y, \psi(y))$. For the sake of convenience, we denote $F(z,y)$ by $f(z,y)$; with the above conditions, this should not cause any misunderstanding. We shall show that the function $\psi(y)$ and the function $f(z,y)$, for $|z| \leq \rho < 1$, are regular in the disc $\mathfrak{U}$.

The proof will be rather lengthy, although it is really quite simple. Let $v = (v_1, \cdots, v_{2n-1})$ be an arbitrary vector. Assuming that $\psi(y)$ possesses some definite a priori smoothness properties, we shall construct a variational formula for $f(z, y + \epsilon v)$, considering it as a variation of the function $f(z,y)$. This variational formula will not be the same as the Taylor formula for $f(z,y)$ considered as a function of y. The whole of the remainder of this section will be devoted to the construction of this variational formula.

We suppose that a function $\varphi(\epsilon)$ is defined for $0 \leq \epsilon \leq \epsilon_0$, and that

$$\varphi(\epsilon) = \varphi_0 + \epsilon\varphi_1 + \cdots + \epsilon^m \varphi_m + \epsilon^{m+1}\varphi_{m+1}(\epsilon),$$

where φ_l, $0 \leq l \leq m$, are constants. The number $l!\,\varphi_l$ is called the lth generalized derivative of φ at the point $\epsilon = 0$. If φ is real, then the upper and lower limits of $(m+1)!\,\varphi_{m+1}(\epsilon)$ at $\epsilon = 0$ are called the $(m+1)$th generalized derivatives. In what follows we shall assume that the constants φ_l satisfy

$$(33.1)\qquad |\varphi_l| \leqq AB^{l-1}l^{-\beta}, \qquad 1 \leqq l \leqq m,$$

and

$$(33.2)\qquad |\varphi_{m+1}(\epsilon)| \leqq AB_1^m(m+1)^{-\beta}, \qquad B_1 \geqq B,$$

for $0 \leqq \epsilon \leqq \epsilon_0$. The constant β is at least 2, and in what follows we take $\beta = 3$.

Let the function $F(w)$ be regular in the disc $|w - \varphi_0| \leqq \Delta$, and bounded there by M. We shall examine the question of the existence of the generalized derivatives of the function $F[\varphi(\epsilon)]$.

Proposition 33.1. *If* $\epsilon \leqq \min(\epsilon_0, 1/2B_1)$ *and*

$$(33.3)\qquad \left(\frac{A}{\Delta B}\right)^{1/2} \leqslant \frac{1}{2\zeta(\beta)},$$

where $\zeta(\beta) = \sum_1^\infty k^{-\beta}$, *then the function* $\Phi(\epsilon) = F[\varphi(\epsilon)]$ *has at the point* $\epsilon = 0$ *generalized derivatives of order up to and including* m. *That is to say,* $\Phi(\epsilon) = \sum_0^m \epsilon^l \Phi_l + \epsilon^{m+1}\Phi_{m+1}(\epsilon)$, *where*

$$|\Phi_l| < \frac{M}{\Delta}\left(1 + \gamma\sqrt{\frac{A}{\Delta B}}\right) AB^{l-1}l^{-\beta}, \qquad 1 \leqslant l \leqslant m,$$

and

$$|\Phi_{m+1}(\varepsilon)| < 2\frac{M}{\Delta}\left(1 + \gamma\sqrt{\frac{A}{\Delta B}}\right) AB_1^m(m+1)^{-\beta}.$$

The constant γ *depends only on* β.

Proof. For any $\epsilon \leqq \epsilon_0$ and for $|\delta| \leqq \delta_0$, the values of $\pi(\delta) = \sum_0^m \varphi_k\delta^k + \varphi_{m+1}(\epsilon)\delta^{m+1}$ lie in the disc $|w - \varphi_0| \leqq \Delta$. Thus the function $\Psi(\delta) = F(\pi(\delta))$ is regular for $|\delta| \leqq \delta_0$. Having set $\Psi(\delta) = \sum_0^m \Psi_k\delta^k + \sum_{m+1}^\infty \Psi_k(\epsilon)\delta^{k+1}$, we find that

$$(33.4)\qquad \Psi_k = F'(\varphi_0)\frac{\pi^{(k)}(0)}{k!} + \tilde{\Psi}_k,$$

where

$$\tilde{\Psi}_k = \frac{1}{k!}\sum_{\substack{\nu+2\nu_2+\ldots=k\\ \nu_1+\nu_2+\ldots>1}} k(\nu_1, \nu_2, \ldots)F^{(\nu_1+\nu_2+\ldots)}(\varphi_0)(1!)^{\nu_1}(2!)^{\nu_2}\ldots\varphi_1^{\nu_1}\varphi_2^{\nu_2}\ldots.$$

Using (33.1) and Cauchy's inequality $|F^{(\mu)}(\varphi_0)| < M\mu!\Delta^{-\mu}$, we get for $k \leqq m$ that

$$|\tilde{\Psi}_k| < \frac{M}{\Delta k!}AB^{k-1}\sum_{\substack{\nu_1+2\nu_2+\ldots=k\\ \nu_1+\nu_2+\ldots>1}} k(\nu_1, \nu_2, \ldots)\frac{(1!)^{\nu_1}(2!)^{\nu_2}\ldots}{1^{\beta\nu_1}2^{\beta\nu_2}\ldots}$$
$$\times(\nu_1+\nu_2+\ldots)!\left(\frac{A}{B\Delta}\right)^{\nu_1+\nu_2+\ldots-1}.$$

Since $(A/B\Delta)^{1/2} = \vartheta \leqq 1/2\zeta(\beta)$, we have

$$|\tilde{\Psi}_k| < \frac{M}{\Delta}AB^{k-1}\frac{\eta\beta^{(k)}(0, \vartheta)}{k!},$$

and on applying Lemma 1.7 we get that

(33.5) $$|\tilde{\Psi}_k| < \gamma \frac{M}{\Delta} \vartheta A B^{k-1} k^{-\beta}.$$

For $k \geqq m+1$ we must change the constant B in this latter inequality to B_1. Hence

$$|\Psi_k| < \frac{M}{\Delta}(1+\gamma\vartheta) A B^{k-1} k^{-\beta}, \quad 1 \leqslant k \leqslant m,$$

and for $k \geqq m+1$

$$|\Psi_k| < \frac{M}{\Delta}(1+\gamma\vartheta) A B_1^{k-1} k^{-\beta}.$$

Set $\Psi_{m+1}(\delta, \epsilon) = \sum_{m+1}^{\infty} \Psi_k(\epsilon) \delta^{k-m-1}$. For $|\delta| \leqq 1/2B_1$, the above inequalities imply that

$$|\Psi_{m+1}(\delta, \varepsilon)| < (2 + 2\gamma\vartheta) \frac{M}{\Delta} A B_1^m (m+1)^{-\beta}.$$

Considering that $\epsilon \leqq \min(\epsilon_0, 1/2B_1)$, we set $\delta = \epsilon$; then $\Phi(\epsilon) = \Psi(\epsilon)$, and, having set $\Phi_l = \Psi_l$, $0 \leqq l \leqq m$, and $\Phi_{m+1}(\epsilon) = \Psi_{m+1}(\epsilon, \epsilon)$, we find that $\Phi(\epsilon)$ $\sum_0^m \Phi_k \epsilon^k + \Phi_{m+1}(\epsilon) \epsilon^{m+1}$. Q.E.D.

Later on we shall have occasion to use formulas (33.4) and (33.5), which we can write in the form

(33.6) $$\Phi_k = F'(\varphi_0) \varphi_k + \tilde{\Phi}_k, \qquad 1 \leqq k \leqq m,$$

and

(33.7) $$|\tilde{\Phi}_k| < \gamma\vartheta \frac{M}{\Delta} A B^{k-1} k^{-\beta}, \qquad 1 \leqslant k \leqslant m.$$

We now consider the problem of constructing a solution of a special differential equation. Let G be a closed set on the v-sphere containing $w=0$ and bounded by an analytic Jordan curve. We shall introduce some geometric properties of G which we shall use later on. Denote by d the diameter of G, by d_1 the length of ∂G, and by d_0 the radius of the largest disc $|v| \leqq d_0$ contained in the closure of G. If $\Delta < d_0$ and $\Delta < d/2$, we denote by G_Δ a subdomain of G containing $w = 0$ such that the distance from ∂G_Δ to ∂G is at least Δ.

Let $F(v, \epsilon)$ be regular in G for $0 \leqq \epsilon \leqq \epsilon_0$, and assume that, as a function of ϵ, $F(v, \epsilon)$ is given by

$$F(v, \varepsilon) = \sum_0^m \varepsilon^l F_l(v) + \varepsilon^{m+1} F_{m+1}(v, \varepsilon).$$

Suppose that in G

(33.8) $$\mathcal{M}_0^{-1} \leqslant |F_0(v)| \leqslant \mathcal{M}_0,$$

(33.9) $$|F_l(v)| \leqslant A B^{l-1} l^{-\beta}, \qquad 1 \leqslant l \leqslant m,$$
$$|F_{m+1}(v, \varepsilon)| \leqslant A B_1^m (m+1)^{-\beta},$$

where $B_1 \geqq B$ and $0 \leqq \epsilon \leqq \epsilon_0$.

We shall construct a solution of the differential equation

$$\frac{F(u,\varepsilon)}{F(v,0)}\sqrt{\frac{v}{u}}\frac{du}{dv}=1,$$

satisfying $u(0)=0$; the branch of the square root is chosen so that $\sqrt{1}=1$.

Set $A_1=2(2+d_1/d_0)\mathscr{M}_0A$. Having defined ϑ by the condition

$$A_1d/B\Delta=\vartheta^2, \tag{33.10}$$

we require that

$$\vartheta\leqslant\frac{1}{2\zeta(\beta)},\qquad 2\gamma\vartheta\mathscr{M}_0(2\mathscr{M}_0+2^{\beta-2})\leqslant 1. \tag{33.11}$$

Proposition 33.2. *If (33.11) is satisfied and if $\epsilon\leqq\min(\epsilon_0,1/2B_1)$, then in G_Δ the differential equation given above has a solution of the form*

$$u=v\left[1+\sum_1^m\varepsilon^k u_k(v)+\varepsilon^{m+1}u_{m+1}(v,\varepsilon)\right].$$

The functions $u_k(v)$ and $u_{m+1}(v)$ satisfy

$$|u_k(v)|<A_1B^{k-1}k^{-\beta},\qquad 1\leqq k\leqq m,$$
$$|u_{m+1}(v,\epsilon)|\leqq 2A_1B_1^m(m+1)^{-\beta},$$

in G_Δ.

Proof. Before beginning the proof we must establish some auxiliary inequalities. Suppose that the disc $|v|\leqq d_0$ is tangent to ∂G at the point v_1. Construct in G the line from $v=0$ to $v=v_1$. In this slit domain the function

$$\widetilde{F}_k(v)=v^{-\frac{1}{2}}\int_0^v F_k(t)\,t^{-\frac{1}{2}}\,dt$$

is regular and single-valued, where the integration is over any rectifiable curve lying in G and not intersecting the slit. After orienting the slit in any way desired, we can define its "left" and "right" sides as we have done before. It is easily seen that $[\widetilde{F}_k(v)]_+=[\widetilde{F}_k(v)]_-$ if v lies on the slit. In this way we have that $\widetilde{F}_k(v)$ is regular and single-valued in G. We now estimate the modulus of this function. We choose the path of integration as follows: join $v=0$ to the point v by a line segment; then replace any open intervals of this segment lying in the complement of G by the corresponding arcs of ∂G which connect the end points of the segment. Integrating along this path, we get that

$$|\widetilde{F}_k(v)|<(2+d_1/d_0)F_k, \tag{33.12}$$

where $F_k=\max_{v\in G}|F_k(v)|$. According to Leibniz' formula,

$$v^{\mu+\frac{1}{2}}\frac{d^{\mu+1}}{dv^{\mu+1}}(\sqrt{v}\,\widetilde{F}_k(v))=\sum_{l=0}^{\mu}(-1)^l C_\mu^l\frac{(2l-1)!!}{2^l}v^{\mu-l}F_k^{(\mu-l)}(v).$$

If $v \in G_\Delta$, then it is easily seen that

$$|F_k^{(l)}(v)| < l!\,\Delta^{-1}F_k.$$

In order to prove this, set $\max_{v\in G_\Delta}|F^{(l)}(v)| = |F^{(l)}(v_0)|$. There exists a disc, centered at v_0 with radius not less than Δ, which lies entirely within G. By Cauchy's inequality $|F_k^{(l)}(v_0)| < l!\,\Delta^{-l}F_k$, and so if $v \in G_\Delta$ we have that

$$\left| v^{\mu+\frac{1}{2}} \frac{d^{\mu+1}}{dv^{\mu+2}} (\sqrt{v}\,\widetilde{F}_k(v)) \right| < \mu!\,F_k \sum_{l=1}^{\mu} \frac{(2l-1)!!}{2^l\,l!} \left(\frac{d}{\Delta}\right)^{\mu-l}.$$

But

$$\sum_{l=0}^{\mu} \frac{(2l-1)!!}{2^l\,l!} \left(\frac{d}{\Delta}\right)^{\mu-l} < \left(\frac{d}{\Delta}\right)^{\mu} \sum_{l=0}^{\infty} \frac{(2l-1)!!}{2^l\,l!} \left(\frac{\Delta}{d}\right)^{l} < 2\left(\frac{d}{\Delta}\right)^{\mu}.$$

Therefore

$$\left| v^{\mu+\frac{1}{2}} \frac{d^{\mu+1}}{dv^{\mu+1}} (\sqrt{v}\,\widetilde{F}_k(v)) \right| < 2\cdot\mu!\left(\frac{d}{\Delta}\right)^{\mu} F_k. \tag{33.13}$$

Having fixed ϵ, we consider the auxiliary differential equation

$$\frac{F_*(u,\delta)}{F(v,0)} \sqrt{\frac{v}{u}}\,\frac{du}{dv} = 1,$$

where $F_*(v) = \sum_0^m \delta^k F_k(v) + \delta^{m+1}F_{m+1}(v,\epsilon)$. For notational ease we have not included a reference to ϵ in the subscript of F_{m+1}. Assuming that δ is sufficiently small, we change the above differential equation via formal manipulation into the equation

$$\sum_0^{m+1} \delta^k \int_0^u F_k(t)\frac{dt}{\sqrt{t}} = \int_0^v F_0(t)\frac{dt}{\sqrt{t}}.$$

Seeking a solution of the form

$$u = v\left(1 + \sum_1^{\infty} \delta^k u_k(v)\right)$$

for this equation, we find after substitution of this series into the above equation, and after combining like powers of δ, that

$$\sum_{r=0}^{\min(l,m+1)} \frac{1}{(l-r)!} \sum_{\nu_1+2\nu_2+\ldots=l-r} k(\nu_1,\nu_2,\ldots)(1!)^{\nu_1}(2!)^{\nu_2}\ldots v^{\nu_1+\nu_2+\ldots} \times [\sqrt{v}\,\widetilde{F}_r(v)]^{(\nu_1+\nu_2+\ldots)} u_1^{\nu_1}u_2^{\nu_2}\ldots = 0, \quad 1 \leq l < \infty. \tag{33.14}$$

For $l = 1$ we get $F_0(v)u_1(v) + \widetilde{F}_1(v) = 0$, and so

$$u_1(v) = -\widetilde{F}_1(v)/F_0(v). \tag{33.15}$$

Hence $u_1(v)$ is regular in G. From (33.14) it follows for $1 \leq l \leq m+1$ that

$$u_l(v) = -\frac{\widetilde{F}_l(v)}{F_0(v)} - \widetilde{u}_l(v), \tag{33.16}$$

where

$$\widetilde{u}_l(v) = \frac{1}{\sqrt{v}\,F_0(v)}\Big\{\frac{1}{l!}\sum_{\substack{\nu_1+2\nu_2+\ldots=l\\ \nu_1+\nu_2+\ldots>1}} k(\nu_1, \nu_2, \ldots)(1!)^{\nu_1}(2!)^{\nu_2}\ldots v^{\nu_1+\nu_2\ldots}$$

$$\times [\sqrt{v}\,\widetilde{F}_0(v)]^{(\nu_1+\nu_2+\ldots)}u_1^{\nu_1}\ldots u_{\nu_{l-1}}^{\nu_{l-1}} + \sum_{r=1}^{\min(l-1,\,m+1)} \frac{1}{(l-r)!}\sum_{\nu_1+2\nu_2+\ldots=l-r} k(\nu_1, \nu_2, \ldots)$$

$$\times (1!)^{\nu_1}(2!)^{\nu_2}\ldots v^{\nu_1+\nu_2+\ldots}[\sqrt{v}\,F_r(v)]^{(\nu_1+\nu_2+\ldots)}u_1^{\nu_1}u_2^{\nu_2}\ldots\Big\}.$$

For $l > m+1$ it is obvious that $u_l(v) = -\widetilde{u}_l(v)$. From these equations it follows that $u_l(v)$, $1 \leq l < \infty$, are regular functions in G. We next estimate $\max_{v\in G_\Delta}|u_l(v)|$. Considering that $l \leq m$ and assuming that $|u_k(v)| < A_1B^{k-1}k^{-\beta}$, $1 \leq k \leq l-1$, $v \in G_\Delta$, we find from (33.13) that

$$|\widetilde{u}_l(v)| < 2A_1\,B^{l-1}|F_0(v)|^{-1}F_0\Big\{\frac{1}{l!}\sum_{\substack{\nu_1+2\nu_2+\ldots=l\\ \nu_1+\nu_2+\ldots>1}} k(\nu_1, \nu_2, \ldots)\frac{(1!)^{\nu_1}(2!)^{\nu_2}}{1^{\beta\nu_1}2^{\beta\nu_2}}$$

$$\times (\nu_1+\nu_2+\ldots)!\left(\frac{A_1d}{B\Delta}\right)^{\nu_1+\nu_2+\ldots-1}\Big\} + 2\frac{\Delta}{d}|F_0(v)|^{-1}\sum_{r=1}^{l-1}\frac{B^{l-r}F_r}{(l-r)!}$$

$$\times \sum_{\nu_1+2\nu_2+\ldots=l-r} k(\nu, \nu_2, \ldots)\frac{(1!)^{\nu_1}(2!)^{\nu_2}\ldots}{1^{\beta\nu_1}2^{\beta\nu_2}\ldots}(\nu_1+\nu_2+\ldots)!\left(\frac{A_1d}{B\Delta}\right)^{\nu_1+\nu_2+\ldots}.$$

From (33.10) it follows that $(A_1d/B\Delta)^{k-1} \leq \vartheta^k$, $k \geq 2$, and so

$$|\widetilde{u}_l(v)| < 2A_1B^{l-1}\frac{F_0}{|F_0(v)|}\cdot\frac{\eta_\beta^{(l)}(0, \vartheta)}{l!} + 2\frac{\Delta\vartheta}{d\,|F_0(v)|}\sum_{r=1}^{l-1}B^{l-r}F_r\frac{\eta^{(l-r)}(0, \vartheta)}{(l-r)!}.$$

From this, via Lemma 1.7 and inequalities (33.8) and (33.9), we conclude that

$$|\widetilde{u}_l(v)| < 2\gamma\vartheta M_0\Big\{A_1B^{l-1}M_0l^{-\beta} + \frac{\Delta\vartheta}{d}AB^{l-1}\sum_{r=1}^{l-1}r^{-\beta}(l-r)^{-\beta}\Big\}.$$

Using the obvious inequality

$$\sum_{r=1}^{l-1}r^{-\beta}(l-r)^{-\beta} < \frac{2^{\beta+1}}{l^\beta}\zeta(\beta),$$

we get that

$$|\widetilde{u}_l(v)| < \gamma\vartheta M_0A_1B^{l-1}[2M_0 + \vartheta 2^{\beta-1}\zeta(\beta)], \tag{33.17}$$

and so on the basis of (33.11) we have

$$|\widetilde{u}_l(v)| < \tfrac{1}{2}A_1B^{l-1}l^{-\beta}.$$

Therefore from (33.12) and (33.16) it follows that $|u_l(v)| < A_1B^{l-1}l^{-\beta}$ if $v \in G_\Delta$. Since $|u_1(v)| \leq A_1$ for $v \in G_\Delta$, as follows from (33.15) and (33.12), we have that $|u_l(v)| \leq A_1B^{l-1}l^{-\beta}$, $1 \leq l \leq m$, and $|u_l(v)| \leq A_1B_1^{l-1}l^{-\beta}$, $l \geq m+1$, provided $v \in G_\Delta$. It follows that for $|\delta| \leq 1/2B_1$ the series $v(1+\sum_1^\infty \delta^k u_k(v))$

converges in G_Δ to a regular function. We note that

$$|u(v)-v|<|v|\sum_1^\infty|\delta|^k A_1B_1^{k-1}k^{-\beta}<2|v||\delta|A_1\leqslant\vartheta^2\frac{\Delta}{d}|v|$$

and so, if $v\in G_\Delta$, then $u(v)\in G$. By construction $u(v)$ is a solution of the auxiliary differential equation. Setting $\delta=\epsilon$, we arrive at a solution of the desired differential equation having the form

$$u(v,\varepsilon)=u\Big(1+\sum_1^\infty\epsilon^k u_k(v)\Big),$$

where the functions $u_k(v)$ depend on ϵ for $k\geqq m+1$. Set $u_{m+1}(v,\epsilon)=\sum_{k=m+1}^\infty\epsilon^{k-m-1}u_k(v)$; using the inequalities derived above, we find for $\epsilon\leqq\min(\epsilon_0,1/2B_1)$ that

$$|u_{m+1}(v,\varepsilon)|<A_1B_1^m(m+1)^{-\beta}\sum_0^\infty(\varepsilon B_1)^k\leqslant 2A_1B_1^m(m+1)^{-\beta},$$

Q.E.D.

We now turn our attention to the construction of the variational formula for the function $f(z,y+\epsilon v)$. Denote by $Q(w,y)(dw/w)^2$ the quadratic differential corresponding to the function $f(z,y)$. We no longer adhere to our previous convention that the vector $\lambda(x)$ must be normalized, and we write

$$\lambda(x)=U\mu(y),$$

where $\mu(y)=(-\operatorname{grad}\psi(y),1)$, and where x is the image of y. If $Q(w,y)=\sum_1^n A_k(y)w^{-k}$, then, according to (10.3),

$$A_k(y)=-\sum_{l=k}^n\bar\lambda_l(x)\,a_l^{(k+1)},\qquad 1\leqslant k\leqslant n.$$

It is easy to see that $a_l^{(k+1)}$ is a polynomial in $a_1,\cdots,a_{l-k}$, of degree at most $l-k$, with integral coefficients, and with absolute value equal to 1. Since $x=x^0+U(y,\psi(y))$,

$$A_k(y)=\sum_{j=1}^{2n-1}\frac{\partial\psi}{\partial y_j}A_{kj}(y,\psi(y))+A_{k0}(y,\psi(y)),$$

where $A_{kj}(y,\psi(y))=A_{kj}(y)$, $0\leqq j\leqq 2n-1$, is a polynomial in $y_1,\cdots,y_{2n-1}$ and $\psi(y)$ of degree $n-k$, and where the coefficients depend only on the elements of the matrix U and on the coordinates of x^0. Hence the coefficients of this polynomial are bounded by some constant depending only on n.

We shall now make an a priori assumption on the smoothness of $\psi(y)$. We assume that $\psi(y)$ is $m+1$ times continuously differentiable, and that its $(m+1)$th derivative on the disc $\mathfrak{U}$ satisfies a Lipschitz condition with constant B_0. Let $v=(v_1,\cdots,v_{2n-1})$ be an arbitrary vector such that $|v_j|\leqq 1$, $1\leqq j\leqq 2n-1$. Denote the kth order derivative in the direction of v by

$d_v^k\psi(y)$. We suppose that for any vector v as above and for any y, $|y| < R$, the conditions

$$\text{(33.18)} \quad \frac{1}{k!}|d_v^k\psi_{y_l}(y)| \leqslant A_* B^{k-1}k^{-\beta}, \quad 1 \leqslant k \leqslant m, \quad 1 \leqslant l \leqslant 2n-1,$$

are satisfied. We shall assume that the constant B is sufficiently large; the specific choice of such a constant will be indicated later. If ϵ is sufficiently small, in particular small enough that $|y+\epsilon v| < R$, then we see by Taylor's formula that

$$A_k(y+\varepsilon v) = \sum_{j=0}^{m} \varepsilon^j \tilde{A}_{kj}(y) + \varepsilon^{m+1}\tilde{A}_{km+1}(y, \varepsilon).$$

Without loss of generality we assume that on $\mathfrak{U}$ we have $|\operatorname{grad}\psi(y)| < C_*$, where C_* is an absolute constant. Then, using (33.18), we can easily prove that

$$|\tilde{A}_{kj}(y)| < A^* B^{j-1} j^{-\beta}, \quad 1 \leqq k \leqq n, \quad 1 \leqq j \leqq m,$$

where $A^* < CA_*$, with the constant C depending only on n. We shall in fact prove this inequality in a slightly more general form than the form stated above.

We begin by taking the domain G and its subdomain G_Δ such that the distance from ∂G to ∂G_Δ is not less than Δ. Assume that $\Phi(w,\epsilon)$ is a regular function of w in G, provided $0 \leqq \delta \leqq \epsilon_0$. Suppose also that $\Phi(w,\epsilon)$ is m times continuously differentiable on the interval $0 \leqq \epsilon \leqq \epsilon_0$, and that

$$\max_{w \in G}\left|\left\{\frac{\partial^k \Phi(w, \varepsilon)}{\partial \varepsilon^k}\right\}_{\varepsilon=0}\right| \leqslant k!\, M_k, \quad 0 \leqslant k \leqslant m.$$

Finally, we suppose that $\varphi(\epsilon)$ is m times continuously differentiable for $0 \leqq \epsilon \leqq \epsilon_0$, that $\varphi(0) \in G_\Delta$, and that

$$\frac{1}{k!}|\varphi^{(k)}(0)| \leqslant AB^{k-1}k^{-\beta}, \quad 1 \leqslant k \leqslant m.$$

Proposition 33.3. *If the constants A, B and Δ satisfy inequality* (33.3), *then, for any function* $\Psi(\epsilon) = \Phi[\varphi(\epsilon)\epsilon]$ *as above,*

$$\frac{1}{k!}|\Psi^{(k)}(0)| \leqslant M_k + \Delta^{-1}\left(1 + \gamma\sqrt{\frac{A}{B\Delta}}\right) AB^{k-1} \sum_{l=0}^{k-1} \frac{M_l B^{-l}}{(k-l)^\beta}, \quad 1 \leqslant k \leqslant m.$$

Proof. We set

$$F_l(w) = \frac{1}{l!}\left\{\frac{\partial^l}{\partial \varepsilon^l}\Phi(w, \varepsilon)\right\}_{\varepsilon=0}, \quad 0 \leqslant l \leqslant m.$$

Let $h > 0$ be sufficiently small. We consider the function $\varphi_h(\epsilon) = h^{-1}\int_0^h \varphi(t+\epsilon)\,dt$, $0 \leqq \epsilon \leqq \epsilon - h$. By Taylor's formula, $\varphi_h(\epsilon) = \sum_0^m \epsilon^k \varphi_{h,k} + \epsilon^{m+1}\varphi_{h,m+1}(\epsilon)$, where $\varphi_{h,k} \to \varphi^{(k)}(0)/k!$, $1 \leqq k \leqq m$, as $h \to 0$. It is easily seen that the quantities A_h, B_h and Δ_h can be determined so that:

1) $|\varphi_{h,k}| \leqq A_h B_h^{k-1} k^{-\beta}$, $1 \leqq k \leqq m$;

2) inequality (33.3) is satisfied;

3) the disc $|w - \varphi_{h,0}| \leqq \Delta_h$ lies in G, and $A_h \to A$, $B_h \to B$ and $\Delta_h \to \Delta$ as $h \to 0$.

Applying Proposition 33.1 to the functions $F_l(w)$ and $\varphi_h(\epsilon)$, and then letting $h \to 0$, we find that

$$\frac{1}{k!}\left|\left\{\frac{d^k}{d\varepsilon^k} F_l\,[\varphi(\varepsilon)]\right\}_{\varepsilon=0}\right| \leqslant \frac{M_l}{\Delta}\left(1+\gamma\sqrt{\frac{A}{B\Delta}}\right) AB^{k-1}k^{-\beta}, \qquad 1 \leqslant k \leqslant m.$$

Since

$$\frac{1}{k!}\,\Psi^{(k)}(0) = \sum_{l=0}^{k} \frac{1}{l!}\left\{\frac{d^l}{d\varepsilon^l}\,\mathcal{F}_{k-l}\,[\varphi(\varepsilon)]\right\}_{\varepsilon=0},$$

the desired inequality now follows from the previous inequalities.

We now apply this proposition to the case under consideration. We first must derive an estimate for the derivatives $d_v^k\psi(y)$. It is easily seen that

$$d_v^k\psi(y) = \sum_{j=1}^{2n-1} v_j d_v^{k-1}\psi_{v_j}(y), \qquad 2 \leqslant k \leqslant m+1.$$

Thus

$$\frac{1}{k!}\,|\,d_v^k\psi(y)| \leqslant (2n-1)\;A_* B^{k-2}\;(k-1)^{-\beta},$$

and therefore

$$\frac{1}{k!}|\,d_v^k\psi(y)| < 2^{\beta}\,(2n-1)\;A_* B^{k-1}k^{-\beta},$$

since the constant B can be chosen without loss of generality to be greater than 1. This latter inequality also holds for $k=1$ in view of our argument concerning $|\operatorname{grad}\psi(y)|$. Thus

$$\frac{1}{k!}\,|\,d_v^k\;\psi(y)\,| < 2^{\beta}(2n-1)\,A_* B^{k-1}k^{-\beta}, \qquad 1 \leqslant k \leqslant m+1.$$

We now apply Proposition 33.3 to the function $A_{kj}(y, \psi(y))$. Below the symbol O will denote constants depending only on n. Recalling the remarks we made concerning the coefficients of the polynomial $A_{kj}(y, w)$, we have

$$\left\{\frac{d^k}{d\varepsilon^k}\,A_{lj}\,(y+\varepsilon v,\,w)\right\}_{\varepsilon=0} = O\,(1),$$

provided $|y| < R$ and $|w| \leqq D_n$, where D_n is the diameter of $\mathscr{V}_n$. Therefore

$$\frac{1}{k!}\left\{\frac{d^k}{d\varepsilon^k}\;A_{lj}\,(y+\varepsilon v,\;\psi)\,(y+\varepsilon v)\right\}_{\varepsilon=0} = O\,(A_* B^{k-1}k^{-\beta}),\;\; 1 \leqslant k \leqslant m,$$

since we have already assumed that A_*/B is less than 1. From this we can easily derive, via Leibniz' formula, that

$$\frac{1}{k!}\;\left\{\frac{d^k}{d\varepsilon^k}\,A_l(y+\varepsilon v)\right\}_{\varepsilon=0} = 0\,(A_* B^{k-1}k^{-\beta}),\;\; 1 \leqslant k \leqslant m,\; 1 \leqslant l \leqslant n.$$

If we set

$$Q(w,\ y+\varepsilon v)=Q(w,\ y)+\sum_{1}^{m}\varepsilon^{k}Q_{k}(w,\ y;\ v)+\varepsilon^{m+1}Q_{m+1}(w,\ y;\ \varepsilon;\ v),$$

then from the estimates derived above it follows that

$$\max_{|w|\geqslant\frac{1}{8}}|wQ_{k}(w,\ y;\ v)|\leqslant AB^{k-1}k^{-\beta},\qquad 1\leqslant k\leqslant m, \tag{33.19}$$

where $A=O(A_*)$. For $0\leqq\epsilon\leqq\epsilon_0$ the inequality

$$\max_{|w|\geqslant\frac{1}{8}}|wQ_{m+1}(w,\ y,\ \varepsilon;\ v)|\leqslant AB_1^m(m+1)^{-\beta} \tag{33.20}$$

is satisfied, where B_1 is determined from the Lipschitz constant B_0. Without loss of generality we assume that $B_1\geqq 2B$.

We now construct a univalent mapping of some neighborhood of the tree $\mathscr{T}_y$ onto a neighborhood of the tree $\mathscr{T}_{y+\epsilon v}$ in a manner such that the image of $\mathscr{T}_y$ will be $\mathscr{T}_{y+\epsilon v}$. For $|y|\leqq R$ the distance from $\mathscr{T}_y$ to the nearest zero of the function $Q(w,y)$ is greater than some constant $C(y)>0$. Let $\mathscr{L}_r=\mathscr{L}_{r,y}$ be the level curve $\mathscr{L}_r=\{w:|f(z,y)|=r\}$, and let $\mathscr{G}_r$ be the complement of $\mathscr{G}(\mathscr{L}_r)$. It is easily verified that there exists r^*, independent of y, such that in $\mathscr{G}(\mathscr{L}_{r^*},\mathscr{T}_y)$ the function $wQ(w,y)$ does not vanish for $|y|\leqq R$. In order to prove this, let $w_1(y)$ be any zero of $Q(w,y)$, and let $z_1(y)$, $|z_1(y)|<1$, be its image. Assume that $\sup_{|y|\leqq R}|z_1(y)|=1$. It follows that there exists a convergent sequence $\{y^k\}$, $\lim_{k\to\infty}y^k=y^0$, such that $\lim_{k\to\infty}|z_1(y^k)|=1$. On $|z|=1$ the quasipolynomial $P(z,y^0)$ has a unique double path. Therefore there exists $\rho<1$ such that in the annulus $\rho\leqq|z|\leqq\rho^{-1}$ the function $P(z,y^0)$ has only one double path. Since $\lim_{k\to\infty}P(z,y^k)=P(z,y^0)$ uniformly in the above annulus, the relation $\lim_{k\to\infty}|z_1(y^k)|=1$ is a contradiction.

We set $r_1=\frac{1}{2}(1+r^*)$. Then

$$\min_{|y|\leqslant R}\min_{w\in\mathscr{G}_{r_1}}|wQ(w,\ y)|>0.$$

Since $\max_{w\in\mathscr{G}_{r_1}}|wQ(w,y)|$ is bounded above by a constant depending only on n, there exists a constant $\mathscr{M}_0$ such that

$$\mathscr{M}_0^{-1}\leqq|wQ(w,y)|\leqq\mathscr{M}_0,$$

if $w\in\mathscr{G}_{r_1}$ and $|y|\leqq R$.

Let $\alpha_0=1,\ \alpha_1,\cdots,\alpha_m$ be real constants, and let $\alpha_{m+1}(\epsilon)$ be a real-valued function of ϵ. Assume that the constants α_k satisfy

$$|\alpha_k|\leqq LAB^{k-1}k^{-\beta},\qquad 1\leqq k\leqq m, \tag{33.21}$$

where L is a constant to be determined below. Also assume that

$$|\alpha_{m+1}(\epsilon)|<LAB_1^m(m+1)^{-\beta}. \tag{33.22}$$

We set

$$\sum_0^m \varepsilon^k \widetilde{Q}_k(w) + \varepsilon^{m+1}\widetilde{Q}_{m+1}(w, \varepsilon) = \left(\sum_0^m \varepsilon^k \alpha_k + \varepsilon^{m+1}\alpha_{m+1}(\varepsilon)\right) Q(w, y+\varepsilon v).$$

From this, upon denoting $Q(w,y)$ by $Q_0(w,y;v)$, we find that

$$\widetilde{Q}_k(w) = \sum_{l=0}^{k} \alpha_{k-l} Q_l(w, y; v) = \alpha_k Q_0(w, y; v) + Q_k(w, y; v) + \hat{Q}_k(w), \ 1 \leqslant k \leqslant m,$$

and

$$\widetilde{Q}_{m+1}(w) = \sum_{k+l \geqslant m+1} \varepsilon^{k+l-m-1} \alpha_k Q_l(w, y; v)$$
$$= \alpha_{m+1}(\varepsilon) Q_0(w, y; v) + Q_{m+1}(w, y, \varepsilon; v) + \hat{Q}_{m+1}(w, \varepsilon).$$

Without loss of generality we can assume that the level curve $\mathscr{L}_{r_1}$ lies in the domain $|w| > 1/8$, since if this is not the case we can always enlarge r^* to make it so. Then in view of (33.19) we have

$$|w\hat{Q}_k(w)| < 2^{\beta+1}\zeta(\beta) LA^2 B^{k-2} k^{-\beta}, \qquad 2 \leqq k \leqq m, \tag{33.23}$$

if $w \in \mathscr{G}_{r_1}$, and so

$$|w\widetilde{Q}_k(w)| < AB^{k-1}(L + 1 + 2^{\beta+1}\zeta(\beta) AB^{-1}L) k^{-\beta}.$$

Set $A_1 = A(L + 1 + 2^{\beta+1}\zeta(\beta) AB^{-1}L)$; we assume that B is so large that the quotient A_1/B is small. The smallest possible value for B will become apparent in the course of the proof. It is clear that for any fixed m the constant B can be defined so as to satisfy all the conditions which we impose upon it. It is important to note that such a choice can be made independent of m. The constant B_1 need not be chosen in such a strict way, and B_1 can depend on m. We need only the fact that $B_1 < \infty$.

It is easy to see that

$$|w\hat{Q}_{m+1}(w, \varepsilon)| < 2^{\beta+1}\zeta(\beta) A^2 B_1^{m-1} L (m+1)^{-\beta} (1 + \varepsilon B_1 + \ldots).$$

Assuming $\epsilon \leqq \min(\epsilon_0, 1/2B_1)$, we get that

$$|w\hat{Q}_{m+1}(w, \epsilon)| < 2^{\beta+2}\zeta(\beta) A^2 B_1^{m+1} L (m+1)^{-\beta},$$

and so

$$|w\widetilde{Q}_{m+1}(w, \epsilon)| < A_1 B_1^m (m+1)^{-\beta}.$$

Set $r_0^* = \frac{1}{4}(1 + 3r_1)$ and $r_0 = \frac{1}{2}(1 + r_1)$. In $\mathscr{G}_{r_0^*}$ we shall seek a solution $w(t)$ of the differential equation

$$\left[\sum_0^m \varepsilon^k \widetilde{Q}_k(w) + \varepsilon^{m+1}\widetilde{Q}_{m+1}(w, \varepsilon)\right] \left(\frac{dw}{w}\right)^2 = \widetilde{Q}_0(t) \left(\frac{dt}{t}\right)^2,$$

with the additional condition that $w(\infty) = \infty$. Since $\widetilde{Q}_0(w) = Q(w,y)$, $w(t)$ maps $\mathscr{T}_y$ to a trajectory of the differential

$$\left[\sum_0^m \varepsilon^k \alpha_k + \varepsilon^{m+1}\alpha_{m+1}(\varepsilon)\right] Q(w, y + \varepsilon v) \left(\frac{dw}{w}\right)^2.$$

But $\sum_0^m \epsilon^k\alpha_k + \epsilon^{m+1}\alpha_{m+1}(\epsilon) > 0$ for small ϵ, and so $w(t)$ maps $\mathscr{T}_y$ to a trajectory of the differential $Q(w, y+\epsilon v)(dw/w)^2$. The constants α_k, $1 \leqq k \leqq m$, and $\alpha_{m+1}(\epsilon)$ will be chosen such that the image of $\mathscr{T}_y$ will be $\mathscr{T}_{y+\epsilon v}$. Below we shall prove that such choices are possible.

We change variables via the formulas $\xi = 1/w$, $\eta = 1/t$. The images of the domains $\mathscr{G}_{r_1}$ and $\mathscr{G}_{r_\delta^*}$ will be denoted by G and G_Δ, respectively. In the proof of Proposition 33.2 we discussed the geometric characteristics of G. Using the notation introduced there, it is clear that $d < 8$, and by the distortion theorem we can estimate the quantities d_0 and d_1 from above and below, respectively, by constants depending only on r_1. According to a theorem of Goluzin, which holds for any $f(z) \in S$,

$$\left|\frac{1}{f(z_1)} - \frac{1}{f(z_2)}\right| \geqslant (1-|z_1|^2)^{1/2}(1-|z_2|^2)^{1/2}\left|\frac{z_1-z_2}{z_1z_2}\right|.$$

Therefore $\Delta \geqq \Delta_{r_1}$, where Δ_{r_1} is a constant depending only on r_1.

After the change of variables, our differential equation becomes

$$\xi^{-1}\left[\sum_0^m \varepsilon^k \widetilde{Q}_k(\xi^{-1}) + \varepsilon^{m+1}\widetilde{Q}_{m+1}(\xi^{-1}, \varepsilon)\right]\frac{(d\xi)^2}{\xi} = \eta^{-1}\widetilde{Q}_0(\eta^{-1})\frac{(d\eta)^2}{\eta}.$$

Set $q_k(\xi) = \xi^{-1}\widetilde{Q}(\xi^{-1})$, $0 \leqq k \leqq m$, $q_{m+1}(\xi, \epsilon) = \xi^{-1}\widetilde{Q}_{m+1}(\xi^{-1}, \epsilon)$. Clearly $q_k(\xi)$ and $q_{m+1}(\xi, \epsilon)$ are polynomials in ξ of degree at most $n-1$. In order to apply Proposition 33.2, we rewrite the above equation as

$$\left(\frac{q_0(\xi)}{q_0(\eta)}\right)^{1/2}\left(1 + \sum_1^m \varepsilon^k \frac{q_k(\xi)}{q_0(\xi)} + \varepsilon^{m+1}\frac{q_{m+1}(\xi, \varepsilon)}{q_0(\xi)}\right)^{1/2}\left(\frac{\eta}{\xi}\right)^{1/2}\frac{d\xi}{d\eta} = 1.$$

We now apply Propositions 33.1 and 33.2 (in that order). All the constants arising from these propositions are either absolute constants, constants depending on r_1, or constants depending on the quantity $\mathscr{M}_0$ used to majorize $q_0(\xi)$:

$$\mathscr{M}_0^{-1} \leqq |q_0(\xi)| \leqq \mathscr{M}_0, \qquad \xi \in G.$$

These constants depend neither on m nor on y. The symbol O will now be used in cases when the constants involved depend only on n, r_1 or $\mathscr{M}_0$. Since for $\epsilon \leqq \epsilon_0$

$$\left|\frac{1}{q_0(\xi)}\left[\sum_1^m \varepsilon^k q_k(\xi) + \varepsilon^{m+1}q_{m+1}(\xi, \varepsilon)\right]\right| \leqslant \frac{1}{2}, \qquad \xi \in G,$$

it follows from Proposition 33.1 that

$$[q_0(\xi)]^{1/2}\left[1 + \frac{1}{q_0(\xi)}\left(\sum_1^m \varepsilon^k q_k(\xi) + \varepsilon^{m+1}q_{m+1}(\xi, \varepsilon)\right)\right]^{1/2} = \sum_0^m \varepsilon^k F_k(\xi) + \varepsilon^{m+1}F_{m+1}(\xi, \varepsilon),$$

where

$$F_k(\xi) = O(A_1 B^{k-1} k^{-\beta}), \quad 1 \leq k \leq m, \quad \text{and} \quad F_{m+1}(\xi, \epsilon) = O(A_1 B^m (m+1)^{-\beta}).$$

By (33.4),

$$F_k(\xi) = \frac{q_k(\xi)}{2\sqrt{q_0(\xi)}} + \tilde{F}_k(\xi), \qquad 1 \leqslant k \leqslant m, \tag{33.24}$$

where, according to (33.5),

$$\tilde{F}_k(\xi) = O(A_1^{3/2} B^{k-3/2} k^{-\beta}), \quad \xi \in G, \quad 1 \leq k \leq m. \tag{33.25}$$

Analogously we have

$$F_{m+1}(\xi, \varepsilon) = \frac{q_{m+1}(\xi, \varepsilon)}{2\sqrt{q_0(\xi)}} + \tilde{F}_{m+1}(\xi, \varepsilon), \tag{33.26}$$

where

$$\tilde{F}_{m+1}(\xi, \epsilon) = O(A_1^{3/2} B^{-1/2} B_1^m (m+1)^{-\beta}). \tag{33.27}$$

Proposition 33.2 now guarantees the existence of a solution of the differential equation under consideration, and moreover the solution has the form

$$\xi = \eta \left(1 + \sum_1^m \varepsilon^k u_k(\eta) + \varepsilon^{m+1} u_{m+1}(\eta, \varepsilon)\right),$$

where $u_k(\eta) = O(A_1 B^{k-1} k^{-\beta})$, $1 \leq k \leq m$, $u_{m+1}(\eta, \epsilon) = O(A_1 B_1^m (m+1)^{-\beta})$ for $\eta \in G_\Delta$. Using (33.16) and (33.24), we see that

$$u_k(\eta) = -\frac{1}{2\sqrt{\eta q_0(\eta)}} \int_0^\eta q_k(\eta) \frac{d\eta}{\sqrt{\eta q_0(\eta)}} + \tilde{u}_k(\eta), \qquad 1 \leqslant k \leqslant m, \tag{33.28}$$

where $\tilde{u}_k(\eta)$, as a consequence of (33.17), (33.25) and (33.12), satisfies

$$\tilde{u}_k(\eta) = O(A_1^{3/2} B^{k-3/2} k^{-\beta}), \qquad \eta \in G_\Delta. \tag{33.29}$$

Since (33.16) holds also for $u_{m+1}(\eta, \epsilon)$, we see as above that

$$u_{m+1}(\eta, \varepsilon) = -\frac{1}{2\sqrt{\eta q_0(\eta)}} \int_0^\eta q_{m+1}(\eta, \varepsilon) \frac{d\eta}{\sqrt{\eta q_0(\eta)}} + \tilde{u}_{m+1}(\eta, \varepsilon),$$

where

$$\tilde{u}_{m+1}(\eta, \epsilon) = O(A_1^{3/2} B^{-1/2} B_1^m (m+1)^{-\beta}), \qquad \eta \in G_\Delta.$$

Changing back to the variables w and t, we find that

$$w = t\left[1 + \sum_1^m \varepsilon^k u_k\left(\frac{1}{t}\right) + \varepsilon^{m+1} u_{m+1}\left(\frac{1}{t}, \varepsilon\right)\right]^{-1}.$$

By Proposition 33.1

$$\left[1 + \sum_1^m \varepsilon^k u_k\left(\frac{1}{t}\right) + \varepsilon^{m+1} u_{m+1}\left(\frac{1}{t}, \varepsilon\right)\right]^{-1} = 1 + \sum_1^m \varepsilon^k h_k(t) + \varepsilon^{m+1} h_{m+1}(t, \varepsilon),$$

where

$$h_k(t) = O(A_1B^{k-1}k^{-\beta}), \quad 1 \leqq k \leqq m, \quad h_{m+1}(t,\epsilon) = O(A_1B_1^m(m+1)^{-\beta}), \quad t \in \mathscr{G}_{r_\delta^*}.$$

Hence we have

$$w = t\left[1 + \sum_1^m \varepsilon^k h_k(t) + \varepsilon^{m+1}h_{m+1}(t,\varepsilon)\right].$$

Using (33.4) and (33.28), we get that

$$h_k(t) = \frac{1}{2}\sqrt{\frac{t}{q_0\left(\frac{1}{t}\right)}}\int_0^{t^{-1}} q_k(\eta)\frac{d\eta}{\sqrt{\eta q_0(\eta)}} + \widetilde{h}_k(t), \tag{33.30}$$

where $\widetilde{h}_k(t)$, in view of (33.5) and (33.29), satisfies

$$\widetilde{h}_k(t) = O(A_1^{3/2}B^{k-3/2}k^{-\beta}), \qquad t \in \mathscr{G}_{r_\delta^*}. \tag{33.31}$$

Analogously

$$h_{m+1}(t,\varepsilon) = \frac{1}{2}\sqrt{\frac{t}{q_0\left(\frac{1}{t}\right)}}\int_0^{t^{-1}} q_{m+1}(t,\varepsilon)\frac{d\eta}{\sqrt{\eta q_0(\eta)}} + \widetilde{h}_{m+1}(t,\varepsilon), \tag{33.32}$$

where

$$\widetilde{h}_{m+1}(t,\epsilon) = O(A_1^{3/2}B^{-1/2}B_1^m(m+1)^{-\beta}), \qquad t \in \mathscr{G}_{r_\delta^*}. \tag{33.33}$$

We now show that the function we have constructed is in fact univalent in $\mathscr{G}_{r_0}$. Let G_0 be the image of $\mathscr{G}_{r_0}$ under the mapping $\eta = t^{-1}$. Then $G_0 \subset G_\Delta$, and the distance from ∂G_0 to ∂G_1 is bounded below by some constant depending only on r_1. Thus $u_k'(\eta) = O(A, B^{k-1}k^{-\beta})$, $1 \leqq k \leqq m$, $u_{m+1}'(\eta,\epsilon) = O(A_1B_1^m(m+1)^{-\beta})$, provided $\eta \in G_0$. Thus, for sufficiently small ϵ, the function $\xi = \xi(\eta,\epsilon)$ is univalent in G_0, and so the function

$$w = \frac{1}{\xi\left[\frac{1}{t},\varepsilon\right]}$$

is univalent in $\mathscr{G}_{r_0}$.

We shall now rewrite (33.30) and (33.32) in a more convenient form. By the definition of the functions $q_k(\eta)$, $1 \leqq k \leqq m$, and $q_{m+1}(\eta,\epsilon)$ we have

$$h_k(t) = \frac{1}{2\sqrt{Q(t,y)}}\int_0^{t^{-1}} [\alpha_k Q(\eta^{-1},y) + Q_k(\eta^{-1},y;v)]\frac{d\eta}{\eta\sqrt{Q(\eta^{-1},y)}} + h_k^*(t), \tag{33.34}$$

where $h_k^*(t)$, on the basis of (33.21) and (33.31), satisfies the inequality

$$h_k^*(t) = O(A_1^{3/2}B^{k-3/2}k^{-\beta}), \qquad t \in \mathscr{G}_{r_\delta^*}. \tag{33.35}$$

Similarly,

$$\begin{aligned} h_{m+1}(t,\varepsilon) = \frac{1}{2\sqrt{Q(t,y)}}\int_0^{t^{-1}} [\alpha_{m+1}(\varepsilon)\,Q(\eta^{-1},y) + \\ + Q_{m+1}(\eta^{-1},y,\varepsilon;v)]\frac{d\eta}{\eta\sqrt{Q(\eta^{-1},y)}} + h_{m+1}^*(t,\varepsilon), \end{aligned} \tag{33.36}$$

where

(33.37) $$h^*_{m+1}(t,\epsilon) = O(A_1^{3/2}B^{-1/2}B_1^m(m+1)^{-\beta}).$$

If we now retrace our calculations from the beginning, it is easy to see that

(33.38) $$h_1^*(t) \equiv 0.$$

We now pass to the determination of the quantities $\alpha_1, \cdots, \alpha_m$ and $\alpha_{m+1}(\epsilon)$. Having set $t=f(z,y)$, $f(z,y)h_k[f(z,y)]=g_k(z)$, $1 \leqq k \leqq m$, $f(z,y)h_{m+1}[f(z,\eta),\epsilon] = g_{m+1}(z,\epsilon)$, we get that the function

$$f(z,y) + \sum_1^m \varepsilon^k g_k(z) + \varepsilon^{m+1} g_{m+1}(z,\varepsilon),$$

is univalent in an annulus $r_0 \leqq |z| < 1$. Thus the function

$$g(z,\varepsilon) = \sum_1^m \varepsilon^k g_k(z) + \varepsilon^{m+1} g_{m+1}(z,\varepsilon)$$

is admissible with respect to $f(z,y)$. Set $\Delta = (1-r_0)/2$; with $K_0 = \{z : r_0 \leqq |z| \leqq r_0 + \Delta\}$, we note that $f(z,y) = O((1-r_0)^{-2})$ if $z \in K_0$, by the distortion theorems, and so if $\epsilon \leqq \epsilon_0$ we have that $g_k(z) = O(A_1B^{k-1}k^{-\beta})$, $1 \leqq k \leqq m$, and $g_{m+1}(z,\epsilon) = O(A_1B_1^m(m+1)^{-\beta})$. Thus we can apply Lemma 1.6 to the admissible function $g(z,\epsilon)$. Via this lemma we get a function

$$F_*(z,\varepsilon) = \sum_0^m \varepsilon^k F_k(z) + \varepsilon^{m+1} F_{m+1}(z,\varepsilon), \qquad F_0(z) = f(z,y),$$

univalent in $|z| < 1$. We shall now define the constants α_k, $1 \leqq k \leqq m$, and $\alpha_{m+1}(\epsilon)$ so that

$$F_k'(0) = 0, \qquad 1 \leqq k \leqq m, \qquad F'_{m+1}(0,\epsilon) = 0.$$

If the function $\psi_k(z)$ (see §1 for notation) satisfies the condition

$$\operatorname{Re} \frac{1}{2\pi i} \int_{|z|=r} \psi_k(z) \frac{dz}{z} = 0, \qquad r_0 + \frac{1}{4}\Delta < r < r_0 + \frac{3}{4}\Delta,$$

then, according to our construction, $F_k'(0) = 0$. The constants α_k will be defined by induction. Since $\psi_1(z) = g_1(z)/Df(z,y)$, we get the equation

$$\operatorname{Re}\left\{\frac{1}{2\pi i} \int_{|z|=r} \frac{g_1(z)}{Df(z,y)} \frac{dz}{z}\right\} = 0$$

for α_1. From (33.34) and (33.38) it follows that

$$\alpha_1 \operatorname{Re} J(y)$$
$$= -\operatorname{Re}\left\{\frac{1}{2\pi i} \int_{|z|=r} \left[\frac{f(z,y)}{Df(z,y)} Q^{-\frac{1}{2}}(f(z,y),y) \int_0^{\frac{1}{f(z,y)}} \frac{Q_1(\eta^{-1},y;v)}{\sqrt{Q(\eta^{-1},y)}} \frac{d\eta}{\eta}\right] \frac{dz}{z}\right\},$$

(33.39)

where

$$J(y)=\frac{1}{2\pi i}\int_{|z|=r}\left[\frac{f(z,y)}{Df(z,y)}Q^{-\frac{1}{2}}(f(z,y),y)\int_0^{\frac{1}{f(z,y)}}Q^{1/2}(\eta^{-1},y)\frac{d\eta}{\eta}\right]\frac{dz}{z}.$$

(33.40)

We must show that $\operatorname{Re}J(y)\neq 0$ if $|y|\leqq R$. To show this, we change the path of integration with respect to z so that it consists of an arc of the circle $|z|=1$ and an arc of the circle $|z-z_0|=\delta$, where z_0 is a double zero of $P(z,y)$ lying on $|z|=1$. For $|z|=1$ the function $(Df(z,y)/f(z,y))Q^{1/2}[f(z,y),y]$ assumes only purely imaginary values, and the integral

$$\int_0^{\frac{1}{f(z,y)}}Q^{1/2}(\eta^{-1},y)\frac{d\eta}{\eta}$$

is real, since in this case the path of integration can be deformed into the image of a trajectory of the differential $Q(w,y)(dw/w)^2$. Therefore

$$\operatorname{Re}J(y)=\operatorname{Re}\left\{\frac{1}{2\pi i}\int_{\mathscr{C}_\delta}\left[\frac{f(z,y)}{Df(z,y)}Q^{-1/2}[f(z,y),y]\int_0^{\frac{1}{f(z,y)}}Q^{1/2}(\eta^{-1},y)\frac{d\eta}{\eta}\right]\frac{dz}{z}\right\},$$

where $\mathscr{C}_\delta$ is an arc of the circle $|z-z_0|=\delta$. Since $(Df(z,y)/f(z,y))Q^{1/2}[f(z,y),y]=iP^{1/2}(z,y)$, we see upon letting $\delta\to 0$ that

$$\operatorname{Re}J(y)=-[-2z_0^2P''(z,y)]^{-1/2}\int_{f(z_0,y)}^{\infty}\left[-\frac{Q(u)}{u^2}\right]^{1/2}du,$$

where the integration is over a branch of the tree $\mathscr{T}_y$. Therefore $\inf_{|y|\leqq R}|\operatorname{Re}J(y)|=I_0>0$. From (33.39) it follows that

$$|\alpha_1|<L_0\max_{w\in\mathscr{G}_{r_0}}|wQ_1(w,y;v)|,$$

where L_0 is a constant depending only on I_0, r_1 and $\mathscr{M}_0$. We now show that the remaining constants $\alpha_2,\cdots,\alpha_m$ can be defined in such a way that we can choose the constant L to be $2L_0$. Let $l\leqq m$, and assume that α_j has already been defined for $j\leqq l-1$, and that $|\alpha_j|\leqq 2L_0AB^{j-1}j^{-\beta}$. Using (1.29) and (1.30), we conclude that α_j must be defined to satisfy

$$\operatorname{Re}\left\{\frac{1}{2\pi i}\int_{|z|=r}\frac{g_l(z)}{Df(z,y)}\frac{dz}{z}\right\}=O(A_1^{3/2}B^{l-3/2}l^{-\beta}).$$

Applying (33.34), (33.35), and Bieberbach's distortion theorem, we find that

$$\alpha_l\operatorname{Re}J(y)=-\operatorname{Re}\left\{\frac{1}{2\pi i}\int_{|z|=r}\left[\frac{f(z,y)}{Df(z,y)}Q^{-1/2}[f(z,y),y]\int_0^{\frac{1}{f(z,y)}}Q_l(\eta^{-1},y;v)\right.\right.$$
$$\left.\left.\times\frac{d\eta}{\sqrt{\eta Q(\eta^{-1},y)}}\right]\frac{dz}{z}\right\}+O(A_1^{3/2}B^{l-3/2}l^{-\beta}).$$

Hence

$$|\alpha_l| \leqq (L_0 + C(A/B)^{1/2}L^{3/2})AB^{l-1}l^{-\beta},$$

where C is a constant depending only on I_0, r_1 and $\mathscr{M}_0$. If $2^{3/2}C(AL_0/B)^{1/2} \leqq 1$, then $|\alpha_l|_r \leqq 2L_0AB^{l-1}l^{-\beta}$. Considering that $B \geqq 8AL_0C^2$, we see that (33.21) holds with $L = 2L_0$. We now define $\alpha_{m+1}(\epsilon)$. In order that $F'_{m+1}(0,\epsilon) = 0$, it is necessary that

$$\operatorname{Re}\left\{\frac{1}{2\pi i}\int_{|z|=r}\sum_{k=m+1}^{\infty}\varepsilon^{k-m-1}\psi_k(z)\frac{dz}{z}\right\} = 0.$$

Note that $\psi_k(z) = \tilde{\psi}_k(z)$ for $k > m+1$, and for $k \geqq m+1$ we must substitute B_1 for B in (1.30). Thus for $\epsilon \leqq 1/2B_1$ the above equation can be written as

$$\operatorname{Re}\left\{\frac{1}{2\pi i}\int_{|z|=r}\psi_{m+1}(z)\frac{dz}{z}\right\} = O(\varepsilon A_1^{3/2}B^{-1/2}B_1^{m+1}(m+1)^{-\beta}).$$

Using (33.36) and (33.37), we find as above that $\alpha_{m+1}(\epsilon)$ is well defined, and that if $B_1 > 2B$ and $\epsilon < \min\{\epsilon_0, 1/2B_1\}$, then

$$|\alpha_{m+1}(\epsilon)| < LAB_1^m(m+1)^{-\beta}.$$

Thus $F_*(z,\epsilon) \in S$ and maps $|z| < 1$ onto the complement of the trajectory of $Q(w, y+\epsilon v)(dw/w)^2$ emanating from $w = \infty$; that is to say, onto the complement of the tree $\mathscr{T}_{y+\epsilon v}$. Hence the function

$$w = t\left[1 + \sum_{1}^{m}\varepsilon^k h_k(t) + \varepsilon^{m+1}h_{m+1}(t, \varepsilon)\right]$$

maps the domain $\mathscr{G}_{r_0}$ univalently onto a neighborhood of $\mathscr{T}_{y+\epsilon v}$ in a manner so that this tree is the image of $\mathscr{T}_y$. The functions $h_k(t)$, $1 \leqq k \leqq m$, and $h_{m+1}(t,\epsilon)$ are regular in $\mathscr{G}_{r_0^*}$. Therefore

$$th_k(t) = \frac{1}{2\pi i}\int_{\mathscr{G}_{r_0^*}}\frac{t^2}{t-\xi}h_k(\xi)\frac{d\xi}{\xi}, \qquad t \in \mathscr{G}_{r_0^*}, \qquad 1 \leqslant k \leqslant m,$$

and the same representation also holds for $h_{m+1}(t,\epsilon)$. Thus the admissible function $g(z,\epsilon)$ can be represented in the form

$$g(z, \varepsilon) = \frac{1}{2\pi i}\int_{|\zeta|=r_0^*}\frac{f^2(z, y)}{f(z, y) - f(\zeta, y)}\left(\frac{Df(\zeta, y)}{f(\zeta, y)}\right)^2\left[\sum_{1}^{m}\varepsilon^k\frac{g_k(\zeta)}{Df(\zeta, y)} + \varepsilon^{m+1}\frac{g_{m+1}(\zeta, \varepsilon)}{Df(\zeta, y)}\right]\frac{d\zeta}{\zeta}. \tag{33.41}$$

Thus we have shown that for fixed ϵ the function $g(z,\epsilon)$ belongs to the class $\mathscr{D}_f^0$. This very important fact will be useful in what follows. We formulate the above results in the following lemma.

Lemma 33.1. *Let $\psi(y)$ be $m+1$ times continuously differentiable in $|y|<R$, and suppose that its $(m+1)$th partial derivatives satisfy a Lipschitz condition with constant B_0. Assume that the derivatives of $\psi(y)$ satisfy inequality* (33.18). *Then, if A_*B^{-1} is sufficiently small, and if $\epsilon \leqq \epsilon_0$, the function $f(z,y)$ has an admissible function $g(z,\epsilon)$ such that $f(z,y)+g(z,y)$ maps $|z|=1$ onto the tree $\mathscr{T}_{y+\epsilon v}$, where $v=(v_1,\cdots,v_{2n-1})$, $|v_j|\leqq 1$, $1\leqq j\leqq 2n-1$. The function $g(z,\epsilon)$ is of the form*

$$g(z,\varepsilon)=\sum_{1}^{m}\varepsilon^k g_k(z)+\varepsilon^{m+1}g_{m+1}(z,\varepsilon),$$

where $g_k(z)=O(A_*B^{k-1}k^{-\beta})$, $1\leqq k\leqq m$, and $g_{m+1}(z,\epsilon)=O(A_*B_1^m(m+1)^{-\beta})$, for $r_0\leqq|z|\leqq r_0+\Delta$. The function $g(z,\epsilon)$ can also be represented in the form (33.41).

In the following lemma we establish a Taylor-like formula for $f(z,y)$ as a function of y.

Lemma 33.2. *Suppose that $\psi(y)$, the constants A_* and B, and the vector v satisfy the conditions in the previous lemma. Then, for $\epsilon\leqq\epsilon_0$ and $|z|\leqq r_0+3\Delta/4$,*

$$f(z,y+\varepsilon v)=f(z,y)+\sum_{1}^{m}\varepsilon^k f_k(z,y;v)+\varepsilon^{m+1}f_{m+1}(z,y,\varepsilon;v), \tag{33.42}$$

where

$$f_k(z,y;v)=O(A_*B^{k-1}k^{-\beta}),\quad 1\leqq k\leqq m,$$

and

$$f_{m+1}(z,y,\epsilon;v)=O(A_*B_1^m(m+1)^{-\beta}).$$

Remark 33.1. Applying Theorem 1 to the admissible function $g(z,\epsilon)$ and choosing the arbitrary constants $c_1,c_2,\cdots$ as in Lemma 1.4, we arrive at the following expansion:

$$f(z,y+\varepsilon v)=\sum_{0}^{\infty}\varepsilon^k f_k(z,\varepsilon),\qquad f_0(z,\varepsilon)=f(z,y). \tag{33.43}$$

Since $g(z,\epsilon)$, in view of (33.41), has the form of the standard type of admissible function (2.2), we can evaluate $f_1(z,\epsilon)$ and $f_2(z,\epsilon)$ by the formulas established in Chapter 1. If we now expand the series (33.43) up through the term of order ϵ^{m+1}, we arrive at (33.42).

§34. In this section we shall prove a rather simple lemma which will be the kernel of the proof of the theorem on the analyticity of $\psi(y)$. Since each tree $\mathscr{T}_y$, $|y|\leqq R$, has no singular vertices, $\mathfrak{W}$ lies in the complement of the set $\mathscr{A}_2$. Hence, according to Lemma 32.3, $\psi(y)$ is twice continuously differentiable.

We shall now refine one of the inequalities used in the proof of Proposition 31.1. Let $(y,0)$ be an arbitrary point in the disc $\mathfrak{U}$, and let x be its image. We consider the variation $f_*(z,\epsilon)$ of the function $f(z,y)$ with respect to an admissible function of the form (2.2), where the measure $\nu(dz)$ is allowed to depend on δ. We require that the measure $\nu(dz)$ and the constant c_1 be chosen such that the first coefficient variations satisfy $\delta a_k = t_k$, $1 \leqq k \leqq n$, where t_k is such that the vector $t=(t_1,\cdots,t_n)$ is connected with a preassigned vector v via formula (31.3). The function $f_*(z,\epsilon)$ determines a point $x^*(\epsilon) \in \mathscr{V}_n$ which, in general, is an interior point. If (y^*,τ^*) are the local coordinates of this point, then $(y^*,\tau^*)=U'(x^*(\epsilon)-x^0)$ and $\tau^*-\psi(y^*)\leqq 0$, where equality can hold only if $x^*(\epsilon)\in\partial\mathscr{V}_n$. From this we get

$$\tau^*-\tau-(\operatorname{grad}\psi(y),y^*-y)-[\psi(y^*)-\psi(y)-(\operatorname{grad}\psi(y),y^*-y)]\leqq 0.$$

But in connection with our condition on the normalization of the vector $\lambda(x)$ we have

$$\tau^*-\tau-(\operatorname{grad}\psi(y),y^*-y)=\operatorname{Re}\sum_1^n\bar\lambda_k(x)(\overset{*}{x}_k-x_k).$$

If L is a linear functional definable by $f(z,y)$, then

$$\operatorname{Re}\sum_1^n\bar\lambda_k(x)(\overset{*}{x}_k-x_k)=L[f_*(z,\varepsilon)-f(z,y)]=\sum_2^\infty\varepsilon^kL[f_k(z,\varepsilon)].$$

Therefore

$$\varepsilon^2L[f_2(z,\varepsilon)]+\varepsilon^3L[f_3(z,\varepsilon)]-[\psi(y^*)-\psi(y)-(\operatorname{grad}\psi(y),\ y^*-y)] + O(\varepsilon^4)\leqslant 0.$$

We now examine the expression in square brackets a bit more closely. As a consequence of our condition on the first variations,

$$y^*=y+\varepsilon v+\frac{\varepsilon^2}{2}\tilde y(\varepsilon).$$

Therefore

$$\psi(y^*)-\psi(y)-(\operatorname{grad}\psi(y),y^*-y)=\psi(y+\varepsilon v)-\psi(y)-\varepsilon d_v\psi(y) +\frac{\varepsilon^2}{2}(\operatorname{grad}(\psi(y+\varepsilon v)-\psi(y)),\tilde y(\varepsilon))+O(\varepsilon^4),$$

and so

$$\psi(y^*)-\psi(y)-(\operatorname{grad}\psi(y),y^*-y)=\psi(y+\varepsilon v)-\psi(y)-\varepsilon d_v\psi(y) +\frac{\varepsilon^3}{2}\sum_1^{2n-1}\tilde y_l(\varepsilon)\,d_v\psi_{y_l}(y)+o(\varepsilon^3).$$

Thus we have

$$\varepsilon^2L(f_2(z,\varepsilon))+\varepsilon^3L(f_3(z,\varepsilon))-[\psi(y+\varepsilon v)-\psi(y)-\varepsilon d_v\psi(y) +\frac{\varepsilon^3}{2}\sum_1^{2n-1}\tilde y_l(\varepsilon)\,d_v\psi_{y_l}(y)]+o(\varepsilon^3)\leqslant 0, \tag{34.1}$$

and equality can hold only when $x^*(\epsilon) \in \partial \mathscr{V}_n$.

Lemma 34.1. *For $|y| < R$ the second derivatives of $\psi(y)$ satisfy a Lipschitz condition.*

Proof. Applying Lemma 33.1 with $m = 0$, we construct an admissible function $g(z,\epsilon)$ corresponding to a vector v, and as in Remark 33.1 we represent $f(z,y+\epsilon v)$ in the form of the series (33.43). Let the measure $\nu(dz;\epsilon)$ be determined by the relation

$$\nu(dz;\epsilon) = \frac{g_1(z,\varepsilon)}{Df(z,y)} \frac{dz}{2\pi iz}, \quad z \in \mathscr{C}_0 = \{z : |z| = r_0^*\},$$

and suppose the constant $c_1(\epsilon)$ is determined by

$$\operatorname{Im}\left\{\int_{\mathscr{C}_0} \nu(dz;\varepsilon) + ic_1(\varepsilon)\right\} = 0.$$

Applying (34.1) (equality holds in this case), we get

$$\varepsilon^2 L(f_2(z,\varepsilon)) - [\psi(y+\varepsilon v) - \psi(y) - \varepsilon d_v \psi(y)] + O(\varepsilon^3) = 0.$$

We divide this relation by ϵ^2 and then take the limit as $\epsilon \to 0$. We first note, however, that, since $\psi(y)$ is twice continuously differentiable, $\lim_{\epsilon\to 0} Q_1(t,y,\epsilon;v) = Q_1(t,y;v)$, $t \neq 0$, and so $\lim_{\epsilon\to 0}\alpha_1(\epsilon) = \alpha_1$, where α_1 is determined by (33.39). We thus find that

$$\lim_{\varepsilon\to 0} h_1(t;\varepsilon) = h_1(t), \qquad \lim_{\varepsilon\to 0} \frac{g_1(z,\varepsilon)}{Df(z,y)} = \frac{g_1(z)}{Df(z,y)},$$

and $g_1(z) = f(z,y)h_1[f(z,y)]$. Determine the measure $\nu(dz)$ with support $\mathscr{C}_0$ by the relation

$$\nu(dz) = \frac{g_1(z)}{Df(z,y)} \frac{dz}{2\pi iz}, \qquad z \in \mathscr{C}_0. \tag{34.2}$$

Clearly $\lim_{\epsilon\to 0} c_1(\epsilon) = c_1$, where c_1 is defined by

$$\operatorname{Im}\left\{\int_{\mathscr{C}_0} \nu(dz) + ic_1\right\} = 0. \tag{34.3}$$

Since $L(f_2(z,\epsilon)) = \frac{1}{2}\delta^2 L$, where the second variation is defined by (4.18) with the measure $\nu(dz;\epsilon)$ and constant $c_1(\epsilon)$, we have $\lim_{\epsilon\to 0} L(f_2(z,\epsilon)) = \frac{1}{2}\delta^2 L[f(z,y);\nu,c_1]$, where $\nu(dz)$ and c_1 are defined by (34.2) and (34.3).

In this way we have

$$d_v^2\psi(y) = \delta^2 L[f(z,y);\nu,c_1].$$

Let $|y^1| < R$; define the first and second variation of the function $f(z,y^1)$ by the measure $\nu(dz)$ and the constant c_1. Let $t_1^1,\cdots,t_n^1$ be the first coefficient variations of $f(z,y^1)$, and let v^1 be the vector determined from $t^1 = (t_1^1,\cdots,t_n^1)$ via (31.3). From (32.8) it follows that $d_{v^1}^2\psi(y^1) = (D(y^1)v^1,v^1)$, and on the other hand we have by the definition of the function $(D(y)v,v)$ that $\delta^2 L[f(z,y^1);\nu,c_1] \leqq (D(y^1)v^1,v^1)$.

In other words, we have

$$\delta^2 L\,[f\,(z,\,y^1);\ \ \nu,\,c_1) - \delta^2 L\,[f\,(z,\,y);\ \ \nu_1,\,c_1] \leqslant d^2_{v^1}\psi\,(y^1) - d^2_v\psi\,(y).$$

From this we get that

$$d^2_v\psi\,(y^1) - d^2_{v^1}\psi\,(y^1) + \delta^2 L\,[f\,(z,\,y^1);\,\nu,\,c_1] - \delta^2 L\,[f\,(z,\,y);\,\nu,\,c_1] \leqslant d^2_v\psi\,(y^1) - d^2_v\psi\,(y).$$

Let $V(\nu)$ denote the total variation of the measure $\nu(dz)$, and set $\omega_0 = \max_{|z|\leq r_0}|f(z,y) - f(z,y^1)|$. From (34.3) it follows that $|c_1| < V(\nu)$. From (31.3) it follows that $|v - v^1| < |t - t^1|$, and from Proposition 3.1 and formula (2.13) we deduce that $|t - t^1| < \alpha^n\omega_0(V(\nu) + |c_1|)$, and so $|t - t^1| < 2\alpha^n\omega_0 V(\nu)$. From Proposition 5.1 and formula (4.18) it follows that

$$|\delta^2 L\,[f\,(z,\,y);\ \ \nu,\,c_1] - \delta^2 L\,[f\,(z,\,y^1);\ \ \nu,\,c_1]\,| < \alpha_1^n\,[V\,(\nu) + |\,c_1\,|\,]^2\,(\omega_0 + \omega_1),$$

where $\omega_1 = |\operatorname{grad}\psi(y) - \operatorname{grad}\psi(y^1)|$. Since $V(\nu)$ is bounded by a constant independent of y, the above results imply that

$$d^2_v\psi(y) - d^2_v\psi(y^1) \leq N_0[\omega_0 + |y - y^1|],$$

where N_0 is independent of both y and y^1. Interchanging the roles of y and y^1, we obtain the above inequality with the sign changed on the left-hand side, so that

$$|d^2_v\psi(y) - d^2_v\psi(y^1)| \leq N_0[\omega_0 + |y - y^1|].$$

To conclude the proof, it suffices to show that, as a function of y, $f(z,y)$ satisfies a Lipschitz condition. This will be proved below as a separate proposition. Then we shall have

$$|d^2_v\psi(y) - d^2_v(y^1)| \leq N|y - y^1|,$$

which will complete the proof of the present lemma.

PROPOSITION 34.1. *For $|z| < 1$ and $|y| < R$, $f(z,y)$ is continuously differentiable as a function of y. Its derivative in the direction of the vector v satisfies the inequality $|d_v f(z,y)| \leq N_1(1 - |z|)^{-3}$.*

PROOF. From (33.43) we have for $|z| \leq r_0 + 3\Delta/4$ that

$$f(z, y + \epsilon v) = f(z,y) + \epsilon f_1(z,\epsilon) + O(\epsilon^2), \tag{34.4}$$

with $f_1(z,\epsilon)$ defined via (2.11) and the measure $\nu(dz;\epsilon)$ and constant $c_1(\epsilon)$. We know that $\lim_{\epsilon\to 0} f_1(z,\epsilon) = \delta f(z;\nu,c_1)$, where δf is the first variation of $f(z,y)$ determined by the measure (34.2) and the constant c_1. Then

$$\lim_{\varepsilon\to 0}\frac{1}{\varepsilon}\,[f\,(z,\,y + \varepsilon v) - f\,(z,\,y)] = \delta f\,(z;\,\nu,\,c_1).$$

But $\delta f(z;\nu,c_1) = f_1(z,y;v)$, and so $d_v f(z,y) = f_1(z,y;v)$. It is easy to see that

$$g_1(z)/Df(z,y) = g_1(z,y)/Df(z,y)$$

is a continuous function of y if $|y| \leqq R$. Therefore $f(z,y)$ is continuously differentiable for $|z| \leqq r_0 + 3\Delta/4$. Applying the distortion theorem, we deduce easily that

$$|f_1(z,y;v)| < C_{r_0} V(\nu)(1-|z|)^{-3} \leqq N_1(1-|z|)^{-3}, \tag{34.5}$$

where the constant C_{r_0} depends only on r_0, and so N_1 is independent of y. Thus the differentiability of $f(z,y)$ has been proved for $|z| \leqq r_0 + 3\Delta/4$. In the derivation of Lemmas 33.1 and 33.2 the constant r_0 was chosen as small as possible. However, it is always possible to choose it arbitrarily close to 1, and so the formula

$$d_v f(z,y) = f_1(z,y;v)$$

thus holds for $|z| < 1$. From (34.5) it follows that

$$|f(z,y) - f(z,y^1)| < N_1(1-|z|)^{-3}|y-y^1|,$$

which completes the proof of the proposition.

We shall next show that $\psi(y)$ is three times continuously differentiable, and in the process a formula for $d_v^3\psi(y)$ will be given. Let $\nu_1(dz)$ be any measure with support $\mathscr{C}_0$, and set $\tilde{\nu}(dz,\epsilon) = \nu(dz) + \epsilon\nu_1(dz)$. We construct the variation $f_*(z,\epsilon) \in S$ of the function $f(z,y)$ with respect to the admissible function

$$\varepsilon \int_{\mathscr{C}_0} \frac{f^2(z,y)}{f(z,y) - f(\zeta,y)} \left(\frac{Df(\zeta,y)}{f(\zeta,y)}\right)^2 \tilde{\nu}(d\zeta,\varepsilon).$$

In the determination of this first variation we choose the arbitrary constant to be the constant defined via (34.3). According to Theorem 1

$$\tilde{f}_*(z,\varepsilon) = \sum_0^\infty \varepsilon^k \tilde{f}_k(z,\varepsilon), \qquad \tilde{f}_0(z,\varepsilon) = f(z,y).$$

It is clear that $\tilde{f}_1(z,\epsilon) = f_1(z,y;v) + \epsilon f_{11}(z)$, where $f_{11}(z)$ is the first variation of $f(z,y)$ determined from the measure $\nu_1(dz)$ and with the constant $c_{11} = 0$. We note that

$$f_1(z,y;v) = \sum_1^n t_k z^{k+1} + \cdots.$$

Set

$$f_{11}(z) = \sum_1^n t_k^1 z^{k+1} + \cdots.$$

It is clear that $\tilde{f}_2(z,\epsilon) = f_2(z) + \epsilon f_{21}(z) + \epsilon^2 f_{22}(z)$, where $f_2(z) = \frac{1}{2}\delta^2 f(z,y)$, and where the second variation of $f(z,y)$ is determined via the measure $\nu(dz)$ and the constant c_1. The function $f_{21}(z)$ is defined by the measures $\nu(dz)$ and $\nu_1(dz)$, and obviously depends linearly (over the real numbers) on $\nu_1(dz)$.

Similarly, $f_3(z,\epsilon)=f_3(z)+\epsilon f_{31}(z)+\cdots$. It is clear that $f_3(z)$ does not depend on $\nu_1(dz)$. If we denote the coefficients of the power series of $f_*(z,\epsilon)$ by $\tilde{a}_k(\epsilon)$, then

$$\tilde{a}_k(\varepsilon)=a_k(y)+\varepsilon t_k+\varepsilon^2\left(t_k^1+\frac{1}{2}\delta^2 a_k(y)\right)+\ldots,\qquad 1\leqslant k\leqslant n,$$

where $a_k(y)$ are the coefficients of the power series of $f(z,y)$ and where $\delta^2 a_k(y)$ are the second variations of these coefficients defined via $\nu(dz)$ and c_1.

Applying inequality (34.1), we get that

$$\varepsilon^2 L(f_2)+\varepsilon^3 L(f_{21}+f_3)-[\psi(y+\varepsilon v)-\psi(y)-\varepsilon d_v\psi(y) \tag{34.6}$$
$$+\frac{\varepsilon^3}{2}\sum_1^{2n-1}\tilde{y}_l(\varepsilon)\,d_v\psi_{y_l}(y)]+o(\varepsilon^3)\leqslant 0.$$

By what we have already shown, $L(f_2)=\frac{1}{2}\delta^2 L(f;\nu,c_1)=\frac{1}{2}d_v^2\psi(y)$. Set $x^*(\epsilon)=(\tilde{a}_1(\epsilon),\cdots,\tilde{a}_n(\epsilon))$; the local coordinates of $x^*(\epsilon)$ are determined from the relation $(y^*,\tau^*)=U'(x^*(\epsilon)-x)$, and so $y^*=y+\epsilon v+\epsilon^2 U_0(t^1+\frac{1}{2}\delta^2 x)+O(\epsilon^3)$, where U_0 is the rectangular $(2n-1)\times(2n)$ matrix gotten by eliminating the last row of U', and where $\delta^2 x=(\delta^2 a_1,\cdots,\delta^2 a_n)$. Thus we have $\tilde{y}(\epsilon)=U_0(2t^1+\delta^2 x)+O(\epsilon)$. We set $\tilde{v}=U_0 t^1$ and $\tilde{y}=U_0\delta^2 x$. If $\tilde{y}=(\tilde{y}_1,\cdots,\tilde{y}_{2n-1})$, we find that

$$\tilde{y}_l=\operatorname{Re}\sum_{k=1}^{n}\bar{\mu}_{lk}\delta^2 a_k,\qquad 1\leqslant l\leqslant 2n-1, \tag{34.7}$$

where the quantities μ_{lk} are the elements of the matrix U_0. It is easy to verify that $\sum_{k=1}^n|\mu_{lk}|^2=1$, $1\leq l\leq 2n-1$. Substituting in (34.6), we find that

$$\varepsilon^3 L(f_{21}+f_3)-\frac{\varepsilon^3}{2}\sum_1^{2n-1}(2\tilde{v}_l+\tilde{y}_l)\,d_v\psi_{y_l}(y)$$
$$-\left[\psi(y+\varepsilon v)-\psi(y)-\varepsilon d_v\psi(y)-\frac{\varepsilon^2}{2}d_v^2\psi(y)\right]+o(\varepsilon^3)\leqslant 0.$$

We divide this inequality by ϵ^3 and take the limit as $\epsilon\to+0$. We find that

$$L(f_{21}+f_3)-\frac{1}{2}\sum_1^{2n-1}(2\tilde{v}_l+\tilde{y}_l)\,d_v\psi_{y_l}(y)\leqslant\frac{1}{3!}\underline{D}_v^3\psi(y),$$

where $\underline{D}_v^3\psi(y)$ is the lower de la Vallée-Poussin third derivative. Since $L(f_{21})$ and $\sum_1^{2n-1}\tilde{v}_l d_v\psi_{y_l}(y)$ depend linearly on $\nu_1(dz)$, we conclude as a consequence of the arbitrariness of this measure that

$$L(f_3)-\frac{1}{2}\sum_1^{2n-1}\tilde{y}_l d_v\psi_{y_l}(y)\leqslant\frac{1}{3!}\underline{D}_v^3\psi(y). \tag{34.8}$$

§35. We shall show that the left-hand side of (34.8) is a continuous function of y. Toward this end we evaluate $L(f_3)$. For this we require a process of regularization for the function

$$F(z,\varepsilon) = f(z,y) + \varepsilon \int\limits_{\mathscr{C}_0} \frac{f^2(z,y)}{f(z,y)-f(t,y)} \left(\frac{Df(t,y)}{f(t,y)}\right)^2 \nu(dt). \tag{35.1}$$

We set $\mu(z,y) = g_1(z,y)/Df(z,y)$ and consider the piecewise analytic function

$$G(z) = \frac{1}{2\pi i} \int\limits_{|t|=r_0^*} \frac{f^2(z,y)}{f(z,y)-f(t,y)} \left(\frac{Df(t,y)}{f(t,y)}\right)^2 \mu(t) \frac{dt}{t}.$$

It is clear that if $|z| < r_0^*$, then $G(z) \equiv 0$, and if $|z| > r_0^*$, then $G(z) = g_1(z,y)$. According to Theorem 1 there exists a change of variable $z = \zeta \exp\{\varphi(\zeta,\epsilon)\}$ which regularizes the function (35.1). In the process of finding an expansion $\varphi(z,\epsilon) = \sum_1^\infty \epsilon^k \varphi_k(z)$, the functions $\varphi_k(z)$ will be defined to be piecewise analytic. As usual, we set

$$F_*(z,\varepsilon) = F[z\exp(\varphi(z,\varepsilon)),\varepsilon] = \sum_0^\infty \varepsilon^k F_k(z), \qquad F_0(z) = f(z,y).$$

The function $F_1(z)$ is determined by formula (1.5). From (2.8) it follows that

$$\varphi_1(z) = \int\limits_{|t|=r_0^*} \frac{t}{t-z} \mu(t,y) \frac{dt}{2\pi i t} + \int\limits_{|t|=r_0^*} \frac{\bar{t}z}{1-\bar{t}z} \overline{\mu(t,y) \frac{dt}{2\pi i t}} + i_1 c.$$

In view of the construction of $g_1(z,y)$ we have $\operatorname{Re} \int_{\mathscr{C}_1} \nu(dz) = 0$. Therefore from (34.3) we obtain

$$\int\limits_{|t|=r_0^*} \mu(t,y) \frac{dt}{2\pi i t} + ic_1 = 0,$$

and so

$$\varphi_1(z) = \int\limits_{|t|=r_0^*} \frac{z}{t-z} \mu(t,y) \frac{dt}{2\pi i t} + \int\limits_{|t|=r_0^*} \frac{\bar{t}z}{1-\bar{t}z} \overline{\mu(t,y) \frac{dt}{2\pi i t}}. \tag{35.2}$$

From (1.5) and (35.2) we get that $F_1(z) = Df(z,y)\varphi_1(z) + G(z)$, and so

$$F_1(z) = \int\limits_{|t|=r_0^*} \left[\frac{f^2(z,y)}{f(z,y)-f(t,y)} \left(\frac{Df(t,y)}{f(t,y)}\right)^2 + \frac{zDf(z,y)}{t-z}\right] \mu(t,y) \frac{dt}{2\pi i t}$$
$$+ Df(z,y) \int\limits_{|t|=r_0^*} \frac{z\bar{t}}{1-z\bar{t}} \overline{\mu(t,y) \frac{dt}{2\pi i t}}.$$

By (2.16)

$$F_2(z) = -\frac{1}{2} D^2 f(z,y) \varphi_1^2(z) + DF_1(z)\varphi_1(z) + Df(z,y)\psi(z), \tag{35.3}$$

and from (1.4) with $n = 3$ we get the simple formula

$$F_3(z) = D\left[\frac{1}{2} DF_1(z)\varphi_1^2(z) - \frac{1}{3} D^2 f(z,y)\varphi_1^3(z)\right] + DF_1(z)\psi(z) + Df(z,y)\chi(z), \tag{35.4}$$

where $\chi(z) = \varphi_3(z) - \varphi_2(z) D\varphi_1(z) - \frac{1}{2}\varphi_1^2(z) D\varphi_1(z)$. Since $\operatorname{Re}\varphi_3(z) = 0$ for $|z| = 1$, it is clear that $\operatorname{Re}\chi(z) = 0$ for $|z| = 1$. Considering $\psi(z)$ to be piecewise analytic in $|z| < 1$, we shall look for a $\psi(z)$ of the form

$$\psi(z) = \int\limits_{|t|=r_0^*} \frac{t}{t-z} \psi^*(t) \frac{dt}{2\pi i t} + \int\limits_{|t|=r_0^*} \overline{\frac{\bar{t}z}{1-\bar{t}z} \psi^*(t) \frac{dt}{2\pi i t}} + ic_2.$$

The limiting values of $\psi(z)$ on $|z| = r_0^*$ will be denoted by ψ_+ and ψ_-, the "+" sign denoting approach from the inside. Determining $\psi(z)$ by the condition that $F_2(z)$ be regular in $|z| < 1$, we get that

$$\psi^* = \frac{1}{2} \frac{D^2 f}{Df} \mu [(\varphi_1)_+ + (\varphi_1)_-] - \mu \frac{DF_1}{Df}.$$

Since $F_1(z) = Df(z,y)\varphi_1(z) + G(z)$, we have $F_1(z) = Df(z,y)[\varphi_1(z)]_+$ on $|z| = r_0^*$, and so $(\varphi_1)_+ + (\varphi_1)_- = 2F_1/Df - \mu$. From this it follows that

$$\psi^* = -\mu D\left[\frac{F_1}{Df}\right] - \frac{\mu^2}{2} \frac{D^2 f}{Df}. \tag{35.5}$$

Seeking a $\chi(z)$ of the form

$$\chi(z) = \int\limits_{|t|=r_0^*} \frac{t}{t-z} \chi^*(t) \frac{dt}{2\pi i t} + \int\limits_{|t|=r_0^*} \overline{\frac{\bar{t}z}{1-\bar{t}z} \chi^*(t) \frac{dt}{2\pi i t}} + ic_3,$$

we find, as above, that

$$\chi^* = -\frac{1}{Df} D\left\{\frac{1}{2} DF_1\mu [(\varphi_1)_+ + (\varphi_1)_-] - \frac{1}{3} D^2 f\mu [(\varphi_1)_+^2 + (\varphi_1)_+(\varphi_1)_- + (\varphi_1)_-^2]\right\} - \frac{DF_1}{Df}\psi^*,$$

and so

$$\chi^* = -\frac{1}{Df} D\left\{\mu \left[\left(F_1 - \frac{1}{2}\mu Df\right) D\left(\frac{F_1}{Df}\right) + \frac{1}{2}\mu \frac{D^2 f}{Df} F_1 - \frac{1}{3}\mu^2 D^2 f\right]\right\} - \frac{DF_1}{Df}\psi^*. \tag{35.6}$$

We now normalize the function $F_*(z,\epsilon)$. We set

$$f_*(z,\varepsilon) = \frac{F_*(z,\varepsilon)}{F_*'(0,\varepsilon)} = \sum_0^\infty \varepsilon^k f_k(z), \qquad f_0(z) = f(z,y).$$

Since $F_1'(0) = 0$, we have $f_1(z) = F_1(z) = f_1(z,y;v)$ and

$$f_2(z) = F_2(z) - F_2'(0)f(z,y) \quad \text{and} \quad f_3(z) = F_3(z) - F_3'(0)f(z,y) - F_2'(0)F_1(z).$$

Since $f_2(z)$ and $f_3(z)$ depend linearly on the constant c_2, the same is true of $\delta^2 a_k$ and $L(f_3)$; $L(f_3)$ is independent of c_3. Therefore the left-hand side of (34.8), considered as a function of c_2, has the form $\alpha + \beta c_2$, and since this function is bounded uniformly for all c_2 we have that $\beta = 0$. Hence we can set $c_2 = c_3 = 0$. For $|z| < r_0^*$ we have $F_1(z) = Df(z,y)\varphi_1(z)$, and since $F_1(z) = f_1(z,y;v)$, we see that $\varphi_1(z) = f_1(z,y;v)[Df(z,y)]^{-1}$. Therefore

$$F_2(z) = -\frac{1}{2} D^2 f \left(\frac{f_1}{Df}\right)^2 + Df_1 \frac{f_1}{Df} + Df\psi,$$

from which it follows that

$$(35.7) \qquad F_2(z) = \frac{1}{2} D \left[\frac{f_1^2(z, y; v)}{Df(z, y)}\right] + Df(z, y)\psi(z), \qquad |z| < r_0^*.$$

Similarly we get from (35.4) that for $|z| < r_0^*$

$$(35.8) \qquad F_3(z) = \frac{1}{6} D^2 \left[\frac{f_1^3(z, y; v)}{(Df(z, y))^2}\right] + Df_1(z, y; v)\psi(z) + Df(z, y)\chi(z).$$

We shall now calculate the values of the quantities $\tilde{y}_l$, $1 \leq l \leq 2n-1$. Prior to this we introduce the following quasipolynomials:

$$(35.9) \qquad \begin{aligned} P_l(z, y) = \sum_{k=1}^{n} \Big[\bar{\mu}_{lk} \sum_{r=1}^{k} (k - r + 1) a_{k-r}(y) z^{-r} \\ + \mu_{lk} \sum_{r=1}^{k} (k - r + 1) \overline{a_{k-r}(y)} z^r + k\mu_{lk} a_k(y)\Big], \end{aligned}$$

where $a_k(y)$, $k \geq 0$, are the coefficients of the power series for the function $f(z, y)$. Set

$$(35.10) \qquad R_l(z) = \sum_{k=1}^{n} (k+1)\bar{\mu}_{lk} z^{-k-1}, \qquad 1 \leq l \leq 2n-1.$$

Since $f_2(z) = \frac{1}{2}\sum_1^\infty \delta^2 a_k z^{k+1}$, it follows from (34.7) and (35.7) that

$$(35.11) \qquad \tilde{y}_l = \operatorname{Re}\left\{\frac{1}{2\pi i} \int_{|z|=r_0^*} \left[\frac{f_1^3(z, y; v)}{Df(z, y)} \mathrm{R}_l(z) + 2\psi^*(z) P_l(z, y)\right] \frac{dz}{z}\right\}, \qquad 1 \leq l \leq 2n-1.$$

In order to evaluate $L(f_3)$, we first note that

$$f_3(z) = F_3(z) - \chi(0) f(z, y) - \psi(0) F_1(z).$$

Thus for $|z| < r_0^*$ we have

$$\begin{aligned} f_3(z) = \frac{1}{6} D^2 \left[\frac{f_1^3(z, y; v)}{(Df(z, y))^2}\right] + Df_1(z, y; v)\psi(z) - f_1(z, y; v)\psi(0) \\ + Df(z, y)\chi(z) - f(z, y)\chi(0). \end{aligned}$$

We set

$$(35.12) \qquad R(z, y) = \frac{1}{6} \sum_1^n (k+1)^2 \bar{\lambda}_k z^{-k-1}.$$

It is easily seen that

$$(35.13) \quad J_1(y) = L\left[\frac{1}{6} D^2 \left(\frac{f_1^3(z, y; v)}{(Df(z, y))^2}\right)\right] = \operatorname{Re}\left\{\frac{1}{2\pi i} \int_{|z|=r_0^*} R(z, y) \frac{f_1^3(z, y; v)}{(Df(z, y))^2} \frac{dz}{z}\right\}.$$

Denote by $P(z,f_1)$ the quasipolynomial (2.20) constructed in connection with the function $f_1(z,y;v)$. Since $f_1(z,y;v)=\sum_1^\infty t_k z^{k+1}+\cdots$, we have

(35.14)
$$P(z,f_1)=\sum_{k=2}^{n}\Big(\bar\lambda_k\sum_{r=1}^{k-1}(k-r+1)t_{k-r}z^{-r}+\lambda_k\sum_{r=1}^{k-1}(k-r+1)\bar t_{k-r}z^r\Big)+\sum_{k=1}^{n}k\bar\lambda_k t_k.$$

Performing some elementary calculations, we see that

(35.15)
$$\begin{aligned}J_2(y)&=L\,[Df_1(z,y;v)-f_1(z,y;v)\,\psi(0)]\\&=\operatorname{Re}\Big\{\frac{1}{2\pi i}\int_{|z|=r_0^*}\psi^*(z)\,P(z;f_1)\frac{dz}{z}\Big\}.\end{aligned}$$

Similarly,

(35.16)
$$J_3(y)=L\,[Df(z,y)\,\chi(z)-f(z,y)\,\chi(0)]=\operatorname{Re}\Big\{\frac{1}{2\pi i}\int_{|z|=r_0^*}\chi^*(z)\,P(z,y)\frac{dz}{z}\Big\}.$$

Upon summing, we find that

(35.17)
$$L(f_3)=J_1(y)+J_2(y)+J_3(y).$$

§36. It is now an easy matter to show that the left-hand side of (34.8) is a continuous function of y. In order to prove the continuity of the functions $\psi^*(z,y)$ and $\chi^*(z,y)$, it suffices to prove the continuity of the function $\mu(z,y)$ for $r_0^*\leqq|z|\leqq r_0$. But since

$$\mu(z,y)=\frac{f(z,y)}{Df(z,y)}h_1[f(z,y),y],$$

and since $h_1(t,y)$, in view of (33.32), (33.37), and (33.38), is both a regular function of t in $\mathscr{G}_{r_1}$ and a continuous function of y, it follows that $\mu(z,y)$ is a continuous function of y if $r_0^*\leqq|z|<1$. It now follows trivially that both terms of the left-hand side of (34.8) are continuous. Since $\partial^2\psi(y)/\partial y_k\partial y_l$, $1\leqq k,\ l\leqq 2n-1$, satisfy a Lipschitz condition, it follows that $d_v^3\psi(y)$ exists except possibly on a set of measure zero, and moreover this exceptional set does not depend on v. Thus the inequality

$$L(f_3)-\frac{1}{2}\sum_1^{2n-1}\tilde y_l\,d_v\psi_{y_l}(y)\leqslant\frac{1}{3!}d_v^3\psi(y)$$

is satisfied almost everywhere. However, both sides of this inequality are third degree forms in the variables $v_1,\cdots,v_{2n-1}$. Thus the relation

(36.1)
$$L(f_3)-\frac{1}{2}\sum_1^{2n-1}\tilde y_l\,d_v\psi_{y_l}(y)=\frac{1}{3!}d_v^3\psi(y)$$

is satisfied almost everywhere. We set

$$L(f_3) - \frac{1}{2}\sum_{1}^{2n-1} \tilde{y}_l\, d_v\psi_{y_l}(y) = \frac{1}{3!}\sum_{i,j,k=1}^{2n-1} c_{ijk}(y)\, v_i v_j v_k,$$

where the coefficients $c_{ijk}(y)$ are invariant under all permutations of the indices i, j, k. Then $c_{ijk}(y)$, $1 \leqq i, j, k \leqq 2n-1$, is a continuous function of y.

Let $K(y) \in C^\infty$ be a nonnegative function with support equal to the sphere $|y| \leqq 1$. As in §32 we introduce the kernel $K_\delta(y)$,

$$K_\delta(y) = K\left(\frac{y}{\delta}\right)\left(\int_{|y|\leqslant\delta} K\left(\frac{y}{\delta}\right)dy\right)^{-1}.$$

The average of a function by means of the kernel $K_\delta(y)$ will be denoted in the same manner as before. Since $d_v^2\psi(y)$ satisfies a Lipschitz condition, the sequence $\{(1/\epsilon)[d_v^2\psi(y+\epsilon v) - d_v^2\psi(y)]\}$ is bounded and converges as $\epsilon \to 0$ almost everywhere in the sphere $|y| \leqq R_1 < R$. Thus we have from well-known theorems that

$$\sum_{i,j,k=1}^{2n-1} [c_{ijk}(y)]_\delta\, v_i v_j v_k = \lim_{\varepsilon\to+0} \varepsilon^{-1}[d_v^2\psi_\delta(y+\varepsilon v) - d_v^2\psi_\delta(y)] = d_v^3\psi_\delta(y),$$

provided δ is sufficiently small. From this it follows that

$$\frac{\partial^3\psi_\delta(y)}{\partial y_i\,\partial y_j\,\partial y_k} = [c_{ijk}(y)]_\delta, \qquad 1 \leqslant i, j, \quad k \leqslant 2n-1. \tag{36.2}$$

Let $|y^1| \leqq R_1$, $|y^2| \leqq R_1$, $v^1 = (y^2 - y^1)/|y^2 - y^1|^{-1}$ and $v^1 = (v_1^1, \cdots, v_{2n-1}^1)$. It follows from (36.2) that

$$d_v^2\psi_\delta(y^2) - d_v^2\psi_\delta(y^1) = \int_0^h (\operatorname{grad} d_v^2\psi_\delta(y^1 + \chi v^1), v^1)\, d\chi$$

$$= \int_0^h \sum_{i,j,k=1}^{2n-1} [c_{ijk}(y^1 + \chi v^1)]_\delta\, v_i^1 v_j v_k\, d\chi, \qquad h = |y^2 - y^1|.$$

Since $\psi(y)$ is twice continuously differentiable, and since $c_{ijk}(y)$, $1 \leqq i, j, k \leqq 2n-1$, is a continuous function, we see upon taking the limit as $\delta \to 0$ above that

$$d_v^2\psi(y^2) - d_v^2\psi(y^1) = \int_0^h \sum_{i,j,k=1}^{2n-1} c_{ijk}(y^1 + \chi v^1)\, v_i^1 v_j v_k\, d\chi.$$

Since R_1 is arbitrary subject to the condition that $R_1 < R$, we see that the latter relation must in fact hold for any y^1 and y^2 such that $|y^1| < R$ and $|y^2| < R$. From this it follows that $\psi(y)$ is three times continuously differentiable, and at the same time we see that (36.1) must hold everywhere in $|y| < R$.

It is now almost obvious that $\psi(y)$ is an analytic function. To prove this, note that the left-hand side of (36.1) depends only on the first and second derivatives of $\psi(y)$, and at the same time the right-hand side is a third derivative. Equation (36.1) is the characteristic differential equation for $\psi(y)$. In order to conclude the proof, we must first overcome some relatively minor technical difficulties. The proof will proceed by induction. We assume that $\psi(y)$ is continuously differentiable through order $m+2$, and that the derivatives satisfy inequality (33.18) for $1 \leqq k \leqq m+1$. We shall show that these statements are also true with $m+3$ and $m+2$ replacing $m+2$ and $m+1$, respectively. In the process of this proof, we shall suppose that the constant B satisfies the conditions placed on it in §33.

PROPOSITION 36.1. *Let $F(y)$ be continuous for $|y| < R$, and for any y and any vector v such that $|v_j| \leqq 1$, $1 \leqq j \leqq 2n-1$, suppose that*

$$F(y+\varepsilon v) = F(y) + \sum_1^m \frac{\varepsilon^k}{k!} F_k(y; v) + \frac{\varepsilon^{m+1}}{(m+1)!} F_{m+1}(y, \varepsilon; v), \tag{36.3}$$

where $F_k(y; v)$, $1 \leqq k \leqq m$, is a kth degree form in the variables $v_1, \cdots, v_{2n-1}$ which is bounded for $|y| < R$: $|F_k(y; v)| \leqq M_k$, $1 \leqq k \leqq m$, and $|F_{m+1}(y, \epsilon; v)| \leqq M_{m+1}$ for $\epsilon \leqq \epsilon_y$. Assume that for any $R_1 < R$ we have $\epsilon_y \geqq \epsilon_{R_1}$ for $|y| < R_1$. Then $F(y)$ is m times continuously differentiable.

PROOF. Let $|y| \leqq R_1 < R$, and let δ be sufficiently small. Then for (36.3) we have for $\epsilon \leqq \epsilon_{R_1}$ that

$$F_\delta(y+\varepsilon v) = F_\delta(y) + \sum_1^m \frac{\varepsilon^k}{k!} F_{k,\delta}(y; v) + \frac{\varepsilon^{m+1}}{(m+1)!} F_{m+1,\delta}(y, \varepsilon; v).$$

Since $F_\delta(y)$ is infinitely differentiable, we have $F_{k,\delta}(y; v) = d_v^k F_\delta(y)$, $1 \leqq k \leqq m$, and $\lim_{\epsilon \to 0} F_{m+1,\delta}(y, \epsilon; v) = d_v^{m+1} F_\delta(y)$. Thus we have constructed a sequence $\{F_\delta(y)\}$ whose $(m+1)$th partial derivatives are bounded by a constant independent of δ. It follows that each of the sequences $\{d_v^k F_\delta(y)\}$, $1 \leqq k \leqq m$, is uniformly bounded and equicontinuous for $|y| \leqq R_1$. Thus from the sequence $\{\operatorname{grad} F_\delta(y)\}$ we can extract a subsequence which converges to some continuous function $H(y)$. Let y^1 and y^2 be arbitrary points of the sphere $|y| \leqq R_1$. Since

$$F_\delta(y^2) - F_\delta(y^1) = \int_0^h (\operatorname{grad} F_\delta(y^1 + \chi v^1), v^1)\, d\chi,$$

where $v^1 = h^{-1}(y^2 - y^1)$, $h = |y^2 - y^1|$, we have

$$F(y^2) - F(y^1) = \int_0^h (H(y^1 + \chi v^1), v^1)\, d\chi,$$

from which it follows that $F(y)$ is continuously differentiable in $|y| \leqq R_1$. This latter line of reasoning applies to the sequence $[F_{y_i}(y)]_\delta$, where $F_{y_i}(y) = \partial F/\partial y_i$, provided that $|y| \leqq R_2 < R$. We get that $F(y)$ is twice continuously differentiable in the sphere $|y| \leqq R_2$. Repeating the same process, we eventually find that $F(y)$ is m times continuously differentiable for $|y| \leqq R_m$. Since R_m can be taken as close to R as desired, $F(y)$ is m times continuously differentiable in $|y| < R$. Q.E.D.

Below, in the course of the proof of the theorem, the symbol O will be used to denote constants not depending on either m or y, but possibly depending on either absolute constants or on n, r_0, $\mathscr{M}_0$, or I_0, where r_0, $\mathscr{M}_0$ and I_0 are the constants defined in §33. By C^m we shall denote the set of functions m times continuously differentiable for $|y| < R$. We shall also say that the function $F(y)$, $|y| < R$, belongs to $C^m(A_*, B)$, if $F \in C^m$ and if its derivatives satisfy

$$\frac{1}{k!} d_v^k F(y) = O(A_* B^{k-1} k^{-\beta}), \qquad 1 \leqslant k \leqslant m.$$

$C_*^m(A_*, B)$ will denote the subset of C^m consisting of those functions whose derivatives satisfy

$$\frac{1}{k!} d_v^k F(y) = O(A_* B^k (k+1)^{-\beta+1}), \qquad 0 \leqslant k \leqslant m.$$

Below we shall assume that $\beta = 3$.

Proposition 36.2. *If $\psi(y)$ is $m+2$ times continuously differentiable and if its derivatives satisfy* (33.18) *for* $1 \leqq k \leqq m+1$, *then* $f(z,y) \in C^m(A_*, B)$, *provided* $|z| \leqq r_0 + 3\Delta/4$.

Proof. If $|y| \leqq R_1 < R$, then Lemma 33.2 holds for $\epsilon \leqq \min(\epsilon_0, 1/2B_1)$, where ϵ_0 is independent of y. The constant B_1 depends only on the maximum of the $(m+2)$th derivative of $\psi(y)$, and so is independent of y. Thus the proposition follows from Lemma 33.2 and the previous proposition.

Proposition 36.3. *Let $\varphi(\epsilon)$ be m times continuously differentiable for* $0 \leqq \epsilon \leqq \epsilon_0$, *and suppose that the quantities* $\varphi_l = \varphi^{(l)}(0)/l!$, $1 \leqq l \leqq m$, *satisfy inequality* (33.1). *Suppose $F(w)$ satisfies the conditions of Proposition* 33.1. *Then the quantities* $\Phi_l = (1/l!)\{d^l F[\varphi(\epsilon)]/d\epsilon^l\}_{\epsilon=0}$, $1 \leqq l \leqq m$, *satisfy the same inequalities as those stated in Proposition* 33.1.

Proof. This proposition is a special case of Proposition 33.3, namely when $\Phi(w,\epsilon) = F(w)$ is independent of ϵ.

Proposition 36.4. *Let* $F(y), G(y) \in C^m$, *and for* $|y| < R$ *suppose that* $(1/k!)|d_v^k F(y)| \leqq M_k$, $(1/k!)|d_v^k G(y)| \leqq N_k$, $0 \leqq k \leqq m$. *Then*

$$\frac{1}{k!} | d_v^k (FG) | \leqslant \sum_{l=0}^{k} M_l N_{k-l}, \qquad 0 \leqslant k \leqslant m. \tag{36.4}$$

Proof. By Leibniz' formula,

$$d_v^k (FG) = \sum_{l=0}^{k} C_k^l d_v^l F_v^{k-l} G, \qquad 0 \leqslant k \leqslant m,$$

and (36.4) now follows immediately.

COROLLARY 36.1. *If $|z| \leqq r_0$, then $Df(z,y), D^2f(z,y)/Df(z,y), Df^3(z,y)/Df(z,y)$, and $f(z,y)/Df(z,y)$ belong to $C^m(A_*,B)$. If $r_0^* \leqq |z| \leqq r_0$, then $1/Df(z,y)$ belongs to $C^m(A_*,B)$.*

The results stated above follow from the relations

$$D^l f(z, y) = \frac{1}{2\pi i} \int\limits_{|t|=r_0+\frac{3}{4}\Delta} f(t, y)\, D_z^l \frac{1}{(t-z)}\, dt, \qquad |z| \leqslant r_0, \qquad 1 \leqslant l \leqslant 3,$$

the distortion theorems of Bieberbach, and the following three propositions.

PROPOSITION 36.5. *If the conditions of Proposition 36.2 are satisfied and if $r_0^* \leqq |z| \leqq r_0$, then $\mu(z,y) \in C_*^m(A_*,B)$.*

PROOF. If $t \in \mathscr{G}_{r_1}$, then from (33.34) and (33.38) it follows that

$$h_1(t) = \frac{1}{2Q^{1/2}(t, y)} \left[\alpha_1(y) \int\limits_0^{t-1} Q^{1/2}(\eta^{-1}, y) \frac{d\eta}{\eta} + \int\limits_0^{t-1} \frac{Q_1(\eta^{-1}, y; v)}{\eta Q^{1/2}(\eta^{-1}, y)}\, d\eta\right].$$

We shall now consider the differentiability of the integral

$$j_1(t, y) = \int\limits_0^{t-1} \frac{Q_1(\eta^{-1}, y; v)}{\eta Q^{1/2}(\eta^{-1}, y)}\, d\eta = 2 \int\limits_0^{t-1} d_v \left[(\eta^{-1} Q(\eta^{-1}, y))^{1/2}\right] \frac{d\eta}{\eta^{1/2}}.$$

By construction the function $[wQ(w,y)]^{1/2}$ is regular in $\mathscr{G}_{r_1}$, and by hypothesis $[\eta^{-1}Q(\eta^{-1},y)]^{1/2}$ is $m+1$ times continuously differentiable. Thus $d_v[(\eta^{-1}Q(\eta^{-1},y))^{1/2}]$ is m times continuously differentiable. By assumption, inequality (33.19) is satisfied for $1 \leqq k \leqq m+1$; we write this inequality in the form

$$\frac{1}{k!} d_v^k [\eta^{-1} Q(\eta^{-1}, y)] = O(A_* B^{k-1} k^{-3}), \qquad 1 \leqslant k \leqslant m+1,$$

from which by (36.3) we get

$$\frac{1}{k!} d_v^k [(\eta^{-1} Q(\eta^{-1}, y))^{1/2}] = O(A_* B^{k-1} k^{-3}), \qquad 1 \leqslant k \leqslant m+1,$$

and hence

$$\frac{1}{k!} d_v^k \{d_v [(\eta^{-1} Q(\eta^{-1}, y))^{1/2}]\} = O(A_* B^k (k+1)^{-2}), \qquad 0 \leqslant k \leqslant m.$$

Thus we have that $j_1(t,y) \in C^m$ and its derivatives satisfy the inequalities

$$\frac{1}{k!} d_v^k j_1(t, y) = O(t^{-1/2} A_* B^k (k+1)^{-2}), \qquad 0 \leqslant k \leqslant m. \tag{36.5}$$

Similarly we can consider the function

$$j_0(t, y) = \int\limits_0^{t-1} [\eta^{-1} Q(\eta^{-1}, y)]^{1/2} \frac{d\eta}{\eta^{1/2}},$$

which is clearly $m+1$ times continuously differentiable, and in place of (36.5) we find that its derivatives satisfy

$$\frac{1}{k!} d_v^k j_0(t, y) = O(t^{-1/2} A_* B^{k-1} k^{-3}), \quad 1 \leqslant k \leqslant m. \tag{36.6}$$

By the results of §33 the functions $Q^{-1/2}(t,y)j_l(t,y)$, $l=0,1,\cdots$, are regular in $\mathcal{G}_{r_1}$. We now must estimate their derivatives. By Proposition 36.3 we have for $t \in \mathcal{G}_{r_1}$ that

$$\frac{1}{k!} d_v^k [(tQ(t, y))^{-1/2}] = O(A_* B^{k-1} k^{-3}), \quad 1 \leqslant k \leqslant m. \tag{36.7}$$

From (36.4) – (36.7) it follows that for $t \in \mathcal{G}_{r_1}$,

$$\frac{1}{k!} | d_v^k I_l(t, y) | \leqslant C \left[A_* B^{k-1+l} (k+l)^{l-3} + A_*^2 B^{k-2+l} \sum_{p=1}^{k-1+l} p^{-3} (k-p+l)^{-3+l} \right],$$

where $I_l(t,y) = Q^{-1/2}(t,y) j_l(t,y)$, $l=0,1$, and so

$$\frac{1}{k!} d_v^k I_l(t, y) = O(A_* B^{k-1+l} (k+l)^{l-3}), \quad 1-l \leqslant k \leqslant m, \quad l=0,1. \tag{36.8}$$

Recalling that $r_0^* \leqq |z| \leqq r_0$, we apply Proposition 33.3 to the functions $I_l[f(z,y),y]$, $l=0,1$. We find that $I_l[f(z,y),y] \in C^m$, and its derivatives satisfy

$$\frac{1}{k!} d_v^k I_l [f(z, y), y] = O(A_* B^{k-1+l} (k+l)^{l-3}), \quad 1-l \leqslant k \leqslant m, \quad l=0,1. \tag{36.9}$$

Applying Bieberbach's distortion theorem, Corollary 36.1, Proposition 36.4, and inequality (36.9), we find that

$$\frac{f(z, y)}{Df(z, y)} I_0 [f(z, y), y] \in C^m (A_*, B),$$

and

$$\frac{f(z, y)}{Df(z, y)} I_1 [f(z, y), y] \in C_*^m (A_*, B), \quad r_0^* \leqslant |z| \leqslant r_0.$$

Since the function

$$I(y) = \frac{1}{2\pi i} \int\limits_{|z|=r} \frac{f(z, y)}{Df(z, y)} I_0 [f(z, y), y] \frac{dz}{z}, \quad r_0^* \leqslant |z| \leqslant r_0,$$

satisfies the inequality $|\operatorname{Re} I(y)| \geqq I_0 > 0$ (as we saw earlier), it follows from Propositions 36.3 and 36.4 that $\alpha_1(y) \in C_*^m(A_*, B)$. Since

$$\mu(z, y) = \frac{f(z, y)}{2Df(z, y)} [\alpha_1(y) I_0 [f(z, y), y] + I_1 [f(z, y), y]],$$

it follows from Proposition 36.4 and the above that for $r_0^* \leqq |z| \leqq r_0$ we have $\mu(z,y) \in C_*^m(A_*, B)$. Q.E.D.

COROLLARY 36.2. *The derivatives of the functions* $\mu^l(z,y)$, $l=2,3$, *satisfy*

$$\frac{1}{k!} d_v^k \mu^l(z, y) = O(A_*^l B^k (k+1)^{-2}), \quad 0 \leqslant k \leqslant m, \quad l=2,3. \tag{36.10}$$

To prove this, we note that from (36.4) we have

$$\frac{1}{k!}|d_v^k\mu^2(z, y)| < CA_*^2B^k\sum_{l=0}^{k}(l+1)^{-2}(k-l+1)^{-2} = O(A_*^2B^k(k+1)^{-2}).$$

Applying (36.4) to the functions μ^2 and μ, we get (36.10) for $l=3$.

PROPOSITION 36.6. *If* $|z| \leqq \frac{1}{2}(r_0^* + r_0)$, *then* $f_1(z,y;v) \in C_*^m(A_*, B)$.

PROOF. We shall use the results of §35; since $F_1(z) = f_1(z,y;v)$, we have

$$f_1(z, y; v) = \int\limits_{|t|=r_0^*} \left[\frac{f^2(z, y)}{f(z, y) - f(t, y)}\left(\frac{Df(t, y)}{f(t, y)}\right)^2 + \frac{zDf(z, y)}{t-z}\right]\mu(t, y)\frac{dt}{2\pi it}$$

$$+ Df(z, y)\int\limits_{|t|=r_0^*}\overline{\frac{zt}{1-zt}\mu(t, y)\frac{dt}{2\pi it}}.$$

Thus

$$f_1(z, y; v) = \int\limits_{|t|=r_0} \left[\frac{f^2(z, y)}{f(z, y) - f(t, y)}\left(\frac{Df(t, y)}{f(t, y)}\right)^2 + \frac{zDf(z, y)}{t-z}\right]\mu(t, y)\frac{dt}{2\pi it}$$

$$+ Df(z, y)\int\limits_{|t|=r_0}\frac{z\bar{t}}{1-z\bar{t}}\overline{\mu(t, y)\frac{dt}{2\pi it}}.$$

If $|z| < \frac{1}{2}(r_0^* + r_0)$, then

$$f_1(z, y; v) = Df(z, y)\left[\int\limits_{|t|=r_0}\frac{z}{t-z}\mu(t, y)\frac{dt}{2\pi it} + \int\limits_{|t|=r_0}\frac{z\bar{t}}{1-z\bar{t}}\overline{\mu(t, y)\frac{dt}{2\pi it}}\right].$$

Applying Corollary 36.1, Proposition 36.5, and inequality (36.4), we find that $f_1(z,y;v) \in C_*^m(A_*, B)$, as required. Q.E.D.

COROLLARY 36.3. $Df_1(z,y;v) \in C_*^m(A_*, B)$ *for* $|z| \leqq r_0^*$.

To prove this, we note that

$$Df_1(z, y; v) = \frac{1}{2\pi i}\int\limits_{|t|=\frac{1}{2}(r_0^*+r_0)} f_1(t, y; v)\frac{z}{(t-z)^2}\,dt.$$

The desired result now follows from the above equality and Proposition 36.6.

COROLLARY 36.4. *For* $|z| \leqq r_0^*$ *the derivatives of the functions* $[f(z,y;v)]^l$, $l = 2, 3$, *satisfy*

$$\frac{1}{k!}d_v^k\{[f(z, y)]^l\} = O(A_*^lB^k(k+1)^{-2}), \qquad 0 \leqslant k \leqslant m, \qquad l = 2,3.$$

This result follows from repeated applications of (36.4).

PROPOSITION 36.7. *For* $|z| = r_0^*$ *the function* $\psi^*(z,y)$ *belongs to* C^m, *and its derivatives satisfy*

(36.11) $$\frac{1}{k!}d_v^k\psi^*(z, y) = O(A_*^2B^k(k+1)^{-2}),\ 0\leqslant k\leqslant m.$$

Proof. By (35.5),

$$\psi^*(z, y) = -\mu(z, y)D\left[\frac{f_1(z, y; v)}{Df(z, y)}\right] - \frac{1}{2}\mu^2(z, y)\frac{D^2f(z, y)}{Df(z, y)}.$$

Hence $\psi^*(z,y)\in C^m$. From Corollaries 36.1 – 36.3 and Bieberbach's distortion theorem we conclude that (36.11) holds.

Proposition 36.8. *For* $1\leqq l\leqq 2n-1$ *the function* $\tilde{y}_l(y)$ *belongs to* C^m, *and its derivatives satisfy*

$$\frac{1}{k!}d_v^k\tilde{y}_l(y) = O(A_*^2B^k(k+1)^{-2}),\quad 0\leqslant k\leqslant m,\quad 1\leqslant l\leqslant 2n-1.$$

Proof. The fact that $\tilde{y}_l\in C^m$ for $1\leqq l\leqq 2n-1$ follows from (35.11). The inequalities for the derivatives follow directly from Corollaries 36.1 and 36.4, and Propositions 36.4 and 36.7.

Proposition 36.9. $\Psi(y)=\sum_1^{2n-1}\tilde{y}_l(y)d_v\psi_{y_l}(y)\in C^m$, *and*

$$\frac{1}{m!}d_v^m\Psi(y) = O(A_*^3B^m(m+1)^{-2}).$$

Proof. The fact that $\Psi(y)\in C^m$ is obvious. Applying inequality (36.4), we find that

$$\frac{1}{m!}|d_v^m\Psi(y)| < C\sum_{l=1}^{2n-1}A_*^3B^m\sum_{p=0}^{m}(m-p+1)^{-2}(p+1)^{-2} = O(A_*^3B^m(m+1)^{-2}),$$

as required.

Proposition 36.10. *The functions* $R_1(z,y)$ *(see* (36.12)), $P(z;f_1)$ *(see* (35.14)), $P(z,y)$ *and* $DP(z,y)$ *belong to* $C^m(A_*, B)$, *provided* $|z|=r_0^*$.

Proof. The basic structure of $R_1(z,y)$ and $P(z;f_1)$ is the same, and so the proof will only be given in detail for $R_1(z,y)$. It follows from (35.12) that $R_1(z,y)$ depends linearly on the functions $\psi_{y_l}(y)$, $1\leqq l\leqq 2n-1$. In particular, $R_1(z,y)=\sum_1^{2n-1}\mathscr{R}_l(z)\psi_{y_l}(y)+\mathscr{R}_0(z)$. It now follows trivially that $R_1(z,y)\in C^m(A_*,B)$. The proof for $P(z,y)$ is a bit more complicated, since

$$P(z, y) = \sum_1^{2n-1}\mathscr{P}_l(z, y)\psi_{y_l}(y) + \mathscr{P}_0(z, y),$$

where $\mathscr{P}_l(z,y)$ is a linear form of y_l, $1\leqq l\leqq 2n-1$, and $\psi(y)$. However, in this case the fact that $P(z,y)\in C^m(A_*,B)$ follows trivially from Proposition 36.4.

Proposition 36.11. $J_l(y)\in C^m$, $l=1,2$, *and*

$$\frac{1}{m!}d_v^mJ_l(y) = O(A_*^3B^m(m+1)^{-2}),\qquad l=1,2.$$

PROOF. That $J_l(y) \in C^m$ is a trivial consequence of (35.13) and (35.15). The inequality follows from the previous proposition, Propositions 36.4 and 36.7, and Corollaries 36.1 and 36.4.

PROPOSITION 36.12. *$J_3(y) \in C^m$, and*

$$\frac{1}{m!} d_v^m J_3(y) = O(A_*^3 B^m (m+1)^{-2}).$$

PROOF. We use formulas (35.16) and (35.6), after having replaced $F_1(z)$ by $f_1(z, y; v)$ in the latter. Integrating by parts, we find that

$$J_3(y) = \mathrm{Re}\left\{\frac{1}{2\pi i} \int_{|z|=r_0^*} \left[\left(\mu f_1 - \frac{1}{2}\mu^2 Df\right) D\left(\frac{f_1}{Df}\right) + \frac{1}{2}\mu^2 f_1 \frac{D^2 f}{Df}\right.\right.$$
$$\left.\left. - \frac{1}{3}\mu^3 D^2 f\right] D\left(\frac{P}{Df}\right)\frac{dz}{z} - \frac{1}{2\pi i}\int_{|z|=r_0^*} \frac{Df_1}{Df}\psi^* P\frac{dz}{z}\right\}$$
$$= \mathrm{Re}\left\{J_4(y) - \frac{1}{2\pi i}\int_{|z|=r_0^*}\frac{Df_1}{Df}\psi^* P \frac{dz}{z}\right\}.$$

Thus $J_3(y) \in C^m$. It is not difficult to estimate the derivatives of

$$\Phi(z, y) = \frac{1}{Df(z, y)}\left\{\left[\mu(z, y) f_1(z, y; v) - \frac{1}{2}\mu^2(z, y) Df(z, y)\right] D\left(\frac{f_1(z, y; v)}{Df(z, y)}\right)\right.$$
$$\left. + \frac{1}{2}\mu^2(z, y) f_1(z, y)\frac{D^2 f(z, y)}{Df(z, y)} - \frac{1}{3}\mu^3(z, y) D^2 f(z, y)\right\}, \quad |z| = r_0^*,$$

since we may use Corollaries 36.1–36.4, and Propositions 36.4–36.6. It is easily seen that for $0 \leqq k \leqq m$ we have

$$\frac{1}{k!} d_v^k \Phi(z, y) = O(A_*^3 B^k (k+1)^{-2}).$$

Since

$$J_4(y) = \frac{1}{2\pi i}\int_{|z|=r_0^*} \Phi(z, y)\left[DP(z, y) - P(z, y)\frac{D^2 f(z, y)}{Df(z, y)}\right]\frac{dz}{z},$$

it follows from the above inequalities, Propositions 36.4 and 36.10, and Corollary 36.1 that

$$\frac{1}{m!} d_v^m J_4(y) = O(A_*^3 B^m (m+1)^{-2}).$$

It is also easy to prove that the derivatives of the integral

$$\frac{1}{2\pi i}\int_{|z|=r_0^*}\frac{Df_1}{Df}\psi^* P\frac{dz}{z}$$

satisfy the same inequality, and the desired estimate now follows immediately.

From Propositions 36.9, 36.11 and 36.12, and relation (36.1) it follows that $d_v^3 \psi(y) \in C^m$, so $\psi(y) \in C^{m+3}$. From the results derived so far we have

$$\frac{1}{m!}\, d_v^{m+3}\psi(y) = O\,(A_*^3 B^m\,(m+1)^{-2}),$$

and so

$$\text{(36.12)} \qquad \frac{1}{(m+3)!}\,|\, d_v^{m+3}\psi(y)\,| < C A_*^3 B^m\,(m+3)^{-2}\,(m+2),$$

where the constant C depends only on r_0, M_0, I_0, n, and absolute constants. We note that $d_v^{m+3}\psi(y)/(m+3)!$ is a form of degree $m+3$ in the variables v_j, $1 \leqq j \leqq 2n-1$. It is easy to see that

$$\frac{1}{(m+2)!}\, d_v^{m+2}\psi_{y_l}(y) = \frac{1}{(m+3)!}\,\frac{\partial}{\partial v_l}\,[d_v^{m+3}\psi(y)].$$

Since inequality (36.12) holds under the conditions $|v_l| \leqq 1$, $1 \leqq l \leqq 2n-1$, it follows from a classical inequality of V. A. Markov that

$$\left|\frac{\partial}{\partial v_l}\left[\frac{1}{(m+3)!}\, d_v^{m+3}\psi(y)\right]\right| < C A_*^3 B^m\,(m+2)^{-3}, \qquad 1 \leqslant l \leqslant 2n-1.$$

Requiring that $CA_*^2 B^{-2} \leqq 1$, we have

$$\frac{1}{(m+2)!}\,|\, d_v^{m+2}\psi_{y_l}(y)| \leqslant A_* B^{m+2}\,(m+2)^{-3}, \qquad 1 \leqslant l \leqslant 2n-1.$$

Since $\psi(y) \in C^3$, in the same way we have that $\psi(y) \in C^\infty$, and then inequality (33.18) holds for all $k \geqq 1$. Since $d_v^k\psi(y) = \sum_1^{2n-1} v_l d_v^{k-1}\psi_{y_l}(y)$, we have

$$\frac{1}{k!}\,|\, d_v^k\psi(y)\,| < (2n-1)\,A_*\,B^{k-2}\,(k-1)^{-4}, \qquad k > 2,$$

and this in turn implies that $\psi(y)$ is analytic for $|y| < R$. Hence we have proved

Theorem 15. *Let* $\mathfrak{W}$ *be the image of the disc* $\mathfrak{U} = \{(y,0):|y| < R\}$, *where* R *is sufficiently small. If* $\mathfrak{W} \subseteq \mathscr{R}_1^*$, *then* $\psi(y)$ *is analytic in the sphere* $|y| < R$.

From the intermediate results proved above it is easy to deduce the following theorem.

Theorem 16. *If the conditions of Theorem* 15 *are satisfied, then the function* $f(z,y)$ *is analytic in the set* $\{z:|z| < 1\} \times \{y:|y| < R\}$.

Proof. It suffices to establish the analyticity of $f(z,y)$, as a function of y, under the condition that $|z| \leqq r$, where $r < 1$ is arbitrary. According to the proof of the previous theorems, the quantity r_0 is fixed, and is chosen to be as small as possible. It follows from Theorem 15 that the hypotheses of Lemma 33.1 are satisfied for all m. The function $g(z,\epsilon)$ constructed in this lemma is thus admissible with respect to $f(z,y)$ in the annulus $r_0 \leqq |z| \leqq 1$. Therefore the construction given in Lemma 1.6 can be carried out in any annulus $r^* \leqq |z| \leqq r^* + \Delta$ chosen so that $r^* + 3\Delta/4 = r$. Thus (33.42) will hold in the disc $|z| \leqq r$, provided that $\epsilon \leqq \epsilon(r)$, and by Lemma 1.6 there exist constants $A(r)$ and $B(r)$ such that

$$|f_k(z, y; v)| \leqq A(r) B^{k-1}(r) k^{-3}, \qquad 1 \leqq k \leqq m.$$

From Proposition 36.1 it follows that $d_v^k f(z, y)/k! = f_k(z, y; v)$. Since m is arbitrary,

$$\frac{1}{k!} | d_v^k f(z, y) | \leqslant A(r) B^{k-1}(r) k^{-3}, \quad k = 1, 2, \ldots.$$

It now follows that for fixed z, $|z| \leqq r$, $f(z, y)$ is an analytic function of y. Q.E.D.

CHAPTER 6

ON EXTREMAL FUNCTIONS WHICH SATISFY MORE THAN ONE DIFFERENTIAL EQUATION

Our previous investigations of the metric properties of the boundary of the coefficient region have been concerned only with the set $\mathscr{R}\subset\partial\mathscr{V}_n$, and the set $\mathscr{A}_1$ has been completely ignored. The set $\mathscr{A}_1$ is the set of singularities of $\partial\mathscr{V}_n$, since at each point $x\in\mathscr{A}_1$ the tangent hyperplane to $\partial\mathscr{V}_n$ does not exist. As we saw earlier, $\mathscr{A}_1$ is a set of small dimension, and hence one would think that in general it could be excluded from our considerations. However, a series of interesting extremal problems leads to extremal functions $f(z,x)$ with $x\in\mathscr{A}_1$, and for this reason it is appropriate to study in some detail both the structure of $\mathscr{A}_1$ and the extremal functions $f(z,x)$, $x\in\mathscr{A}_1$.

§37. Since $\mathscr{A}_1=\{x:x\in\partial\mathscr{V}_n, d_{n-1/2}(x)=0\}$, the linear manifold of associated vectors of a function $f(z,x)$ is at least two-dimensional. In what follows this subspace will be denoted by $\Lambda_n(x)$. For any $\lambda\in\Lambda_n(x)$, equation (3.15) holds, or in a different notation, equation (3.17) holds, and we write this equation in the form

$$\left(\frac{Df(z,x)}{f(z,x)}\right)^2 Q[f(z,x),\lambda]+P(z,\lambda)=0. \tag{37.1}$$

In this form the associated vector λ determining the equation is explicitly displayed. Below we shall be dealing with a fixed point $x\in\mathscr{A}_1$, and therefore we shall not include the argument x as a subscript to f. Since on $|z|=1$ there is at least one zero of $P(z,\lambda)$, it follows from (3.19) that on $|z|=1$

$$\operatorname{Im}P(z,\lambda)=0. \tag{37.2}$$

PROPOSITION 37.1. *The function $f(z)$ is algebraic. Each branch of $f(z)$ is regular at the points $z=0$ and $z=\infty$, and has an expansion of the form $f(z)=\eta z+\cdots$, $|\eta|=1$, in a neighborhood of $z=0$, and an expansion $f(z)=\eta z^{-1}+\cdots$, $|\eta|=1$, in a neighborhood of $z=\infty$. If $z\neq 0,\infty$, then $f(z)\neq 0$.*

PROOF. In actual fact this proposition has already been proved during the proof of Lemma 3.2. However, we shall reconstruct the proof for the convenience of the reader. Let $\lambda^1,\lambda^2\in\Lambda_n(x)$, and suppose that λ^1 and λ^2 are linearly independent (over the reals). In (37.2) we take $\lambda=\lambda^1$ and $\lambda=\lambda^2$ and divide these equations to get

$$Q(f,\lambda^1)/Q(f,\lambda^2)=P(z,\lambda^1)/P(z,\lambda^2). \tag{37.3}$$

But $P(z,\lambda^1)\,[P(z,\lambda^2)]^{-1}\neq\text{const}$, since if the quotient were constant, then

$\lambda^1 = C\lambda^2$, and on the basis of (37.2) C would be real, which would contradict the linear independence of λ^1 and λ^2. Hence $f(z)$ is an algebraic function. Assume that at $z=0$ we have some branch of $f(z)$ that does not vanish. Then either $f(z) = w_0 + C_1 z^\alpha + \cdots$, where $\alpha > 0$, or $f(z) = C_1 z^{-\alpha} + \cdots$, where $\alpha > 0$. Thus we have that

$$\left(\frac{Df(z)}{f(z)}\right)^2 = \left(\frac{C_1 \alpha z^\alpha}{w_0}\right)^2 + \cdots, \quad \text{or} \quad \left(\frac{Df(z)}{f(z)}\right)^2 = \alpha^2 + \cdots.$$

Then either $Q(f(z), \lambda^{-1})(Df(z)/f(z))^2 = O(z^{2\alpha})$, or $Q(f(z), \lambda^1)(Df(z)/f(z))^2 = O(z^\alpha)$; on the other hand,

$$Q(f(z), \lambda^1)\left(\frac{Df(z)}{f(z)}\right)^2 + P(z, \lambda^1) = 0, \tag{37.4}$$

and if for any $\lambda \in \Lambda_n(x)$ we have that $\lambda_{m+1} = \cdots = \lambda_n = 0$, then we can assume that $\lambda_m^1 \neq 0$, from which it follows that

$$P(z, \lambda^1) = \overline{\lambda_m^1} z^{-m} + O(z^{-m+1}), \tag{37.5}$$

and hence $Q(f(z), \lambda^1)(Df(z)/f(z))^2 = -\lambda_m^1 z^{-m} + O(z^{-m+1})$, which is a contradiction. Thus we have proved that $f(z) \to 0$ as $z \to 0$. Analogously we can prove that $f(z) \to 0$ as $z \to \infty$. We can now suppose that in a neighborhood of $z=0$, $f(z)$ has an expansion $f(z) = \eta z^\alpha + \cdots$. From (10.3) and (10.4) it follows that

$$Q(w, \lambda^1) = -\lambda_m^{-1} w^{-m} + O(w^{-m+1}). \tag{37.6}$$

Then from (37.5), (37.6) and (37.4) we have $\alpha = 1$ and $\eta^m = 1$. If in a neighborhood of $z = \infty$ we have $f(z) = \eta z^{-\alpha} + \cdots$, then, using the fact that $P(z, \lambda^1) = \lambda_m^1 z^m + O(z^{m-1})$, we conclude as above that $\alpha = 1$ and $\eta^m = \overline{\lambda}_m^1 / \lambda_m^1$. Having defined γ by $\gamma \overline{\lambda}_m^1 + \overline{\lambda}_m^2 = 0$, we get $\gamma Q(w, \lambda^1) + Q(w, \lambda^2) = \sum_1^p \widetilde{A}_k w^{-k}$, where $p < m$. From the equation

$$\sum_1^p \widetilde{A}_k f^{-k} \left(\frac{Df}{f}\right)^2 + \gamma P(z, \lambda^1) + P(z, \lambda^2) = 0$$

and the above expansions for $f(z)$ about $z=0$ and $z=\infty$ it follows that

$$\gamma P(z, \lambda^1) + P(z, \lambda^2) = \sum_{-p}^p \widetilde{B}_k z^{-k}.$$

Therefore $\lambda_m^1 \gamma + \lambda_m^2 = 0$, and hence γ is a real constant. Therefore in $\Lambda_n(x)$ there exists a vector $\lambda^3 = \gamma\lambda^1 + \lambda^2$ with the form $\lambda^3 = (\lambda_1^3, \cdots, \lambda_p^3, 0, \cdots, 0)$, where $p \leq m-1$. If $Q(w, \lambda^1) = \sum_1^m A_k w^{-k}$, $P(z, \lambda^1) = \sum_{-m}^m B_k z^{-k}$, then

$$\frac{\sum_1^p \widetilde{A}_k f^{-k}}{\sum_1^m A_k f^{-k}} = \frac{\sum_{-p}^p \widetilde{B}_k z^{-k}}{\sum_{-m}^m B_k z^{-k}}.$$

Thus in a neighborhood of $z=0$ we have the equation

$$f^{m-p}(\alpha_0+\alpha_1 f+\ldots)=z^{m-p}(\beta_0+\beta_1 z+\ldots), \qquad \alpha_0\neq 0, \qquad \beta_0\neq 0,$$

and it now follows easily that any branch of $f(z)$ is regular in a neighborhood of $z=0$. Similarly it can be shown that any branch of f is regular in a neighborhood of $z=\infty$. To complete the proof, we must show that $f(z_0)\neq 0$ for $z_0\neq 0,\infty$. Suppose not, so that for some $z_0\neq 0,\infty$ we have $f(z_0)=0$, and thus $f(z)=C(z-z_0)^\alpha+\cdots$. Then

$$Q(f,\lambda^1)\left(\frac{Df}{f}\right)^2=-\alpha^2 z_0^2 c^{-m}\overline{\lambda_m^1}(z-z_0)^{-m\alpha-2}+0((z-z_0)^{-m\alpha-2}).$$

On the other hand, from (37.4) we find that $Q(f,\lambda^1)(Df/f)^2=O(1)$. This contradiction shows that $f(z)\neq 0$ if $z\neq 0,\infty$, and the proof is complete.

Proposition 37.2. *The subspace $\Lambda_n(x)$ has a basis $\lambda^1,\cdots,\lambda^k$ of the form*

$$\lambda^l=(\lambda_1^l,\ldots,\lambda_{n_l}^l,0,\ldots,0),\ 1\leqslant l\leqslant k,$$

where $n\geqq n_1>n_2>\cdots>n_k$, and where $\lambda_{n_l}^l\neq 0,\ 1\leq l\leq k$.

Proof. If λ and $\tilde\lambda$ are any two linearly independent associated vectors with $\lambda=(\lambda_1,\cdots,\lambda_m,0,\cdots,0)$ and $\tilde\lambda=(\tilde\lambda_1,\cdots,\tilde\lambda_m,0,\cdots,0)$, then, as was shown in the proof of Proposition 37.1, there exists a real constant γ such that $\gamma\lambda+\tilde\lambda=(\gamma\lambda_1+\tilde\lambda_1,\cdots,\gamma\lambda_{m-1}+\tilde\lambda_{m-1},0,\cdots,0)$. We now begin with an arbitrary basis, and, upon applying the above process a finite number of times, we arrive at a basis with the desired properties. Q.E.D.

Corresponding to the vector λ^l we define $Q_l(w)=Q(w,\lambda^l)$ and $P_l(z)=P(z,\lambda^l)$, $1\leq l\leq k$. We shall determine the greatest common divisor of the functions $P_{k-1}(z)$ and $P_k(z)$. First we shall show that the number of common zeros of these quasipolynomials must be even. In view of (37.2) the zeros of these quasipolynomials are symmetric with respect to the circumference $|z|=1$. Thus it is sufficient to consider zeros lying on $|z|=1$. Denote by $f_0(z)$ that branch of $f(z)$ which is regular and univalent in $|z|<1$ with $f_0'(0)=1$. We separate the zeros of $P_k(z)$ lying on $|z|=1$ into two sets $\mathfrak{A}$ and $\mathfrak{B}$, where $\mathfrak{A}$ consists of those zeros which are the image of some vertex of the tree $\mathscr{T}$ under the mapping $z=f_0^{-1}(w)$. Let $z_1,\cdots,z_l$ be the images of some vertex $w_j\in\mathscr{T}$. Denote by s_{p_1} the multiplicity of the zero z_p, $1\leq p\leq l$, of the function $P_k(z)$, and by s_{p_2} the multiplicity of this same point as a zero of $P_{k-1}(z)$. Let κ_1 be the multiplicity of this same point as a zero of $Q_k(w)$, and let κ_2 be the multiplicity of w_j as a zero of $Q_{k-1}(w)$. In a neighborhood of z_p the function $f_0(z)$ has an expansion $f(z)=w_j+\alpha_p(z-z_p)^{r_p}+\cdots$, where $r_p=(s_{p_1}+2)/(\kappa_1+2)=(s_{p_2}+2)/(\kappa_2+2)$. Hence if $\kappa_1<\kappa_2$ we have $s_{p_1}<s_{p_2}$, $1\leq p\leq l$, and if $\kappa_1\geqq\kappa_2$ we have $s_{p_1}\geqq s_{p_2}$. Since $\sum_1^l r_p=2$, we have that $\sum_1^l(s_{p_1}+2)=2(\kappa_1+2)$ and $\sum_1^l(s_{p_2}+2)=2(\kappa_2+2)$. If $w_j=\infty$, then in these formulas we must make the obvious changes. It now follows that there must be an even number of common zeros of $P_{k-1}(z)$ and $P_k(z)$ which

belong to the set $\mathfrak{A}$. We now consider the zeros belonging to $\mathfrak{B}$. If $z_j \in \mathfrak{B}$ is a zero of $P_k(z)$, then $w_j = f_0(z_j)$ is not a vertex of $\mathscr{T}$. Since in a neighborhood of z_j the function $f_0(z)$ is regular, and since $f_0'(z_j) \neq 0$, we have that w_j is a zero of $Q_k(w)$ with the same multiplicity as z_j; this follows from the differential equation $(Df_0/f_0)^2 Q_k(f_0) + P_k(z) = 0$. Since on $|z| = 1$ there are precisely two images of w_j, say the points z_j and z_j', we see that z_j' is a zero of $P_k(z)$ of the same multiplicity as z_j. Hence there must be an even number of common zeros of $P_{k-1}(z)$ and $P_k(z)$ which belong to $\mathfrak{B}$, and therefore from the above we see that the number of common zeros of these quasipolynomials must be even. If $z_1, \cdots, z_{2r}$ are the common zeros, then we set

$$p_0(z) = Cz^{-r} \prod_1^{2r} (z - z_j),$$

where the constant C is defined so that for $|z| = 1$

$$\operatorname{Im} p_0(z) = 0. \tag{37.7}$$

Set $P_l(z) = p_0(z) P_l^*(z)$, $l = k-1, k$. Then the quasipolynomials $P_{k-1}^*(z)$ and $P_k^*(z)$ have no common zero, and in a sense we can say that $p_0(z)$ is the greatest common divisor of $P_{k-1}(z)$ and $P_k(z)$. Denote by $q_0(w)$ the greatest common divisor of $Q_{k-1}(w)$ and $Q_k(w)$, and set $Q_l(w) = q_0(w) Q_l^*(w)$, $l = k-1, k$. Clearly $q_0(w)$ is a polynomial in $1/w$ of degree s, and $q_0(w)$ is well defined up to multiplication by constants. Setting $\lambda = \lambda^{k-1}$ and $\lambda = \lambda^k$ in (37.1) and dividing the two equations, we find that $Q_{k-1}(f)[Q_k(t)]^{-1} = P_{k-1}(z)[P_k(z)]^{-1}$ or

$$\frac{Q_{k-1}^*(f)}{P_{k-1}^*(z)} = \frac{Q_k^*(f)}{P_k^*(z)}. \tag{37.8}$$

We shall show that if $z \neq 0, \infty$, then $Q_{k-1}^*(f)[P_{k-1}^*(z)]^{-1} \neq 0, \infty$. To prove this, we first note that the statement is obvious at any point for which $P_{k-1}^*(z) \neq 0$ and $Q_{k-1}^*(f) \neq 0, \infty$. If $P_{k-1}^*(z) = 0$, then $P_k^*(z) \neq 0$, and if in this case $Q_k^*(f) = 0$, then $Q_{k-1}^*(f) \neq 0$, and relation (37.8) would be violated. If $Q_{k-1}^*(f) = 0$, then $Q_k^*(f) \neq 0$, and then $P_k^*(z) \neq 0$. When $Q_k^*(f) = \infty$, then either $z = 0$ or $z = \infty$, as was shown in Proposition 37.1. On the Riemann surface R of the function $f(z)$ the fraction $Q_{k-1}^*(f)[P_{k-1}^*(z)]^{-1}$ is a meromorphic function. From the remarks made above it follows that the zeros and poles of this function can be only at $z = 0, f = 0$ and $z = \infty, f = 0$. From Proposition 37.1 it follows that $Q_{k-1}^*(f)[P_{k-1}^*(z)]^{-1}$ has a pole for $z = 0$, $f = 0$ or $z = \infty$, $f = 0$, if $r > s$, and a zero if $r < s$. Thus the meromorphic function under consideration assumes neither the value 0 nor ∞, and hence it must be constant. Thus

$$\frac{Q_{k-1}^*(f)}{P_{k-1}^*(z)} = \frac{Q_k^*(f)}{P_k^*(z)} = C_0.$$

Since

$$\left(\frac{Df}{f}\right)^2 Q_k(f) + P_k(z) = 0,$$

we have

$$C_0\left(\frac{Df}{f}\right)^2 q_0(f) + p_0(z) = 0. \tag{37.9}$$

From (37.7) it follows that equation (37.9) determines some vector $\lambda^0 \in \Lambda_n(x)$, and if $P_k^*(z) \not\equiv \text{const}$, then $\lambda^0 = (\lambda_1^n, \cdots, \lambda_{n_0}^0, 0, \cdots, 0)$, where $n_0 < n_k$. It is obvious that this vector λ^0 and the vectors $\lambda^1, \cdots, \lambda^k$ are linearly independent. But this is a contradiction. Therefore $P_k^*(z) = C_1$, $Q_k^*(w) = C_2$, and we can take $C_1 = C_2 = 1$. This implies $C_0 = 1$, and so $Q_{k-1}^*(f) - P_{k-1}^*(z) = 0$. Set $q(w) = Q_{k-1}^*(w) - Q_{k-1}^*(\infty)$ and $p(z) = P_{k-1}^*(z) - Q_{k-1}^*(\infty)$. Since $f_0(x)$ has a pole z_0 with $|z_0| = 1$, we have that $Q_{k-1}^*(\infty) = P_{k-1}^*(z_0)$, and so $Q_{k-1}^*(\infty)$ is a real constant. We set

$$q(w) = \sum_1^r \alpha_l w^{-l}, \qquad p(z) = \sum_{-r}^r \beta_l z^{-l},$$

where by construction $\beta_l = \overline{\beta}_{-l}$, $l = 0, \cdots, r$. In our new notation the equation for f can be written in the form

$$q(f) - p(z) = 0. \tag{37.10}$$

We never required the specific functions $Q_{k-1}(f)$ and $P_{k-1}(z)$ in the above discussion, and so for any $\lambda \in \Lambda_n(x)$ we have

$$Q(w, \lambda) = Q^*(w, \lambda)\, Q_k(w,) \qquad P(z, \lambda) = P^*(z, \lambda)\, P_k(z)$$

and

$$Q^*(f, \lambda) - P^*(z, \lambda) = 0. \tag{37.11}$$

Denote by s the degree of $Q^*(w, \lambda)$ as a polynomial in $1/w$. We shall show that r divides s. Define an integer j by the condition $jr \leq s < (j+1)r$. From (37.10) and (37.11) we have

$$\frac{Q^*(f, \lambda)}{[q(f)]^j} - \frac{P^*(z, \lambda)}{[p(z)]^j} = 0.$$

Note that it is possible to find a quasipolynomial $v(z)$ satisfying the following conditions:

1) $\operatorname{Im} v(z) = 0$ for $|z| = 1$.

2) $P^*(z, \lambda) = [p(z)]^j v(z) + \mathfrak{W}(z)$, where $\mathfrak{W}(z) = \sum_{-jr}^{jr} C_l z^{-l}$ and $\overline{C_l} = C_{-l}$, $l = 0, \cdots, jr$.

We divide $Q^*(w, \lambda)$ by $[q(w)]^j$, so that $Q^*(w, \lambda) = [q(w)]^j u(w) + \mathfrak{r}(w)$. Thus

$$u(f) - v(z) + \frac{\mathfrak{r}(f)}{[q(f)]^j} - \frac{\mathfrak{W}(z)}{[p(z)]^j} = 0.$$

On the Riemann surface R the function $u(f) - v(z)$ is meromorphic, and by (37.1) it can have a pole only at $z = 0$, $f = 0$ or $z = \infty$, $f = 0$. On the other hand

$$u(f)-v(z)=-\left[\frac{\mathfrak{r}(f)}{[q(f)]^{j}}-\frac{\mathfrak{W}(z)}{[p(z)]^{j}}\right],$$

and by construction the meromorphic function $\mathfrak{r}(f)/[q(f)]^j-\mathfrak{W}(z)/[p(z)]^j$ has no poles at $z=0$, $f=0$ or at $z=\infty$, $f=0$. Thus $u(f)-v(z)=C$. As previously noted, the branch of $f(z)$ (univalent in $|z|<1$) which we have denoted by $f_0(z)$ has a pole at $z=z_0$, $|z_0|=1$. Therefore $u(\infty)-v(z_0)=C$; setting $u_0(w)=u(w)-u(\infty)$ and $v_0(z)=v(z)-v(z_0)$, we find that $u_0(f)-v_0(z)=0$. Therefore

$$u_0(f)\,Q_k(f)\left(\frac{Df}{f}\right)^2+v_0(z)\,P_k(z)=0.$$

We put $u_0(w)Q_k(w)=Q(w,\lambda^0)$ and $v_0(z)P_k(z)=P(z,\lambda^0)$. Since $r=n_{k-1}-n_k$, and since the degree of $Q(w,\lambda^0)$ as a polynomial in $1/w$ is less than n_{k-1}, we have in view of our construction of a basis that $\lambda^0=\alpha\lambda_k$. Therefore $u(w)=$ const and $v(z)=$ const, and this implies that $s=jr$, as desired.

From (37.10) it follows that $Dq(f)Df/f-Dp(z)=0$, and so

$$(Dq(f))^2\left(\frac{Df}{f}\right)^2-(Dp(z))^2=0.$$

According to what we have proved, $(Dq(w))^2=\widetilde{Q}(w)Q_k(w)$, $(Dp(z))^2=\widetilde{P}(z)P_k(z)$. The function $\widetilde{Q}(w)$, as a polynomial in $1/w$, has degree equal to either r or 0. In the first case

$$(Dq(w))^2=\beta Q_{k-1}(w)+\gamma Q_k(w),\quad (Dp(z))^2=-[\beta P_{k-1}(z)+\gamma P_k(z)].$$

In the second case $(Dq(w))^2=\beta Q_k(w)$, $(Dp(z))^2=-\beta P_k(z)$. Since $P_k(z)$ and $Q_k(w)$ are determined up to multiplication by real constants, we have that either

$$(Dq(w))^2=(q(w)+\alpha)\,Q_k(w),\qquad (Dp(z))^2=-(p(z)+\alpha)\,P_k(z),$$

or

$$(Dq(w))^2=Q_k(w),\qquad (Dp(z)^2=-P_k(z).$$

It is obvious that the first case occurs only when the functions $q(w)+\alpha$ and $p(z)+\alpha$ do not have finite simple zeros. In what follows we shall apply the term "minimal" to equation (37.1), with λ replaced by λ_k. From the above discussion it follows that one of two possibilities may occur:

1) There exists a real quantity α such that the functions $q(w)+\alpha$ and $p(z)+\alpha$ divide $(Dq(w))^2$ and $(Dp(z))^2$ respectively. Then the minimal equation has the form

$$\frac{(Dq(f))^2}{q(f)+\alpha}\left(\frac{Df}{f}\right)^2-\frac{(Dp(z))^2}{p(z)+\alpha}=0. \tag{37.12}$$

2) A choice of a constant α as above is impossible. Then the minimal equation has the form

$$(Dq(f))^2\left(\frac{Df}{f}\right)^2-(Dp(z))^2=0. \tag{37.13}$$

It is easy to see that $f(z)$ will satisfy an equation either of the form

$$(37.14)\qquad \frac{(Dq(f))^2}{q(f)+\alpha}\sum_0^s \gamma_l\,[q(f)]^l\left(\frac{Df}{f}\right)^2-\frac{(Dp(z))^2}{p(z)+\alpha}\sum_0^s \gamma_l\,[p(z)]^l=0,$$

or of the form

$$(37.15)\qquad (Dq(f))^2\sum_0^s \gamma_l\,[q(f)]^l\left(\frac{Df}{f}\right)^2-(Dp(z))^2\sum_0^s \gamma_l\,[p(z)]^l=0,$$

where $\gamma_0,\cdots,\gamma_s$ are arbitrary real constants. According to what we have shown, each of the numbers $n_1-n_2,\cdots,n_{k-1}-n_k$ is a multiple of r, and in fact from (37.14) and (37.15) these numbers must actually each be equal to r. Since each of the systems (37.14) or (37.15) determines k linearly independent vectors $\tilde{\lambda}^1,\cdots,\tilde{\lambda}^{k-1}$, $\tilde{\lambda}^k$, $\tilde{\lambda}^k=\lambda^k$, and since $\dim\Lambda_n(x)=k$, the vectors $\tilde{\lambda}^1,\cdots,\tilde{\lambda}^k$ form a basis of $\Lambda_n(x)$, and so equation (37.1) for $\lambda\in\Lambda_n(x)$ can be written in either the form of (37.14) or the form of (37.15). In both cases $f(z)$ satisfies equation (37.13), and since $-(Dp(z))^2\geqq 0$ for $|z|=1$, we have that

$$\left(Dq(w)\frac{dw}{w}\right)^2\geqslant 0$$

on $\mathscr{T}$. Therefore $\operatorname{Im}[q'(w)\,dw]=0$, $w\in\mathscr{T}$, and this in turn implies that

$$(37.16)\qquad \operatorname{Im} q(w)=0,\qquad w\in\mathscr{T}.$$

We shall summarize the above results in the following lemma.

Lemma 37.1. *If $x\in\mathscr{A}_1$, then $f(z,x)$ is an algebraic function satisfying equation* (37.10). *The minimal differential equation satisfied by $f(z,x)$ has either the form* (37.12) *or the form* (37.13). *The differential equation of any associated vector $\lambda\in\Lambda_n(x)$ can be expressed in either the form* (37.14) *or the form* (37.15). *On $\mathscr{T}$ relation* (37.16) *holds.*

§38. Let $q(w)$ be any polynomial in $1/w$ of the form

$$(38.1)\qquad q(w)=\sum_1^r \alpha_l w^{-l}.$$

On the w-sphere we shall examine the structure of the set of solutions of the equation

$$(38.2)\qquad \operatorname{Im} q(w)=0;$$

in other words, we shall examine the structure of the zero set of the function $\operatorname{Im} q(w)$. We first need some terminology.

A point w_0 will be called a saddle point of multiplicity κ if w_0 is a zero of $q'(w)$ of multiplicity κ. Saddle points and poles of $q(w)$ will be called critical points. The point $w=\infty$ will be called a critical point if the polynomial $q(1/s)$ has $s=0$ as a saddle point. For each point of the w-sphere it is possible to define a canonical neighborhood in which the structure of the level lines of the function $\operatorname{Im} q(w)$ can be very simply described. Let w_0 be any point; we set $q(w) = q(w_0) + Ct^{\kappa+1}$, $\operatorname{Im} q(w_0) = \alpha$, where κ is the multiplicity of w_0 as a zero of the derivative of $q(w)$; the case $\kappa = 0$ is not excluded. Then

$$\sum_{\varkappa+1}^{\infty} \beta_l (w - w_0)^l = Ct^{\varkappa+1}.$$

We set $C = \beta_{\kappa+1}$; taking the $(\kappa+1)$th root, we get that

$$(w - w_0)\left(1 + \sum_1^{\infty} \gamma_l (w - w_0)^l\right) = t,$$

and by the implicit function theorem we have $w - w_0 = t + \delta_2 t^2 + \cdots$. The function $w(t)$ constructed in this way is regular in a neighborhood of $t = 0$; the radius of the largest disc in which $w(t)$ is univalent will be denoted by ρ. It is clear that there exists a largest ρ_0 such that the image of the disc $|w - w_0| \leqq \rho_0$ lies in the disc $|t| \leqq \rho$. The α level curves of $\operatorname{Im} q(w)$ are the images of the level curves $\operatorname{Im}(Ct^{\kappa+1}) = 0$. In turn these are the rays $\arg t = (1/(\kappa+1))(l\pi - \arg C)$, $0 \leqq l \leqq 2\kappa + 1$. Thus in the disc $|w - w_0| < \rho_0$ we have $2\kappa + 2$ analytic arcs as the α level curves of $\operatorname{Im} q(w)$. It is clear that two adjacent arcs form the angle $\pi/(\kappa+1)$ at w_0. We note that each of the arcs under consideration has a parametric representation

$$w = w\left[\tau \exp\left\{\frac{i}{\varkappa+1}(l\pi - \arg C)\right\}\right], \quad \tau > 0.$$

If we extend the domain of definition of this function by allowing τ to assume negative values, then we obtain an analytic arc, which is the union of two level curves forming an angle π at the point w_0.

Consider the case of a pole. Set $q(w) = \alpha_r t^{-r}$. From this we have

$$w^r\left(1 + \frac{\alpha_{r-1}}{\alpha_r} w + \ldots + \frac{\alpha_1}{\alpha_r} w^{r-1}\right)^{-1} = t^r;$$

taking the rth root, we get that

$$w\left(1 + \sum_1^{\infty} \beta_l w^l\right) = t,$$

and by the implicit function theorem $w = t + \sum_2^{\infty} \gamma_l t^l$. The function constructed in this way will be regular and univalent in some maximal disc $|t| < \rho$, and, as above, there exists a disc $|w| \leqq \rho_0$ of maximal radius ρ_0, whose image will be contained in the disc $|t| \leqq \rho$. In the canonical neighborhood (the disc $|w| < \rho_0$) the zero level lines of $\operatorname{Im} q(w)$ will be the images of the rays $\operatorname{Im}\{\alpha_r t^{-r}\} = 0$, an equation which can clearly be written in the form

$$\arg t = \frac{1}{r}[\arg \alpha_r - k\pi], \qquad k = 0, \ldots, 2r - 1.$$

Thus in the disc $|w| < \rho_0$ we have $2r$ analytic arcs, the zero level curves of the function $\operatorname{Im} q(w)$. Two adjacent arcs form an angle π/r at $w = 0$.

We now consider the structure in the entire set of the zero level lines of the harmonic function $\operatorname{Im} q(w)$. We construct canonical neighborhoods for all the critical points and for the point $w = \infty$, even if it is not a critical point. Let $\mathscr{D}$ be the complement with respect to the w-sphere of the union of the above canonical neighborhoods. If $w \in \mathscr{D}$ satisfies equation (38.2), then there exists a corresponding canonical neighborhood. The diameter of this canonical neighborhood will be defined to be the zero level curve of $\operatorname{Im} q(w)$ which lies inside the neighborhood. A canonical neighborhood of the point w_0 will be denoted by $U(w_0) = \{w : |w - w_0| \leq \rho(w_0)\}$. The set of all points satisfying equation (38.2) can be covered by a finite number of canonical neighborhoods $U(w_k)$ of points $w_k \in \mathscr{D}$, $k = 1, \cdots, N$. In this way we see that the set of all points lying in $\mathscr{D}$ and satisfying equation (38.2) is the intersection of $\mathscr{D}$ with the union of a finite number of diameters. Hence it consists of a finite number of open analytic arcs. The end points of these arcs lie on the boundary of the canonical neighborhoods of the critical points.

To each such arc we join two corresponding arcs of zero level lines of $\operatorname{Im} q(w)$ which lie in canonical neighborhoods of critical points. If $w = \infty$ is not a critical point, then in the same way we agree to say that the set of points in this neighborhood which satisfy equation (38.1) consists of two arcs with the common boundary point $w = \infty$.

As a result of this construction we get maximal arcs of the zero level curve of $\operatorname{Im} q(w)$. The set of points satisfying (38.1) consists of a finite number of maximal arcs of the zero level curves, and of boundary points which are either critical points or the point $w = \infty$. Hence this set is a planar graph $\mathscr{G}$. The complement of $\mathscr{G}$ with respect to the w-sphere consists of finitely many domains $\mathscr{D}_1, \mathscr{D}_2, \cdots$. The boundary of each of these domains contains $w = 0$. To prove this, suppose the opposite were true. Then there would exist a domain $\mathscr{D}_j$ on whose boundary the harmonic function $\operatorname{Im} q(w) = V(w)$ equals zero, and $\sup_{w \in \mathscr{D}_j} |V(w)| < \infty$. Then $V(w) \equiv 0$, and this contradiction proves that each boundary contains $w = 0$. In particular, the graph $\mathscr{G}$ is connected. Since $2r$ arcs of zero level curves emanate from $w = 0$, the graph $\mathscr{G}$ divides the w-sphere into at least $2r$ domains.

We shall consider the various trees which can be formed from the branches of $\mathscr{G}$, assuming that each time the base of the tree is located at $w = \infty$. The terminal vertices of such trees need not in general coincide with the vertices of the graph $\mathscr{G}$. We limit ourselves to the consideration of those trees whose complements contain $w = 0$ and which have conformal radius 1 at $w = 0$.

Let $\mathscr{T}$ be one of these trees, and let $f(z)$ be a univalent function mapping $|z| < 1$ onto the complement of $\mathscr{T}$, where we have normalized f by the conditions $f(0) = 0$ and $f'(0) > 0$. Then $f'(0) = 1$, and so $f(z) \in S$. We shall prove that $f(z)$ is an extremal function. To begin the proof, we note that in the disc $|z| < 1$ the function $q[f(z)]$ has a unique singular point at $z = 0$, which is a pole of order r. With an appropriate choice of constants the function

$$g(z) = q\,[f(z)] - \left(\sum_{1}^{r} \beta_k z^{-k} + \sum_{-r}^{-1} \bar{\beta}_k z^{-k}\right)$$

is regular in $|z| < 1$. Since $f(z)$ maps $|z| < 1$ onto the tree $\mathscr{T}$, which is the sum of finitely many analytic arcs, the function $f(z)$ is regular everywhere on $|z| = 1$ with the exception of finitely many points. Thus $\operatorname{Im} q(w) = 0$ everywhere on $|z| = 1$, with the exception of the above-mentioned finite set. Since $\sup_{|z|<1} |g(z)| < \infty$, we have $g(z) = \beta_0$, where β_0 is real. Hence

$$q[f(z)] - p(z) = 0, \tag{38.3}$$

where

$$p(z) = \sum_{-r}^{r} \beta_k z^{-k},$$

and moreover $\beta_{-k} = \bar{\beta}_k$, $0 \leqq k \leqq r$. Thus $f(z)$ is an algebraic function. From (38.3) we see that $Dq(f)Df/f - Dp(z) = 0$, and so

$$(Dq(f))^2 \left(\frac{Df}{f}\right) - (Dp(z))^2 = 0. \tag{38.4}$$

Since $\operatorname{Im} p(z) = 0$ for $|z| = 1$, we have $-(Dp(z))^2 \geqq 0$ for $|z| = 1$, and thus $f(z) \in S_{2r}$. If there exists a real constant α such that the functions $q(w) + \alpha$ and $p(z) + \alpha$ divide $(Dq(w))^2$ and $(Dp(z))^2$ respectively, then $f(z)$ will satisfy

$$\frac{(Dq(f))^2}{q(f) + \alpha} \left(\frac{Df}{f}\right)^2 - \frac{(Dp(z))^2}{p(z) + \alpha} = 0 \tag{38.5}$$

in addition to (38.4). In this case $f(z)$ clearly satisfies an equation of the form (37.14). If a choice of α as above is not possible, $f(z)$ will satisfy (37.15). Thus, for $r \leqq n/2$ in the first case and $r \leqq n/3$ in the second case, the function $f(z)$ determines on $\partial \mathscr{V}_n$ a point $x \in \mathscr{A}_1$. These results are summarized in the following lemma.

Lemma 38.1. *Let $q(w)$ be any polynomial in $1/w$ of the form* (38.1), *and let $\mathscr{G}$ be a corresponding graph of zero level lines of* $\operatorname{Im} q(w)$. *Let the tree $\mathscr{T} \in \mathscr{G}$ with base $w = \infty$ be such that the complement of $\mathscr{T}$ has conformal radius* 1 *at $w = 0$. Then the function $f(z)$, mapping $|z| < 1$ onto the complement of $\mathscr{T}$, such that $f(0) = 0$ and $f'(0) = 1$, is an algebraic extremal function. It satisfies equation* (38.3), *and also either the differential equations* (38.5), (37.14), *or* (38.4), (37.15). *If $f(z)$ satisfies equations* (38.5) *and* (37.14) [(38.4) *and* (37.15)] *then for $r \leqq n/2$ $(n/3)$ it defines on $\partial \mathscr{V}_n$ a point $x \in \mathscr{A}_1$.*

Lemmas 37.1 and 38.1 characterize completely those extremal functions which satisfy the conditions $d_{n-1/2}(f) = 0$ and $d_n(f) = 0$ for some n. We note, however, that equation (38.4) or (38.5) need not be minimal for the function $f(z)$ constructed in Lemma 38.1. We can say that equation (38.3) will be minimal for the function f if $d_{r-1/2}(f) \neq 0$. Clearly, if equation (38.3) is minimal, then either (38.4) or (38.5) will be the minimal differential equation.

§39. We now apply the results proved above to obtain an upper estimate for the dimension of the set $\mathscr{A}_1$. In order to do this, it suffices to estimate the dimension of the set of univalent functions satisfying the minimal equation (38.3), where $q(w)$ is a polynomial in $1/w$ of fixed degree r. Since a univalent extremal function is uniquely determined by its tree, we must estimate the dimension of the set of trees (with conformal radius 1 at the point $w=0$) included in all the possible graphs formed by the zero level curves of the function $\operatorname{Im} q(w)$. For this we parametrize the sets of trees under consideration. We first parametrize the set of graphs, since after this it is relatively simple to parametrize the set of all the trees contained in a given graph. In the end it will not be difficult to determine the number of important parameters which determine the trees in the set under consideration.

For the sake of convenience we change from the variable w to the variable $\zeta=1/w$, and henceforth we shall consider the image of the graph $\mathscr{G}$ on the ζ-sphere. Set $q_0(\zeta)=q(1/\zeta)$ and let $\zeta_0=0, \zeta_1, \cdots, \zeta_\nu$ be the different roots of $q_0'(\zeta)$, with respective multiplicities $\kappa_0, \kappa_1, \cdots, \kappa_\nu$. Since the polynomial $q_0(\zeta)$ is uniquely determined up to multiplication by real constants, we set

$$q_0'(\zeta)=\eta\zeta^{\kappa_0}\prod_{j=1}^{\nu}(\zeta-\zeta_j)^{\kappa_j},$$

where $|\eta|=1$ and $\sum_0^\nu \kappa_j=r-1$. We suppose that the zeros $\zeta_1, \cdots, \zeta_\mu$ lie on the graph of a zero level curve, and that the remaining zeros do not. We now fix the quantities $\kappa_0, \cdots, \kappa_\nu$ and μ. Denote by κ the vector $(\kappa_0, \cdots, \kappa_\nu)$, and let $G_{\kappa\mu}$ be the set of polynomials $q_0(\zeta)$ determined by the method above. Clearly any polynomial normalized so that $q_0(0)=0$ and $\lim_{\zeta\to\infty} q_0(\zeta)\zeta^{-r}=\eta/r$, $|\eta|=1$, belongs to $\bigcup_{\kappa\mu} G_{\kappa\mu}$, where the union is taken over all possible vectors κ and integers μ. In this way the problem is reduced to the parameterization of those graphs defined by polynomials of the set $G_{\kappa\mu}$. We set $\zeta_j=\xi_j+i\xi_{j+\nu}$, $1\le j\le \nu$, $\eta=e^{i\theta}$. According to hypothesis, the roots $\zeta_1, \cdots, \zeta_\mu$ are vertices of the graph, and the variables $\xi_1, \cdots, \xi_{2\nu}, \theta$ are not independent, but are subject to the conditions

$$\operatorname{Im} q_0(\zeta_j)=0, \qquad 1\le j\le \mu. \tag{39.1}$$

It is easy to see that $\operatorname{Im} q_0(\zeta_j)$, $1\le j\le \nu$, is a homogeneous polynomial in $\xi_1, \cdots, \xi_{2\nu}$, whose coefficients are linear forms in $\cos\theta$ and $\sin\theta$. Thus equation (39.1) for fixed θ determines an algebraic manifold which is a cone. To determine its dimension, we shall show that conditions (39.1) are independent. Set $\mathscr{P}_j=\mathscr{P}_j(\xi_1, \cdots, \xi_{2\nu}, \theta)=\operatorname{Im} q_0(\zeta_j)$, and let $\mathscr{J}$ be the Jacobi matrix

$$\mathscr{J}=\begin{pmatrix} \dfrac{\partial\mathscr{P}_1}{\partial\xi_1} & \cdots\cdots & \dfrac{\partial\mathscr{P}_1}{\partial\xi_{2\nu}} \\ \cdot & \cdots\cdots & \cdot \\ \dfrac{\partial\mathscr{P}_\mu}{\partial\xi_1} & \cdots\cdots & \dfrac{\partial\mathscr{P}_\mu}{\partial\xi_{2\nu}} \end{pmatrix}$$

PROPOSITION 39.1. *If the ζ_j are pairwise distinct, then the rank of the matrix $\mathscr{J}$ is equal to μ.*

Proof. We shall show that the rows of $\mathscr{J}$ are linearly independent; that is, that the equations

$$\sum_{l=1}^{\mu} \alpha_l \frac{\partial \mathscr{P}_l}{\partial \xi_k} = 0, \qquad k = 1, \ldots, 2\nu, \tag{39.2}$$

imply $\alpha_1 = \cdots = \alpha_\mu = 0$. Since $\mathscr{P}_l = \operatorname{Im}\{\int_0^{\zeta_l} q_0'(\zeta)\,d\zeta\}$, it is easily seen that

$$\frac{\partial \mathscr{P}_l}{\partial \xi_k} = -\operatorname{Im}\left\{\int_0^{\zeta_l} \frac{\varkappa_k}{\zeta-\zeta_k} q_0'(\zeta)\,d\zeta\right\}, \qquad \frac{\partial \mathscr{P}_l}{\partial \xi_{\nu+k}} = -\operatorname{Re}\left\{\int_0^{\zeta_l} \frac{\varkappa_k}{\zeta-\zeta_k} q_0'(\zeta)\,d\zeta\right\}.$$

Therefore, in view of the fact that the α_l, $1 \leqq l \leqq \mu$, are real, equation (39.2) implies that

$$\sum_{l=1}^{\mu} \alpha_l \int_0^{\zeta_l} \frac{q_0'(\zeta)}{\zeta-\zeta_k}\,d\zeta = 0, \qquad k = 1, \ldots, \nu. \tag{39.3}$$

Setting $q_*(\zeta) = \prod_1^\nu (\zeta - \zeta_k)$, we see from (39.3), upon using the well-known relation

$$\frac{\zeta^s}{q_*(\zeta)} = \sum_{k=1}^{\nu} \frac{\zeta_k^s}{q_*'(\zeta_k)(\zeta-\zeta_k)}, \qquad s = 0, \ldots, \nu - 1,$$

that

$$\sum_{l=1}^{\mu} \alpha_l \int_0^{\zeta_l} \prod_{j=1}^{\nu} (\zeta-\zeta_j)^{\varkappa_j - 1} \zeta^{\varkappa_0 + s}\,d\zeta = 0, \qquad s = 0, \ldots, \nu - 1. \tag{39.4}$$

We shall construct a polynomial $P(\zeta)$ with the following properties:

1) The quantities ζ_j, $2 \leqq j \leqq \nu$, are zeros of $P(\zeta)$ of order κ_j, respectively.
2) $\zeta = 0$ is a zero of $P(\zeta)$ of order $\kappa_0 + 1$.
3) $P(\zeta_1) = 1$, and $P'(\zeta)$ has a zero of order $\kappa_1 - 1$ at ζ_1.

We shall seek $P(\zeta)$ as above in the form $P(\zeta) = \zeta^{\kappa_0+1}\prod_{j=2}^{\nu}(\zeta-\zeta_j)^{\kappa_j}P_0(\zeta)$, where $P_0(\zeta)$ is such that condition 3) is satisfied. Toward this end we expand the function $\{\zeta^{\kappa_0+1}\prod_{j=2}^{\nu}(\zeta-\zeta_j)^{\kappa_j}\}^{-1}$ in a power series about ζ_1 and we take the sum of the first κ_1 terms of this series; this will be our polynomial $P_0(\zeta)$. It is clear that the degree of $P_0(\zeta)$ equals $\kappa_1 - 1$, and the degree of $P(\zeta)$ equals $\kappa_0 + 1 + \sum_{j=2}^{\nu}\kappa_j + \kappa_1 - 1 = r - 1$. By construction

$$P'(\zeta) = \zeta^{\varkappa_0} \prod_1^{\nu} (\zeta - \zeta_j)^{\varkappa_j - 1} P_1(\zeta),$$

where $P_1(\zeta)$ is a polynomial of degree $r - 2 - \kappa_0 - \sum_{j=1}^{\nu}(\kappa_j - 1) = \nu - 1$. Taking a suitable linear combination of equations (39.4), we get that

$$\sum_{l=1}^{\mu} \alpha_l \int_0^{\zeta_l} P'(\zeta)\,d\zeta = 0,$$

and so $\alpha_1 = 0$; $\alpha_2 = \cdots = \alpha_\mu = 0$ now follows by repeated applications of the above line of reasoning. Thus the dimension of the algebraic cone under consideration does not exceed $2\alpha - \mu$. Denote this cone by $M_{\kappa\mu\theta}$. In what follows we shall consider that part of this set on which the inequalities

$$|\operatorname{Im} q_0(\zeta_j)| \geqq \epsilon, \qquad \mu+1 \leqq j \leqq \nu, \tag{39.5}$$

$$\begin{gathered} \epsilon \leqq |\xi_j| \leqq \epsilon^{-1}, \qquad 1 \leqq j \leqq 2\nu, \\ |\xi_j - \xi_k| \geqq \epsilon, \qquad j, k = 1, \cdots, 2\nu, \quad j \neq k, \end{gathered} \tag{39.6}$$

are satisfied. The resulting compact set will be denoted by $M_{\kappa\mu\theta\epsilon}$. Let $\mathscr{C} = \{\eta : |\eta| = 1, \eta = e^{i\theta}\}$. Consider the set of pairs $\mathfrak{y} = (\xi,\eta)$, where $\xi = (\xi_1, \cdots, \xi_{2\nu})$, $\xi \in M_{\kappa\mu\theta}$ and $\eta \in \varphi$, and define a distance by

$$\rho(\mathfrak{y}_1, \mathfrak{y}_2) = [|\xi^1 - \xi^2|^2 + |\eta_1 - \eta_2|^2]^{1/2},$$

$\mathfrak{y}_j = (\xi^j, \eta_j)$, $j = 1, 2$. Then the set $\mathfrak{M}_{\kappa\mu\epsilon}$ of those pairs for which $\xi \in M_{\kappa\mu\theta\epsilon}$ and $\eta = e^{i\theta}$ is a compact set. Let $\mathfrak{y}_0 = (\xi^0, \eta_0)$ be an arbitrary point of the compact set $\mathfrak{M}_{\kappa\mu\epsilon}$. By Proposition 39.1 there exists a minor, with order μ, of the matrix $\mathscr{I}$, which is nonnull for $\xi_j = \xi_j^0$, $1 \leqq j \leqq 2\nu$, $\theta = \theta_0$. Let $j_1, \cdots, j_\mu$ be the numbers of those columns of $\mathscr{I}$ which are used to make up this minor. Denote the variables $\xi_{j_1}, \cdots, \xi_{j_\mu}$ by $y_1, \cdots, y_\mu$, and denote the remaining variables by $x_1, \cdots, x_{2\nu-\mu}$. By the implicit function theorem there exist segments $a_j \leqq x_j \leqq b_j$, $1 \leqq j \leqq 2\nu - \mu$, $\theta_1 \leqq \theta \leqq \theta_2$, such that on these segments y_j, $1 \leqq j \leqq \mu$, will be analytic functions of the variables x_j, $1 \leqq j \leqq 2\nu - \mu$, and of θ:

$$y_j = \varphi_j(x_1, \cdots, x_{2\nu-\mu}, \theta), \qquad 1 \leqq j \leqq \mu. \tag{39.7}$$

This segment can be taken so that the image belongs to the set $\mathfrak{M}_{\kappa\mu\epsilon/2}$. In this way we have covered the compact set $\mathfrak{M}_{\kappa\mu\epsilon}$ by closed sets, and so we can extract a finite subcover. This cover will be composed of finitely many segments $I_1, \cdots, I_N$. On each segment I_l, $1 \leqq l \leqq N$, the variables $x_1, \cdot\cdot, x_{2\nu-\mu}$ and θ can be considered as parameters determining the corresponding polynomial $q_0(\zeta) = q(1/\zeta)$, since all the roots of $q_0'(\zeta)$ can be determined with the aid of (39.7). Thus we have parametrized the set of graphs corresponding to $q(w)$; we denote this set by $\mathscr{I}_l$. To each graph $\mathscr{G} \in \mathscr{I}_l$ there corresponds an infinite set of trees (recall that we consider only those trees whose complement has conformal radius 1 at $w = 0$). We shall determine the number of parameters on which those trees belonging to the same graph depend. Let $\mathscr{G}_0$ be any fixed graph of the set $\mathscr{I}_l$, with $w_0 = \infty$, $w_1^0, \cdots, w_\mu^0$ as its vertices. It is possible to choose the segments I_l so small that the vertices of each graph $\mathscr{G} \in \mathscr{I}_l$ will lie in the discs $K_j = \{w : |w - w_j^0| \leqq \rho\}$, $1 \leqq j \leqq \mu$, and moreover these discs will be mutually disjoint and will not contain any of the remaining zeros of $q'(w)$. We now index the vertices and branches of $\mathscr{G}_0$. This induces a numbering of the vertices and branches of any graph $\mathscr{G} \in \mathscr{I}_l$, and moreover vertices with the same index have equal degrees and lie in the same disc K_j, $1 \leqq j \leqq \mu$, and branches with the same index emenate from vertices with the same index. Any tree $\mathscr{T} \subset \mathscr{G}$ will be uniquely determined by some vector of integers—the indices of these vertices of the graph, which will be the vertices of the tree, and the indices of those branches of the graph which intersect $\mathscr{T}$ (the branches are considered as open arcs). It is clear that each vector of integers determines a class of trees, and the entire set of trees is partitioned into a finite number of classes. Trees of the same class, which are contained in one and the same graph $\mathscr{G}$, will differ among themselves according to the location of their terminal vertices. Suppose some class of trees has τ terminal vertices. According to our method of indexing the vertices and branches of the trees, the terminal vertices of the tree will be given an index only when they are vertices of the graph. Among the τ terminal vertices suppose there are τ_0 vertices which are not vertices of the graph. To each such vertex we assign the same index possessed by the terminal branch of the tree emanating from it. Thus for all terminal vertices of a given class of trees the number of the corresponding terminal vertices will be one and the same. Let these numbers be $m_1 < m_2 < \cdots < m_{\tau_0}$. By m we denote the vector of numbers assigned to the vertices and the branches of a given class of trees. The position of the terminal vertex with index m_j will be uniquely determined by the integral

$$\int \left| Dq'(w) \frac{dw}{w} \right|, \tag{39.8}$$

taken over the terminal branch with index m_j. Denote this integral by σ_j. Then the terminal vertices will be fixed if we preassign the quantities $\sigma_1, \cdots, \sigma_{\tau_0 - 1}$, since the vertex with index m_{τ_0} will be uniquely determined by the condition that the conformal radius of the complement of the tree must be 1. Denote by σ_j^*, $1 \leqq j \leqq \tau_0$, the length of the branch of the graph with index m_j, where the length is computed via the metric defined by the integral (39.8). Then, generally speaking, we have

$$0 < \sigma_j < \sigma_j^*, \qquad 1 \leqq j \leqq \tau_0.$$

Below we shall not consider all the trees of a given class, but only those for which

$$\epsilon \leqq \sigma_j \leqq \sigma_j^* - \epsilon, \qquad 1 \leqq j \leqq \tau_0. \tag{39.9}$$

The set of trees of various classes, defined by the graphs of the set $\mathscr{T}_l$ and subject to conditions (39.9), will be denoted by $\mathscr{P}_{\kappa\mu l\epsilon}$. It is clear that the vector of numbers m and the variables $x_1, \cdots, x_{2\nu-\mu}$, θ, $\sigma_1, \cdots, \sigma_{\tau_0 - 1}$ are parameters which determine a tree of a given set. The above variables define a point of $(2\nu - \mu + \tau_0)$-dimensional Euclidean space, and the class of trees determined by the vector m will correspond to a set $\mathfrak{N}_m$. Clearly $\mathfrak{N}_m$ is compact. Using the construction outlined above, we can deduce an upper bound for the dimension of the set $\mathscr{A}_1 \subset \partial \mathscr{V}_n$. In order to obtain a sharper result, we require a fairly simple proposition. Let $\mathfrak{z} = (x_1, \cdots, x_{2\nu-\mu}, \theta, \sigma_1, \cdots, \sigma_{\tau_0 - 1})$,

$\mathfrak{z} \in \mathfrak{N}_m$. Then $\mathfrak{z}$ determines a tree $\mathscr{T}_\mathfrak{z}$, and we denote by $f(z, \mathfrak{z})$ the function in S which maps $|z| < 1$ onto the complement of $\mathscr{T}_\mathfrak{z}$.

Proposition 39.2. *If $\mathfrak{z}, \mathfrak{z}^0 \in \mathfrak{N}_m$, then* $\max_{|z| \leq 1/2} |f(z; \mathfrak{z}) - f(z; \mathfrak{z}^0)| \leq C_m |\mathfrak{z} - \mathfrak{z}^0|$.

Proof. Let $\mathfrak{z}$ be any point of $\mathfrak{N}_m$, with $q_0(\zeta; \mathfrak{z})$ its corresponding polynomial, and set $q(w; \mathfrak{z}) = q_0(1/w; \mathfrak{z})$. Denote by $\mathscr{G}_\mathfrak{z}^*$ the graph of the quadratic differential $(Dq(w; \mathfrak{z})dw/w)^2$, and by $\mathscr{G}_\mathfrak{z}$ denote the graph formed by the zero level curves of $\operatorname{Im} q(w; \mathfrak{z})$. Let $\zeta_1(\mathfrak{z}), \cdots, \zeta_\nu(\mathfrak{z})$ be the zeros of $q_0^1(\zeta, \mathfrak{z})$, and put $w_j(\mathfrak{z}) = 1/\zeta_j(\mathfrak{z})$, $1 \leq j \leq \nu$; the quantities $w_j(\mathfrak{z})$ are the finite zeros of the function $q^1(w, \mathfrak{z})$. It is clear that the vertices of $\mathscr{G}_\mathfrak{z}^*$ are located at the points $w_j(\mathfrak{z})$, $1 \leq j \leq \nu$. We shall estimate the distance from the tree $\mathscr{T}_\mathfrak{z}$ to each vertex of $\mathscr{G}_\mathfrak{z}^*$ not lying on $\mathscr{T}_\mathfrak{z}$. If $\mu + 1 \leq j \leq \nu$, then $w_j(\mathfrak{z})$ does not belong to the tree, and by the construction of the segment I_l we have $|\operatorname{Im} q(w_j)| \geq \epsilon/2$. Since for $\mathfrak{z} \in \mathfrak{N}_m$ the functions $q(w, \mathfrak{z})$ are uniformly bounded in the domain $|w| \geq 1/8$, it follows from the previous inequality that the distance from $w_j(\mathfrak{z})$, $\mu + 1 \leq j \leq \nu$, to $\mathscr{T}_\mathfrak{z}$ is not less than some constant depending only on ϵ. If $j \leq \mu$, then $w_j(\mathfrak{z})$ is a vertex of the graph $\mathscr{G}_\mathfrak{z}$, but if $w_j(\mathfrak{z})$ does not belong to $\mathscr{T}_\mathfrak{z}$, it follows from (39.9) that in the metric

$$ds^2 = |Dq(w; \mathfrak{z})dw/w|^2$$

the distance from $w_j(\mathfrak{z})$ to $\mathscr{T}_\mathfrak{z}$ is not less than ϵ. From this it follows that the distance from such a vertex to $\mathscr{T}_\mathfrak{z}$ is not less than some constant depending only on ϵ. Therefore the distance from $\mathscr{T}_\mathfrak{z}$ to those vertices of $\mathscr{G}_\mathfrak{z}^*$ not belonging to $\mathscr{T}_\mathfrak{z}$ is not less than C_ϵ. From this we conclude that there exists a quantity $r_0 < 1$, independent of $\mathfrak{z}$, such that in the domain G_{r_0} bounded by $\mathscr{T}_\mathfrak{z}$ and the level curve $\mathscr{L}_{r_0} = \{w : |f^{-1}(w; \mathfrak{z})| = r_0\}$ there are no vertices of $\mathscr{G}_\mathfrak{z}^*$.

Recalling relation (39.7), it is easy to prove that the corresponding zeros of the functions $q'(w, \mathfrak{z})$ and $q'(w, \mathfrak{z}^1)$ satisfy the inequality $|w_j(\mathfrak{z}) - w_j(\mathfrak{z}^1)| \leq C_0 |\mathfrak{z} - \mathfrak{z}^1|$, where C_0 depends only on ϵ. Therefore there exists $\delta_0 > 0$ such that for any points $\mathfrak{z}$ and $\mathfrak{z}^0$ with $|\mathfrak{z} - \mathfrak{z}^0| \leq \delta_0$ the vertices of the graph $\mathscr{G}_{\mathfrak{z}^0}^*$ lying in the domain G_{r_1}, where $r_1 = \frac{1}{2}(1 + r_0)$, will also be the vertices of $\mathscr{T}_{\mathfrak{z}^0}$. Let $l_k(\mathfrak{z})$ be the length of a nonterminal branch of $\mathscr{T}_\mathfrak{z}$; it is easy to see that $|l_k(\mathfrak{z}) - l_k(\mathfrak{z}^0)| \leq C_1 |\mathfrak{z} - \mathfrak{z}^0|$, where C_1 is a constant depending only on ϵ. It is also easy to prove that if $\sigma_j^*(\mathfrak{z}) < \infty$, then $|\sigma_j^*(\mathfrak{z}) - \sigma_j^*(\mathfrak{z}^0)| < C|\mathfrak{z} - \mathfrak{z}^0|$, and thus if $|\mathfrak{z} - \mathfrak{z}^0| \leq \delta_0$ and $\sigma_j^*(\mathfrak{z}) < \infty$, then

$$|\sigma_j^*(\mathfrak{z}) - \sigma_j^*(\mathfrak{z}^0)| \leq \epsilon/4. \tag{39.10}$$

Henceforth we shall assume that $|\mathfrak{z} - \mathfrak{z}^0| \leq \delta_0$, and we shall not write this assumption each time. For any two trees $\mathscr{T}_\mathfrak{z}$ and $\mathscr{T}_{\mathfrak{z}^0}$ with parameters σ_j and σ_j^0, $1 \leq j \leq \tau_0 - 1$, there are well-defined quantities σ_{τ_0} and $\sigma_{\tau_0}^0$ which determine the vertex with index m_{τ_0}. Suppose that all the vertices of the trees $\mathscr{T}_\mathfrak{z}$ and $\mathscr{T}_\mathfrak{z}^0$ coincide, with the exception of a terminal vertex of number m_{τ_0}, and suppose that for the tree $\mathscr{T}_\mathfrak{z}^0$ this vertex corresponds to the parameter value $\sigma_{\tau_0}^0$. From (39.9) and (39.10) it follows that if $\sigma_{\tau_0}^*(\mathfrak{z}) < \infty$, then

$$\epsilon \leqq \sigma_{\tau_0}^0 \leqq \sigma_{\tau_0}^0(\mathfrak{z}) - \epsilon/2.$$

All the statements of the concluding part of §16, where we constructed a solution of equation (16.12), apply to the trees $\mathscr{T}_\mathfrak{z}^0$ and $\mathscr{T}_{\mathfrak{z}^0}$, and to the quadratic differentials $(Dq(w;\mathfrak{z})dw/w)^2$ and $(Dq(w;\mathfrak{z}^0)dw/w)^2$. To see that this is in fact the case, we use Proposition 16.5 to construct the function $\xi(w)$, and we consider the equation

$$\left(D(t;\mathfrak{z})\frac{dt}{t}\right)^2 = \left[D(w;\mathfrak{z}^0)(1+\delta\xi(w))\frac{dw}{w}\right]^2,$$

where $\delta = |\mathfrak{z} - \mathfrak{z}^0|$. Since this latter equation is a special case of (16.12), we conclude that in the domain G_{r_2}, $r_2 = \frac{1}{2}(1+r_1)$, there exists a univalent function $\varphi(w,\delta)$ mapping G_{r_2} to some neighborhood of $\mathscr{T}_\mathfrak{z}^0$. Since $\varphi(w) = w + \delta h(w,\delta)$, where $|h(w,\delta)| \leqq C|w|$, the function $h[f(z;\mathfrak{z}^0),\delta]$ will be admissible with respect to $f(z;\mathfrak{z}^0)$ in the annulus $r_2 \leqq |z| < 1$. Applying Theorem 1 and choosing the arbitrary constants as in Lemma 1.4, we get a function $F(z,\delta)$ mapping $|z|<1$ conformally onto the complement of $\mathscr{T}_\mathfrak{z}^0$; and $F(z,\delta)$ is normalized by the conditions $F(0,\delta) = 0$ and $F'(0,\delta) > 0$. According to Theorem 1, $|F'(0,\delta) - 1| \leqq C\delta$ and $|F(z,\delta) - f(z;\mathfrak{z}^0)| \leqq C\delta$ for $|z| \leqq r_2$. Note that the constant C is independent of δ and $\mathfrak{z}^0$. If the conformal radius at $w = 0$ of the complement of $\mathscr{T}_\mathfrak{z}^0$ is greater than 1, then the function $t = \psi(z) = f^{-1}[F(z,\delta),\mathfrak{z}]$ is regular and univalent in $|z|<1$, and maps $|z|<1$ onto the disc $|t|<1$ slit by some Jordan curve. Since $\psi(0) = 0$, Schwarz' lemma implies $|\psi(z)| < |z|$. By Schwarz' formula

$$\log\frac{\psi(z)}{z} = \frac{1}{2\pi}\int_0^{2\pi}\log|\psi(e^{i\vartheta})|\frac{e^{i\vartheta}+z}{e^{i\vartheta}-z}d\vartheta,$$

and so $|\log(\psi(z)/z)| \leqq 3\log(1/F'(0,\delta))$ for $|z| \leqq \frac{1}{2}$. Therefore $|\psi(z) - z| \leqq (3/2)C(1-C\delta)^{-4}\delta$ for $|z| \leqq \frac{1}{2}$. Since

$$|f(z,\mathfrak{z}) - f(z,\mathfrak{z}^0)| \leqslant |f[\psi(z);\mathfrak{z}] - f(z,\mathfrak{z})| + |F(z,\delta) - f(z,\mathfrak{z}^0)|,$$

it follows for $|z| \leqq \frac{1}{2}$ that $|f(z,\mathfrak{z}) - f(z,\mathfrak{z}^0)| \leqq C_2\delta$, where C_2 is a constant depending only on ϵ. The case where $F'(0,\delta) > 1$ is treated analogously. If $|\mathfrak{z} - \mathfrak{z}^0| > \delta_0$, then, using the distortion theorem, we get that $|f(z;\mathfrak{z}) - f(z;\mathfrak{z}^0)| \leqq 4 < 4\delta/\delta_0$ for $|z| \leqq \frac{1}{2}$. Thus the inequality of Proposition 39.2 follows by taking $C_m = C_2 + 4/\delta_0$.

We have established that the trees of the set $\mathscr{P}_{\kappa\mu l\epsilon}$ are determined by $2\nu - \mu + \tau_0$ parameters. We shall now give a more easily handled upper bound on this number of parameters. Suppose the tree has l vertices, of which τ are terminal by assumption. If m is the number of nonterminal branches, then $m = l - 1 - \tau$, since the total number of branches is $l - 1$. Denoting by k_j the degree of the jth vertex, we have

$$\sum{}' k_j = 2m + \tau,$$

where the sum is over all nonterminal vertices of the tree. Each nonterminal vertex is a vertex of the graph, and its degree as a vertex of the graph is $2\kappa_j + 2$. Thus $2\sum{}' (\kappa_j + 1) \geqq 2m + \tau$, and so

$$\sum{}' \varkappa_j \geqslant \frac{\tau - 2}{2}.$$

Therefore

$$\sum_{j=0}^{\mu} \varkappa_j \geqslant \frac{\tau - 2}{2}. \tag{39.11}$$

From the obvious inequality $\sum_{j=0}^{\mu} (\kappa_j - 1) \leqq 2\sum_{j=0}^{\nu} (\kappa_j - 1) + 1 - \kappa_0$ it follows that $2\nu - \mu + \sum_{j=0}^{\mu} \kappa_j \leqq 2r - 2 - \kappa_0$, and so upon using (39.11) we get that

$$2\nu - \mu + \tau \leqq 2r + \tau/2 - 1 - \kappa_0.$$

We now estimate the number of terminal vertices of the tree. Let $f(z) \in S$ be the function mapping $|z| < 1$ onto the complement of $\mathscr{T}$. If w_j is a terminal vertex of the tree and if z_j is its image, $|z_j| = 1$, then $f'(z_j) = 0$. Hence z_j is a zero of the quasipolynomial $Dp(z)$, since $f(z)$ satisfies (38.4). Since $Dp(z)$ has at most $2r$ zeros on $|z| = 1$, we have $\tau \leqq 2r$, and thus we finally conclude that

$$2\nu - \mu + \tau_0 \leqq 3r - 1. \tag{39.12}$$

Denote by $\mathfrak{A}_r$ the set of algebraic functions of class S which satisfy the minimal equation (38.3). If we denote by $\mathfrak{A}_{\kappa\mu l\epsilon}$ the set of all univalent functions of class S which are determined by trees of the set $\mathscr{T}_{\kappa\mu l\epsilon}$, and if we set

$$\mathfrak{A}_{\varkappa\mu\varepsilon} = \bigcup_{l} \mathfrak{A}_{\varkappa\mu l\varepsilon}, \qquad \mathfrak{A}_{r\varepsilon} = \bigcup_{\varkappa, \mu} \mathfrak{A}_{\varkappa\mu\varepsilon},$$

then

$$\mathfrak{A}_r \subseteq \bigcup_{l \geqslant 1} \mathfrak{A}_{r\frac{1}{l}}.$$

Lemma 39.1. *The set $\mathfrak{A}_r$ has $3r$-dimensional Hausdorff measure zero;* $\dim \mathfrak{A}_r \leqq 3r - 1$.

Proof. Recall that on S our metric is given by

$$\rho(f, g) = \max_{|z| \leqslant \frac{1}{2}} |f(z) - g(z)|.$$

Let $h > 0$ be arbitrarily small. Since the set $\mathscr{T}_{\kappa\mu l\epsilon}$ consists of finitely many classes, determined by the vector of indices m, we have that $\mathfrak{A}_{\kappa\mu l\epsilon}$ is likewise divided into a finite number of classes $\mathfrak{B}_m$. Each class of trees is determined by the set $\mathfrak{N}_m$, which in view of inequality (39.12) is a compact set in $(3r - 1)$-dimensional Euclidean space. We shall take a partition of the compact set $\mathfrak{N}_m$ into sets $A_1, \cdots, A_M$ of diameter at most $h(2C_m)^{-1}$, where C_m is the constant determined in Proposition 39.2. It is clear that this partition can be taken so that $M \leqq C_m^* h^{-3r+1}$. This partition defines a partition of the class $\mathfrak{B}_m$ into sets $B_1, \cdots, B_M$, if we associate to the set B_l those functions which correspond to points of A_l. By Proposition 39.2,

$$\delta(B_l) \leqq h/2 < h,$$

where $\delta(B_l)$ is the diameter of B_l. Thus $\sum_1^M [\delta(B_l)]^{3r} < C_m^* h$, and hence it follows that $m_{3r}(\mathfrak{B}_m) = 0$, and so $m_{3r}(\mathfrak{A}_{\kappa\mu l\epsilon}) = 0$, since there are only finitely many classes. This latter relation implies that

$$m_{3r}(\mathfrak{A}_{r,\epsilon}) = 0$$

for any $\epsilon > 0$. Since the sequence $\{\mathfrak{A}_{r\frac{1}{l}}\}$ is an expanding sequence, and since the sets $\mathfrak{A}_{r\frac{1}{l}}$, $l \geqq 1$, are regular with respect to the measure m_{3r}, it follows by a well-known theorem ([**13**], Chapter II, Theorem 6.1) that

$$m_{3r}(\lim_{p\to\infty} \mathfrak{A}_{r\frac{1}{p}}) = \lim_{p\to\infty} m_{3r}(\mathfrak{A}_{r\frac{1}{p}}) = 0.$$

Since $\lim_{l\to\infty} \mathfrak{A}_{r\frac{1}{l}} = \bigcup_{l \geqq 1} \mathfrak{A}_{r\frac{1}{l}} \supseteq \mathfrak{A}_r$, we have $m_{3r}(\mathfrak{A}_r) = 0$. Hence by a theorem of Szpilrajn ([**14**], Theorem VII, 3) $\dim \mathfrak{A}_r \leqq 3r - 1$. Q.E.D.

With respect to the set of algebraic functions satisfying (38.5), this latter estimate is very crude. Denote by $\mathfrak{A}_r^*$ the set of algebraic functions satisfying (38.3) and (38.5). It is very easy to parametrize this set; we shall not give the details of how this is done, since the line of reasoning is precisely the same as the one given previously.

If $\alpha \neq 0$, then the polynomial $q_0(\zeta) + \alpha = q(1/\zeta) + \alpha$ has no simple zeros. For $\alpha = 0$ the zero $\zeta = 0$ may be simple, and the remaining roots must be multiple. Denote these remaining roots by $\zeta_1, \cdots, \zeta_\nu$, and let their respective multiplicities be $\kappa_1, \cdots, \kappa_\nu$. If we set $\zeta_j = \xi_j + i\xi_{\nu+j}$, $1 \leqq j \leqq \nu$, then the graph corresponding to the polynomial $q_0(\zeta) + \alpha$ will be determined by the quantities ξ_j, $1 \leqq j \leqq 2\nu$, and by the vector $(\kappa_1, \cdots, \kappa_\nu) = \kappa$. In order to fix the tree, it is necessary to give the position of its terminal vertices. If τ_0 is the number of terminal vertices which are not vertices of the graph, then $\tau_0 - 1$ parameters must still be specified. Thus the total number of parameters will not exceed $2\nu + \tau_0 - 1$ if $\alpha \neq 0$, and $2\nu + \tau_0$ if $\alpha = 0$. We now estimate the number of terminal vertices of the tree. In view of equation (38.4), any terminal vertex of the tree which is not a vertex of the graph must correspond to a zero of $Dp(z)$ which is not a zero of $p(z) + \alpha$ and which lies on $|z| = 1$. Since the quasipolynomial $p(z) + \alpha$ has only multiple zeros, we have $\tau_0 \leqq r$. Since $2\nu \leqq r$ for $\alpha \neq 0$ and $2\nu \leqq r - 1$ for $\alpha = 0$, it follows that in general the number of parameters does not exceed $2r - 1$. From this fact we obtain, as above,

LEMMA 39.2. *The set* $\mathfrak{A}_r^*$ *has* $2r$*-dimensional Hausdorff measure zero*; $\dim \mathfrak{A}_r^* \leqq 2r - 1$.

With the aid of the results proved above it is easy to estimate the dimension of the set $\mathscr{A}_1 = \{x : x \in \partial \mathscr{V}_n, d_{n-1/2}(x) = 0\}$.

Theorem 17. *Let* $p = \max(2[n/2], 3[n/3])$. *Then* $m_p(\mathscr{A}_1) = 0$ *and* $\dim \mathscr{A}_1 \leqq p - 1$.

Proof. If $f(z) \in \mathfrak{A}_r$, then $d_{n-1/2}(f) = 0$ only when $3r \leqq n$. Therefore for a mapping $F: S \to R^{2n}$ the image of $\mathfrak{A}_r$ belongs to $\mathscr{A}_1$ if $3r \leqq n$. Denote the image of $\mathfrak{A}_r$ by $\mathscr{A}_{r_1}$. As above it follows that the image of $\mathfrak{A}_r^*$ belongs to $\mathscr{A}_1$ if $2r \leqq n$; denote this image by $\mathscr{A}_{r_1}^*$. Then, according to Lemma 37.1,

$$\mathscr{A}_1 = \Big(\bigcup_{3r \leqslant n} \mathscr{A}_{r1}\Big) \cup \Big(\bigcup_{2s \leqslant n} \overset{*}{\mathscr{A}}_{s1}\Big).$$

We note that if $F(f) = x$ and $F(g) = y$, then $|x - y| \leqq \alpha^n \rho(f, g)$, where α is an absolute constant. It follows easily from this fact that $m_{3r}(\mathscr{A}_{r_1}) = 0$ and $m_{2r}(\mathscr{A}_{r_1}^*) = 0$. Therefore $m_p(\mathscr{A}_1) = 0$, and so $\dim \mathscr{A}_1 \leqq p - 1$. Q.E.D.

Note that $p \leqq n$, and hence

$$\dim \mathscr{A}_1 \leqq n - 1. \tag{39.13}$$

Later on we shall require a slight generalization of Theorem 17. Let $l_1, \cdots, l_q$ be linearly independent linear forms of the type

$$l_j(x) = \operatorname{Re} \sum_{k=1}^{n} \overline{l}_{jk} x_k, \qquad 1 \leqslant j \leqslant q.$$

Consider a mapping $\Phi_q: S \to R^q$ defined in the following way: if $f \in S$ and $f(z) = \sum_0^n x_k z^{k+1} + \cdots$, then

$$\Phi_q(f) = \xi, \quad \xi = (\xi_1, \cdots, \xi_q), \quad \xi_j = l_j(x), \qquad 1 \leqq j \leqq q.$$

Denote the image of S by $\mathfrak{W}_q$. It is easy to show that $\mathfrak{W}_q$ is a compact set homeomorphic to a q-dimensional sphere. Consider the subset of $\mathfrak{W}_q$ consisting of the image of the set $(\bigcup_{3r \leqq n} \mathfrak{A}_r) \cup (\bigcup_{2s \leqq n} \mathfrak{A}_s^*)$. Denote this subset by $\mathscr{A}_{nq}$.

Lemma 39.3. *Let* $p = \max(2[n/2], 3[n/3])$, *and suppose* $q \geqq p$. *Then* $m_p(\mathscr{A}_{nq}) = 0$ *and* $\dim \mathscr{A}_{nq} \leqq p - 1$.

Proof. If $\Phi(f) = \xi$, $\Phi(g) = \eta$, then $|\xi - \eta| < C\rho(f, g)$, where C depends only on n and the coefficients of the forms l_j, $1 \leqq j \leqq q$. It now follows that $m_p(\mathscr{A}_{nq}) = 0$ and $\dim \mathscr{A}_{nq} \leqq p - 1$.

CHAPTER 7

ON EXTREMAL FUNCTIONS YIELDING A LOCAL EXTREMUM OF A LINEAR FUNCTIONAL

It follows in the obvious manner from the results of Chapter 5 that for any point $x^0 \in \mathscr{R}$ the function $f(z, x^0)$ gives the absolute extremum for some functional which, in general, is nonlinear. To see that this is true, we consider the local coordinate system with origin at x^0, and in a neighborhood of x^0 we obtain the equation of $\partial \mathscr{V}_n$ in the form $\tau = \psi(y)$. Since $\psi(y) = O(|y|^2)$, there exists a sphere contained in the complement of $\mathscr{V}_n$ which is tangent to $\partial \mathscr{V}_n$ at the single point x^0. If x^* is the center of this sphere, then the functional $J(f) = |x - x^*|^{-1}$ has its absolute extremum at the function $f(z, x^0)$. On the other hand, $f(z, x)$, with $x \in \mathscr{R}$, is not distinguished from the set S_n by anything.

The situation is quite different, however, if $f(z, x)$ yields a local extreme of a linear functional. Such univalent functions possess a whole sequence of important properties, and above all their boundary trees are structured in a very simple way. We shall investigate the structure of these boundary trees.

§40. Suppose the linear functional L has a local maximum at the function $f(z)$. Then for any measure $\nu(dz)$ and constant c_1

$$\delta^2 L(f; \nu, c_1) \leqq 0, \tag{40.1}$$

where the second variation is defined by (4.18). In Chapter 4 we introduced the space $\mathscr{A}$ of first variations, and we mentioned the importance of the subspaces $\mathscr{R}_j$ which characterize the properties of the tree $\mathscr{T}$ and thus the properties of the function $f(z)$. It is true that if the subspace $\mathscr{R}_j$ is nonempty, then it contains elements for which $\delta^2 L(f, \zeta) > 0$. Therefore, in the case under consideration above, the subspace $\mathscr{R}_j$ must be empty. The proof that we shall give below does not resort to a consideration of the subspaces $\mathscr{R}_j$, since the study of the structure of $\mathscr{R}_j$ can become quite complicated.

Let z_0, $|z_0| = 1$, be a singular point of $f(z)$, or the image of a singular point of $f^{-1}(w)$. We take $\mathscr{L}$ to be

$$\mathscr{L} = \{z : z = z_0 e^{-u}, 0 \leqq a\epsilon \leqq u \leqq b\epsilon < \infty\},$$

and we take a measure, with support $\mathscr{L}$, of the form

$$\nu(dz) = i^{\rho+1} \epsilon^{-1} g(u/\epsilon)\, du,$$

where $\rho = 0, 1$, where $\epsilon > 0$ is an arbitrary small quantity, where a and b are arbitrary, and where g is a bounded real-valued function. We choose the constant c_1 and the measure so that

$$\int_{\mathscr{C}} \nu(dz) + ic_1 = 0.$$

Setting $\mathfrak{A}_0(z,t) = \mathfrak{A}(z,t) - \mathfrak{C}(z) - \mathfrak{C}(t) + \sum_1^n k^2 \overline{\lambda}_k a_k$, we find from (4.18) and (4.21), in accordance with the choice of ν and c_1, that

$$\delta^2 L(f; \nu, c_1) = \operatorname{Re}\left\{\int_{\mathscr{C}}\int_{\mathscr{C}} \mathfrak{A}_0(z,t)\,\nu(dz)\,\nu(dt) - \int_{\mathscr{C}}\int_{\mathscr{C}} \mathfrak{B}(z,\overline{t})\,\nu(dz)\,\overline{\nu(dt)}\right\},$$

where

$$\mathfrak{A}_0(z,t) = \mathscr{K}(z,t)P(z) + \mathscr{K}(t,z)P(t) + \frac{zP^*(t) - tP^*(t)}{z-t}$$
$$+ \left(\frac{Df(z)}{f(z)}\right)^2 Q^*[f(z), t] + \left(\frac{Df(t)}{f(t)}\right)^2 Q^*[f(t), z] - \overline{R\left(\frac{1}{\overline{z}}\right)} - \overline{R\left(\frac{1}{\overline{t}}\right)}.$$

Set

$$A(u,v) = (-1)^\rho\,[\mathscr{K}(z,t)P(z) + \mathscr{K}(t,z)P(t)]$$
$$(40.2)\qquad + \frac{z\overline{t}}{(1 - z\overline{t})^2}[P(z) + \overline{P(t)}], \quad z = z_0e^{-u},\ t = z_0e^{-v},$$

$$A_1(u,v) = (-1)^\rho \frac{zP^*(t) - tP^*(z)}{z-t} - \frac{z\overline{t}}{1 - z\overline{t}}[P^*(z) + \overline{P^*(t)}] + 2\left(\frac{Df(z)}{f(z)}\right)^2$$
$$(40.3)\qquad \times\left\{(-1)^\rho Q^*[f(z), t] - Q^*\left[f(z), \frac{1}{\overline{t}}\right]\right\} + 2\,[1 - (-1)^\rho]\,\overline{R\left(\frac{1}{\overline{z}}\right)},$$
$$z = z_0e^{-u}, \quad t = z_0e^{-v}.$$

From (4.18) – (4.20) and our choice of the measure we find that

$$\delta^2 L(f; \nu, C_1) = -\varepsilon^{-2}\operatorname{Re}\left\{\int_{a\varepsilon}^{b\varepsilon}\int_{a\varepsilon}^{b\varepsilon}[A(u,v) + A_1(u,v)]\,g\left(\frac{u}{\varepsilon}\right)g\left(\frac{v}{\varepsilon}\right)du\,dv\right\}.$$

Assuming that z_0 is a zero of $P(z)$ of multiplicity $2r$, we shall pick out the main terms, with respect to ϵ, of the right-hand side of this latter relation. We impose upon $g(u)$ the conditions

$$(40.4)\qquad \int_a^b g(u)\,u^{2l+1-\rho}du = 0, \quad 0 \leqslant l \leqslant r - 2 + \rho.$$

Since $P^*(z) = -\overline{P^*(1/\overline{z})}$, we see that for $z = z_0e^{-u}$ and $t = z_0e^{-v}$ the first two terms of the right-hand side of (40.3) are

$$\frac{1}{2}\left\{\frac{(-1)^\rho}{\tanh\frac{u-v}{2}}[P^*(z_0e^{-u}) - P^*(z_0e^{-v})] - \frac{1}{\tanh\frac{u+v}{2}}[P^*(z_0e^{-u}) - P^*(z_0e^{v})]\right.$$
$$\left. + (-1)^\rho P^*(z_0e^{-v}) - P^*(z_0e^{v})\right\} + \frac{1}{2}[P^*(z_0e^{-u}) - \overline{P^*(z_0e^{-u})}]$$
$$- \frac{(-1)^\rho}{2}P^*(z_0e^{-u}) - \frac{1}{2}P^*(z_0e^{u}).$$

The expression in the curly brackets is an odd function of v when $\rho=0$, and an even function of v when $\rho=1$. The expression in the square brackets is purely imaginary. For $z=z_0e^{-u}$ and $t=z_0e^{-v}$ the function $(-1)^\rho Q^*[f(z),t] - Q^*[f(z),1/t]$ is an odd function of v if $\rho=0$, and an even function of v if $\rho=1$; the last term on the right-hand side of (40.3) either vanishes when $\rho=0$ or is a function of u. Therefore the contribution to $\delta^2L(f;\nu,C_1)$ made by the term $A_1(u,v)$ is, in view of (40.4), at most $O(\epsilon^{2r-2})$. Thus

$$\delta^2 L(f;\nu,C_1) = -\varepsilon^{-2}\operatorname{Re}\left\{\int_{a\varepsilon}^{b\varepsilon}\int_{a\varepsilon}^{b\varepsilon} A(u,v)\, g\left(\frac{u}{\varepsilon}\right) g\left(\frac{v}{\varepsilon}\right) du\, dv\right\} + O(\varepsilon^{2r-2}).$$

The study of the remaining integral forms the most complex part of the proof. Set $w_0=f(z_0)$, and let κ be the multiplicity of w_0 as a zero of $Q(w)$; the possibility that $w_0=\infty$ is not excluded. From equation (5.13) it follows that for $|z-z_0|<r_0$

$$f(z) = f(z_0) + \sum_{l=1}^{\kappa+2} (z-z_0)^{ls}\varphi_l(z),$$

where $s=(2r+2)/(\kappa+2)$, and $\varphi_l(z)$, $1\leq l\leq \kappa+2$, are functions regular in the disc $|z-z_0|<r_0$, with $\varphi_1(z_0)\neq 0$. If $f(z_0)=w_0=\infty$, then

$$\frac{1}{f(z)} = \sum_{l=1}^{\kappa} (z-z_0)^{ls}\varphi_l(z),$$

where $s=(2r+2)/\kappa$, and $\varphi_l(z)$, $1\leq l\leq\kappa$, are regular in $|z-z_0|<r_0$, with $\varphi_1(z_0)\neq 0$. Set $F(u)=f(z_0e^{-u})$, $0<u<\infty$. Then from the previous expressions we get that for $|u|\leq\delta_0$ either

$$F(u) = w_0 + \sum_{l=1}^{\kappa+2} u^{ls}\psi_l(u), \tag{40.5}$$

or

$$\frac{1}{F(u)} = \sum_{l=1}^{\kappa} u^{ls}\psi_l(u), \tag{40.6}$$

where $\psi_l(u)$ are functions regular for $|u|<\delta_0$, with $\psi_1(0)\neq 0$. We set $\mathscr{K}_0(u,v) = \mathscr{K}(z_0e^{-u}, z_0e^{-v})$.

From (4.7) we find that

$$\mathscr{K}_0(u,v) = \left(\frac{F'(v)}{F(u)-F(v)}\right)^2 - \left(\frac{F'(v)}{F(v)}\right)^2 - \frac{1}{4\sinh^2\dfrac{u-v}{2}}, \quad \text{if} \quad w_0\neq\infty,$$

and

$$\mathscr{K}_0(u,v) = \left(\frac{G(u)}{G(v)}\right)^2\left(\frac{G'(v)}{G(u)-G(v)}\right)^2 - \left(\frac{G'(v)}{G(v)}\right)^2 - \frac{1}{4\sinh^2\dfrac{u-v}{2}}, \quad \text{if} \quad w_0=\infty,$$

where $G(u)=1/F(u)$.

Set $u=x+y$ and $v=x-y$. Since $a\epsilon\leq u,\ v\leq b\epsilon$, we have $|y/x|\leq 1-\delta$, where $\delta > 0$ depends only on a and b. Hence

$$y\frac{u^{ls}+v^{ls}}{u^{ls}-v^{ls}} = \frac{1}{ls}x + O(y^2x^{-1}), \qquad l \geqslant 1, \tag{40.7}$$

$$\frac{u^{ls}-v^{ls}}{u^s-v^s} = lx^{(l-1)s} + O(y^2x^{(l-1)s-2}), \quad l \geqslant 2. \tag{40.8}$$

Using (40.5) and expanding $\psi_l(u)$ and $\psi_l(v)$ in powers of y, we find from (40.7) that

$$F(u)-F(v) = \sum_{l=1}^{\varkappa+2}(u^{ls}-v^{ls})\left[\psi_l(x) + \frac{1}{ls}x\psi_l'(x) + O(y^2x^{-1})\right].$$

and so by (40.8)

$$F(u)-F(v) = (u^s-v^s)\left\{\sum_{l=1}^{\varkappa+2} lx^{(l-1)s}\left[\psi_l(x) + \frac{1}{ls}x\psi_l'(x)\right] + O[y^2(x^{-1}+x^{s-2})]\right\}.$$

Similar computations give that

$$F'(v) = sv^{s-1}\sum_{l=1}^{\varkappa+2} lx^{(l-1)s}\left[\psi_l(x) + \frac{1}{ls}x\psi_l'(x)\right]\{1 + y\varphi(x) + O[y^2(x^{-1}+x^{s-2})]\},$$

where $\varphi(x) = O(x^{s-1})$ if $s < 1$ and $\varphi(x) = O(1)$ if $s \geqq 1$.

Since $\psi_1(0) \neq 0$, we have

$$\left(\frac{F'(v)}{F(u)-F(v)}\right)^2 = \frac{s^2v^{2s-2}}{(u^s-v^s)^2}[1 + 2y\varphi(x)] + O(x^{-1}+x^{s-2}).$$

We note that $(F'(v)/F(v))^2 = O(v^{2s-2}) = O(x^{2s-2})$ since $u^{-1} = O(x^{-1})$ and $v^{-1} = O(x^{-1})$. It is easily seen that $(G'(v)/G(v))^2 = s^2/v^2 + O(x^{-1}+x^{s-2})$. From the relations proved above it follows that for $w_0 \neq \infty$

$$\mathscr{K}_0(u,v) = \frac{s^2v^{2s-2}}{(u^s-v^s)^2}(1 + 2y\varphi_1(x)) - \frac{1}{(u-v)^2} + O(x^{-1}+x^{s-2}).$$

If $w_0 = \infty$, then, taking into account the presence of the factor $(G(u)/G(v))^2$, we get that

$$\mathscr{K}_0(u,v) = \frac{s\cdot u^{2s}v^{-2}}{(u^s-v^s)^2}(1 + 2y\varphi(x)) - \frac{s^2}{v^2} - \frac{2}{(u-v)^2} + O(x^{-1}+x^{s-2}),$$

where $\varphi_1(x) = O(x^{s-1})$ if $s < 1$ and $\varphi_1(x) = O(y)$ for $s \geqq 1$. Set $P(z_0e^{-u}) = u^{2r}p(u)$. Expanding $p(u)$ and $p(v)$ in powers of y and noting that $y^2\mathscr{K}_0(u,v) = O(1)$, $y[u^{2r}\mathscr{K}_0(u,v) - v^{2r}\mathscr{K}_0(v,u)] = O(x^{2r-1}+x^{2r+s-2})$, we find that

$$P(z_0e^{-u})\mathscr{K}_0(u,v) + P(z_0e^{-v})\mathscr{K}_0(v,u) = p(x)[u^{2r}\mathscr{K}_0(u,v) + v^{2r}\mathscr{K}_0(v,u)] + O(x^{2r-1}+x^{2r+s-2})$$

Set

$$k_0(u,v) = \frac{s^2v^{2s-2}}{(u^s-v^s)^2} - \frac{1}{(u-v)^2}, \qquad \text{if} \quad w_0 \neq \infty,$$

and

$$k_0(u, v) = \frac{s^2u^{2s}v^{-2}}{(u^s - v^s)^2} - \frac{s^2}{v^2} - \frac{1}{(u-v)^2}, \quad \text{if} \quad w_0 = \infty.$$

Then

$$P(z_0e^{-u})\mathscr{K}_0(u, v) + P(z_0e^{-v})\mathscr{K}_0(v, u) = p(0)[u^{2r}k_0(u, v) + v^{2r}k_0(u, v)] + O(x^{2r-1} + x^{2r+s-2}).$$

Put

$$k(u,v) = u^{2r}k_0(u,v) + v^{2r}k_0(v,u).$$

Thus we have

$$\delta^2L(f; \nu, C_1) = -\varepsilon^{2r-2}p(0)\int_a^b\int_a^b\left[(-1)^\rho k(u, v) + \frac{u^{2r}+v^{2r}}{(u+v)^2}\right]g(u)g(v)\,du\,dv + o(\varepsilon^{2r-2}).$$

Since $P(z) \geqq 0$ for $|z| = 1$, it follows that $(-1)^r\rho(0) > 0$. Thus from (40.1) it follows that

$$(-1)^r\int_a^b\int_a^b\left[(-1)^\rho k(u, v) + \frac{u^{2r}+v^{2r}}{(u+v)^2}\right]g(u)g(v)\,du\,dv \geqslant 0. \tag{40.9}$$

We now reformulate the kernel in the integral (40.9). Set

$$k_*(u, v) = \begin{cases} 2\left[s^2\dfrac{(uv)^{r+s-1}}{(u^s-v^s)^2} - \dfrac{(uv)^r}{(u-v)^2}\right], & \text{if } \kappa \text{ is even,} \\[2ex] 2\left[s^2\dfrac{(uv)^{r+s-1}}{(u^s-v^s)^2} - \dfrac{(uv)^r}{(u-v)^2}\right] + \dfrac{(uv)^{r+\frac{s}{2}-1}}{(u^{s/2}+v^{s/2})^2}, & \text{if } \kappa \text{ is odd.} \end{cases}$$

It is not difficult to see that the following identity holds:

$$k(u, v) = k_*(u, v) + s^2\sum_{\nu=1}^{\leqslant\frac{r+1}{s}}{}' (\nu\mp1)(u^{2r-\nu s}v^{\nu s-2} + u^{\nu s-2}v^{2r-\nu s}) - \sum_{\nu=0}^{r-1}{}'(\nu+1)(u^{2r-2-\nu}v^\nu + u^\nu v^{2r-2-\nu}).$$

The prime on the summation sign means that those terms with $\nu = (r+1)/s$ and $\nu = r-1$ are to be taken with coefficients equal to $\frac{1}{2}$. In the expression $\nu\mp1$ the upper sign is used when $w_0 \neq \infty$, and the lower when $w_0 = \infty$. We shall now verify one of these identities. Let κ be even and $w_0 \neq \infty$. It is easy to see that

$$k(u, v) - k_*(u, v) = \frac{s^2}{(uv)^{2-2s}}\left[\frac{u^{r+1-s} - v^{r+1-s}}{u^s - v^s}\right]^2 - \left[\frac{u^r - v^r}{u-v}\right]^2.$$

If $\mu = \frac{1}{2}\kappa$ and so $\kappa + 2 = 2(\mu+1)$, then $r^{r+1} = u^{s(\mu+1)}$. Therefore

$$\left[\frac{v^{r+1-s}-v^{r+1-s}}{u^s-v^s}\right]^2=\sum_{\nu=0}^{\mu-1}{}'(\nu+1)(u^{s(2\mu-2-\nu)}v^{s\nu}+u^{s\nu}v^{s(2\mu-2-\nu)}),$$

and so

$$k(u,v)-k_*(u,v)=s^2\sum_{\nu=0}^{\mu-1}{}'(\nu+1)(u^{s(2\mu-\nu)-2}v^{s(\nu+2)-2}+u^{s(\nu+2)-2}v^{s(2\mu-\nu)-2})$$

$$-\sum_{\nu=0}^{r-1}{}'(\nu+1)(u^{2r-2-\nu}v^{\nu}+u^{\nu}v^{2r-2-\nu}).$$

Changing the index of summation in the first sum from ν to $\nu+2$ and noting that $\mu+1=(r+1)/s$, we arrive at the desired identity. The remaining cases of these identities can be verified in a similar way, although the case where κ is odd requires a somewhat longer calculation. We also use another simple identity:

$$\frac{u^{2r}+v^{2r}}{(u+v)^2}=2(-1)^r\frac{(uv)^r}{(u+v)^2}+\sum_{\nu=0}^{r-1}(-1)^{\nu}(\nu+1)(u^{2r-2-\nu}v^{\nu}+u^{\nu}v^{2r-2-\nu}).$$

The verification is elementary.

Set $k_\rho(u,v)=(-1)^\rho k_*(u,v)+2(-1)^r(uv)^r/(u+v)^2$ and

$$I_\rho(g,g)=\int_a^b\int_a^b k_\rho(u,v)\,g(u)\,g(v)\,du\,dv,\qquad g_l=\int_a^b g(u)\,u^l du,\qquad -\infty<l<\infty.$$

From the above identities and (40.4) it follows that (40.9) can be written in the form

$$(-1)^r\Big\{I_\rho(g,g)+2(-1)^\rho s^2\sum_{\nu=1}^{\leqslant\frac{r+1}{s}}(\nu\mp1)\,g_{2r-\nu s}g_{\nu s-2}\Big\}\geqslant0. \tag{40.10}$$

At this point we make a few remarks which will prove useful later on. First of all we note that via Schur's theorem ([15], Theorem 319)

$$|I_\rho(g,g)|<A\int_0^\infty u^{2r-1}g^2(u)\,du, \tag{40.11}$$

where

$$A=\int_0^\infty\left[x^{-r}|k_*(1,x)|+\frac{2}{(1+x)^2}\right]dx<\infty.$$

It is easily verified that A depends only on s.

Condition (40.4), which was used to establish (40.10), can be relaxed in some cases. For example, if $w_0\neq\infty$, $\rho=0$ and $s=2$, then (40.10) involves only the moments g_l of the function $g(u)$ for which $2\leq l\leq 2r-4$. But in addition to these moments, (40.4) requires unnecessarily that $g_1=g_{2r-3}=0$.

Similar situations arise for other values of the parameter s. It is clear that (40.10) continues to be true in the case when condition (40.4) is imposed only on those moments g_l for which $2s-2 \leq l \leq 2r-2s$ or $s-2 \leq l \leq 2r-s$. It is also possible to relax condition (40.4) in other ways. For example, suppose $g(u)$ satisfies (40.4) for $0 \leq l \leq r-3$, but $g_{2r-3} \neq 0$. It is not difficult to construct a sequence $\{g^k(u)\}$ such that $g^k(u)$, $1 \leq k < \infty$, satisfies the following conditions.

1) $g^k(u)$ is bounded and nonzero for $a_k \leq u \leq b_k$,

$$0 < a_k < b_k < \infty, \qquad \lim_{k\to\infty} a_k = 0, \qquad \lim_{k\to\infty} b_k = \infty.$$

2) Condition (40.4) is satisfied for $g^k(u)$.

3) $$\lim_{k\to\infty} g^k_{2l} = g_{2l}, \qquad 1 \leqslant l \leqslant r-2, \qquad \lim_{k\to\infty} I(g^k, g^k) = I(g, g).$$

Since (40.10) holds for each $g^k(u)$, this inequality continues to hold for $g(u)$. Upon beginning with this result and repeating the above argument, we have removed the restriction $g_1 = 0$. Similar arguments can be applied to other cases.

§41. Suppose the linear functional L has a local maximum at the function $f(z)$.

Proposition 41.1. *If z_0, $|z| = 1$, is the image of the base of the tree, then $P(z_0) \neq 0$ and $s = 2$. The function $Q(w)$ has a simple zero at $w = \infty$.*

Proof. Suppose z_0 is a zero of $P(z)$ of multiplicity $2r > 0$. From (40.10) we have

$$(-1)^r \Big\{ I_\rho(g, g) + 2(-1)^\rho s^2 \sum_{\nu=1}^{\leqslant \frac{r+1}{s}}{}' (\nu+1) g_{2r-\nu s} g_{\nu s-2} \Big\} \geqslant 0. \tag{41.1}$$

Since $s \leq 2$, the moments g_{2r-3} and g_{s-2} will always be present on the left-hand side. If $r > 1$, then $2r - s \neq s - 2$. Take $\rho = 0$ if $s \neq 1$ and $\rho = 1$ if $s = 1$. It is not difficult to construct a function $g(u, \epsilon)$ such that $g(u, \epsilon)$ does not vanish for $0 < a_\epsilon < u < b_\epsilon < \infty$, such that $|g_{s-2}| > \epsilon^{-1}$ and $|g_{2r-s}| > \epsilon^{-1}$, and at the same time such that $|g_{\nu s-2}| \leq C$ for $2 \leq \nu \leq (r+1)/s$, $|g_{2r-\nu s}| \leq C$ for $2 \leq \nu \leq (r+1)/s$, $g_{2l+1} = 0$, $0 \leq l \leq r-2$, and

$$\int_0^\infty u^{2r-1} g^2(u, \varepsilon)\, du < C,$$

where C is independent of ϵ. Since for such a function it is easy to arrange that the derivative of $g_{s-r} g_{2r-s}$ has any desired sign, it follows that (41.1) cannot hold for $g(u, \epsilon)$, provided ϵ is sufficiently small. The contradiction implies that $r \leq 1$. But for $r = 1$ and $s < 2$, the above argument applies again. Thus we must consider the case $r = 1$, $s = 2$. We now take $\rho = 0$, and so (41.1) takes the form

$$-\{I_0(g,g)+8g_0^2\}\geqq 0.$$

But this is absurd, since it implies via (40.11) the inequality

$$\left(\int_0^\infty g(u)\,du\right)^2 \leqslant \frac{1}{8}A\int_0^\infty ug^2(u)\,du,$$

which obviously cannot hold for any bounded finite function $g(u)$. Thus we have $r=0$, and we must now show that $s=2$. Take $\rho=1$; since $r=0$, condition (40.4) is automatically eliminated from consideration. Inequality (41.1) takes the form

(41.2) $$I_1(g,g)-2s^2\sum_{\nu=1}^{\leqslant\frac{1}{s}}{}' (\nu+1)g_{-\nu s}g_{\nu s-2}\geqslant 0.$$

If $s=1$, then it follows via (40.11) that

$$\left(\int_0^\infty g(u)\frac{du}{u}\right)^2 \leqslant \frac{1}{2}A\int_0^\infty g^2(u)\,\frac{du}{u},$$

which is an obvious contradiction. For $s<1$ we have $s-2\neq -s$; applying the arguments as above, we find that (41.2) cannot be satisfied. Since $s=2/\kappa$, the only possibility left is that $\kappa=1$, and so $s=2$. Q.E.D.

PROPOSITION 41.2. *Any nonterminal vertex of the tree $\mathcal{T}_f$ is a simple zero of the function $Q(w)$. If a terminal vertex w_0 is a zero of $Q(w)$, then w_0 is a simple zero.*

PROOF. If w_0 is not a terminal vertex, then it has an image z_0, $|z_0|=1$, such that $s\leqq 1$. We shall use inequality (40.10) with $\rho=0$ if $s\neq\frac{1}{2}$, and with $\rho=1$ if $s=\frac{1}{2}$. We find that

(41.3) $$(-1)^r\left\{I_\rho(g,g)+2(-1)^\rho s^2\sum_{\nu=2}^{\leqslant\frac{r+1}{s}}{}'(\nu-1)g_{2r-\nu s}g_{\nu s-2}\right\}\geqslant 0.$$

Suppose that w_0 is a zero of $Q(w)$ of multiplicity $\kappa\geqq 2$. Then $(r+1)/s=(\kappa+2)/2\geqq 2$. If $2r-2s\neq 2s-2$, then it is easy to find a function $g(u)$ for which the left-hand side of (41.3) is negative. If $2r-2s=2s-2$, then $s=(r+1)/2$, and so $\kappa=2$. If $r=1$, it follows from (41.3) that

$$-\{I_0(g,g)+g_0^2\}\geqq 0,$$

which in turn yields, on the basis of (40.11), the obvious contradiction

$$\left(\int_0^\infty g(u)\,du\right)^2 \leqslant A\int_0^\infty ug^2(u)\,du.$$

Hence it remains to consider the case $r=0$, $\kappa=2$, $s=\frac{1}{2}$. From the condition $\rho=1$ we get the inequality

$$I_1(g, g) - \frac{1}{4} g_{-1}^2 \geqslant 0,$$

and this in turn gives

$$\left(\int_0^\infty g(u) \frac{du}{u}\right)^2 \leqslant 4A \int_0^\infty g^2(u) \frac{du}{u}.$$

This latter inequality cannot be satisfied for any finite function $g(u)$, and so w_0 must be a simple zero of $Q(w)$. Suppose that w_0 is a terminal vertex and z_0 is its image. In this case we have $s = 2$. For $r = 3$ it follows from (41.3) that

$$-\{I_0(g,g) + 4g_2^2\} \geqq 0,$$

which in turn implies via (40.11) that

$$\left(\int_0^\infty u^2 g(u)\, du\right)^2 \leqslant \frac{A}{4} \int_0^\infty u^5 g^2(u)\, du,$$

which is an obvious contradiction. If $r > 3$, then $2r - 4 \neq 2$, and it is easy to find a function $g(u)$ for which the left-hand side of (41.3) is negative. Thus we have that $r \leqq 2$, and hence w_0 can be a zero of $Q(w)$ of multiplicity at most one. This completes the proof of the proposition. From the results proved above we obtain the following theorem.

THEOREM 18. *Suppose that a linear functional has a maximum at the function $f(z)$. Then the vertices of the tree $\mathscr{T}_f$ can be zeros of $Q(w)$ of order at most one. The base of the tree is a simple zero of $Q(w)$.*

To prove Theorem 18 we do not really use the fact that the functional L has a local maximum at $f(z)$, but rather the following consequences of this fact: a) the inequality (40.1), and b) the fact that $f(z) \in S_n$. Thus we can generalize Theorem 18 as follows. We assume that the linear functional L and $f(z) \in S$ satisfy the following conditions:

a) The first variation of L vanishes for the class of admissible function $\mathscr{D}_f$.

b) The second variation of L with respect to this same class of admissible functions satisfies inequality (40.1).

From condition a) it follows that $f(z)$ satisfies (4.5), where the function $P(z)$ is real-valued on $|z| = 1$. From condition b) and Lemma 5.3 it follows that $P(z) \geqq 0$ for $|z| = 1$. Hence $f(z) \in S_n$. Thus from condition b) it follows, as shown above, that the tree $\mathscr{T}_f$ has the properties stated in Theorem 18. Thus we have

THEOREM 18′. *If the linear functional L and the function $f(z) \in S$ satisfy conditions* a) *and* b), *then $f(z) \in S_n$, and its boundary tree has the properties listed in Theorem* 18.

§42. We now pause for a moment to examine questions on the nonlocal theory of extremal problems, even though strictly speaking these questions do not fall within the scope of this monograph.

If the linear functional L has an absolute maximum at $f(z)$, then we can say a good deal more about the structure of the tree $\mathscr{T}_f$. The following important lemma was first proved by Schiffer for a special type of functional, and it was extended to the general case by Schaeffer and Spencer.

LEMMA 42.1. *Suppose that the linear functional L has an absolute maximum at $f(z)$. If $Q(w)$ has a zero $w_0 \neq \infty$ with $w_0 \in \mathscr{T}_f$, then the tree $\mathscr{T}_f$ is located on the line* $\operatorname{Im}(w/w_0) = 0$.

The proof of this lemma will not be given here, and the reader is referred to [2]. From Theorem 18 and Lemma 42.1 it follows that the tree $\mathscr{T}_f$ of $f(z)$ contains one branch, and moreover the terminal vertex $w_0 \in \Gamma_j$ is not a zero of $Q(w)$ unless $f(z) = z(1 - e^{i\theta}z)^{-2}$. The fact that the tree of the univalent function under consideration has only one branch was established first by Schaeffer and Spencer, although we have shown considerably more, since by Theorem 18 the base of the tree is a simple zero of $Q(w)$.

DEFINITION 42.1. A point $x \in \partial \mathscr{V}_n$ is called a *support point* if there exists a supporting hyperplane of $\mathscr{V}_n$ passing through it. The set of all support points will be called the *supporting set* and will be denoted by $\mathfrak{O}$.

Clearly, if $x \in \mathfrak{O}$, then $j(z, x)$ gives the absolute extremum for some linear functional, and the converse is also true. From Theorem 18 and Lemma 42.1 we obtain the important

LEMMA 42.2. $\mathfrak{O} \cap \mathscr{R} \subseteq \mathscr{R}_1^*$.

CHAPTER 8

ON SUFFICIENT CONDITIONS FOR A LOCAL EXTREMUM, AND ON THE KOEBE FUNCTION

The results given in the preceding chapters allow us to prove some elegant results within the framework of the second variation. If we ignore on $\partial\mathscr{V}_n$ a set of small dimension, namely $\mathscr{A}_1 \cup \mathscr{A}_2$, then with the aid of the second variation we can give a sufficient condition for a local extremum of a functional. Earlier we limited ourselves to a consideration of functionals of the form $J(f) = F(x)$, where x is the image of f under a mapping of S to R^{2n}, and where F is twice continuously differentiable on some open set $K \supset \mathscr{V}_n$. A sufficient condition for a local maximum of the functional $J(f)$ has the classical form, and, as is easily seen, involves the nonpositivity of the second variation. If however we suspect that the functional in question has a maximum at $f(z, x)$, with $x \in \mathscr{A}_1$, then the question of stating a sufficient condition for a local maximum becomes significantly more complicated. We shall not investigate this problem to any great extent; we shall limit ourselves to functionals of the type

$$J(f) = \operatorname{Re} \sum_{k=1}^{n} \lambda_k a_k, \qquad \operatorname{Im} \lambda_k = 0, \qquad 1 \leqslant k \leqslant n,$$

which are closely connected with the well-known Bieberbach conjecture.

§43. Let $F(x)$ be a twice continuously differentiable function on an open set $K \supset \mathscr{V}_n$ such that $\operatorname{grad} F(x) \neq 0$ for $x \in \mathscr{V}_n$. Suppose that the functional $J(f) = F(x)$ is stationary at the function $f(z, x^0)$, $x^0 \in \mathscr{V}_n$. By a stationary value of the functional $J(z)$, we mean the following: consider the variations of a given $f(z) \in S$ determined by the admissible functions of the form (2.2); if for any such variation $f_\epsilon = f(z, \epsilon) \in S$ we have that $J(f_\epsilon) = J(f) + O(\epsilon^2)$, then we say that the functional is stationary at $f(z)$. Upon combining the fact that $J(f)$ is stationary at $f(z, x^0)$ with the condition $\operatorname{grad} F(x) \neq 0$, $x \in \mathscr{V}_n$, we see as before that $f(z, x^0)$ satisfies equation (4.5), with $P(z)$ real-valued on $|z| = 1$. Suppose that there exists a constant $\theta > 0$ for which

$$\delta^2 J(f) \leqq -\theta, \tag{43.1}$$

if the first coefficient variations of $f(z, x^0)$ are subjected to the condition

$$\sum_{1}^{n} |\delta a_k|^2 = 1. \tag{43.2}$$

It follows from Lemma 5.3 that $P(z) \geqq 0$ for $|z| = 1$, and so $x^0 \in \partial\mathscr{V}_n$ and

$f(z, x^0) \in S_n$. In §4 we showed that the associated vector λ (in general, not normalized) determining equation (4.5) has the form $\lambda = \operatorname{grad} F(x^0)$. Suppose that $x^0 \in \mathscr{R}$. Then λ will be an outward pointing normal to $\partial\mathscr{V}_n$ at x^0. In a neighborhood of x^0 we introduce the local coordinate system; as before we denote these coordinates by (y, τ). Then $x = x^0 + U\mathfrak{y}$, where $\mathfrak{y} = (y, \tau)$. In a neighborhood of x^0 the equation of the boundary of $\mathscr{V}_n$ has the form $\tau = \psi(y)$, and moreover $\psi(0) = \operatorname{grad}\psi(0) = 0$. Set $F(x^0 + U\mathfrak{y}) = \Phi(y, \tau)$. Clearly $\operatorname{grad}\Phi(0,0) = U'\lambda$, and so

$$\frac{\partial\Phi(0,0)}{\partial y_j} = 0, \quad 1 \leqslant j \leqslant 2n-1, \qquad \frac{\partial\Phi(0,0)}{\partial\tau} = |\operatorname{grad} F(x^0)|.$$

Taking into account these relations and recalling that $\psi(y) = O(|y|^2)$ (see Theorem 13), we find that

$$\Phi(y, \psi(y)) = \Phi(0,0) + \frac{\partial\Phi(0,0)}{\partial\tau}\psi(y) + \frac{1}{2}\sum_{k,l=1}^{2n-1}\frac{\partial^2\Phi(0,0)}{\partial y_k \partial y_l} y_k y_l + o(|y|^2).$$

Put $v = y|y|^{-1}$, $\mathfrak{v} = (v, 0)$, and $t = U\mathfrak{v}$. Denote by H the Hessian matrix of $F(x)$, computed at $x = x^0$. It is easy to see that

$$\sum_{k,l=1}^{2n-1}\frac{\partial^2\Phi(0,0)}{\partial y_k \partial y_l} v_k v_l = (Ht, t).$$

Suppose that $\mathscr{A}_{x^0} = \mathfrak{H}_{x^0}$; that is to say, $x^0 \notin \mathscr{A}_2$. Then it follows from Lemma 32.8 that at $y = 0$ the function $\psi(y)$ has a second derivative in the direction of any vector v, and that $\psi(y) = (|y|^2/2) d_v^2\psi(0) + o(|y|^2)$. From the above we find that

$$\Phi(y, \psi(y)) = \Phi(0,0) + \frac{|y|^2}{2}\left[|\operatorname{grad} F(x^0)|\, d_v^2\psi(0) + (Ht, t)\right] + o(|y|^2)$$

On the other hand, we showed in §4 that

$$\delta^2 J(f) = \delta^2 L(f) + (Ht, t),$$

where L is a linear functional of the type $L(f) = \operatorname{Re}\sum_1^n \bar{\lambda}_k a_k$, and moreover the second variation of L can be evaluated by a measure satisfying $\delta a_k = t_k$, $1 \leqq k \leqq n$, $(t_1, \cdots, t_n) = t$. From (43.1), (43.2), (32.8) and the assumption that $\mathscr{A}_{x^0} = \mathfrak{H}_{x^0}$ it follows that

$$|\operatorname{grad} F(x^0)|\, d_v^2\psi(0) + (Ht, t) < -0. \tag{43.3}$$

The presence of the factor $|\operatorname{grad} F(x^0)|$ is necessary because we proved (32.8) with a normalized vector λ, but we have now dropped this normalization hypothesis. From (43.3) it follows that

$$\Phi(y, \psi(y)) < \Phi(0,0),$$

provided $|y| < \epsilon$ and $y \neq 0$, and this inequality in turn clearly implies that $F(x) < F(x^0)$ for $|x - x^0| < \delta$, $x \neq x^0$. Hence we have proved the following theorem.

THEOREM 19. *Suppose the functional* $J(f) = F(x)$ *is stationary at* $f(z, x^0)$. *If under the condition* (43.2) *the second variation satisfies equation* (43.1), *then* $x^0 \in \partial\mathscr{V}_n$ *and if* $x^0 \in \mathscr{R}$, $x^0 \notin \mathscr{A}_2$, *then* $J(f)$ *has a local maximum at* $f(z, x^0)$.

For linear functionals the condition $x^0 \notin \mathscr{A}_2$ is redundant.

LEMMA 43.1. *Suppose that the linear functional* $L(f)$ *is stationary at* $f(z, x^0)$, *and that if* (43.2) *holds, then the second variation of* $L(f)$ *satisfies* (43.1). *Then* $f(z, x^0) \in S_n$, *and if* $x^0 \in \mathscr{R}$, *then* $L(f)$ *has a local maximum at* $f(z, x^0)$.

PROOF. Since the conditions of Theorem 18′ are satisfied, the tree $\mathscr{T}_{x^0}$ has no singular vertices. From this and Lemma 25.3 it follows that $\mathscr{A}_{x^0} = \mathfrak{H}_{x^0}$, and thus the previous theorem applies. This proves the lemma.

§44. In the theory of extremal problems within the class S, the Koebe functions $f_\theta(z) = z(1 - e^{i\theta}z)^{-2}$, where θ is real, play a central role. It is well known that for fixed n the linear manifold of associated vectors for the functions $f_\theta(z)$ has dimension equal to n. Using Lemma 37.1, it is not difficult to see that the dimension of the above-mentioned linear manifold is always $\leq n$, and equality can hold only for the Koebe functions. In the following we shall be examining the function $f_0(z)$. Suppose that the linear functional L is stationary at $f_0(z)$ and that, under condition (43.2), the inequality (43.1) holds. Let $L(f) = \operatorname{Re}\sum_1^n \lambda_k a_k$. Since $L(f)$ is stationary at $f_0(z)$, it follows that $f_0(z)$ satisfies equation (4.5), and from Lemma 37.1 we see that $\operatorname{Im}\lambda_k = 0$, $1 \leq k \leq n$. Equation (4.5) can be written in the form

$$Q(f_0; \lambda)\left(\frac{Df_0}{f_0}\right)^2 + P(z; \lambda) = 0.$$

From (43.1) and Lemma 5.3 it follows that $P(z; \lambda) \geq 0$ for $|z| = 1$, and Theorem 18′ implies that $P(z; \lambda)$ can have zeros of multiplicity at most 4 on $|z| = 1$. We shall suppose that $P(z_j, \lambda)$ has only double zeros on $|z| = 1$, or, in other words, that $Q(-1/4; \lambda) \neq 0$. We shall assume that in any arbitrarily small neighborhood of $f_0(z)$ there exists $f(z) \in S$, $f(z) \neq f_0(z)$, such that $L(f) \geq L(f_0)$.

Let σ_α be the set of univalent functions $f(z) \in S$, $f(z) = \sum_0^\infty a_k z^{k+1}$, $a_0 = 1$, which satisfy the condition $\operatorname{Re} a_1 \geq 2 - \alpha$. For $0 \leq \alpha \leq 4$ we have $\sigma_\alpha \neq \emptyset$. By assumption

$$\sup_{f \in \sigma_\alpha} L(f) = l(\alpha) > L(f_0).$$

In view of the compactness of σ_α, there exists a function $f_\alpha(z)$ for which $L(f_\alpha) = l(\alpha)$. If $\alpha_1 < \alpha_2$, then $\sigma_{\alpha_1} \subseteq \sigma_{\alpha_2}$, and so $l(\alpha_1) \leq l(\alpha_2)$.

The extremal problem under consideration easily can be visualized geometrically. To do so, we consider the mapping of S to R^{2n}. We also note that the image of σ_α lies to one side of the hyperplane defined by $\operatorname{Re} a_1 = 2 - \alpha$, namely the one for which $\operatorname{Re} a_1 \geq 2 - \alpha$. If we denote by Σ_α the image of σ_α, then the extremal problem is reduced to the construction of a supporting hyperplane for Σ_α, with the given normal vector $\lambda = (\lambda_1, \cdots, \lambda_n)$. Naturally we assume that λ does not coincide with the normal vector of the hyperplane $\mathrm{Re}\, a_1 = 2 - \alpha$. If x^0 is a point of Σ_α which is contained in the desired supporting hyperplane, then $x^0 \in \partial\Sigma_\alpha$, and in view of our previous remarks we have that $x^0 \in \partial\mathscr{V}_n$. Therefore $f_\alpha(z) \in S_n$. Consider the variation of $f_\alpha(z)$ determined by an admissible function of the type (2.2). Denote this variation by $f_\alpha(z,\epsilon) = \sum_0^\infty a_k(\epsilon) z^{k+1}$. Having set $\alpha_\epsilon = 2 - \mathrm{Re}\, a_1(\epsilon)$, we see that $f_\alpha(z,\epsilon) \in \sigma_{\alpha_\epsilon}$. We recall that

$$\mathrm{Re}\, a_1(\varepsilon) = 2 - \alpha + \varepsilon\, \mathrm{Re}\, \delta a_1 + \frac{\varepsilon^2}{2} \mathrm{Re}\, \delta^2 a_1 + \cdots,$$

and by formulas (2.13) – (2.15)

$$\mathrm{Re}\, \delta a_1 = \mathrm{Re} \left\{ \iint_{\mathscr{C}} \left[-\left(\frac{Df_\alpha(z)}{f_\alpha(z)} \right)^2 \frac{1}{f_\alpha(z)} + z^{-1} + a_1 + z \right] \nu(dz) + ic_1 a_1 \right\}.$$

Choose the measure $\nu(dz)$ and the constant c_1 such that $\mathrm{Re}\, \delta a_1 > 0$. Such a choice is always possible if

$$\frac{(Df_\alpha(z))^2}{f_\alpha^3(z)} - z^{-1} - a_1 - z \not\equiv 0.$$

If this latter condition does not hold, then $f_\alpha(z)$ satisfies a differential equation which we can solve easily upon noting that $z^{-1} + a_1 + z$ must have a double zero on $|z| = 1$. This implies $a_1 = 2$, and so $f_\alpha(z) \equiv f_0(z)$. Hence, if $f_\alpha(z) \not\equiv f_0(z)$, then $\mathrm{Re}\, \delta a_1$, as a function of the measure, can be either positive or negative. Note that $L(f_\alpha(z,\epsilon)) \leqq l(\alpha_\epsilon)$, and so

$$\varepsilon L(\delta f_\alpha(z)) + \frac{\varepsilon^2}{2} L(\delta^2 f_\alpha(z)) + \cdots \leqslant l(\alpha_\varepsilon) - l(\alpha).$$

If $\mathrm{Re}\, \delta a_1 > 0$, we have $\alpha_\epsilon < \alpha$ provided ϵ is sufficiently small. Since $l(\alpha_\epsilon) - l(\alpha) \leqq 0$, we have

$$L(\delta f_\alpha(z)) \leqq 0, \tag{44.1}$$

as soon as $\mathrm{Re}\, \delta a_1 > 0$. It is clear from continuity considerations that inequality (44.1) also holds when $\mathrm{Re}\, \delta a_1 \geqq 0$. Using (2.13), we find that

$$L(\delta f_\alpha(z)) = \mathrm{Re} \left\{ \iint_{\mathscr{C}} \sum_1^n (\lambda_k \mathfrak{A}_k(z) + \lambda_k \overline{\mathfrak{B}_k(\bar{z})}) \nu(dz) + ci_1 \sum_1^n k\lambda_k a_k \right\},$$

and from (2.14) and (2.15) it follows that

$$\sum_1^n (\lambda_k \mathfrak{A}_k(z) + \lambda_k \overline{\mathfrak{B}_k(\bar{z})}) = Q_\alpha(f_\alpha; \lambda) \left(\frac{Df_\alpha(z)}{f_\alpha(z)} \right)^2 + P_\alpha(z; \lambda);$$

where $Q_\alpha(w; \lambda)$ is defined by (3.16), and $P_\alpha(z; \lambda)$ by (2.30). Since (44.1) holds for any measure for which $\mathrm{Re}\, \delta a_1 \geqq 0$, it is easy to see that

$$Q_\alpha(f_\alpha; \lambda) \left(\frac{Df_\alpha(z)}{f_\alpha(z)} \right)^2 + P_\alpha(z; \lambda) = -\mu(\alpha) \left[-\frac{(Df_\alpha(z))^2}{f_\alpha^3(z)} + z^{-1} + a_1 + z \right],$$

where $\mu(\alpha) \geqq 0$. Put $Q_\alpha^*(w;\lambda) = Q_\alpha(w;\lambda) - \mu(\alpha)w^{-1}$,

$$P_\alpha^*(z;\lambda) = P_\alpha(z;\lambda) + \mu(\alpha)(z^{-1} + a_1 + z).$$

Since $P_\alpha^*(r;\lambda)$ has a double zero on $|z| = 1$, we have $\operatorname{Im} P_\alpha^*(z;\lambda) = 0$ for $|z| = 1$. We can also show that $P_\alpha^*(z;\lambda) \geqq 0$ for $|z| = 1$. However, in order to avoid lengthy calculations, we shall only consider small α; the above inequality is then obvious, since for sufficiently small α the coefficients of $P_\alpha(z;\lambda)$ are arbitrarily close to the corresponding coefficients of $P(z;\lambda)$. Since $\mu_\alpha \geqq 0$ and $P(z;\lambda)$ has a unique double zero on $|z| = 1$, we see that for small α the quasipolynomial $P_\alpha^*(z;\lambda)$ has only one double zero on $|z| = 1$. Therefore $P_\alpha^*(z;\lambda)$ has constant sign on $|z| = 1$, and so $P_\alpha^*(z;\lambda) \geqq 0$ for $|z| = 1$.

Consider the differential $Q_\alpha^*(w;\lambda)(dw/w)^2$. Set $Q_\alpha(w;\lambda) = \sum_1^n A_{\alpha k} w^{-k}$, $Q(w;\lambda) = \sum_1^n A_k w^{-k}$. If α is small, then $\sum_1^n |A_{\alpha k} - A_k|^2$ is small, and upon setting $A_{\alpha k} = B_{\alpha k} + iC_{\alpha k}$ we see from the fact that A_k, $1 \leqq k \leqq n$, is real, that $\lim_{\alpha\to 0} \sum_1^n C_{\alpha k}^2 = 0$. Set $\epsilon = \{\sum_1^n C_{\alpha k}^2\}^{1/2}$. Then

$$Q_\alpha^*(w;\lambda) = Q_{r\alpha}(w) + i\varepsilon Q_{i\alpha}(w;\varepsilon), \quad \text{where } \varepsilon Q_{i\alpha}(w;\varepsilon) = \sum_1^n C_{\alpha k} w^{-k}.$$

Since $wQ(w;\lambda)$ by hypothesis does not vanish for $-\infty < w \leqq -1/4$, we see for sufficiently small α that

$$\inf_{-\infty < w \leqslant -1/4} |wQ_{r\alpha}(w)| \geqslant 2C_0 > 0.$$

Thus there exists $\rho < 1$ such that outside the level curve $\mathscr{L}_\rho = \{w : |f_0^{-1}(w)| = \rho\}$ we have $|wQ_{r\alpha}(w)| \geqq C_0$, provided $\alpha \leqq \alpha_0$. It might happen that as $\alpha \to 0$ the quantity $\mu(\alpha)$ will become unbounded. If so, we shall consider a sequence $\alpha_k \downarrow 0$ such that $\mu(\alpha_k) \uparrow \infty$. We also consider $(1/\mu(\alpha))Q_\alpha^*(w;\lambda)$, and so as not to introduce more notation, we shall continue to denote this function by $Q_\alpha^*(w;\lambda)$. Thus without loss of generality we can suppose that for $|z| \geqq \rho$ and $w = f_0(z)$ we have

$$C_0 \leqq |wQ_{r\alpha}(w)| \leqq C_0^*,$$

where the constants C_0 and C_0^* are independent of α (again provided $\alpha \leqq \alpha_0$). Let $\mathscr{G}_\rho$ be the domain containing ∞ and bounded by the level curve $\mathscr{L}_\rho$. In $\mathscr{G}_\rho$ we consider the solution of the equation

$$(1 + \varepsilon\gamma(\varepsilon))[Q_{r\alpha}(w) + i\varepsilon Q_{i\alpha}(w;\varepsilon)]\left(\frac{dw}{w}\right)^2 = Q_{r\alpha}\left(\frac{dt}{t}\right)^2$$

such that $w(\infty) = \infty$; we take $\gamma(\epsilon)$ so that the image of the boundary tree of the function $f_0(z)$ is the boundary tree of $f_\alpha(z)$. This is the situation encountered in §33, and we shall apply the results of that section. Upon doing so we obtain in the domain $\mathscr{G}_{\rho_0}$, $\rho_0 = \frac{1}{2}(1 + \rho)$, a solution of the above equation such that

$$w = t(1 + \epsilon h_1(t,\epsilon)).$$

The function $h_1(t, \epsilon)$ is defined by (33.36) with $m = 0$. Recalling (33.38), we find that

$$h_1(t, \varepsilon) = \frac{1}{2\sqrt{Q_{r\alpha}(t)}} \int_0^{t-1} [\gamma(\varepsilon) Q_{r\alpha}(\eta^{-1}) + iQ_{i\alpha}(\eta^{-1}; \varepsilon)] \frac{d\eta}{\eta \sqrt{Q_{r\alpha}(\eta^{-1})}} + O(\varepsilon).$$

(44.2)

The quantity $\gamma(\epsilon)$ is determined with the aid of the relations established in §33. Thus for $f_0(z)$ we have constructed an admissible function of the form

$$g(z, \varepsilon) = \frac{1}{2\pi i} \int_{|\zeta|=\rho_0} \frac{f_0^2(z)}{f_0(z) - f_0(\zeta)} \left(\frac{Df_0(\zeta)}{f_0(\zeta)}\right)^2 \frac{h[f_0(\zeta), \varepsilon]}{Df_0(\zeta)} \frac{d\zeta}{\zeta}, \tag{44.3}$$

where $h(t, \epsilon) = th_1(t, \epsilon)$ is such that $f_0(z) + \epsilon g(z, \epsilon)$ is regular in $\rho \leqq |z| < 1$ and maps $|z| = 1$ onto the tree $\mathcal{T}_{f_\alpha}$. Applying Lemma 1.5, we get that $f_\alpha(z) = f_0(z, \epsilon)$, where $f_0(z, \epsilon)$ is the variation of $f_0(z)$ defined by the admissible function (44.3). According to Theorem 1,

$$f_0(z, \epsilon) = f_0(z) + \epsilon f_1(z; \epsilon) + \epsilon^2 f_2(z; \epsilon) + \cdots.$$

Thus

$$L(f_\alpha) = L(f_0) + \epsilon^2 L(f_2) + O(\epsilon^3).$$

If

$$\nu(dz; \varepsilon) = \frac{h[f_0(z); \varepsilon]}{Df_0(z)} \frac{dz}{2\pi iz}, \quad z \in \mathscr{C} = \{z: |z| = \rho_0\},$$

then $L(f_2) = \frac{1}{2}\delta^2 L(f_0; \nu, c_1)$, where $c_1 = c_1(\epsilon)$ is determined by the relation

$$\operatorname{Im}\left\{\int_{\mathscr{C}} \nu(dz; \varepsilon) + ic_1\right\} = 0. \tag{44.4}$$

Since $Q_{r\alpha}(w)$ and $Q_{i\alpha}(w; \epsilon)$ are uniformly bounded for $|w| \geqq 1/8$ and $\alpha \leqq \alpha_0$, we can extract a subsequence $\{\alpha_m\}$ and a corresponding subsequence $\{\epsilon_m\}$ such that the corresponding functions converge uniformly to polynomials in $1/w$; denote these limit functions by $Q_r(w)$ and $Q_i(w)$, respectively. From the definition of $Q_{i\alpha}(w; \epsilon)$ it follows that $Q_i(w) \not\equiv 0$. Without loss of generality we can assume that $\gamma(\epsilon_m) \to \gamma$ and $c_1(\epsilon_m) \to c_1^0$, since $\gamma(\epsilon)$ and $c_1(\epsilon)$ are bounded. From (44.2) it follows that

$$\lim_{\varepsilon_m \to 0} h(t, \varepsilon_m) = \frac{t}{2\sqrt{Q_r(t)}} \int_0^{t-1} [\gamma Q_r(\eta^{-1}) + iQ_i(\eta^{-1})] \frac{d\eta}{\eta\sqrt{Q_r(\eta^{-1})}} = h_0(t),$$

and upon defining the measure

$$\nu_0(dz) = \frac{h_0[f_0(z)]}{Df_0(z)} \frac{dz}{2\pi iz}, \quad z \in \mathscr{C},$$

and the constant c_1^0 by equation (44.4) with $\nu(dz; \epsilon)$ replaced by $\nu_0(dz)$, we find that

$$\lim_{\varepsilon_m \to 0} \delta^2 L(f_0; \nu(dz; \varepsilon_m), c_1(\varepsilon_m)) = \delta^2 L(f_0; \nu_0, c_1^0).$$

Since $L(f_\alpha) \geqq L(f_0)$, we have

$$\delta^2 L(f_0; \nu_0, c_1^0) \geqq 0. \tag{44.6}$$

Let $\delta f_0(z)$ be the variation of $f_0(z)$ defined by the measure $\nu_0(dz)$ and the constant c_1^0. By hypothesis, (43.1) holds, subject to condition (43.2). Thus $\delta a_k(\nu_0, c_1^0) = 0$, $1 \leqq k \leqq n$. But then from (44.6) and (23.2) it follows that

$$\delta f_0(z) \equiv 0. \tag{44.7}$$

We now recall the representation for the first variation given in §19. Setting

$$u_k = \int_{\mathscr{C}} z^k \nu_0(dz), \qquad k = 1, 2, \ldots,$$

we consider, in accord with (19.2), the function

$$u_k = \sum_1^\infty (u_k z^{-k} - \bar{u}_k z^k) - ic_1^0.$$

If $\alpha(w) = Df_0(z) u(z)$, $z = f_0^{-1}(w)$, we have for $|z| < \rho_0$ that

$$\delta f_0(z) = \frac{1}{2\pi i} \int_{\mathscr{C}_{\rho_0}} \frac{f_0^2(z)}{f_0(z) - w} \frac{\alpha(w)}{w^2} dw.$$

From (44.7) it follows that $\alpha(w)$ is regular in $\mathscr{G}_{\rho_0}$. Thus $\alpha_+(w) = \alpha_-(w)$ for $-\infty < w \leqq -1/4$. This latter condition is equivalent to the condition that $Df_0(z) u(z) = Df_0(\bar{z}) u(\bar{z})$, where $z = e^{i\vartheta}$, $0 < \vartheta \leqq \pi$. Noting that $Df_0(\bar{z}) = -Df_0(\bar{z})$, we see that $u(z) = -u(z)$. Therefore for $|z| = 1$ we have

$$\sum_1^\infty (u_k - \bar{u}_k)(z^k + z^{-k}) - 2ic_1^0 = 0.$$

Hence $c_1^0 = 0$ and $\operatorname{Im} u_k = 0$, $1 \leqq k < \infty$. Put

$$h_{0r}(t) = \frac{\gamma t}{2\sqrt{Q_r(t)}} \int_0^{t^{-1}} \sqrt{Q_r(\eta^{-1})} \frac{d\eta}{\eta}, \qquad h_{0i}(t) = \frac{t}{2\sqrt{Q_r(t)}} \int_0^{t^{-1}} \frac{Q_i(\eta^{-1})\, d\eta}{\eta \sqrt{Q_r(\eta^{-1})}}.$$

If $|z| < 1$, then $\overline{f_0(z)} = f_0(\bar{z})$, and so

$$h_{0r}[f_0(\bar{z})] = \overline{h_{0r}[f_0(z)]}, \qquad h_{0i}[f_0(\bar{z})] = \overline{h_{0i}[f_0(z)]}.$$

Upon putting

$$u_k = \int_{|z|=\rho_0} z^k \frac{h_{0r}[f_0(z)]}{Df_0(z)} \frac{dz}{2\pi iz} + i \int_{|z|=\rho_0} z^k \frac{h_{0i}[f_0(z)]}{Df_0(z)} \frac{dz}{2\pi iz} = u_{rk} + iu_{ik},$$

$1 \leqq k < \infty$, we deduce from the above that u_{rk} and u_{ik}, $1 \leqq k < \infty$, are all real. Therefore $u_{ik} = 0$, $1 \leqq k < \infty$. Hence $h_{0i}[f_0(z)]/Df_0(z)$ is regular in $|z| < 1$. Since on $|z| = 1$ the limits of this function are purely imaginary, and since $\lim_{z \to -1} (z+1)(h_{0i}[f_0(z)]/Df_0(z)) = -8h_{0i}[-1/4]$, we have that

$$\frac{h_{0i}[f_0(z)]}{Df_0(z)} = A\frac{1-z}{1+z},$$

where A is a real constant. But $\operatorname{Im}\{\int_{\mathscr{C}}\nu(dz)+ic_1^0\}=0$, and so $c_1^0=0$ implies

$$\int_{|z|=\rho_0} \frac{h_{0i}[f_0(z)]}{Df_0(z)}\frac{dz}{2\pi iz}=0,$$

and therefore $A=0$. Hence $h_{0i}(t)\equiv 0$, and thus $Q_i(w)\equiv 0$, which is a contradiction. Hence for $\alpha \leqq \alpha_0$ we have $l(\alpha)=l(0)$ and $f_\alpha(z)\equiv f_0(z)$. These results are summarized in the following lemma.

Lemma 44.1. *Suppose the linear functional $L(f)$ is stationary at $f_0(z)$, and suppose that subject to condition (43.2), its second variation satisfies inequality (43.1). If $Q(-1/4;\lambda)\neq 0$, then $L(f)$ has a local maximum at $f_0(z)$. If $\rho(f,f_0)<\delta$, where δ is sufficiently small and $f\neq f_0$, then $L(f)<L(f_0)$.*

§45. In the preceding section we showed that inequality (43.1) holds for some linear functionals. We shall consider the particular linear functional $L(f)=\operatorname{Re}a_n$. In accord with our previous results, $L(f)$ is stationary at $f_0(z)$. In this particular case the Euler-Schiffer differential equation has the form

$$q_n(f_0)\left(\frac{Df_0(z)}{f_0(z)}\right)^2+P_n(z)=0,$$

where $q_n(w)$ is defined by

$$\frac{f_0^2(z)}{f_0(z)-w}=\sum_{k=1}^{\infty} q_k(w)z^{k+1}.$$

Putting $w=f_0(\zeta)$ and performing some simple calculations, we find that

$$q_n(w)=n+1+\frac{\zeta}{1-\zeta^2}(\zeta^{n+1}-\zeta^{-n-1}). \tag{45.1}$$

We now use the results of Chapter 4. Consider the space $\mathscr{A}$ of first variations. We define the scalar product by the bilinear form (23.1), after first having replaced the constant $1/4$ by a smaller constant, say $1/8$. This does not affect any of the results in Chapter 4. A function $\xi(w)$ will belong to $\mathfrak{H}$ if it satisfies the following conditions:

1) $\xi(w)$ is regular outside $(-\infty,-1/4)$, and $\xi(0)=\xi'(0)=0$.
2) $\|\xi\|<\infty$.
3) $\operatorname{Im}\{\xi_+(w)-\xi_-(w)\}=0$ for $-\infty<w<-1/4$.

The latter condition is equivalent to (21.4), since $\operatorname{Im}\sqrt{q_n(w)/w^2}=0$ for $-\infty<w<-1/4$. Since $q_n(w)$ has no zeros of order more than one at the vertices of $\mathscr{T}$ (the points $w=-1/4$ and $w=\infty$), it follows that the subspace $\mathfrak{R}$ is empty, and hence $H=\mathscr{A}=\mathfrak{H}$. Therefore, in the computations below, we can use elements of the linear manifold H_0. In our particular case, $\xi(w)$

$\in H_0$ if

$$\xi(w) = \frac{1}{2\pi i} \int_{-\infty}^{-1/4} \frac{w^2 \psi(t)\, dt}{(w-t)\, t^2},$$

where $\psi(t)$ is a real-valued twice continuously differentiable function vanishing in some neighborhood of both $w = -1/4$ and $w = -\infty$.

We shall now present a formula for the second variation of a linear functional which differs somewhat from the formulas used previously. Let $\mathcal{T}$ be the boundary tree of any extremal function $f(z) \in S_n$. Suppose that $\xi(w) \in H_0$ is arbitrary. We compare formulas (20.2) and (20.9). Taking (17.17) and (19.17) into account, we find that

$$\delta^2 L(f; \xi) = -\operatorname{Re}\left\{\frac{1}{2\pi i} \int_{\mathcal{T}} \frac{Q(w)}{w^2} \psi(w)\, dw \frac{1}{\pi i} \int_{\mathcal{T}} \frac{w \psi'(t)}{t(t-w)} dt\right\},$$

if

$$\xi(w) = \frac{1}{2\pi i} \int_{\mathcal{T}} \frac{w^2 \psi(t)}{(w-t)\, t^2} dt.$$

Interchanging the order of integration, we find that

$$\delta^2 L(f; \xi) = -\operatorname{Re}\left\{\frac{1}{2\pi^2} \int_{\mathcal{T}} \frac{\psi'(t)}{t} dt \int_{\mathcal{T}} \frac{Q(w)\psi(w)}{w(w-t)} dw\right\}. \tag{45.2}$$

Upon applying this formula to our particular case we get

$$\delta^2 L(f_0; \xi) = -\frac{1}{2\pi^2} \int_{-\infty}^{-1/4} \frac{\psi'(t)}{t} dt \int_{-\infty}^{-1/4} \frac{q_n(w)\psi(w)}{w(w-t)} dw.$$

Denote by J the double integral on the right-hand side of the above formula. In order to evaluate it, we change variables by setting $t = f_0(z)$, where $z = e^{i\theta}$, $0 < \theta \leq \pi$. Put $\psi(t)/t = h(\theta)$. Then

$$J = \int_0^\pi \frac{d\,[f_0(e^{i\theta})\, h(\theta)]}{f_0(\theta)} \int_0^\pi \frac{q_n[f_0(e^{i\varphi})]\, h(\varphi)}{f_0(e^{i\varphi}) - f_0(e^{i\theta})} df_0(e^{i\varphi}).$$

Thus

$$J = \int_0^\pi \left(h'(\theta) - \frac{h(\theta)}{\tan\frac{\theta}{2}}\right) d\theta \int_0^\pi \frac{q_n[f_0(e^{i\varphi})]\, h(\varphi) \sin^2\frac{\theta}{2}}{\left(\sin^2\frac{\varphi}{2} - \sin^2\frac{\theta}{2}\right)\tan\frac{\varphi}{2}} d\varphi$$

$$= \int_0^\pi h'(\theta)\, d\theta \int_0^\pi \frac{q_n[f_0(e^{i\varphi})]\, h(\varphi) \sin^2\frac{\theta}{2}}{\left(\sin^2\frac{\varphi}{2} - \sin^2\frac{\theta}{2}\right)\tan\frac{\varphi}{2}} d\varphi$$

$$- \int_0^\pi h(\theta) \frac{d\theta}{\tan\frac{\theta}{2}} \int_0^\pi \frac{q_n[f_0(e^{i\varphi})]\, h(\varphi) \sin^2\frac{\theta}{2}}{\left(\sin^2\frac{\varphi}{2} - \sin^2\frac{\theta}{2}\right)\tan\frac{\varphi}{2}} d\varphi.$$

Denote the first double integral by J_1, and the second by J_2; then $J = J_1 - J_2$. We now reformulate J_1 and J_2 slightly. It is easy to see that

$$J_1 = -\int_0^\pi h'(\theta)\,d\theta \int_0^\pi q_n[f_0(e^{i\varphi})]\,h(\varphi)\frac{d\varphi}{\tan\frac{\varphi}{2}} + \int_0^\pi h'(\theta)\,d\theta \int_0^\pi \frac{q_n[f_0(e^{i\varphi})]\,h(\varphi)\sin\varphi}{\cos\theta - \cos\varphi}\,d\varphi.$$

By assumption $h(0) = h(\pi) = 0$, and so the first term vanishes. In the second integral we reverse the order of integration, so that

$$J_1 = \int_0^\pi q_n[f_0(e^{i\varphi})]\,h(\varphi)\sin\varphi\,d\varphi \int_0^\pi \frac{h'(\theta)}{\cos\theta - \cos\varphi}\,d\theta.$$

We now operate on J_2. By (45.1) we have $q_n[f_0(e^{i\varphi})] = n+1 - \sin(n+1)\varphi/\sin\varphi$, and so

$$J_2 = (n+1)\int_0^\pi h(\theta)\frac{d\theta}{\tan\frac{\theta}{2}} \int_0^\pi \frac{h(\varphi)\sin^2\frac{\theta}{2}\,d\varphi}{\left(\sin^2\frac{\varphi}{2} - \sin^2\frac{\theta}{2}\right)\tan\frac{\varphi}{2}}$$

$$- \int_0^\pi h(\theta)\sin\theta\,d\theta \int_0^\pi \frac{h(\varphi)\sin(n+1)\varphi\,d\varphi}{\sin\varphi\,(\cos\theta - \cos\varphi)\tan\frac{\varphi}{2}}.$$

The first double integral is easily seen to be equal to

$$-\frac{1}{2}\left(\int_0^\pi h(\theta)\frac{d\theta}{\tan\frac{\theta}{2}}\right)^2.$$

Let $D^*_{n+1}(\varphi)$ be the modified Dirichlet kernel:

$$D^*_{n+1}(\varphi) = \frac{1}{2} + \sum_1^n \cos k\varphi + \frac{1}{2}\cos(n+1)\varphi.$$

It is well known that

$$D^*_{n+1}(\varphi) = \frac{\sin(n+1)\varphi}{2\tan\frac{\varphi}{2}}.$$

Thus

$$J_2 = -\frac{n+1}{2}\left(\int_0^\pi h(\theta)\frac{d\theta}{\tan\frac{\theta}{2}}\right)^2 - 2\int_0^\pi h(\theta)\sin\theta\,d\theta \int_0^\pi \frac{h(\varphi)\,D^*_{n+1}(\varphi)\,d\varphi}{\sin\varphi\,(\cos\theta - \cos\varphi)}.$$

Denote by J_3 the double integral on the right-hand side of this last equation. It is not difficult to verify that

$$2J_3 = \int_0^\pi h(\theta)\sin\theta\,d\theta \int_0^\pi \frac{h(\varphi)\,[D^*_{n+1}(\varphi) - D^*_{n+1}(\theta)]}{\sin\varphi\,(\cos\theta - \cos\varphi)}\,d\varphi$$

$$- \int_0^\pi \frac{h(\theta)}{\sin\theta}\,d\theta \int_0^\pi \frac{h(\varphi)}{\sin\varphi}(\cos\theta + \cos\varphi)\,D^*_{n+1}(\varphi)\,d\varphi.$$

Hence

$$J_2 = -\frac{n+1}{2}\left(\int_0^\pi h(\theta)\frac{d\theta}{\tan\frac{\theta}{2}}\right)^2 + \int_0^\pi h(\theta)\sin\theta\, d\theta \int_0^\pi \frac{h(\varphi)\,[D_{n+1}^*(\varphi) - D_{n+1}^*(\theta)]}{\sin\varphi\,(\cos\varphi - \cos\theta)}\,d\varphi$$

(45.3)

$$+\int_0^\pi h(\theta)\frac{d\theta}{\sin\theta}\int_0^\pi (\cos\theta + \cos\varphi)\,D_{n+1}^*(\varphi)\,h(\varphi)\frac{d\varphi}{\sin\varphi}.$$

We now expand $h(\theta)$ in a Fourier series; having extended this function to $(0, 2\pi)$ as an odd function, we find that

$$h(\theta) = \sum_1^\infty \alpha_k \sin k\theta.$$

We now extend $h(\theta)/\sin\theta$ as an even function, and we find that

$$\frac{h(\theta)}{\sin\theta} = \frac{\beta_0}{2} + \sum_1^\infty \beta_k \cos k\theta.$$

Note that the quantities α_k and β_k, $1 \leqq k < \infty$, satisfy the following easily verified identities:

(45.4) $$2\sum_{\nu=1}^{k} \alpha_{m-k+2\nu-1} = \beta_{m-k} - \beta_{m+k}, \quad k \leqslant m, \quad 1 \leqslant m < \infty,$$

(45.5) $$\sum_{\nu=1}^{k} (k+1-\nu)\,\alpha_\nu = \frac{k+1}{2}(\beta_0 + \beta_1) - \sum_{\nu=0}^{k+1}{}'' \beta_\nu, \quad 1 \leqslant k < \infty,$$

where the double prime on the summation sign means that the terms corresponding to $\nu = 0$ and $\nu = k+1$ are to have the coefficient $\frac{1}{2}$. We now evaluate the integrals J_1 and J_2. Using the well-known integration formula

(45.6) $$\int_0^\pi \frac{\cos k\theta}{\cos\theta - \cos\varphi}\,d\theta = \pi\frac{\sin k\varphi}{\sin\varphi}, \quad k = 0, 1, \ldots,$$

we see, since $h(\theta)$ is twice continuously differentiable, that

(45.7) $$\int_0^\pi \frac{h'(\theta)\,d\theta}{\cos\theta - \cos\varphi} = \sum_1^\infty k\alpha_k \int_0^\pi \frac{\cos k\theta\, d\theta}{\cos\theta - \cos\varphi} = \frac{\pi}{\sin\varphi}\sum_1^\infty k\alpha_k \sin k\varphi.$$

Using (45.6) it is not difficult to prove that for $k = 0, 1, \cdots$

$$\int_0^\pi \frac{\cos k\theta\,[D_{n+1}^*(\theta) - D_{n+1}^*(\varphi)]}{\cos\theta - \cos\varphi}\,d\theta = \begin{cases} 0, & \text{if } k \geqslant n+1, \\ \pi \displaystyle\sum_{\nu=1}^{n+1-k}{}^{*} \frac{\sin \nu\varphi}{\sin\varphi}, & \text{if } k \leqslant n, \end{cases}$$

where the star on the summation sign means that the term corresponding to $\nu = n+1-k$ is to have the coefficient $\frac{1}{2}$. Thus

$$\int_0^{\pi} \frac{h(\theta)\,[D^*_{n+1}(\theta) - D^*_{n+1}(\varphi)]\,d\theta}{\sin\theta\,(\cos\theta - \cos\varphi)} = \pi \sum_{k=0}^{n}{}' \beta_k \sum_{\nu=1}^{n+1-k}{}^{*} \frac{\sin\nu\varphi}{\sin\varphi}, \tag{45.8}$$

where the prime means that the term corresponding to $k=0$ is to have the coefficient $\frac{1}{2}$. Using (45.7), we find that

$$J_1 = \pi \int_0^{\pi} q_n\,[f_0(e^{i\varphi})]\,h(\varphi) \sum_{k=1}^{\infty} k\alpha_k \sin k\varphi\,d\varphi,$$

and so

$$J_1 = \frac{\pi^2}{2}(n+1) \sum_{k=1}^{\infty} k\alpha_k^2 - \frac{\pi^2}{4} \sum_{k=1}^{\infty} k\alpha_k\,[\beta_{|k-n-1|} - \beta_{k+n+1}].$$

In view of the awkwardness of the formula defining J_2 we denote the various terms on the right-hand side of (45.3) by J_{21}, J_{22} and J_{23}. Then

$$J_{21} = -\frac{n+1}{2}\left(\int_0^{\pi} \frac{h(\theta)}{\sin\theta}(1+\cos\theta)\,d\theta\right)^2 = -\frac{\pi^2(n+1)}{8}(\beta_0+\beta_1)^2.$$

Applying (45.8), we find that

$$J_{22} = \frac{\pi^2}{2} \sum_{k=0}^{n}{}' \beta_k \sum_{\nu=1}^{n+1-k}{}^{*} \alpha_\nu.$$

Simple calculations give that

$$J_{23} = \frac{\pi^2}{8}\beta_0\left(\sum_{k=0}^{n+2}{}'' \beta_k + \sum_{k=0}^{n}{}'' \beta_k\right) + \frac{\pi^2}{4}\beta_1 \sum_{k=0}^{n+1}{}'' \beta_k.$$

We shall now further refine the expressions we have derived for J_1 and J_2. Using (45.4), we find that

$$J_1 = \frac{\pi^2}{2}(n+1) \sum_{k=1}^{\infty} k\alpha_k^2 - \frac{\pi^2}{2} \sum_{k=1}^{\infty} k\alpha_k \sum_{\nu=1}^{\min(k,\,n+1)} \alpha_{|k-n-1|+2\nu-1}.$$

Using the easily verified identity

$$\sum_{k=1}^{\infty} \sum_{\nu=1}^{\min(k,\,n+1)} k\alpha_k\alpha_{|k-n-1|+2\nu-1} = \frac{1}{2} \sum_{k=1}^{\infty} \sum_{\nu=1}^{\min(k,\,n+1)} k\,[\alpha_k^2 + \alpha^2_{|k-n-1|+2\nu-1}] - \frac{1}{2} \sum_{k=1}^{\infty} \sum_{\nu=1}^{\min(k,\,n+1)} k\,(\alpha_k - \alpha_{|k-n-1|+2\nu-1})^2,$$

it is not difficult to verify the identity

$$\sum_{k=1}^{\infty} \sum_{\nu=1}^{\min(k,\,n+1)} k\,[\alpha_k^2 + \alpha^2_{|k-n-1|+2\nu-1}] = \sum_{k=1}^{\infty} k\,[n+1+\min(k,\,n+1)]\,\alpha_k^2.$$

Therefore

$$J_1 = \frac{\pi^2}{4}\sum_{k=1}^{n} k(n+1-k)\alpha_k^2 + \frac{\pi^2}{4}\sum_{\nu=1}^{n+1}\sum_{k=\nu}^{\infty} k(\alpha_k - \alpha_{|k-n-1|+2\nu-1})^2.$$

Applying (45.5), we find that

$$J_{32} = \frac{\pi^2}{4}(\beta_0+\beta_1)\left[\frac{n+1}{2}(\beta_0+\beta_1) - \sum_{\nu=1}^{n}(n+1-\nu)\alpha_\nu\right] - \frac{\pi^2}{8}\alpha_{n+1}\beta_0.$$

We now consider J_{22}. Reversing the order of summation, we have

$$J_{22} = \frac{\pi^2}{2}\sum_{\nu=1}^{n}\alpha_\nu \sum_{k=0}^{n+1-\nu}{}'' \beta_k + \frac{\pi^2}{8}\alpha_{n+1}\beta_0.$$

In view of (45.5) it follows that

$$J_{22} = \frac{\pi^2}{4}(\beta_0+\beta_1)\sum_{\nu=1}^{n}(n+1-\nu)\alpha_\nu - \frac{\pi^2}{2}\sum_{k=1}^{n-1}\alpha_k\sum_{\nu=1}^{n-k}(n+1-k-\nu)\alpha_\nu + \frac{\pi^2}{8}\alpha_{n+1}\beta_0.$$

Adding the formulas derived above, we find that

$$J_2 = -\frac{\pi^2}{2}\sum_{k=1}^{n-1}\alpha_k\sum_{\nu=1}^{n-k}(n+1-k-\nu)\alpha_\nu.$$

Thus

$$J = \frac{\pi^2}{4}\sum_{k=1}^{n} k(n+1-k)\alpha_k^2 + \frac{\pi^2}{4}\sum_{\nu=1}^{n+1}\sum_{k=\nu}^{\infty} k(\alpha_k - \alpha_{|k-n-1|+2\nu-1})^2 + \frac{\pi^2}{2}\sum_{k=1}^{n-1}\alpha_k\sum_{\nu=1}^{n-k}(n+1-k-\nu)\alpha_\nu.$$

We now carry out a fairly simple, but still nontrivial, rearrangement of the latter double sum. It is clear that

$$S = \sum_{k=1}^{n-1}\alpha_k\sum_{\nu=1}^{n-k}(n+1-k-\nu)\alpha_\nu = \sum_{k=1}^{n}\alpha_k\sum_{\nu=1}^{n+1-k}(n+1-k-\nu)\alpha_\nu.$$

We have

$$S = \sum_{k=1}^{n}\alpha_{n+1-k}\sum_{\nu=1}^{k}(k-\nu)\alpha_\nu = \frac{1}{n+1}\sum_{k=1}^{n} k\alpha_{n+1-k}\sum_{\nu=1}^{k}(n+1-\nu)\alpha_\nu - \frac{1}{n+1}\sum_{k=1}^{n}(n+1-k)\alpha_{n+1-k}\sum_{\nu=1}^{k}\nu\alpha_\nu.$$

From this the relation

$$S = \frac{1}{n+1}\sum_{k=1}^{n} k\alpha_{n+1-k}\sum_{\nu=1}^{n}(n+1-\nu)\alpha_\nu - \frac{1}{n+1}\sum_{k=1}^{n}(n+1-k)\alpha_{n+1-k}\sum_{\nu=1}^{k}\nu\alpha_\nu -$$

$$-\frac{1}{n+1}\sum_{k=1}^{n} k\alpha_{n+1-k}\sum_{\nu=k+1}^{n}(n+1-\nu)\,\alpha_\nu$$

is easily deduced, and from this it follows that

$$S=\frac{1}{n+1}\Big(\sum_{k=1}^{n}(n+1-k)\,\alpha_k\Big)^2-S_1,$$

where

$$S_1=\frac{1}{n+1}\sum_{k=1}^{n}\sum_{\nu=1}^{k}{}^{*}(n+1-k)\,\nu\,(\alpha_{n+1-k}\alpha_\nu+\alpha_k\alpha_{n+1-\nu}).$$

Using the obvious identity

$$S_1=\frac{1}{2(n+1)}\sum_{k=1}^{n}\sum_{\nu=1}^{k}{}^{*}(n+1-k)\,\nu\,[\alpha_k^2+\alpha_{n+1-k}^2+\alpha_\nu^2+\alpha_{n+1-\nu}^2]$$
$$-\frac{1}{2(n+1)}\sum_{k=1}^{n}\sum_{\nu=1}^{k}{}^{*}(n+1-k)\,\nu\,[(\alpha_{n+1-k}-\alpha_\nu)^2+(\alpha_k-\alpha_{n+1-\nu})^2]$$

it is not difficult to prove that

$$\frac{1}{2(n+1)}\sum_{k=1}^{n}\sum_{\nu=1}^{k}{}^{*}(n+1-k)\,\nu\,[\alpha_k^2+\alpha_{n+1-k}^2+\alpha_\nu^2+\alpha_{n+1-\nu}^2]$$
$$=\frac{1}{2}\sum_{k=1}^{n}k\,(n+1-k)\,\alpha_k^2.$$

Thus we have

$$S=\frac{1}{n+1}\Big(\sum_{1}^{n}(n+1-k)\,\alpha_k\Big)^2+\frac{1}{2(n+1)}\sum_{k=1}^{n}\sum_{\nu=1}^{k}{}^{*}(n+1-k)\,\nu\,[(\alpha_{n+1-k}-\alpha_\nu)^2$$
$$+(\alpha_k-\alpha_{n+1-\nu})^2]-\frac{1}{2}\sum_{k=1}^{n}k\,(n+1-k)\,\alpha_k^2.$$

Upon using this formula, we arrive at our final expression for J, and hence for $\delta^2L(f_0;\xi)$. We have

$$\delta^2L(f_0,\xi)=-\frac{1}{8(n+1)}\sum_{k=1}^{n}\sum_{\nu=1}^{k}{}^{*}(n+1-\nu)\,\nu\,[(\alpha_{n+1-k}-\alpha_\nu)^2+(\alpha_k-\alpha_{n+1-\nu})^2]$$
$$(45.9)\qquad -\frac{1}{8}\sum_{k=1}^{n+1}\sum_{\nu=k}^{\infty}\nu\,(\alpha_\nu-\alpha_{|\nu-n-1|+2k-1})^2-\frac{1}{4(n+1)}\Big(\sum_{k=1}^{n}(n+1-k)\,\alpha_k\Big)^2.$$

This latter reasonably elegant formula answers all questions about the sign of the second variation, and it follows that for any element $\xi(w)\in H_0$, $\xi\not\equiv 0$, we have

$$\delta^2L(f_0;\xi)<0.$$

Since H_0 is dense in H, and since $H=\mathscr{A}=\mathfrak{H}$, we deduce

Lemma 45.1. *The Koebe function $f_0(z)$ satisfies the following necessary conditions for a local maximum of the functional $L(f) = \mathrm{Re}\, a_n$:*

$$\delta L(f_0) = 0, \quad \delta^2 L(f_0) \leqq 0.$$

We shall establish a connection between the first coefficient variations of $f_0(z)$ and the quantities α_k, $1 \leqq k < \infty$, introduced above. Since $\delta f_0(z) = \xi[f_0(z)]$, we have

$$\delta a_k = \frac{1}{2\pi i} \int_{-\infty}^{-\frac{1}{4}} \psi(t) \frac{q_k(t)}{t^2}\, dt.$$

Changing variables via the formula $t = f_0(e^{i\theta})$, $0 < \theta \leqq \pi$, and recalling that $\psi(t)/t = h(\theta)$, we get that

$$\delta a_k = -\frac{1}{2\pi i} \int_0^\pi h(\theta) \left[k + 1 - \frac{\sin(k+1)\theta}{\sin\theta}\right] \frac{d\theta}{\tan\frac{\theta}{2}}.$$

Therefore

$$\delta a_k = -\frac{k+1}{2\pi i} \int_0^\pi \frac{h(\theta)}{\sin\theta}(1 + \cos\theta)\, d\theta + \frac{1}{2\pi i} \int_0^\pi \frac{h(\theta)}{\sin\theta} \frac{\sin(k+1)\theta}{\tan\frac{\theta}{2}}\, d\theta,$$

and so

$$\delta a_k = -\frac{k+1}{4i}(\beta_0 + \beta_1) + \frac{1}{2i} \sum_{\nu=0}^{k+1}{}'' \beta_\nu.$$

Applying (45.5), we find that

$$\delta a_k = \frac{i}{2} \sum_{\nu=1}^{k} (k + 1 - \nu)\alpha_\nu, \quad 1 \leqslant k < \infty. \tag{45.10}$$

We must now consider the question of when condition (43.1) is satisfied. We shall examine this problem in the next section.

§46. We have seen that the vector $\lambda = (\lambda_1, \cdots, \lambda_n)$ will be an associated vector for $f_0(z)$ when $\mathrm{Im}\,\lambda_k = 0$, $1 \leqq k \leqq n$. The linear manifold of associated vectors will be identified with R^n. By $\lambda^1, \cdots, \lambda^n$ we denote the usual basis of R^n; that is, $\lambda^k = (\delta_{k1}, \cdots, \delta_{kk}, \cdots, \delta_{kn})$, $1 \leqq k \leqq n$, where δ_{kn} is the Kronecker delta. For each associated vector λ we construct the corresponding functions $Q(w;\lambda)$ and $P(z;\lambda)$ and the linear functional $L(f) = \mathrm{Re} \sum_1^n \lambda_k a_k$. We define two sets in R^n: the set K_l of those vectors λ for which

$$P(z;\lambda) \geqq 0, \quad |z| = 1,$$

and the set K_m of those vectors λ for which

$$\delta^2 L(f_0;\xi) \leqq 0$$

for any first variation $\delta f_0(z) = \xi[f_0(z)]$. We have already seen that K_l and

K_m are convex cones. From Lemma 5.3 we have that $K_m \subseteq K_l$. It is easily seen that in fact this inclusion is strict, provided $n > 2$. To prove this, let l be even, with $1 < l < n$. Calculation shows that $f_0(z)$ satisfies the equation

$$-w^{-l}\left(\frac{Dw}{w}\right)^2 + (1-z)^{2l-2}(1+z)^2 z^{-l} = 0.$$

Set $(1-z)^{2l-2}(1+z)^2 z^{-l} = P(z;\lambda_0)$; since $P(z;\lambda_0) \geq 0$ if $|z| = 1$, then $\lambda_0 \in K_l$. On the other hand, $\lambda_0 \notin K_m$ by Theorem 18′. For each vector $\lambda \in K_l$ we construct the corresponding Hilbert space $\mathscr{A}_\lambda$ of first variations. The scalar product in $\mathscr{A}_\lambda$ is denoted by $(\xi,\eta)_\lambda$, $\xi,\eta \in \mathscr{A}_\lambda$. We make a slight change in the definition of the scalar product given in Chapter 4, by changing the constant 1/4 in formula (23.2) to 1/8. Thus we have

$$(46.1)\quad (\xi,\eta)_\lambda = \operatorname{Re}\left\{\frac{1}{\pi}\int\limits_{|w|>1/8} \frac{d\xi^*}{dw}\frac{\overline{d\eta^*}}{dw}\,d\sigma - \frac{1}{2\pi i}\int\limits_{|w|=1/8} \xi^*\,\overline{d\eta^*} + \vartheta|\lambda|\sum_1^n \xi_k\overline{\eta_k}\right\},$$

where $\xi^*(w) = \sqrt{Q(w;\lambda)/w^2}\,\xi(w)$, and where an analogous formula holds for $\eta^*(w)$. The constant v is still chosen via Proposition 26.2, which continues to hold with 1/4 replaced by 1/8. The factor $|\lambda|$ arises since the vector λ is not assumed to be normalized. In order to study the cone K_m we must first prove some simple preliminary results. Let $Q(w;\lambda)$ satisfy

$$(46.2)\qquad \inf_{-\infty<w\leqslant-1/4} |wQ(w;\lambda)| > 0.$$

Then from Lemma 25.2 we have that $\mathscr{A}_\lambda = \mathfrak{H}_\lambda$. Since the subspace $\mathscr{R}$ is empty, Lemma 25.4 yields that $\mathscr{A}_\lambda = H_\lambda$. Let $\xi(w) \in \mathscr{A}_\lambda$ be arbitrary. We shall show that

$$\xi(w) = \frac{1}{2\pi i}\int_{-\infty}^{-\frac{1}{4}} \frac{w^2\psi(t)}{(w-t)\,t^2}\,dt,$$

where $\psi(t)$ is real-valued, and where the integral converges absolutely if $w \notin \mathscr{T}_{f_0}$. Note that $\xi(w) = \lim_{m\to\infty}\xi_m(w)$, where $\xi_m(w) \in H_0$, $1 \leq m < \infty$. It is quite obvious that Lemma 23.1 continues to hold in spite of the changes we have made in the definition of the scalar product. Put

$$\xi_m(w) = \frac{1}{2\pi i}\int_{-\infty}^{-1/4} \frac{w^2\psi_m(t)\,dt}{(w-t)\,t^2}.$$

From Lemma 23.1 and inequality (23.5) it follows that the sequence $\{|t|^{-3/2-\alpha/p}\psi_m(t)\}$, $\alpha > 3/2$, converges in the metric of $L^p(-\infty,-1/4)$, where $p > 1$, to some function $\{|t|^{-3/2-\alpha/p}\psi(t)\}$. Taking p sufficiently large and applying Hölder's inequality, we obtain

$$\left|\int_{-\infty}^{-\frac{1}{4}} \frac{w^2\psi(t)\,dt}{(w-t)\,t^2}\right| < |w|^2\left(\int_{-\infty}^{-1/4} |\psi(t)\,t^{-3/2}|^p\,\frac{dt}{|t|^\alpha}\right)^{\frac{1}{p}}\left(\int_{-\infty}^{-\frac{1}{4}} \frac{dt}{|w-t|^q\,|t|^{q/2-\alpha q/p}}\right)^{1/q},$$

where $q = p/(p-1)$. From this, with $d = \inf_{t\in\mathcal{T}_{f_0}} |w-t|$, we have that

(46.3) $$\left|\frac{1}{2\pi i}\int_{-\infty}^{-\frac{1}{4}} \frac{w^2\psi(t)\,dt}{(w-t)\,t^2}\right| < C_p |w|^2 d^{-\frac{1}{p}} \left(\int_{-\infty}^{-1/4} \left|\sqrt{\frac{Q(t,\lambda)}{t^2}}\,\psi(t)\right|^2 \frac{dt}{|t|^\alpha}\right)^{1/p}.$$

Thus the sequence $\{\xi_m(w)\}$ converges in the complement of $\mathcal{T}_{f_0}$ to the function

$$\frac{1}{2\pi i}\int_{-\infty}^{-\frac{1}{4}} \frac{w^2\psi(t)}{(w-t)\,t^2}\,dt.$$

From (23.6) it follows that the latter integral equals $\xi(w)$, as asserted above. Q.E.D.

On the basis of (23.5) and (46.3) we have

(46.4) $$|\xi(w)| < C|w|^2 d^{-1/p} \|\xi\|_\lambda.$$

Let $\lambda_0 \in K_l$ be arbitrary. We shall show that for any function $\xi(w) \in \mathcal{A}_\lambda$

(46.5) $$\|\xi\|_{\lambda_0} < C(\lambda_0, \lambda)\,\|\xi\|_\lambda,$$

provided (46.2) is satisfied. For this it suffices to show that for any sequence $\{\xi_m(w)\} \subset \mathcal{A}_\lambda$ converging to zero we have

$$\lim_{m\to\infty} \|\xi_m\|_{\lambda_0} = 0.$$

Using Green's Theorem, we write (46.1) as

$$\|\xi\|_{\lambda_0}^2 = \operatorname{Re}\left\{\frac{1}{\pi}\int_{\mathcal{G}_\rho} \left|\frac{d\xi^*}{dw}\right|^2 d\sigma - \frac{1}{2\pi i}\int_{\mathcal{C}_\rho} \xi^*\,\overline{d\xi^*} + \vartheta|\lambda_0|\sum_1^n |\xi_k|^2\right\},$$

where $\mathcal{C}_\rho = \{w : |f_0^{-1}(w)| = \rho\}$, $\rho < 1$, and where $\mathcal{G}_\rho$ is the domain containing ∞ and bounded by the level curve $\mathcal{C}_\rho$. We choose ρ so that

$$\inf_{w\in\mathcal{G}_\rho} |wQ(w,\lambda)| = C_0 > 0.$$

This is possible by (46.2). Since $\|\xi_m\|_\lambda \to 0$ as $m\to\infty$, Lemma 23.1 gives that

$$\operatorname{Re}\left\{-\frac{1}{2\pi i}\int_{\mathcal{C}_\rho} \xi_m(w)\sqrt{\frac{Q(w;\lambda_0)}{w^2}}\,d\overline{\left[\sqrt{\frac{Q(w;\lambda_0)}{w^2}}\,\xi_m(w)\right]} + \vartheta|\lambda_0|\sum_{k=1}^n |\xi_{mk}|^2\right\} = o(1);$$

where ξ_{mk}, $1 \le k \le n$, are the coefficients of the power series of $\xi_m(w)$. Thus

$$\|\xi_m\|_{\lambda_0}^2 = \frac{1}{\pi}\int_{\mathcal{G}_\rho} \left|\frac{d}{dw}\left(\sqrt{\frac{Q(w;\lambda_0)}{w^2}}\,\xi_m(w)\right)\right|^2 d\sigma + o(1).$$

Set $q(w) = [Q(w;\lambda_0)/Q(w;\lambda)]^{1/2}$. Since

$$\frac{d}{dw}\left(\sqrt{\frac{Q(w;\lambda_0)}{w^2}}\,\xi(w)\right)=q(w)\frac{d}{dw}\left(\sqrt{\frac{Q(w;\lambda)}{w^2}}\,\xi_m(w)\right)+\sqrt{\frac{Q(w;\lambda)}{w^2}}\,\xi(w)\frac{dq(w)}{dw}$$

and $|q(w)|\leqq C_1$ for $w\in\mathscr{G}_\rho$, it follows that

$$\|\xi_m\|^2_{\lambda_0}<\frac{2C_1^2}{\pi}\int_{\mathscr{G}_\rho}\left|\frac{d}{dw}\left(\sqrt{\frac{Q(w;\lambda)}{w^2}}\,\xi_m(w)\right)\right|^2 d\sigma$$

$$+\frac{2}{\pi}\int_{\mathscr{G}_\rho}\left|\sqrt{\frac{Q(w;\lambda)}{w^2}}\,\xi_m(w)\frac{d}{dw}q(w)\right|^2 d\sigma+O(1).$$

Applying (46.4), we find that

$$\|\xi_m\|^2_{\lambda_0}<C_2\|\xi_m\|^2_\lambda\left[1+\int_{\mathscr{G}_\rho}\left|w^{\frac{1}{2}}\frac{d}{dw}q(w)\right|^2 d^{-2/p}(w)\,d\sigma\right]+O(1).$$

Since $dq(w)/dw=O(w^{-2})$ for large $|w|$, we have

$$\int_{\mathscr{G}_\rho}\left|w^{\frac{1}{2}}\frac{d}{dw}q(w)\right|^2 d^{-2/p}(w)\,d\sigma<C_3,$$

if p is sufficiently large. Therefore $\|\xi\|_{\lambda_0}=O(1)$. Q.E.D.

Let $\lambda\in K_m$ and let $L(f)$ be the corresponding linear functional; $\xi(w)\in\mathscr{A}_\lambda$ and $\xi[f_0(z)]=\sum_1^n t_k z^{k+1}+\cdots$. Suppose that

(46.6) $$\delta^2L(f_0;\xi)\leqq-\theta<0$$

for any $\xi(w)\in\mathscr{A}_\lambda$ satisfying the condition

(46.7) $$\sum_1^n|t_k|^2=1.$$

We shall show that under these hypotheses

(46.8) $$\delta^2L(f_0;\xi)\leqq-C_\theta\|\xi\|^2_\lambda,$$

where $C_\theta>0$. We take any variation $\xi(w)\in\mathscr{A}_\lambda$ satisfying (46.7). Corresponding to $t_1,\cdots,t_n$ we determine the extremal variation $\xi_0(w)$ which solves Problem 2 of §26. Then $\xi(w)=\xi_0(w)+\xi_1(w)$, and moreover $\xi_1[f_0(z)]=t^1_{n+1}z^{n+2}+\cdots$. Therefore

$$\|\xi_0+\xi_1\|^2_\lambda=\|\xi_0\|^2_\lambda+\|\xi_1\|^2_\lambda.$$

Hence, in accord with (23.2),

$$\delta^2L(f_0;\xi)=\delta^2L(f_0;\xi_0)-\|\xi_1\|^2_\lambda.$$

From this, noting that $\|\xi_0\|_\lambda\leqq A<\infty$, we get

$$\delta^2L(f_0;\xi)\leqslant-\frac{\theta}{A^2}\|\xi_0\|^2_\lambda-\|\xi_1\|^2_\lambda<-C_\theta\|\xi\|^2_\lambda.$$

Analogously we can prove that for any $\lambda\in K_l$

(46.9) $$|\delta^2 L(f_0;\xi)| \leq C_\lambda \|\xi\|_\lambda^2.$$

As usual, we denote the interior of the cone K_m by $\operatorname{Int} K_m$. With this notation we have the following lemma.

Lemma 46.1. *A necessary and sufficient condition that* $\lambda \in \operatorname{Int} K_m$ *is that inequality* (46.2) *holds, and that, subject to condition* (46.7), *inequality* (46.6) *also holds.*

Proof. To prove that the condition is necessary, let $\lambda \in \operatorname{Int} K_m$, and suppose that (46.2) does not hold. Then $Q(-1/4;\lambda)=0$ by Theorem 18′. Thus $P(z;\lambda)$ has $z=-1$ as a fourth order zero. As a result there exists a vector λ_0 as near as desired to λ such that $P(z;\lambda_0)$ changes sign in a neighborhood of $|z|=1$. Hence $\lambda \notin \operatorname{Int} K_m$, which is a contradiction. Now we suppose that (46.7) holds but that (46.6) does not. Then there exists a variation

$$\xi(w) \in \mathscr{H}_\lambda, \qquad \xi[f_0(z)] = \sum_1^n t_k z^{k+1} + \dots, \quad \sum_1^n |t_k|^2 = 1,$$

such that $\delta^2 L(f_0;\xi)=0$. Put $\lambda_0 = \lambda + \sum_1^n \epsilon_k \lambda^k$, where ϵ_k, $1 \leq k \leq n$, are arbitrary sufficiently small real constants. By what we have said above, $\xi \in \mathscr{A}_{\lambda_0}$ and $\xi \in \mathscr{A}_{\lambda^k}$, $1 \leq k \leq n$. Denote by $L_k(f)$ the linear functional corresponding to λ^k, $1 \leq k \leq n$, and by L_0 the linear functional corresponding to λ. Clearly $\delta^2 L_0(f_0;\xi) = \delta^2 L(f_0;\xi) + \sum_1^n \epsilon_k \delta^2 L_k(f_0;\xi)$. But $\delta^2 L_0(f_0;\xi) \leq 0$, and so

$$\sum_1^n \varepsilon_k \delta^2 L_k(f_0;\xi) \leqslant 0.$$

From this, in view of the arbitrariness of ϵ_k, we have $\delta^2 L_k(f_0;\xi)=0$, $1 \leq k \leq n$. Then from (45.9) and (45.10) it follows that $t_k=0$, $1 \leq k \leq n$, since the linear manifold H_0 is dense in every space $\mathscr{A}_{\lambda^k}$, $1 \leq k \leq n$. Hence $\sum_1^n |t_k|^2 = 0$, which is a contradiction.

We now prove the sufficiency. From (46.2) it follows that $\lambda \in \operatorname{Int} K_l$. Let the vector λ_0 be defined as above, and let ξ be an arbitrary element of $\mathscr{A}_\lambda$. Then, by the arguments given above, $\xi \in \mathscr{A}_{\lambda^k}$, $1 \leq k \leq n$, and so

$$\delta^2 L_0(f_0;\xi) = \delta^2 L(f_0;\xi) + \sum_1^n \varepsilon_k \delta^2 L_k(f_0;\xi).$$

From this, in view of (46.8) and (46.9), we have that

$$\delta^2 L_0(f_0;\xi) \leqslant -C_\theta \|\xi\|_\lambda^2 + \sum_1^n C_{\lambda k} |\varepsilon_k| \, \|\xi_k\|_{\lambda k}^2,$$

and so it follows from (46.5) that

$$\delta^2 L_0(f_0;\xi) \leqslant -\left[C_\theta - \sum_1^n |\varepsilon_k| C_{\lambda k} C(\lambda, \lambda^k)\right] \|\xi\|_\lambda^2.$$

For sufficiently small ϵ_k, $1 \leq k \leq n$, we get that

$$\delta^2 L(f_0; \xi) \leqslant -\frac{1}{2} C_\theta \|\xi\|_\lambda^2,$$

and so (46.6) is satisfied for $L_0(f)$. Q.E.D.

LEMMA 46.2. *Let $L_k(f)$ be the functionals defined by the vectors λ^k, $1 \leq k \leq n$; that is, $L_k(f) = \operatorname{Re} a_k$. If (46.7) holds, then $L_k(f)$ satisfies (46.6). Therefore $\lambda^k \in \operatorname{Int} K_m$ for odd k, and $\lambda^k \in \partial K_m$ for even k.*

PROOF. Let $\xi(w) \in H_0$ and $\xi[f_0(z)] = \sum_1^\infty t_l z^{l+1}$. In §45 we determined, by starting with $\xi(w)$, quantities α_l, $1 \leq l < \infty$, connected with t_k by relation (45.10). From (45.10) we have that

$$\alpha_1 = \frac{2}{i} t_1,\ \alpha_2 = \frac{2}{i}(t_2 - 2t_1),\ \alpha_l = \frac{2}{i}(t_l - 2t_{l-1} + t_{l-2}),\ l \geqslant 3. \tag{46.10}$$

Suppose that (46.6) does not hold. Then $\sup \delta^2 L_k(f_0; \xi) = 0$, where the supremum is taken subject to the condition $\sum_1^k |t_l|^2 = 1$. It is clear that there exists an element $\xi_0(w) \in \mathscr{A}_{\lambda^k}$, $\xi_0[f_0(z)] = \sum_1^\infty t_l^0 z^{l+1}$, $\sum_1^k |t_l^0|^2 = 1$, for which $\delta^2 L_k(f_0; \xi_0) = 0$. Since H_0 is dense in $\mathscr{A}_{\lambda^k}$, there exists a sequence $\{\xi_m(w)\} \subset H_0$ converging to $\xi_0(w)$. Put $\xi_m[f_0(z)] = \sum_1^\infty t_{ml} z^{l+1}$, $1 \leq m < \infty$. We determine for each t_{ml}, $1 \leq l < \infty$, the corresponding quantities α_{ml}, $1 \leq l < \infty$, as in (46.10). Since $\lim_{m\to\infty} t_{ml} = t_l^0$, $1 \leq l < \infty$, we have that $\lim_{m\to\infty} \alpha_{ml} = \alpha_{0l}$, $1 \leq l < \infty$. Clearly t_l^0 and α_l^0, $1 \leq l < \infty$, are connected via (46.10). Since $\lim_{m\to\infty} \delta^2 L_k(f_0; \xi_m) = \delta^2 L_k(f_0; \xi_0) = 0$, relation (45.9) implies that

$$\alpha_{01} = \cdots = \alpha_{0k},$$

and also that $\sum_{\nu=1}^k (k+1-\nu)\alpha_{0\nu} = 0$. Hence $\alpha_{01} = \cdots = \alpha_{0k} = 0$, and this in turn gives $t_1^0 = \cdots = t_k^0 = 0$, which is a contradiction. To conclude the proof one need only note that $Q(-1/4, \lambda^k) \neq 0$ if k is odd and $Q(-1/4, \lambda^k) = 0$ if k is even.

CHAPTER 9

ON A CONVEXITY PROPERTY OF THE COEFFICIENT REGION

Earlier, in §8, we mentioned that from Teichmüller's Theorem (Theorem 6) an important result follows; we now recall the exact formulation of this result. Let $f(z) \in S_n$, $f(z) = \sum_0^\infty a_k z^{k+1}$, $a_0 = 1$, and let the vector $\lambda = (\lambda_1, \cdots, \lambda_n)$ associated with $f(z)$ correspond to equation (5.13), where $P(z) \geqq 0$ for $|z| = 1$. If $g(z) \in S$ is such that $g(z) = \sum_0^{n-1} a_k z^{k+1} + \sum_n^\infty b_k z^{k+1}$, then

$$\operatorname{Re} \bar{\lambda}_n (a_n - b_n) \geqq 0,$$

and equality holds only if $g(z) \equiv f(z)$. It is easy to give a geometrical interpretation of this result. Toward this end we consider the section of $\mathscr{V}_n$ by the two-dimensional plane

$$x_1 = x_1^0, \ldots, x_{n-1} = x_{n-1}^0 \, (x_1^n = \text{const}, \ldots, x_{n-1}^0 = \text{const}).$$

In view of the above inequality this section will be strictly convex. It is not difficult to check that, conversely, Teichmüller's Theorem can be derived from the fact that the above sections are strictly convex.

The results proved in the preceding chapters lead us to a rather deep generalization of these theorems. We shall prove shortly that the following theorem is true.

THEOREM 20. *The section of $\mathscr{V}_n$ by any plane of the form*

$$x_1 = x_1^0, \ldots, x_m = x_m^0, \qquad m = \begin{cases} \dfrac{n+1}{2} & \textit{for} \quad n \equiv 3 \pmod 6, \\ \left[\dfrac{n}{2}\right] & \textit{for} \quad n \not\equiv 3 \pmod 6 \end{cases}$$

is strictly convex.

Theorem 20 follows fairly easily from the inequalities (32.10), which in fact provide some indication that such a theorem might be true.

§47. Suppose that l and n are natural numbers with $l < n$. We shall consistently denote a point of $\mathscr{V}_l$ by ξ, and a point of $\mathscr{V}_n$ by x. Each point $\xi \in \mathscr{V}_l$ determines in R^{2n} a plane

$$x_1 = \xi_1, \cdots, x_l = \xi_l,$$

which we shall denote by $P_\xi^{l,n}$. The section of $\mathscr{V}_n$ by $P_\xi^{l,n}$ will be denoted by $T_\xi^{l,n}$. Considering $T_\xi^{l,n}$ as a subset of $P_\xi^{l,n}$, we denote the set of its interior points by $\operatorname{Intv} T_\xi^{l,n}$. If $\xi \in \partial \mathscr{V}_l$, then S_ξ contains precisely one function, and hence $T_\xi^{l,n}$ consists of exactly one point. If $\xi \in \operatorname{Int} \mathscr{V}_l$, then S_ξ contains a bounded function, and hence $T_\xi^{l,n}$ has a nonempty intersection with $\operatorname{Int}\mathscr{V}_n$. Therefore $\operatorname{Intv} T_\xi^{l,n} \neq \emptyset$.

PROPOSITION 47.1. *The set* $\operatorname{Intv} T_\xi^{l,n}$ *is linearly connected.*

PROOF. The set $\operatorname{Intv} T_\xi^{n-1,n}$ is connected, since, as we have just noted, it is convex. Thus our proposition will be established if we show that the linear connectedness of the sets $\operatorname{Intv} T_\xi^{l+1,n}$, $l \leqq n-2$, implies the linear connectedness of the set $\operatorname{Intv} T_\xi^{l,n}$, $\xi \in \mathscr{V}_l$. Assuming that $\operatorname{Intv} T_\xi^{l,n} \neq \emptyset$, we take two arbitrary points $x^1, x^2 \in \operatorname{Intv} T_\xi^{l,n}$. If $\xi = (\xi_1, \cdots, \xi_l)$, then $x^j = (\xi_1, \cdots, \xi_l, x_{l+1}^j, \cdots, x_n^j)$, $j = 1, 2$. Clearly in what follows we may assume that $x_{l+1}^1 \neq x_{l+1}^2$, since if $x_{l+1}^1 = x_{l+1}^2$, then by assumption the points x^1 and x^2 can be joined together. We consider the domain $\mathscr{V}_{l+1}$ and its two-dimensional section $T_\xi^{l,l+1}$; having put $\eta^j = (\xi_1, \cdots, \xi_l, x_{l+1}^j)$, $j = 0, 1$, we find that $\eta^j \in \operatorname{Intv} T_\xi^{l,l+1}$, $j = 1, 2$. The straight line segment J which joins η^1 and η^2 belongs to $\operatorname{Intv} T_\xi^{l,l+1}$ by the remarks we made before. Let η be any point of J and let $T_\eta^{l+1,n}$ be the section of $\mathscr{V}_n$ determined by this point. Since $\eta \in \operatorname{Int}\mathscr{V}_{l+1}$, we have $\operatorname{Intv} T_\eta^{l+1,n} \neq \emptyset$. Denote by $\rho(\eta)$ the radius of the largest sphere contained in $\mathscr{V}_n$ which has its center in the set $\operatorname{Intv} T_\eta^{l+1,n}$. It is easily seen that $\inf_{\eta\in J}\rho(\eta) > 0$. From this it follows that on the segment J we can find points η_k, $1 \leqq k \leqq N$, $\eta_1 = \eta^1$, $\eta_N = \eta^2$, such that for each k with $2 \leqq k \leqq N$ there exists a sphere centered at $\tilde{x}_k \in \operatorname{Intv} T_{\eta_k}^{l+1,n}$ and intersecting the set $\operatorname{Intv} T_{\eta_{k+1}}^{l+1,n}$. Denote by $x_{k-1,k}$ any such point of intersection of the above sphere with $\operatorname{Intv} T_{\eta_{k-1}}^{l+1,n}$, and by j_k the straight line segment joining $x_{k-1,k}$ with $\tilde{x}_k$, $2 \leqq k \leqq N$. It follows from the construction that $j_k \subset \operatorname{Intv} T_\xi^{l,n}$. Since $\tilde{x}_k \in \operatorname{Intv} T_{\eta_k}^{l+1,n}$, $x_{k,k+1} \in \operatorname{Intv} T_{\eta_k}^{l+1,n}$, $2 \leqq k \leqq N-1$, it follows by hypothesis that there exists a path $\sigma_k \subset \operatorname{Intv} T_{\eta_k}^{l+1,n}$ which joins $\tilde{x}_k$ and $x_{k,k+1}$. Applying this same argument over again, we see that it is possible to join x^0 and $x_{0,1}$ by a path $\sigma_1 \subset \operatorname{Intv} T_{\eta_1}^{l+1,n}$, and the points x^2 and $\tilde{x}_N$ by a path $\sigma_N \subset \operatorname{Intv} T_{\eta_n}^{l+1,n}$. Thus the path $\sigma_1 \cup j_2 \cup \sigma_2 \cup \cdots \cup j_N \cup \sigma_N \subset \operatorname{Intv} T_\xi^{l,n}$, and it joins the points x^1 and x^2, as required. Q.E.D.

The following lemma is a simple consequence of the above proposition.

LEMMA 47.1. *For any point* $x \in \mathscr{V}_n$ *the set* S_x *is connected.*

Since we shall not use this lemma, its proof will be omitted, despite its simplicity.

The closure of a set $\mathscr{A}$ will be denoted by $\overline{\mathscr{A}}$.

PROPOSITION 47.2. *If* $\operatorname{Intv} T_\xi^{l,n} \neq \emptyset$, *then* $\overline{\operatorname{Intv} T_\xi^{l,n}} = T_\xi^{l,n}$.

PROOF. Suppose $x^0 \in T_\xi^{l,n} \setminus \operatorname{Intv} T_\xi^{l,n}$ and $\operatorname{Intv} T_\xi^{l,n} \neq \emptyset$. Clearly $x^0 \in \partial\mathscr{V}_n$. If $d_l(x^0) \neq 0$, then on the basis of Theorem 2 there exists a function $g(z, \epsilon) \in S$, $g(z) = \sum_{k=0}^{\infty} b_k z^{k+1}$, such that $b_k = \rho^k \xi_k$, $1 \leqq k \leqq l$, $\rho = 1 + \epsilon$, provided $\epsilon > 0$ is sufficiently small. The function $\rho g(z/\rho)$ belongs to S and is bounded in $|z| \leqq 1$, and thus it determines a point $x^\epsilon \in \operatorname{Intv} T_\xi^{l,n}$, and as $\epsilon \to 0$ we have $x^\epsilon \to x^0$.

All that remains is to consider the case when $d_l(x^0) = 0$. If $f(z, x^0) \in S_l$, then $T_\xi^{l,n} = \{x^0\}$ and $\operatorname{Intv} T_\xi^{l,n} = \emptyset$. Thus we may assume that $f(z, x^0) \notin S_l$. Clearly $x^0 \in \mathscr{A}_1$ and $f(z, x^0)$ is algebraic. Define a natural number q by the conditions $d_{q-1}(x^0) \neq 0$ and $d_q(x^0) = 0$, $q \leq l$. By Lemma 37.1, $f(z, x^0)$ satisfies a minimal differential equation of the form (37.12) or (37.13). Therefore $f(z, x^0) \in S_{2q}$, and it is clear that $l < 2q \leq n$. The function $f(z, x^0)$ determines a point $\eta \in \operatorname{Int} \mathscr{V}_{2q-1}$. Consider $T_\eta^{2q-1,2q} \subset \mathscr{V}_{2q}$. Obviously $f(z, x^0)$ corresponds to a point $y^0 \in \partial \mathscr{V}_{2q}$, $y^0 \in T_\eta^{2q-1,2q}$. In view of our construction, $\operatorname{Intv} T_\eta^{2q-1,2q} \neq \emptyset$. We note that if $y \in \operatorname{Intv} T_\eta^{2q-1,2q}$ and $|y^0 - y| < \epsilon$, then $\rho[f(z, x^0), g(z)] < \delta$ for any $g(z) \in S_y$, and also ϵ and δ approach zero simultaneously. If $2q < n$, then $\operatorname{Intv} T_y^{2q,n} \subset \operatorname{Intv} T_\xi^{l,n}$, and $\lim_{y \to y^0} x = x^0$, $x \in \operatorname{Intv} T_y^{2q,n}$. If, however, $2q = n$, then $y^0 = x^0$, $\operatorname{Intv} T_\eta^{n-1,n} \subset \operatorname{Intv} T_\xi^{l,n}$, and x^0 is a limit point of points in $\operatorname{Intv} T_\eta^{n-1,n}$. Q.E.D.

PROPOSITION 47.3. *Let* $\{\xi^\nu\} \subset \mathscr{V}_l$ *and* $\lim_{\nu \to \infty} \xi^\nu = \xi \in \mathscr{V}_l$. *Then*

$$\lim_{\xi^\nu \to \xi} T_{\xi^\nu}^{l,n} = T_\xi^{l,n}.$$

PROOF. Denote by Ls and Li respectively the upper and lower limits of a sequence of sets. Clearly we have $T_\xi^{l,n} \supset \operatorname{Ls}_{\xi^\nu \to \xi} T_{\xi^\nu}^{l,n}$. If $x^0 \in \operatorname{Intv} T_\xi^{l,n}$, then there exists a sphere, with center x^0, lying inside $\mathscr{V}_n$. This sphere will intersect every set $T_{\xi^\nu}^{l,n}$, provided $\nu \geqq \nu_0$. Hence $\operatorname{Li}_{\xi^\nu \to \xi} T_{\xi^\nu}^{l,n} \supseteq \operatorname{Intv} T_\xi^{l,n}$. Since $\operatorname{Li}_{\xi^\nu \to \xi} T_{\xi^\nu}^{l,n} = \operatorname{Li}_{\xi^\nu \to \xi} T_{\xi^\nu}^{l,n}$, in view of the preceding proposition we have $\operatorname{Li}_{\xi^\nu \to \xi} T_{\xi^\nu}^{l,n} \supseteq T_\xi^{l,n}$, and so $\lim_{\xi^\nu \to \xi} T_{\xi^\nu}^{l,n} = T_\xi^{l,n}$. Q.E.D.

The local properties of the boundary of $T_\xi^{l,n}$ are easily studied in any neighborhood of those points $x^0 \in \partial T_\xi^{l,n} = T_\xi^{l,n} \setminus \operatorname{Intv} T_\xi^{l,n}$ for which we have $d_{n-1/2}(x^0) \neq 0$. In this case, as was shown in §12, in a neighborhood of x^0 we can introduce the local coordinates (y, τ) and derive an equation of the boundary of $\mathscr{V}_n$ of the form $\tau = \psi(y)$, $|y| \leq \rho$. The connection between the old and new coordinate systems is given by relation (12.3), from which we obtain

$$y_j = \operatorname{Re} \sum_{k=1}^n \bar{\alpha}_{jk} (x_k - x_k^0), \qquad 1 \leqslant j \leqslant 2n-1, \qquad \tau = \operatorname{Re} \sum_{k=1}^n \alpha_{2nk} (x_k - x_k^0).$$

On the plane $P_\xi^{l,n}$ we have $x_k = x_k^0 = \xi_k$, $1 \leq k \leq l$, and upon setting $u_k = x_k - x_k^0$, $l < k \leq n$, we find that the equation of the boundary of $T_\xi^{l,n}$ in a neighborhood of x^0 has the form

$$\operatorname{Re} \sum_{k=l+1}^n \bar{\alpha}_{2nk} u_k - \psi\left\{\operatorname{Re} \sum_{k=l+1}^n \bar{\alpha}_{1k} u_k, \ldots, \operatorname{Re} \sum_{k=l+1}^n \bar{\alpha}_{2n-1k} u_k\right\} = 0. \tag{47.1}$$

Note that the vector $(\alpha_{2n,1}, \cdots, \alpha_{2n,n})$ is normal to $\partial \mathscr{V}_n$ at x^0. Thus it equals $\lambda(x^0)$, and so we can always assume that $\sum_{k=l+1}^n |\alpha_{2nk}|^2 \neq 0$, since otherwise $x^0 \in \partial \mathscr{V}_l$ and $T_\xi^{l,n} = \{x^0\}$.

We now introduce on the plane $P_\xi^{l,n}$ the Cartesian coordinates $(\mathfrak{z}, \sigma)$, $\mathfrak{z} = (\mathfrak{z}_1, \cdots, \mathfrak{z}_{2n-2l-1})$, with origin at x^0, where the σ axis is the straight line determined by the vector $\lambda^0 = (\alpha_{2n,l+1}, \cdots, \alpha_{2n,n})$, and the remaining axes are allowed to be arbitrary. Then

$$\operatorname{Re} \sum_{k=l+1}^{n} \bar{\alpha}_{2nk} u_k = |\lambda^0| \sigma \text{ and } (\mathfrak{z}, \sigma) = Uu,$$

where $u = (u_{l+1}, \cdots, u_n)$, and where V is an orthogonal matrix. In this new coordinate system equation (47.1) has the form

$$\sigma - \frac{1}{|\lambda^0|} \psi\left\{ \sum_{k=1}^{2n-2l-1} \beta_{1k}\mathfrak{z}_k + \beta_{2n-2l-1}\sigma, \ldots, \sum_{k=1}^{2n-2l-1} \beta_{2n-1k}\mathfrak{z}_k + \beta_{2n-1,\, 2n-2l-1}\sigma \right\} = 0. \tag{47.2}$$

Since $\operatorname{grad} \psi(y) = O(|y|)$, for $(\mathfrak{z}, \sigma) \in \partial T_\xi^{l,n}$ we have

$$\sigma = O(|\mathfrak{z}|^2). \tag{47.3}$$

It is clear that the equation of the hyperplane tangent to $\partial T_\xi^{l,n}$ at x^0 is $\sigma = 0$.

The arguments we present below will be based on a theorem of Tietze on convex sets. A convex set in R^k (and more generally in a linear topological space) can be characterized by local properties. In particular, we can consider its local supporting properties. We introduce the following definitions.

Definition 47.1. Let $\mathfrak{W} \subset R^k$ and $x^0 \in \partial \mathfrak{W}$. Then x^0 is called a *point of weak local convexity* if there exists a hyperplane $(\lambda, x) = (\lambda, x^0)$ and a neighborhood U of x^0 such that $(\lambda, x) > (\lambda, x^0)$ and $x \in U \backslash \{x^0\}$ imply $x \notin \mathfrak{W}$.

Definition 47.2. A set $\mathfrak{W} \subset R^k$ is called *weakly locally convex* if each of its boundary points is a point of weak local convexity.

Theorem of Tietze ([16], [17]). *An open, connected, weakly locally convex set $\mathfrak{W} \subset R^k$ is convex.*

Proposition 47.4. *Let $l \geqq [n/2]$ and $x^0 \in \partial T_\xi^{l,n}$. If $d_{n-1/2}(x^0) \neq 0$, then x^0 is a point of weak local convexity.*

Proof. Introduce the Cartesian coordinates $(\mathfrak{z}, \sigma)$ in a neighborhood of x^0 in the plane $P_\xi^{l,n}$. Let $(\mathfrak{z}, \sigma) \in \partial T_\xi^{l,n}$ and let $\mathfrak{z} = \epsilon \mathfrak{z}^0$, $|\mathfrak{z}^0| = 1$, where $\epsilon > 0$ is a sufficiently small positive quantity. From (47.2) and (47.3) it follows in view of Theorem 13 that

$$\sigma = |\lambda^0|^{-1} \psi\left\{\varepsilon \sum_{k=1}^{2n-2l-1} \beta_{1k}\mathfrak{z}_k^0, \ldots, \varepsilon \sum_{k=1}^{2n-2l-1} \beta_{2n-1k}\mathfrak{z}_k^0\right\} + o(\varepsilon^2). \tag{47.4}$$

Set $v_j = \sum_{k=1}^{2n-2l-1} \beta_{jk}\mathfrak{z}_k^0$, $1 \leqq j \leqq 2n-1$, and $U = (v_2, \cdots, v_{2n-1})$. From (47.4) we find that

$$\overline{\lim_{\varepsilon \to 0}} \frac{\sigma}{\varepsilon^2} = \frac{1}{2} |\lambda^0|^{-1} \overline{D}_v^2 \psi(0). \tag{47.5}$$

Let the vector $\mathfrak{y} \in R^{2n}$ have the form $\mathfrak{y} = \{0, \cdots, 0, V'(\mathfrak{z}^0, 0)\}$, where there are $2l$ zeros. Recalling (12.3), we see that $(v, 0) = U'\mathfrak{y}$. Comparing this with (31.3) and writing the vectors of R^{2n} in complex form, we find that $t = \mathfrak{y}$ and $t_1 = \cdots = t_l = 0$; $|t| = |\mathfrak{z}^0| = 1$. Therefore relation (32.10) gives that $\overline{d}_v^2\psi(0) \leqq -\beta$, and so

$$\overline{\lim_{\varepsilon \to 0}} \frac{\sigma}{\varepsilon^2} < -\frac{1}{2} |\lambda^0|^{-1}\beta^n. \tag{47.6}$$

Since $|\operatorname{grad} \psi(y)| < C|y|$, it follows trivially from (47.6) that for $(\mathfrak{z}, \sigma) \in \partial T_\xi^{l,n}$

$$\sigma < -\frac{1}{4} |\lambda^0|^{-1}\beta^n |\mathfrak{z}|^2,$$

provided $|\mathfrak{z}| \leqq \epsilon_0$. Q.E.D.

COROLLARY 47.1. *Let* $l \geqq [n/2]$. *If* $\inf d_{n-1/2}(x) > 0$, *where the infimum is taken over* $\partial T_\xi^{l,n}$, *then* $T_\xi^{l,n}$ *is a convex set.*

In fact, if (47.1), (47.2) and (47.4) hold, then Tietze's Theorem implies that $T_\xi^{l,n}$ is convex.

Assuming that the hypothesis of Corollary 47.1 is satisfied, we shall examine the question of whether or not $T_\xi^{l,n}$ is strongly convex. First of all we note that in the coordinates $(\mathfrak{z}, \sigma)$ the equation of the boundary of $T_\xi^{l,n}$ in a neighborhood of x^0 can be written in the form $\sigma = \Phi(\mathfrak{z})$, and moreover, as follows from (47.2) and Theorem 13, the function $\operatorname{grad} \Phi(\mathfrak{z})$ satisfies a Lipschitz condition for $|\mathfrak{z}| \leqq \rho_0$. Recalling (47.5), we see easily that

$$\overline{d}_{\mathfrak{z}}^2\Phi(0) = |\lambda^0|^{-1}\overline{d}_v^2\psi(0),$$

and so as a consequence of (32.10) we have

$$\overline{d}_{\mathfrak{z}}^2\Phi(0) < -|\lambda^0|^{-1}\beta^n |\mathfrak{z}|^2. \tag{47.7}$$

In order to estimate the differential of $\Phi(\mathfrak{z})$ at any point at which it exists, we take an arbitrary point $(\mathfrak{z}^0, \sigma_0)$, $\sigma_0 = \Phi(\mathfrak{z}^0)$, on the boundary of $T_\xi^{l,n}$, and in a neighborhood of this point we introduce Cartesian coordinates $(\tilde{\mathfrak{z}}, \tilde{\sigma})$, taking the $\tilde{\sigma}$ axis orthogonal to $T_\xi^{l,n}$. In these new coordinates the equation of the boundary of $T_\xi^{l,n}$ has the form $\tilde{\sigma} = \tilde{\Phi}(\tilde{\mathfrak{z}})$, and since $(\tilde{\mathfrak{z}}, \tilde{\sigma}) = T[(\mathfrak{z}, \sigma) - (\mathfrak{z}^0, \sigma_0)]$, where $T = (\gamma_{jk})$ is an orthogonal matrix, we have

$$\sum_{k=1}^{2n-2l-1} \gamma_{2n-2l,\,k}(\mathfrak{z}_k - \mathfrak{z}_k^0) + \gamma_{2n-2l,\,2n-2l}(\sigma - \sigma_0) = \tilde{\Phi}(\tilde{\mathfrak{z}}_1, \ldots, \tilde{\mathfrak{z}}_{2n-2l-1}), \tag{47.8}$$

where

$$\tilde{\mathfrak{z}}_j = \sum_{k=1}^{2n-2l-1} \gamma_{jk}(\mathfrak{z}_k - \mathfrak{z}_k^0) + \gamma_{j2n-2l}(\sigma - \sigma_0), \qquad 1 \leqslant j \leqslant 2n - 2l - 1.$$

It follows from (47.8) that $\tilde{\Phi}(\tilde{\mathfrak{z}})$ is continuously differentiable if $|\mathfrak{z} - \mathfrak{z}^0| \leqq \rho_0$ and $\operatorname{grad} \tilde{\Phi}(\tilde{\mathfrak{z}})$ satisfies a Lipschitz condition. Denote by $\mathfrak{Z}$ the domain of the variable $\tilde{\mathfrak{z}}$; clearly $\mathfrak{Z}$ is obtained by projecting $T_\xi^{l,n}$ onto the plane $\sigma = 0$. From what we have said it follows that $\tilde{\Phi}(\tilde{\mathfrak{z}})$ is continuously differentiable on Intv $\mathfrak{Z}$ and satisfies a Lipschitz condition on any convex compact set lying in Intv $\mathfrak{Z}$. From (47.8) it follows easily that

$$\overline{d}^2_\eta \Phi(\mathfrak{z}^0) = \frac{1}{\gamma_{2n-2l,\,2n-2l}} \overline{d}^2_{\tilde{\eta}} \tilde{\Phi}(0),$$

where $\tilde{\eta} = (\tilde{\eta}_1, \cdots, \tilde{\eta}_{2n-2l-1})$, $\tilde{\eta}_j = \sum_{k=1}^{2n-2l-1} \gamma_{jk}\eta_k + \gamma_{j2n-2l}(\operatorname{grad}\Phi(\mathfrak{z}^0), \eta)$, $1 \leq j \leq 2n-2l-1$. Since

$$\overset{-1}{\gamma}_{2n-2l,2n-2l} = (1 + |\operatorname{grad}\Phi(\mathfrak{z}^0)|^2)^{1/2} \qquad \text{and} \qquad |\tilde{\eta}| = (|\eta|^2 + (\operatorname{grad}\Phi(\eta^0), \eta)^2)^{1/2},$$

it follows from (47.7) that

$$\overline{d}^2_\eta \Phi(\mathfrak{z}^0) < -\frac{\beta^n}{|\lambda^0(\mathfrak{z}^0)|}(1 + |\operatorname{grad}\Phi(\mathfrak{z}^0)|^2)^{1/2}(|\eta|^2 + (\operatorname{grad}\Phi(\mathfrak{z}^0), \eta)^2), \tag{47.9}$$

where $\lambda^0(\mathfrak{z}^0)$ is the value of the vector λ^0 at the point $(\mathfrak{z}^0, \Phi(\mathfrak{z}^0)) \in \partial T_\xi^{l,n}$.

Denote by $\Delta = \Delta_\xi$ the diameter of the set $T_\xi^{l,n}$, and by θ the greatest lower bound of $|\lambda^0|$ on $\partial T_\xi^{l,n}$.

Proposition 47.5. *Let $x^0 \in \partial T_\xi^{l,n}$ and $x^1 \in T_\xi^{l,n}$, and let δ be the distance from x^1 to the hyperplane tangent to $\partial T_\xi^{l,n}$ at x^0. If the conditions of Corollary* 47.1 *are satisfied, then*

$$\delta \geqslant \min\left(\frac{\beta^n}{8\theta}, \frac{1}{\sqrt{2}\,\Delta}\right)|x^0 - x^1|^2. \tag{47.10}$$

Proof. We introduce the coordinate system $(\mathfrak{z}, \sigma)$, and we construct the two-dimensional plane P passing through the σ axis and the point x^1. This plane intersects the set $T_\xi^{l,n}$ in a convex figure $\mathcal{Y}$. Let (t, σ) be the coordinates in the plane P. With regard to the orientation inherited from the coordinate system $(\mathfrak{z}, \sigma)$, the figure $\mathcal{Y}$ will be situated under the t axis, and the equation of that part of its boundary visible from the t axis will have the form $\sigma = \varphi(t)$, where $\varphi(t) = \Phi(t\mathfrak{z}_1^0, \cdots, t\mathfrak{z}_{2n-2l-1}^0)$, and where $\mathfrak{z}^0 = (\mathfrak{z}_1^0, \cdots, \mathfrak{z}_{2n-2l-1}^0)$ is some well-defined vector. Hence it follows that $\varphi'(t) = (\operatorname{grad}\Phi(t\mathfrak{z}^0), \mathfrak{z}^0)$, and $\overline{d}^2\varphi(t) = \overline{d}^2_{\mathfrak{z}^0}\Phi(t\mathfrak{z}^0)$. We shall examine those points of $\mathcal{Y}$ which lie in the fourth quadrant. We construct a bisector of this quadrant. If a point (σ, t) lies under the bisector, then

$$\sigma = -\frac{|\sigma|}{\rho}\rho \leqslant -\frac{\rho}{\sqrt{2}} \leqslant -\frac{\rho^2}{\sqrt{2}\,\Delta},$$

where $\rho^2 = (\sigma^2 + t^2)^{1/2}$. Suppose that (σ, t) lies above the bisector. First we consider the case when $\sigma = \varphi(t)$. Since $\varphi'(t)$ satisfies a Lipschitz condition, $\varphi'(t)$ is differentiable almost everywhere, and

$$\sigma = \varphi(t) = \int_0^t (t-u)\varphi''(u)\,du.$$

From (47.9) it follows that

$$\sigma < -\frac{\beta^n}{\theta}\int_0^t (t-u)\,du = -\frac{\beta^n}{2\theta}t^2 = -\frac{\beta^n}{2\theta}\rho^2\left(\frac{t}{\rho}\right)^2 \leqslant -\frac{\beta^n}{4\theta}\rho^2.$$

If (t,σ) lies above the bisector but inside $\mathscr{Y}$, it is very simple to show that $\sigma \leqq -\beta^n\rho^2/8\theta$. Thus the proof is complete.

Next we shall examine the set $T_\xi^{l,n}$ in the case when $\partial T_\xi^{l,n} \cap \mathscr{A}_1 \neq \emptyset$.

Proposition 47.6. *If $n \not\equiv 3 \pmod 6$ and $l \geqq [n/2]$, or if $n \equiv 3 \pmod 6$ and $l \geqq (n+1)/2$, then $T_\xi^{l,n}$ is a convex set.*

Proof. We shall apply Lemma 39.3, taking $\mathscr{V}_l$ as the image of S. Denote by $\mathscr{A}_l^n$ the image of the set $\bigcup_{3r\leq n}\mathfrak{A}_r \cup \bigcup_{2s\leq n}\mathfrak{A}_s^*$. The set $\mathscr{A}_l^n$ is closed, and $m_p(\mathscr{A}_l^n) = 0$, where $p = \max(2[n/2], 3[n/3])$. Thus, if $n \not\equiv 3 \pmod 6$ and $l \geqq [n/2]$, or if $n \equiv 3 \pmod 6$ and $l \geqq (n+1)/2$, the set $\mathscr{A}_l^n$ is nowhere dense. For $\xi \in \mathscr{A}_l^n$ we construct a sequence $\{\xi^\nu\} \subset \mathscr{V}_l \backslash \mathscr{A}_l^n$ which converges to ξ. By Corollary 47.1 the sets $T_{\xi^\nu}^{l,n}$, $\nu = 1, 2, \cdots$, are convex, and in view of Proposition 47.3 the set $T_\xi^{l,n}$ is also convex. Q.E.D.

Proof of Theorem 20. We must still examine the question of the strict convexity of the set $T_\xi^{m,n}$. Inequality (47.10) holds for those sets $T_\xi^{m,n}$ for which $\partial T_\xi^{m,n} \cap \mathscr{A}_1 \neq \emptyset$ or for which $d_{n-1/2}(x^0) \neq 0$. Suppose that $d_{n-1/2}(x^0) = 0$. We take a sequence $\{x^k\}$, $\tilde{x}^k \in \partial T_{\xi^k}^{m,n}$, $\xi^k \to \xi$, such that $d_{n-1/2}(\tilde{x}^k) \neq 0$, $k = 1, 2, \cdots$, and $\tilde{x}^k \to x^0$, and we construct a hyperplane P_k which supports $\partial T_{\xi^k}^{m,n}$ at x^k, $k = 1, 2, \cdots$. For each hyperplane P_k and points $\tilde{x}^k, y^k \in T_{\xi^k}^{m,n}$, $k = 1, 2, \cdots$, $y^k \to x^1$, inequality (47.10) will hold. Without loss of generality we assume that the hyperplanes P_k converge to some hyperplane P which supports $T_\xi^{m,n}$ at x^0. Thus (47.10) continues to hold on P. Suppose now that the set $T_\xi^{m,n}$ is not strictly convex. Then on $\partial T_\xi^{m,n}$ there exists a straight line segment (x^1, x^2). Let x^0 be the midpoint of this segment. According to what we showed above, at x^0 the supporting hyperplane P exists, and (47.10) holds. Thus the segments (x^1, x^2) must not lie in P, which is a contradiction. This completes the proof of the theorem.

Remark 47.1. For $n \equiv 3 \pmod 6$ the theorem remains true for the set $T_\xi^{[n/2],n}$. However, in order to prove this, we must study the structure of $\partial\mathscr{V}_n$ in a neighborhood of $\mathscr{A}_1$ in more detail. Due to the lack of space and the lengthy computations involved, we shall not pursue this point any further.

In [**6**] Duren and Schiffer conjectured that the more subtle form of Theorem 20 given above in Remark 47.1 was in fact true. The motivation for this result is the quasi-definite character of the second variation.

§48. Beginning with Theorem 20, we can generalize Teichmüller's theorem on quadratic differentials. On the w-sphere consider the quadratic differential

$$Q(w)\left(\frac{dw}{w}\right)^2 = \sum_{k=1}^n A_k w^{-k}\left(\frac{dw}{w}\right)^2. \tag{48.1}$$

Let $\mathscr{G}$ be the graph of this differential, and let $\mathscr{T} \subset \mathscr{G}$ be any tree with base at $w = \infty$ and having at $w = 0$ exterior conformal radius equal to 1. Denote by $f(z)$ the function mapping $|z| < 1$ onto the complement of $\mathscr{T}$ such that $f(0) = 0$ and $f'(0) > 0$. The function $f(z)$ will be called admissible with respect to the differential (48.1). Clearly $f(z) \in S$. Also, the domain whose complement is $\mathscr{T}$ will be called admissible.

For the proof of our theorem we shall need some information on those extremal functions $f(z) \in S_n$ for which $d_{n-1/2}(f) = 0$. These functions were studied in some detail in Chapter 6. With each such function there is connected a subspace of associated vectors. Of all the associated vectors, the most important are the vectors λ for which the quasipolynomial $P(z; \lambda)$ is nonnegative on $|z| = 1$. The set of all such associated vectors is a convex cone called the normal cone, and denoted by $K_f = K$. As before, we have the set $\operatorname{Intv} K$, and we put $\partial K = K \setminus \operatorname{Intv} K$.

Proposition 48.1. *Suppose $x \in \partial \mathscr{V}_n$ and $d_{n-1/2}(x) = 0$. Assume that $f(z, x) \notin S_{[n/2]}$. If $\lambda \in \partial K_x$, then there exists a sequence $\{x^k\} \subset \mathscr{R}$, $\lim_{k\to\infty} x^k = x$, such that $\lim_{k\to\infty} \lambda(x^k) = \lambda$.*

Proof. First we consider the case when $f(z, x)$ satisfies a minimal differential equation of the form (37.13). Let $q(w)$ be a polynomial in $1/w$ of degree r. Then $3r \leq n$ and $4r > n$, since $f \notin S_{[n/2]}$. The function $f(z, x)$ satisfies

$$(Dq(f))^2 \sum_{l=0}^{1} \gamma_l [q(f)]^l \left(\frac{Df}{f}\right)^2 - (Dp(z))^2 \sum_{l=0}^{1} \gamma_l [p(z)]^l = 0,$$

and if this equation determines the vector $\lambda \in K_x$, it follows that $\sum_0^1 \gamma_l [p(z)]^l \geq 0$ for $|z| = 1$. If $\lambda \in \partial K_x$, then the quasipolynomial $\sum_0^1 \gamma_l [p(z)]^l$ has a zero z_0 on $|z| = 1$. Put $w_0 = f(z_0, x)$, $Q(w) = (Dq(w))^2 \sum_0^1 \gamma_l [q(w)]^l$. Clearly w_0 is a vertex of the graph $\mathscr{G}$ determined by the differential $Q(w)(dw/w)^2$. Also, w^0 can be a vertex of the graph $\mathscr{G}_0$ of the differential $[Dq(w)]^2 (dw/w)^2$, but this can happen only when the degree of this vertex is less than the degree of w_0 as a vertex of $\mathscr{G}$. Take a tree $\mathscr{T}_\epsilon \subset \mathscr{G}$, $\mathscr{T}_\epsilon \supset \mathscr{T}$, where $\mathscr{T}$ is the boundary tree of $f(z, x)$, and moreover so that $\mathscr{T}_\epsilon$ contains portions of all branches of $\mathscr{G}$ emanating from w^0. These portions are taken to be so small that the exterior conformal radius ρ_ϵ of $\mathscr{T}_\epsilon$ will be $1 - \epsilon$. Set $w = \rho_\epsilon t$ and on the t-sphere consider the differential $Q_\epsilon(t)(dt/t)^2 = Q(\rho_\epsilon t)(dt/t)^2$. The image of $\mathscr{T}_\epsilon$ on the t-sphere is a tree $\mathscr{T}_\epsilon^*$ with exterior conformal radius 1. If $f_\epsilon(z) \in S$ maps $|z| < 1$ onto the complement of $\mathscr{T}_\epsilon^*$, then by Theorem 5 we see that $f_\epsilon(z) \in S_n$, and so it determines a point $x_\epsilon \in \partial \mathscr{V}_n$ and an associated vector $\lambda(x_\epsilon)$. We shall show that $x_\epsilon \in \mathscr{R}$. Suppose that $f_\epsilon(z)$ satisfies more than one differential equation. Then by Lemma 37.1 it satisfies the equation

$$q_\epsilon(f_\epsilon) - p_\epsilon(z) = 0,$$

where $q_\epsilon(w)$ is a polynomial in $1/w$ of degree r_ϵ. Since by the construction of $\mathscr{T}_\epsilon$ we have $\operatorname{Im} q_\epsilon[f(z,\epsilon)/\rho_\epsilon]=0$ on $|z|=1$, with the exception of finitely many points, it follows that

$$q_\epsilon[f/\rho_\epsilon]-p_\epsilon^*(z)=0.$$

Thus $f(z)$ satisfies the equation

$$\left(Dq_\varepsilon\left(\frac{f}{\rho_\varepsilon}\right)\right)^2\left(\frac{Df}{f}\right)^2-(Dp_\varepsilon^*(z))^2=0,$$

and so by Lemma 37.1

$$\left(Dq_\varepsilon\left(\frac{w}{\rho_\varepsilon}\right)\right)^2=(Dq\,(w))^2\sum_{l=0}^{s}\gamma_l(\varepsilon)\,[q\,(w)]^l.$$

Note that $[Dq_\epsilon(w/\rho_\epsilon)]^2$ has a root of higher multiplicity at w_0 than does $[Dq(w)]^2$. Since the expression $\sum_{l=0}^{s}\gamma_l(\epsilon)[q(w)]^l$ must be quadratic, we have $s\geqq 2$, since in the opposite case $f(z,x)$ would satisfy a minimal differential equation of the form (37.12), which would contradict our original assumption. Hence $r_\epsilon\geqq 2r$, and so $x_\epsilon\in\mathscr{R}$. As $\epsilon\to 0$ we have $x_\epsilon\to x$, and $Q_\epsilon(w)\to Q(w)$ for $w\neq 0$. Therefore $\lambda(x_\epsilon)\to\lambda\in\partial K_x$. The case when $f(z,x)$ satisfies equation (37.12) is treated completely analogously. This proves the proposition.

Proposition 48.2. *Suppose $x^0\in\partial T_\xi^{m,n}$, where m is defined as in Theorem 20. If $x^0\in\mathscr{A}_1$ and $\lambda=(\lambda_1,\cdots,\lambda_n)\in K_{x^0}$, then there exists a supporting hyperplane for $T_\xi^{m,n}$ at x^0, and $(\lambda_{m+1},\cdots,\lambda_n)$ is a normal vector for this hyperplane.*

Proof. The only interesting case is when $f(z,x^0)\notin S_{[n/2]}$, since otherwise $T_\xi^{m,n}=\{x^0\}$. Thus we have that $\dim K_{x^0}=2$. Let $\lambda^*\in\partial K_{x^0}$, and let $\{x^k\}\subset\mathscr{R}$ and $\{\lambda(x^k)\}$ be the sequence of points and associated vectors constructed in the previous proposition. We construct a sequence $\{\xi^k\}\subset\mathscr{V}_n$ such that $x^k\in\partial T_{\xi^k}^{m,n}$, $k=1,2,\cdots$. Since $x^k\in\mathscr{R}$, the set $T_{\xi^k}^{m,n}$ has a well-defined supporting hyperplane at x^k, and the hyperplane is determined by the vector $\lambda^0(x^k)=(\lambda_{m+1}(x^k),\cdots,\lambda_n(x^k))$. Taking the limit, we find that $T_\xi^{m,n}$ has at x^0 a supporting hyperplane determined by the vector $\lambda^0=(\lambda_{m+1}^*,\cdots,\lambda_n^*)$. Since the cone K_{x^0} is convex, our proposition follows immediately from the facts proved above.

Theorem 21. *Suppose that the function $\varphi(w)=w+\sum_{k=m+1}^{\infty}\alpha_k w^{k+1}$ is regular and univalent in a domain which is admissible with respect to the quadratic differential* (48.1). *Let m be defined as in Theorem* 20. *Then*

$$\operatorname{Re}\sum_{j=m+1}^{n}A_j\alpha_j\geqslant 0,$$

with equality only if $\varphi(w)\equiv w$.

Proof. Let $f(z)\in S$ map $|z|<1$ onto the admissible domain of the theorem. Then $f(z)\in S_n$ by Theorem 5. Set $g(z)=\varphi[f(z)]$. Clearly $g(z)\in S$, and on setting

$$f(z)=\sum_{0}^{\infty} a_k z^{k+1},\quad g(z)=\sum_{0}^{\infty} b_k z^{k+1},$$

we get that

(48.2) $$b_k=a_k,\ 1\leqslant k\leqslant m \text{ and } b_k=a_k+\sum_{j=m+1}^{k} \alpha_j a_k^{(j+1)},\quad m+1\leqslant k.$$

We now take a section $T_\xi^{m,n}$ of the domain $\mathscr{V}_n$, where the section is defined by $x=(a_1,\cdots,a_n)\in\partial\mathscr{V}_n$. By construction, $y=(b_1,\cdots,b_n)\in T_\xi^{m,n}$. The function $f(z)$ satisfies

$$Q(f)(Df/f)^2+P(z)=0,$$

where $P(z)\geqq 0$ for $|z|=1$. This equation uniquely defines an associated vector λ; and, even if $x\in\mathscr{A}_1$, we have that $\lambda=(\lambda_1,\cdots,\lambda_n)\in K_x$. Therefore at x the set $T_\xi^{m,n}$ has a supporting hyperplane determined by $(\lambda_{m+1},\cdots,\lambda_n)$. Thus we have

(48.3) $$\operatorname{Re}\sum_{k=m+1}^{n} \bar\lambda_k b_k\leqslant \operatorname{Re}\sum_{k=m+1}^{n} \bar\lambda_k a_k,$$

and, if equality holds, the strict convexity of $T_\xi^{m,n}$ implies that $b_k=a_k$, $1\leqq k\leqq n$, and so $g(z)\equiv f(z)$. Using (48.2), we get from (48.3) via (10.3) that

$$\operatorname{Re}\sum_{j=m+1}^{n} A_j\alpha_j\geqslant 0.$$

Remark 48.1. Remark 47.1, above, remains in force.

A more general form of Theorem 21 was given by Jenkins [18] as a generalization of Teichmüller's theorem. We obtained the above theorem as a consequence of our theorem on the convexity of corresponding sections of $\mathscr{V}_n$. Also, it is possible to obtain Theorem 20 by combining Jenkins' result with some facts about the structure of the set $\mathscr{V}_n$.

We have considered some aspects of the local theory of extremal problems. Some questions have been investigated reasonably thoroughly, some not very thoroughly, and others not at all. In particular, we have not investigated the structure of $\partial\mathscr{V}_n$ in a neighborhood of a point of $\mathscr{A}_1$. We have touched on the nonlocal theory only in this final chapter. The author intends to study this area more in the very near future.

Of course, we have not mentioned at all the construction of a theory for extremal problems involving functionals of a type more general than those we have considered here. This is a very difficult problem, since one must study mappings from S to infinite-dimensional compact sets, and one must study the structure of these compact sets. However, we consider this to be a very important problem, since the study of variational problems on the nonlinear space S might give valuable insight into the problem of constructing a theory for variational problems on arbitrary nonlinear compact sets.

BIBLIOGRAPHY

[1] M. Schiffer, *A method of variation within the family of simple functions,* Proc. London Math. Soc. (2) **44** (1938), 432-449.

[2] A. C. Schaeffer and D. C. Spencer, *Coefficient regions for schlicht functions,* Amer. Math. Soc. Colloq. Publ., vol. 35, Amer. Math. Soc., Providence, R. I., 1950. MR **12**, 326.

[3] O. Teichmüller, *Ungleichungen zwischen den Koeffizienten schlichter Functionen,* S.-B. Preuss. Akad. Wiss. **1938**, 363-375.

[4] J. Jenkins, *Univalent functions and conformal mappings,* Ergebnisse der Mathematik und ihrer Grenzgebiete, Heft 18, Springer-Verlag, Berlin, 1958; Russian transl., IL, Moscow, 1962. MR **20** #3288.

[5] G. M. Goluzin, *Geometric theory of functions of a complex variable,* GITTL, Moscow, 1952; English transl., Transl. Math. Monographs, vol. 26, Amer. Math. Soc., Providence, R. I., 1969. MR **15**, 112; **40** #308.

[6] P. L. Duren and M. Schiffer, *The theory of the second variation in extremum problems for univalent functions,* J. Analyse Math. **10** (1962/63), 193-252. MR **27** #284.

[7] M. Schiffer, *On the coefficients of simple functions,* Proc. London Math. Soc. (2) **44** (1938), 450-452.

[8] P. R. Garabedian, G. G. Ross and M. M. Schiffer, *On the Bieberbach conjecture for even n,* J. Math. Mech. **14** (1965), 975-989. MR **32** #207.

[9] A. Zygmund, *Trigonometrical series,* Monografie Mat., Band 5, PWN, Warsaw, 1935; Russian transl., ONTI, Moscow, 1939.

[10] I. Verčenko and A. Kolmogorov, *Über Unstetigkeitspunkte von Funktionen zweier Veränderlichen,* C. R. (Dokl.) Acad. Sci. URSS **2** (1934), 106-107.

[11] A. Kolmogorov and I. Verčenko, *Untersuchungen über Unstetigkeitspunkte von Funktionen zweier Veränderlichen,* C. R. (Dokl.) Acad. Sci. URSS **5** (1934), 362-364.

[12] F. Roger, *On the relation between tangential and metric properties of Cartesian spaces,* C. R. Acad. Sci. Paris **201** (1935), 871-873.

[13] S. Saks, *Theory of the integral,* 2nd rev. ed., Monografie Mat., vol. VII, PWN, Warsaw, 1937; Russian transl., IL, Moscow, 1949.

[14] W. Hurewicz and H. Wallman, *Dimension theory,* Princeton Math. Series, vol. 4, Princeton Univ. Press, Princeton, N. J., 1941; Russian transl., IL, Moscow, 1948. MR **3**, 312.

[15] G. H. Hardy, J. E. Littlewood and G. Pólya, *Inequalities,* Cambridge Univ. Press, New York, 1934; Russian transl., IL, Moscow, 1948. MR **18**, 722.

[16] H. Tietze, *Bemerkungen über konvexe und nicht-konvexe Figuren,* J. Reine Angew. Math. **160** (1929), 67-69.

[17] F. A. Valentine, *Convex sets,* McGraw-Hill Ser. in Higher Math., McGraw-Hill, New York, 1964. MR **30** #503.

[18] J. A. Jenkins, *On normalization in the general coefficient theorem,* Proc. Internat. Congr. Math. (Stockholm 1962), Inst. Mittag-Leffler, Djursholm, 1963, pp. 347-350. MR **31** #328.